Vibrations and Waves in Continuous Mechanical Systems

Vibrations and Waves in Continuous Mechanical Systems

Peter Hagedorn
TU Darmstadt, Germany

Anirvan DasGupta
IIT Kharagpur, India

John Wiley & Sons, Ltd

Copyright © 2007 John Wiley & Sons Ltd, The Atrium, Southern Gate, Chichester, West Sussex PO19 8SQ, England

Telephone (+44) 1243 779777

Email (for orders and customer service enquiries): cs-books@wiley.co.uk
Visit our Home Page on www.wileyeurope.com or www.wiley.com

Other Wiley Editorial Offices

John Wiley & Sons Inc., 111 River Street, Hoboken, NJ 07030, USA

Jossey-Bass, 989 Market Street, San Francisco, CA 94103-1741, USA

Wiley-VCH Verlag GmbH, Boschstr. 12, D-69469 Weinheim, Germany

John Wiley & Sons Australia Ltd, 42 McDougall Street, Milton, Queensland 4064, Australia

John Wiley & Sons (Asia) Pte Ltd, 2 Clementi Loop #02-01, Jin Xing Distripark, Singapore 129809

John Wiley & Sons Canada Ltd, 6045 Freemont Blvd, Mississauga, Ontario, L5R 4J3, Canada

Wiley also publishes its books in a variety of electronic formats. Some content that appears in print may not be available in electronic books.

Anniversary Logo Design: Richard J. Pacifico

British Library Cataloguing in Publication Data

A catalogue record for this book is available from the British Library

ISBN: 978-0470-051738-3

Typeset in 10/12 Times by Laserwords Private Limited, Chennai, India

This book is printed on acid-free paper responsibly manufactured from sustainable forestry in which at least two trees are planted for each one used for paper production.

Contents

Preface

This book is a successor to the book written by the first author (with the help of Dr Klaus Kelkel, now at ZF Friedrichshafen), *Technische Schwingungslehre: Lineare Schwingungen kontinuierlicher mechanischer Systeme*, published in 1989 in German. The German book, which has been out of print for many years now, was developed from a course on the vibrations of continuous systems delivered regularly by the first author at the Technische Universität Darmstadt over the last 30 years to fourth and fifth year students of Applied Mechanics, Mechanical Engineering, and other engineering curricula. This course deals exclusively with linear continuous systems and structures, including wave propagation in different media, in particular acoustic waves. The students come from a course on the vibrations of discrete systems, or at least with rudimentary knowledge of discrete vibrations. Over the years, the course content has changed more and more. The plan for a new text came up in 2004 when the second author was spending a year in Darmstadt as an Alexander von Humboldt Research Fellow. It was then that we started to work on the present book. Later, we had a chance to get together again for some time in the Mathematisches Forschungsinstitut Oberwolfach, in the Black Forest, in Germany. In this stimulating and pleasant environment we worked out many details that have found their place in the present book.

From the beginning, in the Darmstadt vibration course we aimed at presenting both the modal solutions and the traveling wave solutions, showing the relations between the two types of representations of solutions. We have found time and again in different engineering problems involving the vibrations of elastic structures, that one and the same problem can be handled in both ways, and this dual approach gives new insights. This is particularly useful whenever the spectra are rather dense, as for example in the vortex-excited vibrations of overhead transmission lines. We believe that stressing the duality between modal representation and a wave-type solution often leads to better understanding of the system's dynamics.

In a time when most of the structural vibrations problems in industry are dealt with by commercial finite-element and/or multi-body codes, often used as black boxes, it may seem that analytical solutions to vibration problems have become superfluous. True, in general it is hopeless to search for analytical solutions for vibrations problems in systems with complex geometry, for example. On the other hand, it can also be extremely dangerous to solve vibration problems using finite-element codes as black boxes without properly checking the applicability and convergence for the problem at hand. Often, for example, gyroscopic terms, non-classical damping and other effects may not be properly handled by the codes if these are used naively. There are worked problems in this book that clearly demonstrate this point. It is therefore important to have benchmark solutions for a large number of vibration problems. Such benchmark results are precisely given by the analytical solutions. Moreover, certain qualitative aspects, such as dependence on parameters, asymptotic behavior, or the basic physics of the problem can be easily recognized from the analytical solutions, and are difficult to find by purely numerical methods. In certain cases, a theoretical/analytical handle can also help in extracting the numerical solution accurately and efficiently. The authors therefore believe that analytical solutions for linear vibrations of continuous systems even today are of great relevance to engineering curricula.

This book deals mainly with the derivation of the linear equations of motion of continuous mechanical systems such as strings, rods, beams, plates and membranes as well as with their solution, both via modal decomposition, and by the wave approach. The equations are derived using the elementary Newton–Euler approach, as well as using variational techniques. Both the free vibrations and forced damped and undamped vibrations are studied. The eigenvalue problems are solved analytically wherever possible, and orthogonality conditions are derived. Problems with non-homogeneous boundary conditions and systems involving simultaneously distributed and lumped parameters are discussed in detail. Eigenvalue problems for systems in which the eigenvalue appears explicitly in the boundary conditions are examined, and the orthogonality of eigenfunctions is also derived for such systems. The forced vibrations are also studied through different solution techniques. Important discretization methods are discussed in a systematic fashion, including the Rayleigh–Ritz and the Galerkin methods. Scattering of waves, and energetics of wave propagation in continuous media are examined in detail. The wave approach is used to explain certain phenomena, such as dispersion, wave propagation during impact and radiation damping.

The dynamics of the aforementioned elastic structural elements are dealt with in the first five chapters. In each of these chapters, a number of free and forced vibration problems are solved, using both exact and also approximate techniques, modal and wave representation. Almost no attention is given to the numerical solution of matrix eigenvalue problems resulting from the discretization of continuous systems, since tools such as MATLAB or Mathematica are readily available for their solution. Among some topics less commonly found in vibration books are dynamics of systems involving continuous and lumped parameters, dynamics and wave propagation in traveling continua, wave propagation during impacts, and the phenomenon of radiation damping.

In Chapter 6, the self-adjoint boundary value problems of continuous elastic systems are dealt with in a somewhat more abstract manner, and general results, such as the expansion theorem and Rayleigh's quotient, are stated and discussed in general form. A formulation for the eigenvalue problem in terms of integral equations using Green's functions is also given. The same chapter also deals with the class of discretization methods in which the solution is written as a series of products of chosen shape-functions with unknown time functions (generalized coordinates). The different ways of minimizing the error then lead to the different methods such as the Rayleigh–Ritz method, the Galerkin method and the collocation method. This also includes finite-element methods, which can be regarded as a particular case of the Rayleigh–Ritz methods.

Chapter 7 is in two parts. The first part is devoted to waves in fluids, including acoustic media, propagation in wave guides and also in slightly viscous fluids. Radiation from membranes and plates is also examined. The second part deals with surface waves in incompressible liquids, sloshing of liquids in partially filled tanks, and surface waves in channels. Chapter 8 deals with elements of wave motion in three-dimensional elastic continua, and includes a short introduction to Rayleigh surface waves.

Three appendices complement the text. The first one is on Hamilton's principle and the variational formulation of dynamics, the second one on harmonic waves, Fourier representation of waves and dispersion, and the third one is on the variational formulation of plate dynamics.

Each chapter comes with a number of problems of different degrees of difficulty, most of which have been used as homework problems in the course. There are many others which are new. Some of the exercise problems are intended to motivate the reader to explore some of the more advanced topics that are available in scientific journals or more advanced texts.

The authors believe that this book will fill a void as a textbook for a course on the linear vibrations of continuous systems. The sections of the book are carefully planned so that they may be used selectively in an undergraduate course, or a post-graduate course. It is hoped that the presence of some of the advanced topics (all of which may not be possible to cover in one course) will inspire the students to explore beyond the limits of a formal course. This book also should be of use to engineers working in the field of structural vibrations and dynamics.

The authors thank the staff of the Dynamics and Vibrations group in Darmstadt, in particular Dr Daniel Hochlenert and Dr Gottfried Spelsberg-Korspeter. They not only participated in the Oberwolfach project and gave important inputs, but also spent some time at IIT Kharagpur with the second author, where they helped in setting up the Latex environment for producing the book. The second author thanks Professor Sandipan Ghosh Moulic for providing useful comments on Chapter 7, and Mr Miska Venu Babu for his help in preparing the figures. The authors also thank the Alexander von Humboldt Foundation, the DAAD (German Academic Exchange Service), which made possible the visit of Darmstadt staff to IIT Kharagpur, the Mathematisches Forschungsinstitut Oberwolfach, as well as Wiley staff, who were extremely helpful in producing this book.

March 2007

Peter Hagedorn
Darmstadt

Anirvan DasGupta
Kharagpur

1

Vibrations of strings and bars

A one-dimensional continuous system, whose configuration at any time requires only one space dimension for description, is the simplest model of a class of continua with boundaries. Strings in transverse vibration, and bars of certain geometries in axial and torsional vibrations may be adequately described by one-dimensional continuous models. In this chapter, we will consider such models that are not only simple to study, but also are useful in developing the basic framework for analysis of continuous systems of one or more dimensions.

1.1 DYNAMICS OF STRINGS AND BARS: THE NEWTONIAN FORMULATION

1.1.1 Transverse dynamics of strings

A string is a one-dimensional elastic continuum that does not transmit or resist bending moment. Such an idealization may be justified even for cable-like components when the ratio of the thickness of the cable to its length (or wavelength of waves in the cable) is small compared to unity. In deriving the elementary equation of motion, it is assumed that the motion of the string is planar, and transverse to its length, i.e., longitudinal motion is neglected. Further, the amplitude of motion is assumed to be small enough so that the change in tension is negligible.

Consider a string, stretched along the x-axis to a length l by a tension T, as shown in Figure 1.1. Arbitrary distributed forces are assumed to act over the length of the string. The transverse motion of any point on the string at the coordinate position x is represented by the field variable $w(x, t)$ where t is the time. Consider the free body diagram of a small element of the string between two closely spaced points x and $x + \Delta x$, as shown in Figure 1.2. Let the element have a mass $\Delta m(x)$, and a deformed length Δs. The tensions at the two ends are $T(x, t)$ and $T(x + \Delta x, t)$, respectively, and the external force densities (force per unit length) are $p(x, t)$ in the transverse direction, and $n(x, t)$ in the longitudinal direction, as shown in the figure. Neglecting the inertia force in the longitudinal direction of the string, we can write the force balance equation for the small element in the longitudinal direction as

$$0 = T(x + \Delta x, t)\cos[\alpha(x + \Delta x, t)] - T(x, t)\cos[\alpha(x, t)] + n(x, t)\Delta s, \tag{1.1}$$

Vibrations and Waves in Continuous Mechanical Systems P. Hagedorn and A. DasGupta

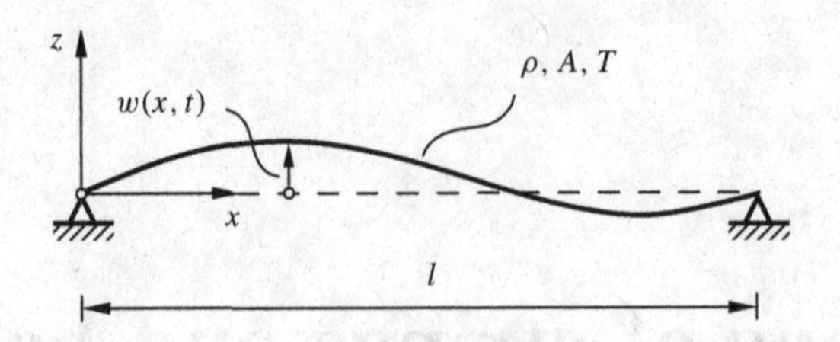

Figure 1.1 Schematic representation of a taut string

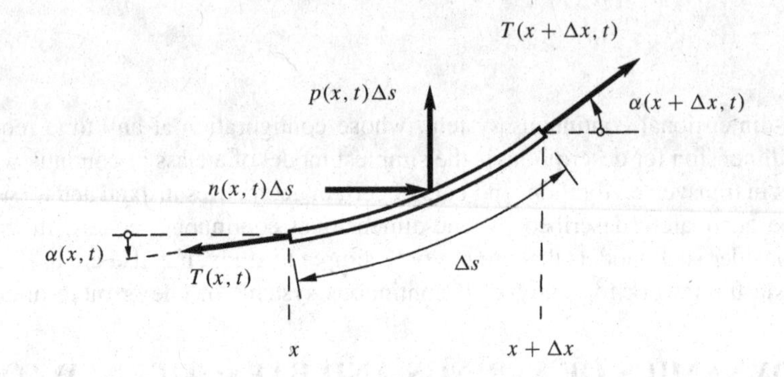

Figure 1.2 Free body diagram of a string element

where $\alpha(x,t)$ represents the angle between the tangent to the string at x and the x-axis, as shown in Figure 1.2. Dividing both sides of (1.1) by Δx and taking the limit $\Delta x \to 0$ yields

$$[T(x,t)\cos\alpha(x,t)]_{,x} = -n(x,t)\frac{\mathrm{d}s}{\mathrm{d}x}, \tag{1.2}$$

where $[\cdot]_{,x}$ represents partial derivative with respect to x. From geometry, one can write

$$\cos\alpha = \frac{1}{\sqrt{1+\tan^2\alpha}} = \frac{1}{\sqrt{1+w_{,x}^2}}, \quad \text{and} \quad \frac{\mathrm{d}s}{\mathrm{d}x} = \sqrt{1+w_{,x}^2}. \tag{1.3}$$

Substituting (1.3) in (1.2), and assuming $w_{,x} \ll 1$, yields on simplification

$$[T(x,t)]_{,x} = -n(x,t). \tag{1.4}$$

Therefore, when $n(x,t) \equiv 0$, (1.4) implies that the tension $T(x,t)$ is a constant. On the other hand, for a hanging string, shown in Figure 1.3, one has $n(x,t) = \rho A(x)g$, where ρ is the density, A is the area of cross-section, and g is the acceleration due to gravity. Then, using the boundary condition of zero tension at the free end, i.e., $T(l,t) \equiv 0$ (for constant ρA), (1.4) yields $T(x,t) = \rho Ag(l-x)$. In general, the tension in a string may also depend on time. However, in the following discussions, it will be assumed to depend at most on x.

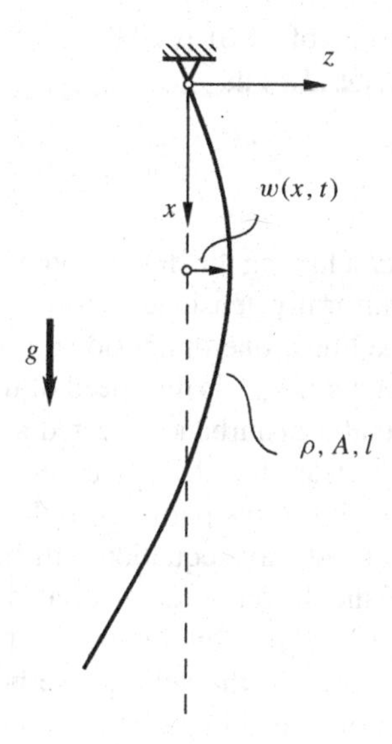

Figure 1.3 Schematic representation of a hanging string

Now, consider the transverse dynamics of the string element shown in Figure 1.1. The equation of motion of the small element in the transverse direction can be written from Newton's second law of motion as

$$\Delta m w_{,tt}(x+\theta\Delta x, t) = T(x+\Delta x)\sin[\alpha(x+\Delta x, t)]$$
$$-T(x)\sin[\alpha(x,t)] + p(x,t)\Delta s, \qquad (1.5)$$

where Δm is the mass of the element, $\theta \in [0, 1]$, and $(\cdot)_{,tt}$ indicates double partial differentiation with respect to time. Again assuming $w_{,x} \ll 1$, one can write $\sin\alpha \approx \tan\alpha = w_{,x}$. Further, $\Delta m = \rho A(x)\Delta s$. Using these expressions in (1.5) and dividing by Δx on both sides, one can write after taking the limit $\Delta x \to 0$

$$\rho A(x) w_{,tt} - [T(x) w_{,x}]_{,x} = p(x,t), \qquad (1.6)$$

where, based on the previous considerations, we have assumed $\mathrm{d}s/\mathrm{d}x \approx 1$. The linear partial differential equation (1.6), along with (1.4), represents the dynamics of a taut string. When the external force is not distributed but a concentrated force acting at, say $x = a$, the forcing function on the right hand side of (1.6) can be written using the *Dirac delta function* as

$$p(x,t) = f(t)\delta(x-a), \qquad (1.7)$$

where $f(t)$ is the time-varying force, and $\delta(\cdot)$ represents the Dirac delta function.

Let us consider the hanging string shown in Figure 1.3 once again. The expression of tension derived earlier was $T(x) = \rho A g(l-x)$. Substituting this expression in (1.6) and assuming $p(x,t) \equiv 0$, one obtains on simplification

$$w_{,tt} - g[(l-x)w_{,x}]_{,x} = 0. \qquad (1.8)$$

This case will be considered again later.

An important particular form of (1.6) is obtained for $p(x,t) \equiv 0$, and T and ρA not depending on x. We can rewrite (1.6) as

$$w_{,tt} - c^2 w_{,xx} = 0, \tag{1.9}$$

where $c = \sqrt{T/\rho A}$ is a constant having the dimension of speed. This represents the unforced transverse dynamics of a uniformly tensioned string. The hyperbolic partial differential equation (1.9) is known as the linear one-dimensional *wave equation*, and c is known as the wave speed. In the case of a taut string, c is the speed of transverse waves on the string, as we shall see later. This implies that a disturbance created at any point on the string propagates with a speed c. It should be clear that the wave speed c is distinct from the transverse material velocity (i.e., the velocity of the particles of the string) which is given by $w_{,t}(x,t)$. The solution and properties of the wave equation will be discussed in detail in Chapter 2.

The complete solution of the second-order partial differential equation (1.6) (or (1.9)) requires specification of two boundary conditions, and two initial conditions. For example, for a taut string shown in Figure 1.1, the appropriate boundary conditions are $w(0,t) \equiv 0$ and $w(l,t) \equiv 0$. For the case of a hanging string, the boundary conditions are $w(0,t) \equiv 0$ and $w(l,t)$ is finite. The initial conditions are usually specified in terms of the initial shape of the string, and initial velocity of the string, i.e., in the forms $w(x,0) = w_0(x)$, and $w_{,t}(x,0) = v_0(x)$, respectively. These will be discussed further later in this chapter.

Boundary conditions are classified into two types, namely *geometric* (or *essential*) boundary conditions, and *dynamic* (or *natural*) boundary conditions. A geometric boundary condition is one that imposes a kinematic constraint on the system at the boundary. The forces at such a boundary adjust themselves to maintain the constraint. On the other hand, a dynamic boundary condition imposes a condition on the forces, and the geometry adjusts itself to maintain the force condition. For example, in Figure 1.4, the right-end boundary condition is obtained from the consideration that the component of the tension in the transverse direction is zero, the roller being assumed massless. This implies $Tw_{,x}(l,t) \equiv 0$, which is a natural boundary condition. As a consequence of this force condition, the slope of the string remains zero. At the left-end boundary, the condition $w(0,t) \equiv 0$ is a geometric boundary condition, and the transverse force from the support point (which can be computed as $Tw_{,x}(0,t)$) will adjust itself appropriately to prevent any transverse motion of the right end of the string. Classification of boundary conditions based on their mathematical structure is discussed in Section 6.1.1.

When a string, in addition to the distributed mass, carries lumped masses (i.e., particles of finite mass) and is subjected to concentrated elastic restoring forces, these can be

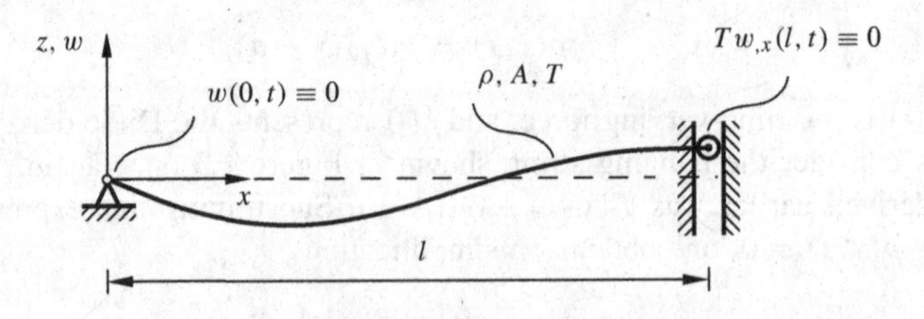

Figure 1.4 A taut string with geometric and natural boundary conditions

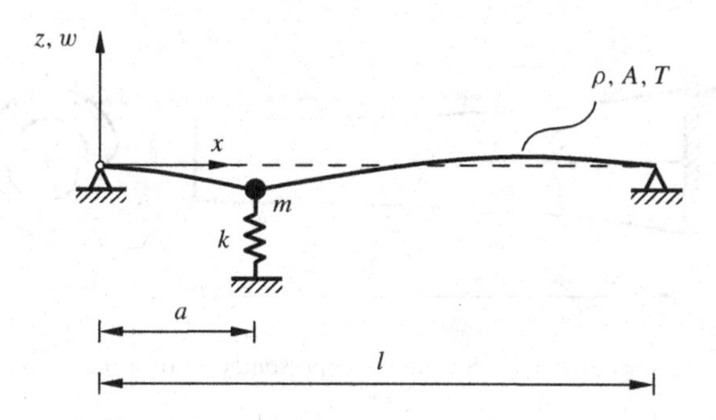

Figure 1.5 A taut string with lumped elements

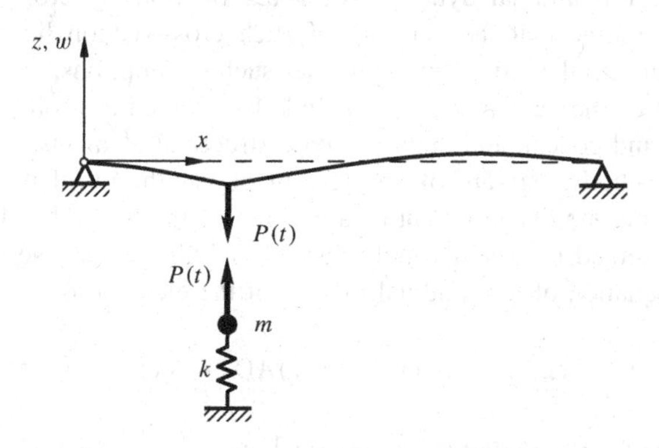

Figure 1.6 The interaction force diagram

easily incorporated into the equation of motion as follows. Consider the system shown in Figure 1.5, and the interaction force diagram shown in Figure 1.6. The force $P(t)$ at the interface between the string and the particle of mass m can be written from Newton's second law for the mass–spring system as $P(t) = mw_{,tt}(a, t) + kw(a, t)$, where $x = a$ is the location of the lumped system. Using the Dirac delta function, one can represent $P(t)$ as a distributed force

$$p(x, t) = mw_{,tt}(x, t)\delta(x - a) + kw(x, t)\delta(x - a). \tag{1.10}$$

Therefore, the equation of motion of the combined system can be written as

$$\rho A(x)w_{,tt} - [T(x)w_{,x}]_{,x} = -p(x, t),$$

or

$$[\rho A(x) + m\delta(x - a)]w_{,tt} - [T(x)w_{,x}]_{,x} + k\delta(x - a)w = 0. \tag{1.11}$$

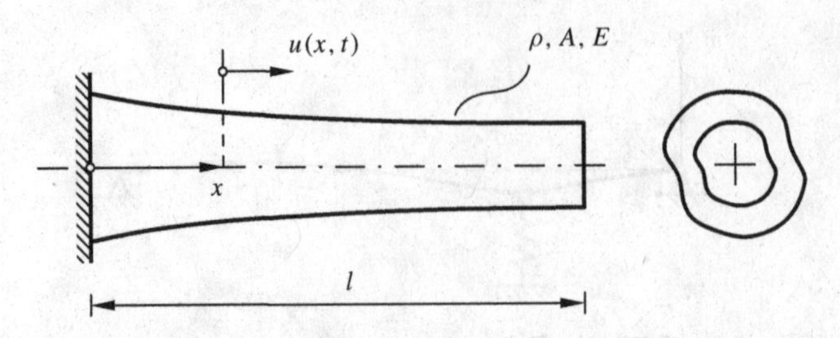

Figure 1.7 Schematic representation of a bar

1.1.2 Longitudinal dynamics of bars

Let us consider the longitudinal dynamics of a bar of arbitrary cross-section, as shown in Figure 1.7. We assume that the centroid of each cross-section lies on a straight line which is perpendicular to the cross-section. Under such assumptions, we can study the pure longitudinal motion of the bar. Such cases include bars which are solids of revolution (for example, cylinders and cones), and other standard structural elements.

Consider the free body diagram of an element of length Δx of the bar, as shown in Figure 1.8. We assume the displacement of any point of the bar to be along the x-axis, so that it can be represented by a single field variable $u(x, t)$. Using Newton's second law, one can write the equation of longitudinal motion of the element as

$$\rho A(x)\Delta x u_{,tt}(x + \theta\Delta x, t) = \sigma_x(x + \Delta x, t)A(x + \Delta x) - \sigma_x(x, t)A(x), \tag{1.12}$$

where ρ is the density, $A(x)$ is the cross-sectional area at x, $\theta \in [0, 1]$, and $\sigma_x(x, t)$ is the normal stress over the cross-section. Dividing (1.12) by Δx, and taking the limit $\Delta x \to 0$, yields

$$\rho A(x)u_{,tt}(x, t) = [\sigma_x(x, t)A(x)]_{,x}. \tag{1.13}$$

From elementary theory of elasticity (see [1]), we can relate the longitudinal strain $\epsilon_x(x, t)$ and the displacement field as $\epsilon_x(x, t) = u_{,x}(x, t)$. Using this strain–displacement relation and Hooke's law, one can write

$$\sigma_x(x, t) = E\epsilon_x(x, t) = Eu_{,x}(x, t), \tag{1.14}$$

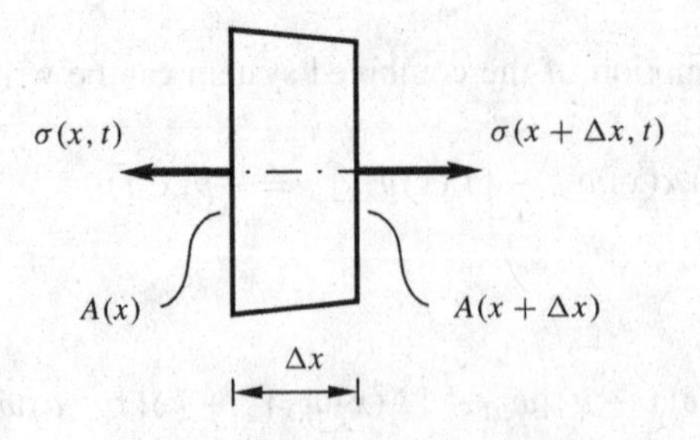

Figure 1.8 Free body diagram of a bar element

where E is the material's Young's modulus. Using (1.14) in (1.13) yields on rearrangement

$$\rho A(x)u_{,tt} - [EA(x)u_{,x}]_{,x} = 0. \tag{1.15}$$

If the bar is homogeneous and has a uniform cross-section, then (1.15) simplifies to

$$u_{,tt} - c^2 u_{,xx} = 0, \tag{1.16}$$

where $c = \sqrt{E/\rho}$ is the speed of the longitudinal waves in a uniform bar.

The boundary conditions for the bar can be written by inspection. For example, in Figure 1.7, the left-end boundary condition is $u(0, t) \equiv 0$, which is a geometric boundary condition. The right end of the bar is force-free, i.e., $EAu_{,x}(l, t) \equiv 0$. Hence, the right end of the bar has a dynamic boundary condition.

1.1.3 Torsional dynamics of bars

In this section, we make the same assumptions regarding the centroidal axis as made for the longitudinal dynamics of bars. The torsional dynamics of a bar depends on the shape of its cross-section. Complications arise due to warping of the cross-section during torsion in bars with non-circular cross-section (see [1]). In general, the torsional vibration of a bar is also coupled with its flexural vibration. Therefore, to keep the discussion simple, we will consider only torsional dynamics of bars with circular cross-section. As is known from the theory of elasticity, for bars with circular cross-section, planar sections remain planar for small torsional deformation. Further, an imaginary radial line on the undeformed cross-section can be assumed to remain straight even after deformation.

Consider a circular bar, as shown in Figure 1.9. A small sectional element of the bar between the centroidal coordinates x and $x + \Delta x$ is shown in Figure 1.10. Let $\phi(x, t)$ be the angle of twist at coordinate x, and $\phi + \Delta\phi(x, t)$ be the twist at $x + \Delta x$. From Figure 1.10, one can write, at any radius r, the kinematic relation

$$r\Delta\phi(x, t) = \Delta x \psi(r, t), \tag{1.17}$$

where $\psi(r, t)$ is the angular deformation of a longitudinal line at r, as shown in the figure. This angular deformation is the shear angle, as shown in Figure 1.11. Then, the shear stress $\tau_{x\phi}(r, t)$ is obtained from Hooke's law as

$$\tau_{x\phi}(r, t) = G\psi(r, t), \tag{1.18}$$

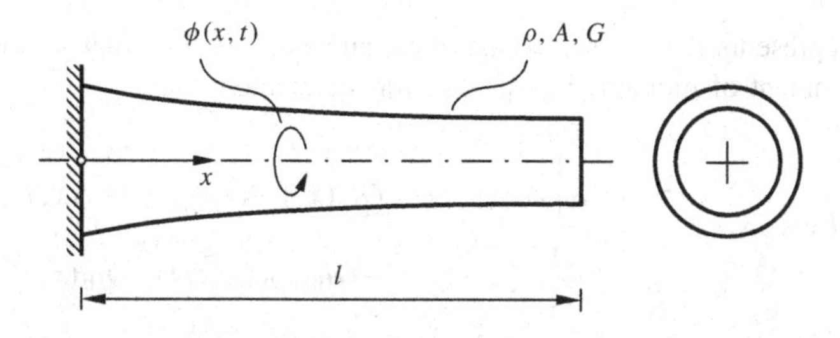

Figure 1.9 Schematic representation of a circular bar

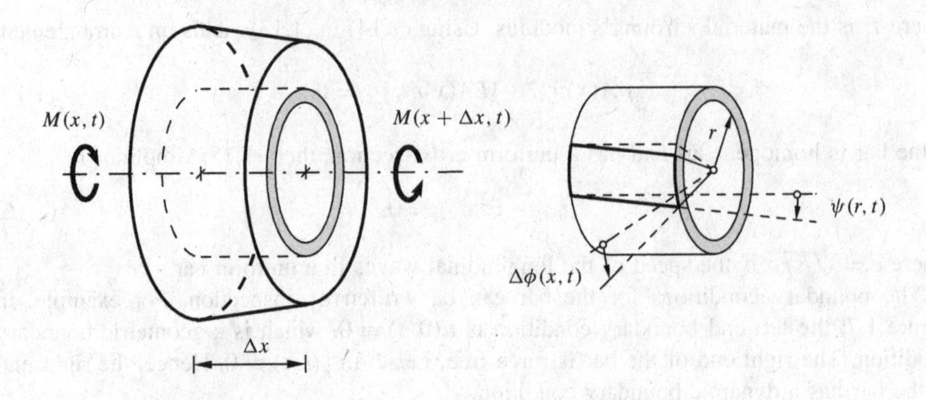

Figure 1.10 Deformation of a bar element under torsion

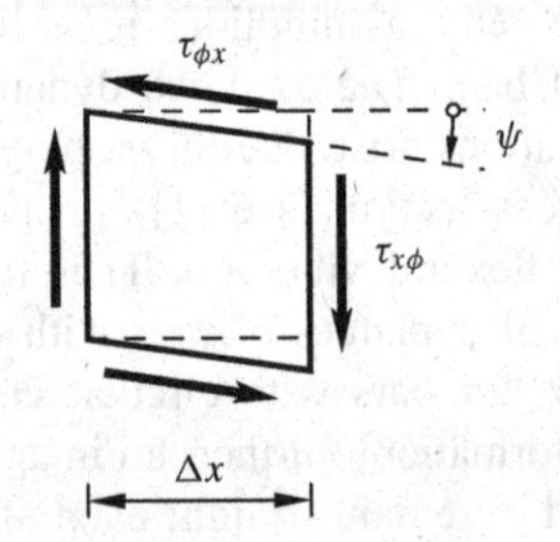

Figure 1.11 State of stress on a bar element under torsion

where G is the shear modulus. Substituting the expression of $\psi(r, t)$ from (1.17) in (1.18), one can write in the limit $\Delta x \to 0$

$$\tau_{x\phi}(r, t) = Gr\phi_{,x}. \tag{1.19}$$

Now, the torque at any cross-section x can be computed as

$$M(x, t) = \int_{A(x)} r\tau_{x\phi}(r, t)\,\mathrm{d}A = G\phi_{,x} \int_{A(x)} r^2\mathrm{d}A = GI_{\mathrm{p}}(x)\phi_{,x}, \tag{1.20}$$

where $A(x)$ represents the cross-sectional area, and $I_{\mathrm{p}}(x)$ is the polar moment of the area. Writing the moment of momentum equation for the element yields

$$\left[\int_{A(x+\theta\Delta x)} \rho r^2 \Delta x\,\mathrm{d}A\right] \phi_{,tt}(x, t) = GI_{\mathrm{p}}(x + \Delta x)\phi_{,x}(x + \Delta x, t)$$
$$-GI_{\mathrm{p}}(x)\phi_{,x}(x, t) + n_{\mathrm{E}}(x, t)\Delta x, \tag{1.21}$$

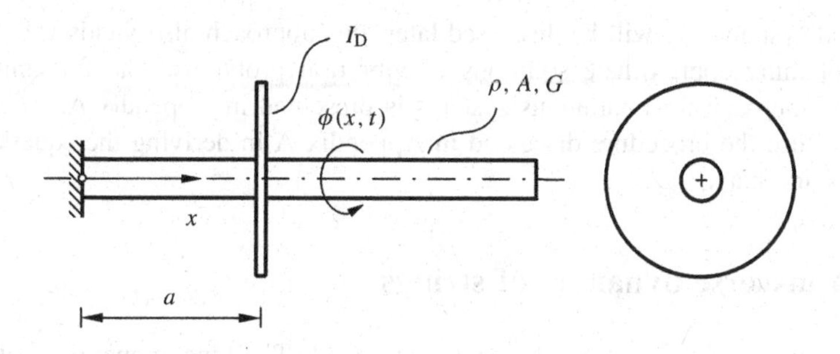

Figure 1.12 A circular bar with a disc

where $n_E(x, t)$ is an externally applied torque distribution. Dividing both sides in (1.21) by Δx and taking the limit $\Delta x \to 0$, we obtain

$$\rho I_p \phi_{,tt} - (G I_p \phi_{,x})_{,x} = n_E(x, t). \tag{1.22}$$

The partial differential equation (1.22) represents the torsional dynamics of a circular bar. For a bar with uniform cross-section (i.e., I_p independent of x), and $n_E(x, t) \equiv 0$, we obtain the wave equation

$$\phi_{,tt} - c^2 \phi_{,xx} = 0, \tag{1.23}$$

where $c = \sqrt{G/\rho}$ is the speed of torsional waves in the bar.

The boundary conditions for the fixed–free bar shown in Figure 1.9 can be written as $\phi(0, t) \equiv 0$, and $M(l, t) = G I_p \phi_{,x}(l, t) \equiv 0$. We can easily identify the first boundary condition as geometric, while the second is a natural boundary condition.

As an example, consider the torsional dynamics of a uniform circular bar with a massive disc at $x = a$, as shown in Figure 1.12. The disc can be considered as having a lumped rotational inertia. Therefore, the bar experiences an external torque due to the rotational inertia of the disc given by $n_E(x, t) = -I_D \phi_{,tt}(x, t)\delta(x - a)$, where I_D is the rotational inertia of the disc. Substituting this expression of external moment in (1.22), the complete equation of torsional dynamics of the bar can then be written as

$$[\rho I_p + I_D \delta(x - a)]\phi_{,tt} - G I_p \phi_{,xx} = 0. \tag{1.24}$$

1.2 DYNAMICS OF STRINGS AND BARS: THE VARIATIONAL FORMULATION

The variational formulation presents an elegant and powerful method of deriving the equations of motion of a dynamical system. Through this formulation, all the boundary conditions of a system are revealed. This is clearly an advantage especially for continuous mechanical systems. As will be discussed later, this approach also yields very useful methods of obtaining approximate solutions of vibration problems. The fundamentals of the variational approach for continuous systems is presented in Appendix A. In the following, we directly use the procedure discussed in Appendix A in deriving the equation of motion for strings and bars.

1.2.1 Transverse dynamics of strings

Consider a string of length l, as shown in Figure 1.1. The kinetic energy $\mathcal{T}$ of the string is

$$\mathcal{T} = \frac{1}{2}\int_0^l \rho A w_{,t}^2 \, \mathrm{d}x. \tag{1.25}$$

The potential energy can be written from the consideration that the unstretched length Δx is stretched to $\Delta s = \sqrt{1 + w_{,x}^2}\,\Delta x$ under a constant tension T. Therefore, the potential energy $\mathcal{V}$ stored in the string is given by

$$\begin{aligned}\mathcal{V} &= \int_0^l T(\mathrm{d}s - \mathrm{d}x) \approx \int_0^l T\left[\left(1 + \frac{1}{2}w_{,x}^2\right) - 1\right]\mathrm{d}x \\ &= \frac{1}{2}\int_0^l T w_{,x}^2 \, \mathrm{d}x.\end{aligned} \tag{1.26}$$

Defining the Lagrangian $\mathcal{L} = \mathcal{T} - \mathcal{V}$, Hamilton's principle can be written as

$$\delta \int_{t_1}^{t_2} \mathcal{L}\, \mathrm{d}t = 0 \tag{1.27}$$

or

$$\delta \int_{t_1}^{t_2} \frac{1}{2}\int_0^l \left[\rho A w_{,t}^2 - T w_{,x}^2\right] \mathrm{d}x. \tag{1.28}$$

As detailed in Appendix A, one obtains from (1.28)

$$\begin{aligned}&\int_0^l \rho A w_{,t} \delta w\Big|_{t_1}^{t_2} \mathrm{d}x - \int_{t_1}^{t_2} T w_{,x} \delta w\Big|_0^l \mathrm{d}t \\ &- \int_{t_1}^{t_2}\int_0^l \left[\rho A w_{,tt} - (T w_{,x})_{,x}\right] \delta w \, \mathrm{d}x \, \mathrm{d}t = 0.\end{aligned} \tag{1.29}$$

The first term in (1.29) is always zero since the variations of the field variable at the initial and final times are zero, i.e., $\delta w(x, t_0) \equiv 0$, and $\delta w(x, t_1) \equiv 0$. Following the arguments in Appendix A, the integrand of the third term in (1.29) has to be zero, i.e.,

$$\rho A w_{,tt} - (T w_{,x})_{,x} = 0, \tag{1.30}$$

which yields the equation of transverse dynamics of the string. The second term in (1.29) is zero if, for example,

$$Tw_{,x}(0,t) \equiv 0 \qquad \text{or} \qquad w(0,t) \equiv 0 \tag{1.31}$$

and

$$Tw_{,x}(l,t) \equiv 0 \qquad \text{or} \qquad w(l,t) \equiv 0, \tag{1.32}$$

which represent possible boundary conditions. For a fixed–fixed string, the conditions $w(0,t) \equiv 0$ and $w(l,t) \equiv 0$ hold, while for a fixed–sliding string (see Figure 1.4), $w(0,t) \equiv 0$ and $Tw_{,x}(l,t) \equiv 0$.

In the case of a string with discrete elements shown in Figure 1.5, the kinetic and potential energies can be written as, respectively,

$$\begin{aligned}\mathcal{T} &= \frac{1}{2}\int_0^l \rho A w_{,t}^2(x,t)\,\mathrm{d}x + \frac{1}{2} m w_{,tt}^2(a,t)\\ &= \frac{1}{2}\int_0^l [\rho A + m\delta(x-a)] w_{,tt}^2(x,t)\,\mathrm{d}x,\end{aligned} \tag{1.33}$$

$$\begin{aligned}\mathcal{V} &= \frac{1}{2}\int_0^l T w_{,x}^2(x,t)\,\mathrm{d}x + \frac{1}{2} k w^2(a,t)\\ &= \frac{1}{2}\int_0^l [T w_{,x}^2(x,t) + k\delta(x-a) w^2(x,t)]\,\mathrm{d}x.\end{aligned} \tag{1.34}$$

Substituting $\mathcal{L} = \mathcal{T} - \mathcal{V}$ in the variational form (1.27) and taking the variation yields on simplification

$$\begin{aligned}&\int_0^l [\rho A + m\delta(x-a)] w_{,t}\delta w\Big|_{t_1}^{t_2}\,\mathrm{d}x - \int_{t_1}^{t_2} T w_{,x}\delta w\Big|_0^l\,\mathrm{d}t\\ &\quad - \int_{t_1}^{t_2}\int_0^l [(\rho A + m\delta(x-a)) w_{,tt} - (T w_{,x})_{,x} + k\delta(x-a) w]\,\delta w\,\mathrm{d}x\,\mathrm{d}t = 0.\end{aligned}$$

The equation of motion is obtained from the third term above which is the same as (1.11). The boundary conditions remain the same as in (1.31)–(1.32). When external forces are present, one can use the extended Hamilton's principle discussed in Appendix A to obtain the equations of motion.

1.2.2 Longitudinal dynamics of bars

In the case of longitudinal vibration of a bar, the kinetic energy is given by

$$\mathcal{T} = \frac{1}{2}\int_0^l \rho A u_{,t}^2\,\mathrm{d}x. \tag{1.35}$$

Defining σ_x and ϵ_x as the longitudinal stress and strain, respectively, the potential energy can be computed from the theory of elasticity as

$$\mathcal{V} = \frac{1}{2}\int_0^l \sigma_x \epsilon_x A \, \mathrm{d}x = \frac{1}{2}\int_0^l EA\epsilon_x^2 \, \mathrm{d}x$$
$$= \frac{1}{2}\int_0^l EAu_{,x}^2 \, \mathrm{d}x. \tag{1.36}$$

Writing the Lagrangian $\mathcal{L} = \mathcal{T} - \mathcal{V}$, Hamilton's principle assumes the form

$$\delta \int_{t_1}^{t_2} \mathcal{L} \, \mathrm{d}t = 0,$$

or

$$\delta \int_{t_1}^{t_2} \frac{1}{2}\int_0^l \left(\rho A u_{,t}^2 - EAu_{,x}^2\right) \mathrm{d}x \, \mathrm{d}t = 0,$$
$$\Rightarrow \int_0^l \rho A \delta u\big|_{t_1}^{t_2} \mathrm{d}x - \int_{t_1}^{t_2} EAu_{,x}\delta u\big|_0^l \, \mathrm{d}t$$
$$- \int_{t_1}^{t_2}\int_0^l [\rho A u_{,tt} - (EAu_{,x})_{,x}]\, \delta u \, \mathrm{d}x \, \mathrm{d}t = 0. \tag{1.37}$$

Since by definition $\delta u(x, t_0) = \delta u(x, t_1) \equiv 0$, the first term in (1.37) vanishes identically. The third term in (1.37) yields the equation of motion

$$\rho A u_{,tt} - (EAu_{,x})_{,x} = 0, \tag{1.38}$$

and the boundary conditions are obtained from the second term. For example, the boundary conditions can be written as

$$EAu_{,x}(0, t) \equiv 0 \qquad \text{or} \qquad u(0, t) \equiv 0, \tag{1.39}$$

and

$$EAu_{,x}(l, t) \equiv 0 \qquad \text{or} \qquad u(l, t) \equiv 0. \tag{1.40}$$

It can be seen that the first condition in both (1.39) and (1.40) is the longitudinal force condition (natural boundary condition) at the two ends of the bar, while the second condition is the displacement condition (geometric boundary condition). Thus, for a fixed–fixed bar, $u(0, t) \equiv 0$, and $u(l, t) \equiv 0$, while for a fixed–free bar, $u(0, t) \equiv 0$ and $EAu_{,x}(l, t) \equiv 0$. In the case of a free–free bar, the boundary conditions are $EAu_{,x}(0, t) \equiv 0$, and $EAu_{,x}(l, t) \equiv 0$.

1.2.3 Torsional dynamics of bars

The kinetic energy of a circular bar undergoing torsional oscillations can be written in the notations used previously in Section 1.1.3 as

$$\begin{aligned}\mathcal{T} &= \frac{1}{2}\int_0^l \int_0^R \int_0^{2\pi} \rho\phi_{,t}^2 r^3\, \mathrm{d}\phi\, \mathrm{d}r\, \mathrm{d}x \\ &= \frac{1}{2}\int_0^l \rho I_{\mathrm{p}}\phi_{,t}^2\, \mathrm{d}x. \end{aligned} \tag{1.41}$$

The potential energy can be written from elasticity theory as

$$\mathcal{V} = \frac{1}{2}\int_0^l \int_0^R \int_0^{2\pi} \tau_{x\phi}\psi r\, \mathrm{d}\phi\, \mathrm{d}r \mathrm{d}x. \tag{1.42}$$

Using the definitions of $\tau_{r\phi}$ and $\psi(x, t)$ from (1.17) and (1.18), respectively, in (1.42), we have

$$\begin{aligned}\mathcal{V} &= \frac{1}{2}\int_0^l \int_0^R \int_0^{2\pi} G\phi_{,x}^2 r^3\, \mathrm{d}\phi\, \mathrm{d}r\, \mathrm{d}x \\ &= \frac{1}{2}\int_0^l GI_{\mathrm{p}}\phi_{,x}^2\, \mathrm{d}x. \end{aligned} \tag{1.43}$$

Hamilton's principle can then be written as

$$\begin{aligned} &\delta \int_{t_1}^{t_2} \frac{1}{2}\int_0^l \left[\rho I_{\mathrm{p}}\phi_{,t}^2 - GI_{\mathrm{p}}\phi_{,x}^2\right] \mathrm{d}x = 0 \\ \Rightarrow &\int_0^l \rho I_{\mathrm{p}}\phi_{,t}\delta\phi\Big|_{t_1}^{t_2} \mathrm{d}x - \int_{t_1}^{t_2} GI_{\mathrm{p}}\phi_{,x}\delta\phi\Big|_0^l \mathrm{d}t \\ &- \int_{t_1}^{t_2}\int_0^l [\rho I_{\mathrm{p}}\phi_{,tt} - (GI_{\mathrm{p}}\phi_{,x})_{,x}]\,\delta\phi\, \mathrm{d}x = 0. \end{aligned} \tag{1.44}$$

The first term in (1.44) is zero by definition of the variational formulation. The third term in (1.44) yields the equation of motion

$$\rho I_{\mathrm{p}}\phi_{,tt} - (GI_{\mathrm{p}}\phi_{,x})_{,x} = 0, \tag{1.45}$$

while the second term provides information on the boundary conditions. For example, the possible boundary conditions could be

$$GI_{\mathrm{p}}\phi_{,x}(0, t) \equiv 0 \quad \text{or} \quad \phi(0, t) \equiv 0, \tag{1.46}$$

and

$$GI_p\phi_{,x}(l,t) \equiv 0 \qquad \text{or} \qquad \phi(l,t) \equiv 0. \tag{1.47}$$

The first condition in (1.46) and (1.47) can be easily identified to be the torque condition (natural boundary condition) at the ends of the bar, while the second condition is on the angular displacement (geometric boundary condition).

1.3 FREE VIBRATION PROBLEM: BERNOULLI'S SOLUTION

Vibration analysis of a system almost always starts with the free or natural vibration analysis. This leads us to the important concepts of natural frequency and mode of vibration of the system. These two concepts form the starting point of any quantitative and qualitative analysis and understanding of a vibratory system.

It was observed in the above discussions that, under certain assumptions of uniformity, the one-dimensional wave equation represents the transverse dynamics of a string, and longitudinal and torsional dynamics of a bar. The wave equation is one of the most important equations that appear in the study of vibrations of continuous systems. The solution and properties of the wave equation are fundamental in understanding vibration and propagation of vibration in continuous media, and will be taken up in detail in later chapters. In this section, we will discuss a simple solution procedure for the one-dimensional wave equation and study some of the solution properties.

Consider the wave equation

$$w_{,tt} - c^2 w_{,xx} = 0, \qquad x \in [0, l], \tag{1.48}$$

with the boundary conditions

$$w(0,t) \equiv 0, \qquad \text{and} \qquad w(l,t) \equiv 0. \tag{1.49}$$

Such a problem corresponds to, for example, a fixed-fixed string or bar.

Let us first look for separable solutions of (1.48) in the form

$$w(x,t) = p(t)W(x). \tag{1.50}$$

Substituting (1.50) in (1.48) yields on rearrangement

$$\frac{\ddot{p}}{p} - c^2\frac{W''}{W} = 0. \tag{1.51}$$

It is easily observed that the first term in (1.51) is solely a function of t, while the second term is solely a function of x. Therefore, (1.51) will hold identically if and only if both the terms are constant, i.e.,

$$\frac{\ddot{p}}{p} = -\omega^2 \qquad \text{and} \qquad c^2\frac{W''}{W} = -\omega^2 \tag{1.52}$$

$$\Rightarrow \ddot{p} + \omega^2 p = 0 \tag{1.53}$$

and

$$W'' + \frac{\omega^2}{c^2} W = 0, \tag{1.54}$$

where ω is an arbitrary constant. It may be noted that the constant in (1.52) is chosen as $-\omega^2$, so that ω later will have the meaning of a circular frequency.

The general solutions of (1.53) and (1.54) can be written as, respectively,

$$p(t) = C \cos \omega t + S \sin \omega t \tag{1.55}$$

and

$$W(x) = D \cos \frac{\omega x}{c} + H \sin \frac{\omega x}{c}, \tag{1.56}$$

where C, S, D, and H are arbitrary constants of integration. The constants C and S are usually determined from the initial position and velocity conditions of the string/bar, while the determination of D and H requires the conditions at the two boundaries of the string/bar. Since the solution (1.50) must satisfy the boundary conditions (1.49), we must have

$$W(0) = 0 \quad \text{and} \quad W(l) = 0, \tag{1.57}$$

which can be written using (1.56) as

$$D + 0 \cdot H = 0,$$

and

$$\left(\cos \frac{\omega l}{c}\right) D + \left(\sin \frac{\omega l}{c}\right) H = 0$$

$$\Rightarrow \begin{bmatrix} 1 & 0 \\ \cos \frac{\omega l}{c} & \sin \frac{\omega l}{c} \end{bmatrix} \begin{Bmatrix} D \\ H \end{Bmatrix} = 0. \tag{1.58}$$

Therefore, for a non-trivial solution of D and H, the determinant of (1.58) must vanish, i.e.,

$$\sin \frac{\omega l}{c} = 0. \tag{1.59}$$

This equation is referred to as the *characteristic equation* of the system (1.48)–(1.49). The characteristic equation (1.59) is satisfied when ω takes any of the discrete values

$$\omega_k = \frac{k \pi c}{l}, \qquad k = 0, 1, \ldots, \infty \tag{1.60}$$

where ω_k is termed as the kth *circular natural frequency* of the system. Thus, there are countably infinitely many natural frequencies of the continuous system (1.48)–(1.49). It may be noted that the negative solutions of ω have been dropped, as they do not yield any new solution to the vibration problem.

To every circular frequency ω_k, there corresponds a solution of (D_k, H_k). On substituting (1.60) in (1.58), one can conclude that $D_k = 0$ for all k, and H_k can be arbitrary. Thus, the solution of (D_k, H_k) can be determined up to an arbitrary (multiplicative) constant. Finally, from (1.56), we have

$$W_k(x) = H_k \sin\frac{\omega_k x}{c} = H_k \sin\frac{k\pi x}{l}, \qquad k = 1, 2, \ldots, \infty, \tag{1.61}$$

corresponding to the circular natural frequencies ω_k, $k = 1, 2, \ldots, \infty$. It may be observed that $k = 0$ has been dropped since $W_0(x) \equiv 0$, which is the trivial solution. We have therefore found infinitely many solutions of the form (1.50), which may be written as

$$\begin{aligned} w_k(x,t) &= p_k(t) W_k(x) \\ &= (C_k \cos\omega_k t + S_k \sin\omega_k t)\sin\frac{k\pi x}{l}, \qquad k = 1, 2, \ldots, \infty, \end{aligned} \tag{1.62}$$

where we have set all $H_k = 1$ without loss of generality, and C_k and S_k are arbitrary constants.

Let us assume that the system is oscillating according to any one of the infinite solutions given by (1.62). Corresponding to this solution, it is clear from (1.62) that all points of the string or bar oscillate with the same circular frequency ω_k. The system is then said to oscillate in the kth *mode*, and the solution $w_k(x,t)$ is known as the *modal solution* of the kth mode. The function $W_k(x)$ is known as the kth *eigenfunction* or *mode-shape-function*. The circular natural frequency ω_k is also referred to as the kth *circular modal frequency* of the system. It may be observed for any modal solution that, whenever $p_k(t) = 0$, the displacement $w(x,t)$ of all points of the string/bar is equal to zero. Thus, when the system is oscillating in a particular mode, all points pass through their equilibrium positions at the same time. Further, any two points on the string/bar have a phase difference of either 0 or π between their motion, i.e., either they move in phase, or in opposite phase. The modal solution has these characteristics only because it is separable.

Since the wave equation is a linear equation and the boundary conditions are assumed homogeneous, linear superposition of the individual modes also gives a solution. Therefore, the general solution of the free vibration problem is of the form

$$w(x,t) = \sum_{k=1}^{\infty} p_k(t) W_k(x) = \sum_{k=1}^{\infty} (C_k \cos\omega_k t + S_k \sin\omega_k t)\sin\frac{k\pi x}{l}. \tag{1.63}$$

To obtain a unique solution, one has to determine the constants C_k and S_k. In practice, a system can vibrate freely when it is released from some non-equilibrium configuration, or started with some non-zero velocity, or both. These *initial conditions* determine the constants C_k and S_k. This problem of determining the free vibration solution uniquely is known as the *initial value problem*. The solution of the initial value problem relies on a very important property of the eigenfunctions, as discussed below.

From the theory of Fourier series (see [2]), (1.63) can be easily identified as the Fourier sine series with time-varying coefficients. It can be easily checked that the eigenfunctions of the string under consideration satisfy the *orthogonality property*

$$\langle W_j(x), W_k(x)\rangle := \int_0^l W_j(x)W_k(x)\,\mathrm{d}x \tag{1.64}$$

$$= \int_0^l \sin\frac{j\pi x}{l}\sin\frac{k\pi x}{l}\,\mathrm{d}x = \frac{l}{2}\delta_{jk}, \tag{1.65}$$

where $\langle W_j(x), W_k(x)\rangle$ is defined as an *inner product* (or *scalar product*) of the two functions $W_j(x)$ and $W_k(x)$, and δ_{jk} is the Kronecker delta symbol, i.e.,

$$\delta_{jk} = \begin{cases} 0, & j \neq k \\ 1, & j = k \end{cases}.$$

Using the orthogonality property, one can *filter* the jth coefficient in (1.63) as

$$\langle w(x,t), W_j(x)\rangle = \int_0^l w(x,t)\sin\frac{j\pi x}{l}\,\mathrm{d}x$$

$$= \frac{l}{2}(C_j\cos\omega_j t + S_j\sin\omega_j t). \tag{1.66}$$

The coefficients C_j and S_j can now be computed easily to match any initial shape and velocity of the string/bar. Let the initial shape and velocity be given by respectively,

$$w(x,0) = w_0(x), \qquad \text{and} \qquad w_{,t}(x,0) = v_0(x). \tag{1.67}$$

Then, from (1.66), one can obtain

$$C_j = \frac{2}{l}\langle w_0(x), W_j(x)\rangle, \qquad j = 1, 2, \ldots, \infty, \tag{1.68}$$

and

$$S_j = \frac{2}{l\omega_j}\langle v_0(x), W_j(x)\rangle, \qquad j = 1, 2, \ldots, \infty, \tag{1.69}$$

where (1.68) is obtained by setting $t = 0$ in (1.66), and (1.69) is obtained by differentiating (1.66) once with respect to t and then setting $t = 0$. This completes the solution (1.63) of the initial value problem of a string/bar defined by (1.48), with the boundary conditions (1.49), and the initial conditions (1.67).

1.4 MODAL ANALYSIS

As observed in the previous section, the solution of the initial value problem requires the mode-shape-functions and the associated circular modal frequencies. In this section, the problem of determination of the mode-shape-functions and the modal frequencies, usually termed *modal analysis*, is formulated as an eigenvalue problem. For convenience, we first introduce the complex notation for representing the general solution of the free vibration problem.

The general solution (1.63) can be compactly represented using the complex notation in the form

$$\begin{aligned} w(x,t) &= \sum_{k=1}^{\infty}\left[\frac{C_k}{2}(\mathrm{e}^{i\omega_k t}+\mathrm{e}^{-i\omega_k t})+\frac{S_k}{2i}(\mathrm{e}^{i\omega_k t}-\mathrm{e}^{-i\omega_k t})\right]W_k(x) \\ &= \sum_{k=1}^{\infty}\left[\frac{F_k}{2}\mathrm{e}^{i\omega_k t}+\frac{F_k^*}{2}\mathrm{e}^{-i\omega_k t}\right]W_k(x) \\ &= \sum_{k=1}^{\infty}\frac{F_k}{2}\mathrm{e}^{i\omega_k t}W_k(x)+\text{c.c.} \\ &= \sum_{k=1}^{\infty}\mathscr{R}\left[F_k\mathrm{e}^{i\omega_k t}W_k(x)\right], \end{aligned} \tag{1.70}$$

where $\mathscr{R}[\cdot]$ denotes real part of a complex number, $F_k = C_k - iS_k$, and the asterisk in the superscript denotes complex conjugate (c.c.). For notational convenience while obtaining the solution, we may write the solution as simply $w(x,t) = W(x)\mathrm{e}^{i\omega t}$, where the unknown mode-shape-function $W(x)$ may be complex in general. It is to be noted that the complex conjugate part can be dropped since the equations considered here are linear. In some cases, such as in the expression of kinetic energy $\rho A w_t^2(x,t)/2$, the complex conjugate part must be written explicitly. One further point to note is that we may also take the imaginary part of $F_k\mathrm{e}^{i\omega_k t}W_k(x)$ in (1.70) as the kth modal solution. The complex representation also allows us to treat problems with non-separable solution, and will be used for studying vibrations in translating strings later in this chapter.

1.4.1 The eigenvalue problem

The equation of motion for the systems discussed above can be represented in the general form

$$\mu(x)w_{,tt} + \mathcal{K}[w] = 0, \tag{1.71}$$

where $\mathcal{K}[\cdot]$ is a linear differential operator. For example, for a taut string

$$\mu(x) = 1 \quad \text{and} \quad \mathcal{K}[\cdot] = -c^2\frac{\partial^2}{\partial x^2}. \tag{1.72}$$

Consider a modal solution for the free vibration problem of (1.71) in the form

$$w(x, t) = W(x)e^{i\omega t}, \tag{1.73}$$

where $W(x)$ is the mode-shape-function and ω is the modal frequency. Substituting (1.73) in (1.71) yields

$$-\omega^2\mu(x)W + \mathcal{K}[W] = 0. \tag{1.74}$$

Only for certain special values of ω, can (1.74) be solved for non-trivial solutions of $W(x)$ satisfying the boundary conditions of the problem. Hence, the differential equation (1.74) along with the boundary conditions on $W(x)$ is known as the *eigenvalue problem* for the system. For a taut string, it can be easily checked that the eigenvalue problem is defined by

$$W'' + \frac{\omega^2}{c^2}W = 0,$$
$$W(0) = 0 \quad \text{and} \quad W(l) = 0.$$

This problem is solved in detail in Section 1.3, and the eigenvalues and eigenfunctions are given by

$$\omega_k = \frac{k\pi c}{l} \quad \text{and} \quad W_k(x) = \sin\frac{k\pi x}{l}, \quad k = 1, 2, \ldots, \infty. \tag{1.75}$$

Thus, the solution of the eigenvalue problem yields the circular modal frequencies, and the corresponding mode-shape-functions, which are also known as the *circular eigenfrequencies* and *eigenfunctions*, respectively. In the following, we consider two slightly more complex eigenvalue problems.

1.4.1.1 The hanging string

Let us consider the unforced dynamics of a hanging string which is described by (1.8). The equation of motion can be represented in the form (1.71) where

$$\mu(x) = 1 \quad \text{and} \quad \mathcal{K}[\cdot] = -g\frac{\partial}{\partial x}\left[(l - x)\frac{\partial}{\partial x}\right]. \tag{1.76}$$

At the free end of the string, the transverse force is zero, i.e., $T(l)w_{,x}(l, t) \equiv 0$. However, since the tension at the free end $T(l) = 0$, it implies that $w_{,x}(l, t)$ can be arbitrary. As we will see shortly, a finiteness condition on the solution is required for the free end. The only boundary condition that can be specified is for the fixed end of the string which is given by

$$w(0, t) \equiv 0. \tag{1.77}$$

Now, the eigenvalue problem for the hanging string can be easily written from (1.74) as

$$\omega^2 W + g[(l-x)W']' = 0, \tag{1.78}$$

with the associated boundary condition obtained from (1.77) as

$$W(0) = 0. \tag{1.79}$$

In the following, the eigenvalue problem (1.78)–(1.79) is now solved to determine the circular eigenfrequencies ω and the corresponding eigenfunctions $W(x)$.

Consider the function

$$s(x) = 2\omega\sqrt{\frac{l-x}{g}}. \tag{1.80}$$

Then, defining $\tilde{W}(s)$ such that $\tilde{W}(s(x)) = W(x)$, one obtains using the chain rule of differentiation

$$\frac{\mathrm{d}W}{\mathrm{d}x} = \frac{\mathrm{d}\tilde{W}}{\mathrm{d}s}\frac{\mathrm{d}s}{\mathrm{d}x} = -\tilde{W}'\frac{\omega}{\sqrt{g(l-x)}} \tag{1.81}$$

$$\begin{aligned}\frac{\mathrm{d}^2W}{\mathrm{d}x^2} &= \frac{\mathrm{d}^2\tilde{W}}{\mathrm{d}s^2}\left(\frac{\mathrm{d}s}{\mathrm{d}x}\right)^2 + \frac{\mathrm{d}\tilde{W}}{\mathrm{d}s}\frac{\mathrm{d}^2s}{\mathrm{d}x^2}\\ &= \tilde{W}''\frac{\omega^2}{g(l-x)} - \tilde{W}'\frac{\omega}{2\sqrt{g(l-x)^3}},\end{aligned} \tag{1.82}$$

where the prime in $\tilde{W}'$ denotes differentiation with respect to s. Using (1.81) and (1.82) in (1.78) yields on simplification

$$\tilde{W}'' + \frac{1}{s}\tilde{W}' + \tilde{W} = 0, \qquad s \in [0, 2\omega\sqrt{l/g}] \tag{1.83}$$

$$\tilde{W}(2\omega\sqrt{l/g}) = 0. \tag{1.84}$$

The differential equation (1.83) is a special case of the Bessel equation (see [2])

$$y''(x) + \frac{1}{x}y'(x) + \left(1 - \frac{n^2}{x^2}\right)y(x) = 0$$

with $n = 0$. Therefore, the general solution of (1.83) can be written as

$$\tilde{W}(s) = DJ_0(s) + EY_0(s), \tag{1.85}$$

where D and E are arbitrary constants, and $J_0(s)$ and $Y_0(s)$ are known as, respectively, zeroth-order Bessel functions of the first and second kind (or Neumann functions). The

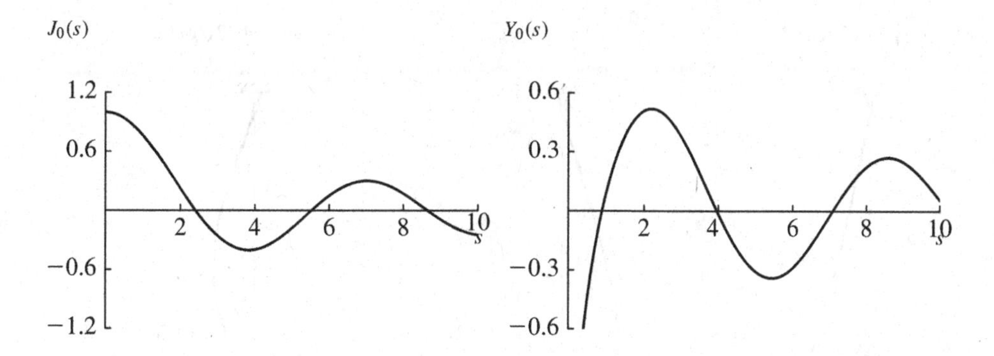

Figure 1.13 Bessel functions $J_0(s)$, and $Y_0(s)$

functions $J_0(s)$ and $Y_0(s)$ are plotted in Figure 1.13. Since $Y_0(s) \to -\infty$ as $s \to 0$ (i.e., $x \to l$), the condition of finiteness of the solution at the free end must imply $E = 0$. Therefore, the solution of (1.83) takes the form $\tilde{W}(s) = DJ_0(s)$. The boundary condition of the fixed end, $W(s(0)) = 0$, then implies, for non-triviality of $\tilde{W}(s)$,

$$J_0(2\omega\sqrt{l/g}) = 0, \tag{1.86}$$

which is the characteristic equation for the problem. The roots of $J_0(\gamma_k) = 0$, yield the eigenfrequencies

$$\omega_k = \frac{\gamma_k}{2}\sqrt{\frac{g}{l}}, \qquad k = 1, 2, \ldots, \infty, \tag{1.87}$$

where $\gamma_1 \approx 2.4048$, $\gamma_2 \approx 5.5201$, $\gamma_3 \approx 8.6537, \ldots$. It is interesting to note that the first eigenfrequency or fundamental frequency of a hanging string, $\omega_1 = 1.2024\sqrt{g/l}$, is about 1.2 times the small-amplitude oscillation frequency of a mathematical pendulum of length l. The kth eigenfunction can now be written as

$$W_k(x) = J_0\left(2\omega_k\sqrt{\frac{l-x}{g}}\right). \tag{1.88}$$

The first three mode-shapes of the hanging string are shown in Figure 1.14. From (1.70), we obtain the general solution of the initial value problem for the hanging string as

$$\begin{aligned} w(x,t) &= \sum_{k=1}^{\infty} \mathscr{R}\left[F_k e^{i\omega_k t} W_k(x)\right] \\ &= \sum_{k=1}^{\infty}\left[(C_k\cos\omega_k t + S_k\sin\omega_k t)J_0\left(2\omega_k\sqrt{\frac{l-x}{g}}\right)\right]. \end{aligned} \tag{1.89}$$

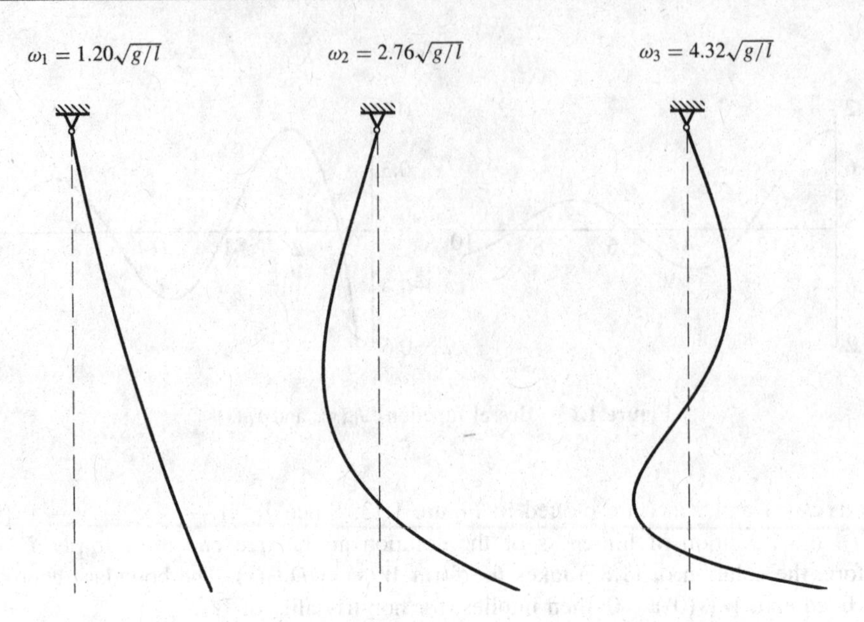

Figure 1.14 First three mode-shapes of a hanging string

The orthogonality property of the eigenfunctions $W_k(x)$ of the hanging string can be obtained from the theory of Bessel functions (see [2]) as

$$\int_0^l W_j(x)W_k(x)\,\mathrm{d}x = lJ_1^2(2\omega_j\sqrt{l/g})\delta_{jk}, \tag{1.90}$$

where $J_1(\cdot)$ is the Bessel function of order one, and δ_{jk} is the Kronecker delta symbol. Later, in Section 1.4.2, we shall see that the eigenfunctions of a large class of eigenvalue problems are always orthogonal in an appropriate sense.

1.4.1.2 Bar with varying cross-section

Consider the longitudinal dynamics of a fixed–free bar with varying cross-section described by (1.15). The equation of motion can be represented by (1.71) where

$$\mu(x) = \rho A(x) \qquad \text{and} \qquad \mathcal{K}[\cdot] = -\frac{\partial}{\partial x}\left(EA(x)\frac{\partial}{\partial x}\right).$$

The boundary conditions are given by

$$u(0,t) \equiv 0 \qquad \text{and} \qquad EA(l)u_{,x}(l,t) \equiv 0. \tag{1.91}$$

The eigenvalue problem in this case can be easily written as

$$A(x)\omega^2 U + c^2[A(x)U']' = 0, \tag{1.92}$$

where $c^2 = E/\rho$, along with the boundary conditions

$$U(0) = 0 \quad \text{and} \quad U'(l) = 0. \tag{1.93}$$

It is not possible to obtain an analytical solution of (1.92) for a general variation of the cross-sectional area $A(x)$. However, for a class of functions $A(x)$, one can solve (1.92) as follows.

Consider the transformation

$$W(x) = h(x)U(x), \tag{1.94}$$

where $h(x)$ is an unknown function. One can then write

$$[h^2U']' = hW'' - h''W. \tag{1.95}$$

Let us choose $h^2(x) = A(x)$. With this choice, one can rewrite (1.92) as

$$h^2\omega^2 U + c^2[h^2U']' = 0,$$

or

$$\omega^2 W + c^2\left(W'' + \frac{h''}{h}W\right) = 0 \quad \text{(using (1.94) and (1.95))}. \tag{1.96}$$

If the variation of the cross-section is such that $h''/h = \alpha$, where α is a constant, one can rewrite (1.96) as

$$W'' + \left(\frac{\omega^2}{c^2} + \alpha\right)W = 0.$$

This differential equation can be easily solved.

As a simple example, let

$$A(x) = A_0\left(1 - \frac{x}{2l}\right)^2.$$

Then, $h(x) = \sqrt{A_0}(1 - x/2l)$, and $h''(x) = 0$. Therefore, (1.96) simplifies to

$$W'' + \frac{\omega^2}{c^2}W = 0, \tag{1.97}$$

and the boundary conditions are given by

$$W(0) = 0 \quad \text{and} \quad W'(l) = \frac{h'(l)}{h(l)}W(l). \tag{1.98}$$

It is not difficult to show that the solutions of (1.97) satisfying the boundary conditions (1.98) can be written as

$$W_k(x) = D \sin \frac{\omega_k x}{c}, \qquad k = 1, 2, \ldots, \infty,$$

where ω_k are the eigenvalues obtained from the characteristic equation

$$\tan \frac{\omega l}{c} + \frac{\omega l}{c} = 0.$$

The first three eigenfrequencies are obtained as $\omega_1 = 2.029c/l$, $\omega_2 = 4.913c/l$, and $\omega_3 = 7.979c/l$. Finally, the eigenfunctions $U_k(x)$ are obtained from (1.94) as

$$U_k(x) = \frac{D \sin \frac{\omega_k x}{c}}{\sqrt{A_0}\left(1 - \frac{x}{2l}\right)}.$$

However, the orthogonality of the eigenfunctions may not be very obvious. Therefore, we need to have a general procedure to determine the orthogonality relations, which is discussed next.

1.4.2 Orthogonality of eigenfunctions

Consider a general eigenvalue problem formed by a differential equation of the type

$$-\lambda\mu(x)W + \mathcal{K}[W] = 0, \qquad x \in [0, l], \tag{1.99}$$

where $\lambda = \omega^2$, along with certain boundary conditions. If W_j and W_k $(j \neq k)$ are solutions of (1.99) corresponding to λ_j and λ_k, respectively, one can write

$$-\lambda_j\mu(x)W_j + \mathcal{K}[W_j] = 0, \tag{1.100}$$

and

$$-\lambda_k\mu(x)W_k + \mathcal{K}[W_k] = 0. \tag{1.101}$$

Multiplying (1.100) with W_k, and (1.101) with W_j, and integrating the difference of the two equations over the length of the string, one can write

$$-(\lambda_j - \lambda_k)\langle\mu(x)W_j, W_k\rangle + \langle W_k, \mathcal{K}[W_j]\rangle - \langle W_j, \mathcal{K}[W_k]\rangle = 0. \tag{1.102}$$

If the operator $\mathcal{K}[\cdot]$ is such that

$$\langle W, \mathcal{K}[\tilde{W}]\rangle = \langle \tilde{W}, \mathcal{K}[W]\rangle, \tag{1.103}$$

for any two functions $W(x)$ and $\tilde{W}(x)$ satisfying the boundary conditions, the operator $\mathcal{K}[\cdot]$ is called a *self-adjoint operator*. Since the eigenfunctions $W_j(x)$ and $W_k(x)$ satisfy the boundary conditions of the problem, one can write for a self-adjoint operator

$$\langle W_k, \mathcal{K}[W_j]\rangle = \langle W_j, \mathcal{K}[W_k]\rangle, \tag{1.104}$$

and (1.102) yields

$$-(\lambda_j - \lambda_k)\langle \mu(x)W_j, W_k\rangle = 0. \tag{1.105}$$

If $\lambda_j \neq \lambda_k$ (which is usually satisfied), we obtain the orthogonality relation

$$\langle \mu(x)W_j, W_k\rangle = 0 \quad \Rightarrow \quad \int_0^l \mu(x)W_j W_k \,\mathrm{d}x = 0. \tag{1.106}$$

It is always possible to normalize the eigenfunctions such that they are orthonormal, i.e,

$$\int_0^l \mu(x)W_j W_k \,\mathrm{d}x = \delta_{jk}, \tag{1.107}$$

where δ_{jk} is the Kronecker delta symbol. A consequence of orthonormality of the eigenfunctions can be obtained from (1.100) as

$$\int_0^l W_k\mathcal{K}[W_j]\,\mathrm{d}x = \lambda_j\delta_{jk}.$$

It can be concluded from the above that the eigenfunctions of a self-adjoint operator are orthogonal with respect to a suitably defined inner product, which may be determined using the above procedure.

1.4.3 The expansion theorem

Let us rework the solution procedure for the free vibration of a taut string presented in Section 1.4.1 in a slightly different manner. Based on our experience thus far, let us assume the solution of (1.71) as an expansion in terms of the eigenfunctions in the form

$$w(x,t) = \sum_{k=1}^{k=\infty} p_k(t)W_k(x), \tag{1.108}$$

where $W_k(x)$ is the kth eigenfunction given by (1.75) and $p_k(t)$ is the corresponding unknown *modal coordinate*. Substituting the expansion (1.108) in (1.71) yields

$$\sum_{k=1}^{k=\infty} \ddot{p}_k(t)W_k(x) + \mathcal{K}\left[\sum_{k=1}^{k=\infty} p_k(t)W_k(x)\right] = 0. \tag{1.109}$$

Using the linearity property of the operator $\mathcal{K}[\cdot]$, one can rewrite (1.109) as

$$\sum_{k=1}^{k=\infty} \mu(x)\ddot{p}_k(t)W_k(x) + \sum_{k=1}^{k=\infty} p_k(t)\mathcal{K}[W_k(x)] = 0,$$

or

$$\sum_{k=1}^{k=\infty} \mu(x)\ddot{p}_k(t)W_k(x) + \sum_{k=1}^{k=\infty} \mu(x)\omega_k^2 p_k(t)W_k(x) = 0 \qquad \text{(using (1.74))},$$

or

$$\sum_{k=1}^{k=\infty} \left[\ddot{p}_k(t) + \omega_k^2 p_k(t)\right] \mu(x)W_k(x) = 0. \tag{1.110}$$

Taking the inner product on both sides with W_j, $j = 1, 2, \ldots, \infty$, and using the orthogonality relations (1.106), one obtains the decoupled differential equations for the modal coordinates as

$$\ddot{p}_j(t) + \omega_j^2 p_j(t) = 0, \qquad j = 1, 2, \ldots, \infty. \tag{1.111}$$

It may be mentioned here that we have exchanged an integral and an infinite sum to arrive at the decoupled equations (1.111). The general solution of the jth modal coordinate is, therefore, obtained as

$$p_j(t) = C_j \cos\omega_j t + S_j \sin\omega_j t,$$

and the general solution of the free vibration problem can be written in the form

$$w(x,t) = \sum_{k=1}^{\infty} (C_k \cos\omega_k t + S_k \sin\omega_k t) W_k(x),$$

which is the same as (1.63) obtained by Bernoulli's method. Thus, we have reconstructed back the solution of the free vibration problem using the eigenfunction expansion (1.108) and the orthogonality relations (1.106).

The fundamental requirement for the expansion method to work is that any physically possible shape of the system, say a string, should be expandable as a linear combination of the eigenfunctions $W_k(x)$ in the form (1.108). In other words, $W_k(x)$ should form a basis of the space of all physically possible shapes of the string. The set of all eigenfunctions of the string is indeed a basis of the function space under consideration, and this follows from the self-adjointness of the differential operator $\mathcal{K}[\cdot]$ in the eigenvalue problem (1.71). This statement is referred to as the *expansion theorem*. The expansion theorem also provides a convenient method for solving forced vibration problems as discussed later.

1.4.4 Systems with discrete elements

A continuous system may interact with discrete elements as discussed in previous sections. For such hybrid systems, the modal analysis can be performed by analyzing the system in parts along with appropriate matching conditions and boundary conditions for each of the parts. Often in these systems, the boundary conditions themselves involve ordinary differential equations, as will be evident in this section.

Let us consider the modal analysis of longitudinal vibrations of a bar with a mass–spring system at the right boundary, as shown in Figure 1.15. This system can be described by one field variable $u(x, t)$ and one discrete variable $y(t)$. The equations of motion are

$$u_{,tt} - c^2 u_{,xx} = 0 \tag{1.112}$$

and

$$M\ddot{y} + Ky = Ku(l, t), \tag{1.113}$$

and the boundary conditions are given by

$$u(0, t) \equiv 0 \quad \text{and} \quad EAu_{,x}(l, t) \equiv K(y - u(l, t)). \tag{1.114}$$

As is evident, the second boundary condition in (1.114) involves the ordinary differential equation (1.113).

Assume a modal solution of the form

$$\left\{ \begin{array}{c} u(x, t) \\ y(t) \end{array} \right\} = \left\{ \begin{array}{c} U(x) \\ Y \end{array} \right\} e^{i\omega t}. \tag{1.115}$$

It may be noted that the modal vector for this problem is given by $(U(x), Y)^T$. Substituting this solution in the equations of motion (1.112)–(1.113) and simplifying, we obtain the eigenvalue problem

$$U'' + \frac{\omega^2}{c^2} U = 0 \tag{1.116}$$

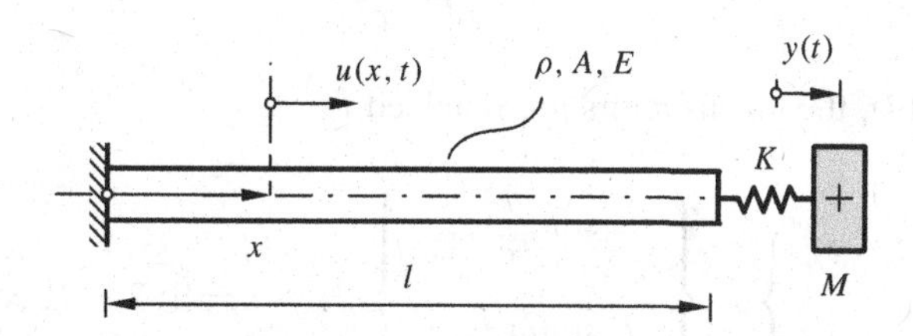

Figure 1.15 A hybrid system formed by a continuous sub-system and lumped elements

and

$$(-M\omega^2 + K)Y = KU(l), \tag{1.117}$$

with the associated boundary conditions given by (1.114) as

$$U(0) = 0 \tag{1.118}$$

and

$$EAU'(l) = K[Y - U(l)] = \frac{KM\omega^2}{K - M\omega^2}U(l) \quad \text{(using (1.117))}. \tag{1.119}$$

Note here that the boundary condition (1.119) also involves the circular frequency ω. Assuming a solution of (1.116) in the form

$$U(x) = C\cos\frac{\omega x}{c} + S\sin\frac{\omega x}{c} \tag{1.120}$$

we have from the boundary conditions (1.118)–(1.119)

$$\begin{bmatrix} 1 & 0 \\ \left(\dfrac{KM\omega^2}{K - M\omega^2}\cos\dfrac{\omega l}{c} + \dfrac{EA\omega}{c}\sin\dfrac{\omega l}{c}\right) & \left(\dfrac{KM\omega^2}{K - M\omega^2}\sin\dfrac{\omega l}{c} - \dfrac{EA\omega}{c}\cos\dfrac{\omega l}{c}\right) \end{bmatrix} \begin{Bmatrix} C \\ S \end{Bmatrix} = 0. \tag{1.121}$$

The non-triviality of the solution of (C, S) implies that the determinant of the matrix in (1.121) must vanish, which yields the characteristic equation

$$\tan\frac{\omega l}{c} - \frac{EA(K - M\omega^2)}{c\omega MK} = 0.$$

This transcendental equation yields infinitely many circular eigenfrequencies ω_k, $k = 1, 2, \ldots, \infty$. Substituting these eigenfrequencies in (1.121), one obtains $(C_k, S_k) = (0, 1)$, and correspondingly

$$U_k(x) = \sin\frac{\omega_k x}{c},$$

so that, using (1.117), the eigenvectors are obtained as

$$\begin{Bmatrix} U_k(x) \\ Y_k \end{Bmatrix} = \begin{Bmatrix} \sin\dfrac{\omega_k x}{c} \\ \dfrac{K\sin(\omega_k l/c)}{-M\omega_k^2 + K} \end{Bmatrix}, \qquad k = 1, 2, \ldots, \infty.$$

It is to be noted that these vectors are formed by the displacement field $U_k(x)$ in the rod, and the discrete coordinate Y_k. They are not vectors in two-dimensional Euclidean space, but rather in an $(\infty + 1)$-dimensional space. Since these infinitely many eigenvectors are all linearly independent, one can conveniently express the solution of (1.112)–(1.113) using the expansion theorem as

$$\begin{Bmatrix} u(x,t) \\ y(t) \end{Bmatrix} = \sum_{k=1}^{\infty} p_k(t) \begin{Bmatrix} U_k(x) \\ Y_k \end{Bmatrix},$$

where $p_k(t)$ is the modal coordinate corresponding to mode k.

The orthogonality relation for the above eigenvectors are obtained from the procedure discussed in Section 1.4.2 as follows. Consider the modes j and k which satisfy the following equations

$$U_j'' + \frac{\omega_j^2}{c^2} U_j = 0, \qquad Y_j = \frac{KU_j(l)}{-M\omega_j^2 + K}, \tag{1.122}$$

$$U_k'' + \frac{\omega_k^2}{c^2} U_k = 0, \qquad Y_k = \frac{KU_k(l)}{-M\omega_k^2 + K}, \tag{1.123}$$

along with appropriate boundary and matching conditions. Multiply the first equation in (1.122) by U_k and the first equation in (1.123) by U_j, and subtract the second product from the first and integrate over the length of the beam to obtain

$$\int_0^l \left(U_k U_j'' + \frac{\omega_j^2}{c^2} U_k U_j \right) \mathrm{d}x - \int_0^l \left(U_j U_k'' + \frac{\omega_k^2}{c^2} U_j U_k \right) \mathrm{d}x = 0$$

$$\Rightarrow \int_0^l \left(U_k U_j'' - U_j U_k'' + \frac{\omega_j^2 - \omega_k^2}{c^2} U_j U_k \right) \mathrm{d}x = 0. \tag{1.124}$$

Integrating by parts the first term in (1.124) twice, and using the boundary and matching conditions from (1.118)–(1.119) yields on simplification

$$(\omega_j^2 - \omega_k^2) \left[\frac{M}{EA} \left(\frac{KU_j(l)}{K - M\omega_j^2} \right) \left(\frac{KU_k(l)}{K - M\omega_k^2} \right) + \frac{1}{c^2} \int_0^l U_j U_k \, \mathrm{d}x \right] = 0$$

$$\Rightarrow MY_jY_k + \rho A \int_0^l U_j U_k \, \mathrm{d}x = 0, \qquad \text{for } j \neq k,$$

where we have used (1.122) and (1.123). These are the orthogonality relations for the system.

1.5 THE INITIAL VALUE PROBLEM: SOLUTION USING LAPLACE TRANSFORM

The Laplace transform method is one of the standard methods of solving initial value problems. Consider the wave equation

$$w_{,tt} - c^2 w_{,xx} = 0, \tag{1.125}$$

with homogeneous boundary conditions $w(0,t) \equiv 0$ and $w(l,t) \equiv 0$, and initial conditions $w(x,0) = w_0(x)$ and $w_{,t}(x,0) = v_0(x)$. Taking the Laplace transform (see [2]) of both sides of (1.125) and the boundary conditions with respect to the variable t yields

$$\tilde{w}'' - \frac{s^2}{c^2}\tilde{w} = -\frac{1}{c^2}\left[s w_0(x) + v_0(x)\right], \tag{1.126}$$

$$\tilde{w}(0,s) \equiv 0 \qquad \text{and} \qquad \tilde{w}(l,s) \equiv 0, \tag{1.127}$$

where $\tilde{w}(x,s)$ represents the Laplace transform of $w(x,t)$, and is defined as

$$\tilde{w}(x,s) = \int_0^\infty w(x,t)\mathrm{e}^{-st}\,\mathrm{d}t. \tag{1.128}$$

The homogeneous solution of (1.126) is obtained as

$$\tilde{w}(x,s) = a\mathrm{e}^{sx/c} + b\mathrm{e}^{-sx/c}. \tag{1.129}$$

Using the boundary conditions (1.127) yields

$$\begin{bmatrix} 1 & 1 \\ \mathrm{e}^{sl/c} & \mathrm{e}^{-sl/c} \end{bmatrix} \begin{Bmatrix} a \\ b \end{Bmatrix} = 0.$$

For non-trivial solutions of (a,b), we must have

$$\mathrm{e}^{2sl/c} - 1 = 0 \qquad \Rightarrow \qquad s = \frac{in\pi c}{l}, \qquad n = 1, 2, \ldots, \infty.$$

For these values of s, one can easily obtain $(a,b) = (1,-1)$, and therefore, the general solution of (1.126)–(1.127) can be written using (1.129) as

$$\tilde{w} = \sum_{n=1}^{\infty} A_n(s)\sin\frac{n\pi x}{l}, \tag{1.130}$$

where $A_n(s)$ are arbitrary constants. Using this solution expansion in (1.126), and taking inner product with $\sin m\pi x/l$ yields on simplification

$$A_m(s) = \frac{s}{s^2+\alpha_m^2}\int_0^l w_0(x)\sin\frac{m\pi x}{l}\,dx + \frac{1}{s^2+\alpha_m^2}\int_0^l v_0(x)\sin\frac{m\pi x}{l}\,dx,$$

where $\alpha_m = m\pi c^2/l$. Substituting this expression in (1.130) and taking the inverse Laplace transform yields

$$w(x,t) = \sum_{n=1}^{\infty}(C_n\cos\alpha_n t + S_n\sin\alpha_n t)\sin\frac{n\pi x}{l}, \tag{1.131}$$

where

$$C_n = \int_0^l w_0(x)\sin\frac{n\pi cx}{l}\,dx \quad \text{and} \quad S_n = \frac{1}{\alpha_n}\int_0^l v_0(x)\sin\frac{n\pi cx}{l}\,dx.$$

The solution (1.131) is the same as obtained in (1.63) before.

1.6 FORCED VIBRATION ANALYSIS

The dynamics of one-dimensional continuous systems discussed above, subjected to an arbitrary distributed forcing $q(x,t)$, can be represented in a general form

$$\mu(x)w_{,tt} + \mathcal{K}[w] = q(x,t). \tag{1.132}$$

When a system is forced at a boundary, one can still convert the problem to the form (1.132), leaving the corresponding boundary condition homogeneous. For example, consider the bar shown in Figure 1.16, forced axially at the boundary. The equation of motion and boundary conditions are given by

$$\rho A u_{,tt} - [EAu_{,x}]_{,x} = 0,$$
$$-EAu_{,x}(0,t) = F(t), \quad \text{and} \quad u(l,t) \equiv 0. \tag{1.133}$$

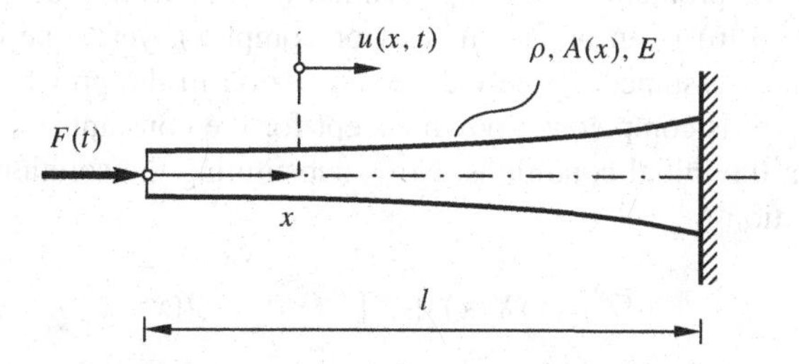

Figure 1.16 A bar forced axially at the boundary

It may be noted that the standard convention of taking compressive stress as negative has been used in (1.133). The dynamics of the bar can also be recast in the form (1.132) as

$$\rho A u_{,tt} - [EAu_{,x}]_{,x} = F(t)\delta(x),$$
$$EAu_{,x}(0,t) = 0, \qquad \text{and} \qquad u(l,t) \equiv 0.$$

Let the eigenvalue problem corresponding to the unforced dynamics in (1.132) be given by

$$-\omega^2 \mu(x) W(x) + \mathcal{K}[W(x)] = 0, \tag{1.134}$$

along with appropriate boundary conditions. We will represent the solutions of (1.134) as $(\omega_k, W_k(x))$, $k = 1, 2, \ldots, \infty$, where ω_k are the eigenvalues and $W_k(x)$ are the corresponding eigenfunctions. The pair (ω_k, W_k) is sometimes also called an *eigenpair*. In the following, we will discuss the solution of (1.132) for different forcing conditions.

1.6.1 Harmonic forcing

Consider a forcing $q(x,t)$ in (1.132) that is separable in space and time, and has a harmonic time function. In particular, consider $q(x,t) = Q(x)\cos\Omega t$, and let us represent the forced dynamics as

$$\mu(x) w_{,tt} + \mathcal{K}[w] = \mathscr{R}[Q(x)\mathrm{e}^{i\Omega t}], \tag{1.135}$$

where Ω is the circular forcing frequency, $Q(x)$ specifies the force distribution, and $\mathscr{R}[\cdot]$ represents the real part. Let us consider a solution of (1.135) in the form

$$\begin{aligned} w(x,t) &= w_{\mathrm{H}}(x,t) + w_{\mathrm{P}}(x,t) \\ &= \sum_{k=1}^{\infty} [C_k \cos\omega_k t + S_k \sin\omega_k t] W_k(x) + \mathscr{R}[X(x)\mathrm{e}^{i\Omega t}], \end{aligned} \tag{1.136}$$

where the first term $w_{\mathrm{H}}(x,t)$ represents the general solution of the homogeneous problem, also simply called the *homogeneous solution*, and the second term $w_{\mathrm{P}}(x,t)$ is a solution of the inhomogeneous problem, also simply called the *particular solution*. The amplitude function $X(x)$ in (1.136) is an unknown (real or complex), yet to be determined. It may be noted that we have assumed, based on the discussion in the previous sections, that the homogeneous solution is completely known except for the constants C_k and S_k, which will be determined from the initial conditions. Now, substituting the solution (1.136) in (1.135) yields on simplification

$$-\Omega^2 \mu(x) X(x) + \mathcal{K}[X(x)] = Q(x). \tag{1.137}$$

This equation along with the boundary conditions forms a *boundary value problem*. In the following, we discuss two methods of solving (1.137), namely the eigenfunction expansion method and Green's function method.

1.6.1.1 Eigenfunction expansion method

Assume the solution of (1.137) as the eigenfunction expansion

$$X(x) = \sum_{k=1}^{\infty} \alpha_k W_k(x), \tag{1.138}$$

where α_k are unknown coefficients. Substituting (1.138) in (1.137) yields

$$-\Omega^2 \mu(x) \sum_{k=1}^{\infty} \alpha_k W_k(x) + \mathcal{K}\left[\sum_{k=1}^{\infty} \alpha_k W_k(x)\right] = Q(x)$$

$$\Rightarrow -\Omega^2 \mu(x) \sum_{k=1}^{\infty} \alpha_k W_k(x) + \sum_{k=1}^{\infty} \alpha_k \mathcal{K}[W_k(x)] = Q(x)$$

$$\Rightarrow \sum_{k=1}^{\infty} (\omega_k^2 - \Omega^2)\alpha_k \mu(x) W_k(x) = Q(x), \qquad \text{(using (1.134))}. \tag{1.139}$$

Taking the inner product on both sides of (1.139) with $W_j(x)$, $j = 1, 2, \ldots, \infty$, and using the orthogonality property, we get

$$(\omega_j^2 - \Omega^2)\alpha_j \langle \mu(x) W_j(x), W_j(x)\rangle = \langle Q(x), W_j(x)\rangle, \qquad j = 1, 2, \ldots, \infty$$

$$\Rightarrow \alpha_j = \frac{\int_0^l Q(x) W_j(x)\,\mathrm{d}x}{(\omega_j^2 - \Omega^2)\int_0^l \mu(x) W_j^2(x)\,\mathrm{d}x}, \qquad j = 1, 2, \ldots, \infty, \tag{1.140}$$

where it has been assumed that the forcing is non-resonant, i.e., $\Omega \neq \omega_j$ for all j. This completes the solution (1.138) of (1.137) for a non-resonant harmonic forcing.

In case $\Omega = \omega_j$ for some j, we have resonance, which is characterized by a very high response amplitude for the jth mode (infinite as far as the linear theory is concerned). To determine the response of the system at resonance, we use the method of variation of parameters in which the particular solution is assumed in the form

$$w_P(x, t) = \mathcal{R}\left[\left(\alpha_j(t) W_j(x) + \sum_{\substack{k=1 \\ k\neq j}}^{\infty} \alpha_k W_k(x)\right) \mathrm{e}^{i\omega_j t}\right]. \tag{1.141}$$

It may be noted that the jth modal coordinate $\alpha_j(t)$ has been taken as a function of time. Substituting this solution form in (1.135) and proceeding as discussed above, one can easily obtain the equation of modal dynamics of the jth mode as

$$\ddot{\alpha}_j + 2i\omega_j \dot{\alpha}_j = \frac{\int_0^l Q(x) W_j(x)\,\mathrm{d}x}{\int_0^l \mu(x) W_j^2(x)\,\mathrm{d}x}.$$

Solving this and substituting in (1.141), the particular solution is finally obtained as

$$w_P(x,t) = \frac{t}{2\omega_j}\frac{\int_0^l Q(x)W_j(x)\,\mathrm{d}x}{\int_0^l \mu(x)W_j^2(x)\,\mathrm{d}x}W_j(x)\sin\omega_j t + \sum_{\substack{k=1\\k\neq j}}^{\infty}\alpha_k W_k(x)\cos\omega_j t,$$

where the constants α_k are obtained from (1.140).

An interesting situation occurs for a resonant forcing with $\Omega = \omega_j$ if $Q(x)$ is such that

$$\int_0^l Q(x)W_j(x)\,\mathrm{d}x = 0,$$

i.e., the forcing amplitude distribution $Q(x)$ is orthogonal to $W_j(x)$. In such a case, the solution is still finite since the jth mode cannot be excited by the force. This situation is referred to as *apparent resonance*.

1.6.1.2 Green's function method

Boundary value problems may be conveniently solved using Green's function method (see [3]). In this method, the solution of (1.137) is obtained in an integral form as discussed in the following.

Let $G(x,\overline{x},\Omega)$ be the solution of (1.137) excited by a concentrated unit force at $x = \overline{x} \in [0,l]$, i.e.,

$$-\Omega^2\mu(x)G(x,\overline{x},\Omega) + \mathcal{K}[G(x,\overline{x},\Omega)] = \delta(x-\overline{x}), \tag{1.142}$$

with all the boundary conditions of (1.137), which are assumed homogeneous. In case the boundary conditions of (1.137) are not homogeneous, one can make them homogeneous by following the procedure discussed in Section 1.9.

Consider the function

$$X(x) = \int_0^l Q(\overline{x})G(x,\overline{x},\Omega)\,\mathrm{d}\overline{x}. \tag{1.143}$$

Substituting (1.143) in the left-hand side of (1.137), one obtains

$$\begin{aligned}
-\Omega^2\mu(x)&\int_0^l Q(\overline{x})G(x,\overline{x},\Omega)\,\mathrm{d}\overline{x} + \mathcal{K}\left[\int_0^l Q(\overline{x})G(x,\overline{x},\Omega)\,\mathrm{d}\overline{x}\right]\\
&= -\Omega^2\mu(x)\int_0^l Q(\overline{x})G(x,\overline{x},\Omega)\,\mathrm{d}\overline{x} + \int_0^l Q(\overline{x})\mathcal{K}[G(x,\overline{x},\Omega)]\,\mathrm{d}\overline{x}\\
&= \int_0^l Q(\overline{x})\big(-\Omega^2\mu(x)G(x,\overline{x},\Omega) + \mathcal{K}[G(x,\overline{x},\Omega)]\big)\,\mathrm{d}\overline{x}\\
&= \int_0^l Q(\overline{x})\delta(x-\overline{x})\,\mathrm{d}\overline{x} \qquad \text{(using (1.142))}\\
&= Q(x).
\end{aligned}$$

Thus, (1.143) is indeed the solution of (1.137) for a general $Q(x)$. It is to be noted that, if $\Omega \neq \omega_k$ for all k, the solution of (1.142) (and also (1.137)) is unique since the homogeneous problem (i.e., with zero right-hand side) has only the trivial solution. The function $G(x, \overline{x}, \Omega)$ is known as *Green's function* for the boundary value problem (1.137). The complete solution of (1.132) can, therefore, be written as

$$w(x,t) = \sum_{k=1}^{\infty} [C_k \cos\omega_k t + S_k \sin\omega_k t]\, W_k(x) + \mathcal{R}\left[e^{i\Omega t}\int_0^l Q(\overline{x})G(x,\overline{x},\Omega)\,\mathrm{d}\overline{x}\right]. \tag{1.144}$$

We now compute Green's function for the forced vibration of a taut string for which $\mu(x) = 1$, and $\mathcal{K}[w] = -c^2 w_{,xx}$. In (1.142), we can consider two regions of the string as follows:

$$-\Omega^2 G - c^2 G_{,xx} = 0, \qquad 0 \leq x < \overline{x}, \tag{1.145}$$

and

$$-\Omega^2 G - c^2 G_{,xx} = 0, \qquad \overline{x} < x \leq l, \tag{1.146}$$

with appropriate matching conditions at $x = \overline{x}$, as discussed below. The solutions of (1.145) and (1.146) can be written as

$$G(x,\overline{x},\Omega) = \begin{cases} A_{\mathrm{L}} \sin\dfrac{\Omega x}{c} + B_{\mathrm{L}} \cos\dfrac{\Omega x}{c}, & 0 \leq x < \overline{x}, \quad (1.147) \\[2ex] A_{\mathrm{R}} \sin\dfrac{\Omega x}{c} + B_{\mathrm{R}} \cos\dfrac{\Omega x}{c}, & \overline{x} < x \leq l, \quad (1.148) \end{cases}$$

where A_{L}, B_{L}, A_{R}, and B_{R} are arbitrary constants. From the requirement of continuity of the solution at $x = \overline{x}$, and satisfaction of all the boundary conditions of the problem, we have

$$G(\overline{x}^-, \overline{x}, \Omega) = G(\overline{x}^+, \overline{x}, \Omega), \tag{1.149}$$

$$G(0, \overline{x}, \Omega) = 0, \qquad \text{and} \qquad G(l, \overline{x}, \Omega) = 0. \tag{1.150}$$

Further, integrating (1.142) over the domain of the string yields

$$-\int_0^l [\Omega^2 G + c^2 G_{,xx}]\,\mathrm{d}x = 1$$

$$\Rightarrow \lim_{\epsilon\to 0} \int_{\overline{x}-\epsilon}^{\overline{x}+\epsilon} [\Omega^2 G + c^2 G_{,xx}]\,\mathrm{d}x = -1 \qquad \text{(using (1.145)–(1.146))}$$

$$\Rightarrow \lim_{\epsilon\to 0} \left[\Omega^2 G(x+\theta\epsilon, \overline{x}, \Omega)\, 2\epsilon + c^2 G_{,x}\Big|_{\overline{x}-\epsilon}^{\overline{x}+\epsilon}\right] = -1 \tag{1.151}$$

$$\Rightarrow c^2 G_{,x}(\overline{x}^+, \overline{x}, \Omega) - c^2 G_{,x}(\overline{x}^-, \overline{x}, \Omega) = -1, \tag{1.152}$$

where, the first term in (1.151) is obtained from the mean value theorem. It can be easily observed that (1.152) provides the force equilibrium condition. Now, (1.149), (1.150), and (1.152) provide four conditions for finding the four constants in (1.148) and (1.148). Finally, after some algebra, Green's function for a taut string is obtained as

$$G(x,\bar{x},\Omega)=\begin{cases}\dfrac{\sin\frac{\Omega}{c}(l-\bar{x})\sin\frac{\Omega x}{c}}{\Omega c\sin\frac{\Omega l}{c}}, & 0\leq x<\bar{x}, \quad (1.153)\\[2ex] \dfrac{\sin\frac{\Omega}{c}(l-x)\sin\frac{\Omega\bar{x}}{c}}{\Omega c\sin\frac{\Omega l}{c}}, & \bar{x}<x\leq l. \quad (1.154)\end{cases}$$

Green's function can also be obtained from (1.142) using the eigenfunction expansion

$$G(x,\bar{x},\Omega)=\sum_{k=1}^{\infty}\alpha_k(\bar{x},\Omega)W_k(x),$$

where $\alpha_k(\bar{x},\Omega)$ are unknown functions. Following the procedure discussed in Section 1.6.1.1, it can be easily checked that

$$G(x,\bar{x},\Omega)=\sum_{k=1}^{\infty}\frac{W_k(\bar{x})W_k(x)}{(\omega_k^2-\Omega^2)\int_0^l\mu(x)W_k^2(x)\,\mathrm{d}x}. \tag{1.155}$$

A collection of Green's functions for various kinds of differential equations can be found in [4].

1.6.2 General forcing

For a general distributed forcing $q(x,t)$ which no longer needs to be separable, let us assume a solution of (1.132) in the form

$$w(x,t)=\sum_{k=1}^{\infty}p_k(t)W_k(x), \tag{1.156}$$

where $W_k(x)$ are the known eigenfunctions and $p_k(t)$ are the unknown modal coordinates. Substituting this solution form in (1.132) yields

$$\mu(x)\sum_{k=1}^{k=\infty}\ddot{p}_k(t)W_k(x)+\mathcal{K}\left[\sum_{k=1}^{k=\infty}p_k(t)W_k(x)\right]=q(x,t). \tag{1.157}$$

Using the linearity property of $\mathcal{K}[\cdot]$ and (1.134), one can simplify (1.157) as

$$\sum_{k=1}^{k=\infty}\left[\ddot{p}_k(t)+\omega_k^2p_k(t)\right]\mu(x)W_k(x)=q(x,t). \tag{1.158}$$

Now, taking the inner product with $W_j(x)$, on both sides of (1.158), and using the orthogonality property yields

$$\ddot{p}_j(t) + \omega_j^2 p_j(t) = f_j(t), \qquad j = 1, 2, \ldots, \infty, \tag{1.159}$$

where

$$f_j(t) = \frac{\langle W_j(x), q(x,t)\rangle}{\langle \mu(x) W_j(x), W_j(x,t)\rangle} = \frac{\int_0^l W_j(x) q(x,t)\,\mathrm{d}x}{\int_0^l \mu(x) W_j^2(x)\,\mathrm{d}x}, \qquad j = 1, 2, \ldots, \infty. \tag{1.160}$$

The second-order ordinary differential equation (1.159) with specified initial conditions $p_j(0) = p_{j0}$ and $\dot{p}_j(0) = v_{j0}$ represents an initial value problem. It can be easily solved using, for example, the method of Laplace transforms or the Duhamel convolution integral.

Consider the example of a constant point force F traveling with a speed v on a stretched string, as shown in Figure 1.17. The equation of motion is given by

$$\rho A w_{,tt} - T w_{,xx} = -F\delta(x - vt).$$

We assume the solution form (1.156), where, as we already know, $W_k(x) = \sin k\pi x/l$. Following the steps as discussed above, we obtain the modal dynamics from (1.159)–(1.160) as

$$\ddot{p}_j(t) + \omega_j^2 p_j(t) = -\frac{2F}{\rho A l} \sin\frac{j\pi v t}{l}, \qquad j = 1, 2, \ldots, \infty, \tag{1.161}$$

where $\omega_j = j\pi c/l$. When the forcing in (1.161) is non-resonant, i.e., $v \neq c$, the solution of the above differential equation can be written as

$$p_j(t) = C_j \cos\omega_j t + S_j \sin\omega_j t - \frac{2Fl}{\rho A j^2 \pi^2 (c^2 - v^2)} \sin\frac{j\pi v t}{l}, \tag{1.162}$$

where C_j and S_j are arbitrary constants of integration. Therefore, the solution for the string is obtained from (1.156) as

$$w(x,t) = \sum_{j=1}^{\infty} \left[C_j \cos\omega_j t + S_j \sin\omega_j t - \frac{2Fl}{\rho A j^2 \pi^2 (c^2 - v^2)} \sin\frac{j\pi v t}{l} \right] \sin\frac{j\pi x}{l}. \tag{1.163}$$

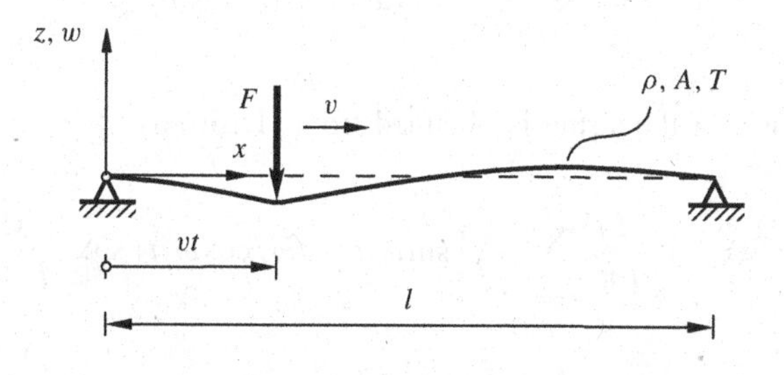

Figure 1.17 Traveling point force on a stretched string

Using the initial conditions of an undeformed string at rest, i.e., $w(x, 0) = 0$ and $w_{,t}(x, 0) = 0$, we get

$$C_j = 0 \quad \text{and} \quad S_j = \frac{2Flv}{\rho Acj^2\pi^2(c^2 - v^2)}, \qquad j = 1, 2, \ldots, \infty. \tag{1.164}$$

Finally, we can write the complete solution of the string as

$$w(x, t) = -\frac{2Fl}{\rho A\pi^2(c^2 - v^2)} \sum_{j=1}^{\infty} \frac{1}{j^2} \left[\sin\frac{j\pi vt}{l} - \frac{v}{c} \sin\frac{j\pi ct}{l} \right] \sin\frac{j\pi x}{l}. \tag{1.165}$$

The above solution is valid as long as the force moves on the string, i.e., $0 \le t \le l/v$. Once the force leaves the string, the string undergoes free vibrations. The shape of the string and its velocity at the onset of the free vibrations can be obtained from (1.165) at $t_f = l/v$. Thus,

$$\begin{aligned} & w(x, t_f^+) = w(x, t_f^-) \quad \text{and} \quad w_{,t}(x, t_f^+) = w_{,t}(x, t_f^-) \\ & \Rightarrow p_j(t_f^+) = p_j(t_f^-) \quad \text{and} \quad \dot{p}_j(t_f^+) = \dot{p}_j(t_f^-), \qquad j = 1, 2, \ldots, \infty, \end{aligned} \tag{1.166}$$

where, for $t \ge t_f$, $p_j(t) = C'_j \cos\omega_j t + S'_j \sin\omega_j t$ with C'_j and S'_j as arbitrary constants. Using the matching conditions (1.166) at $t = t_f$, we obtain the free vibration solution as

$$\begin{aligned} w(x, t) = -\frac{2Flv}{\rho Ac\pi^2(c^2 - v^2)} \sum_{j=1}^{\infty} \frac{(-1)^j}{j^2} \Bigg[& \left(\cos\frac{j\pi c}{v} - (-1)^j \right) \sin\omega_j t \\ & - \sin\frac{j\pi c}{v} \cos\omega_j t \Bigg] \sin\frac{j\pi x}{l}, \end{aligned} \tag{1.167}$$

where $t \ge t_f$. In Figure 1.18, the shape of the string is represented graphically at selected times for a point force traveling at $v = 0.25c$.

Now, we consider the case when the force travels with the resonance speed, i.e., $v = c$ in (1.161). One can obtain the solution of (1.161) as

$$p_j(t) = -\frac{2F}{\rho Al\omega_j^2} \left[\sin\omega_j t - t\omega_j \cos\omega_j t \right]. \tag{1.168}$$

Therefore, the response of the string is obtained from (1.156) as

$$w(x, t) = -\frac{Fl}{T\pi^2} \sum_{j=1}^{\infty} \frac{1}{j^2} \left[\sin\omega_j t - t\omega_j \cos\omega_j t \right] \sin\frac{j\pi x}{l}. \tag{1.169}$$

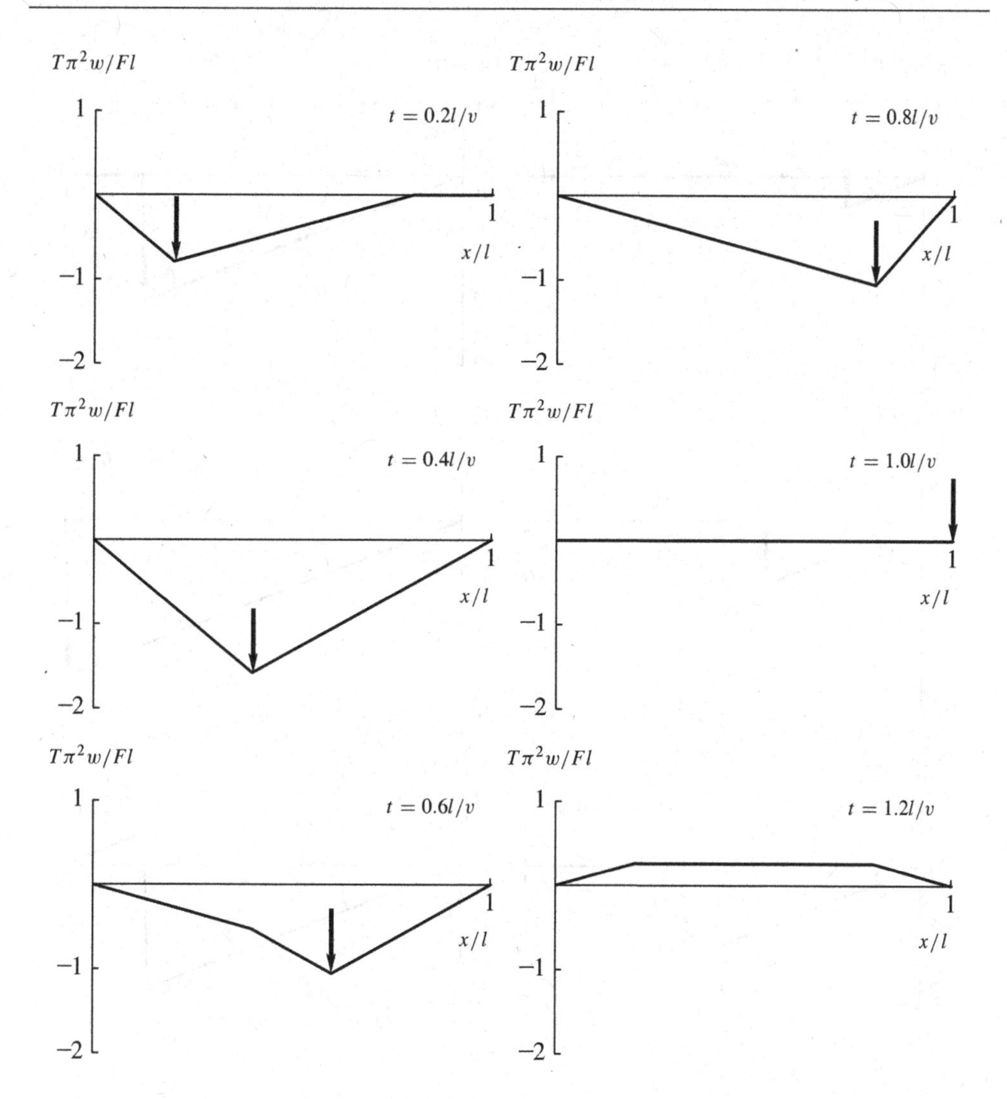

Figure 1.18 Shape of a string at selected times with a traveling point force: $v/c = 0.25$

For $t \geq t_f$, we get, following the procedure discussed above, the free response of the string as

$$w(x, t) = \frac{Fl}{T\pi} \sum_{j=1}^{\infty} \frac{1}{j} \cos \omega_j t \sin \frac{j\pi x}{l}. \tag{1.170}$$

The shape of the string at selected times with the force traveling at resonance speed is shown graphically in Figure 1.19.

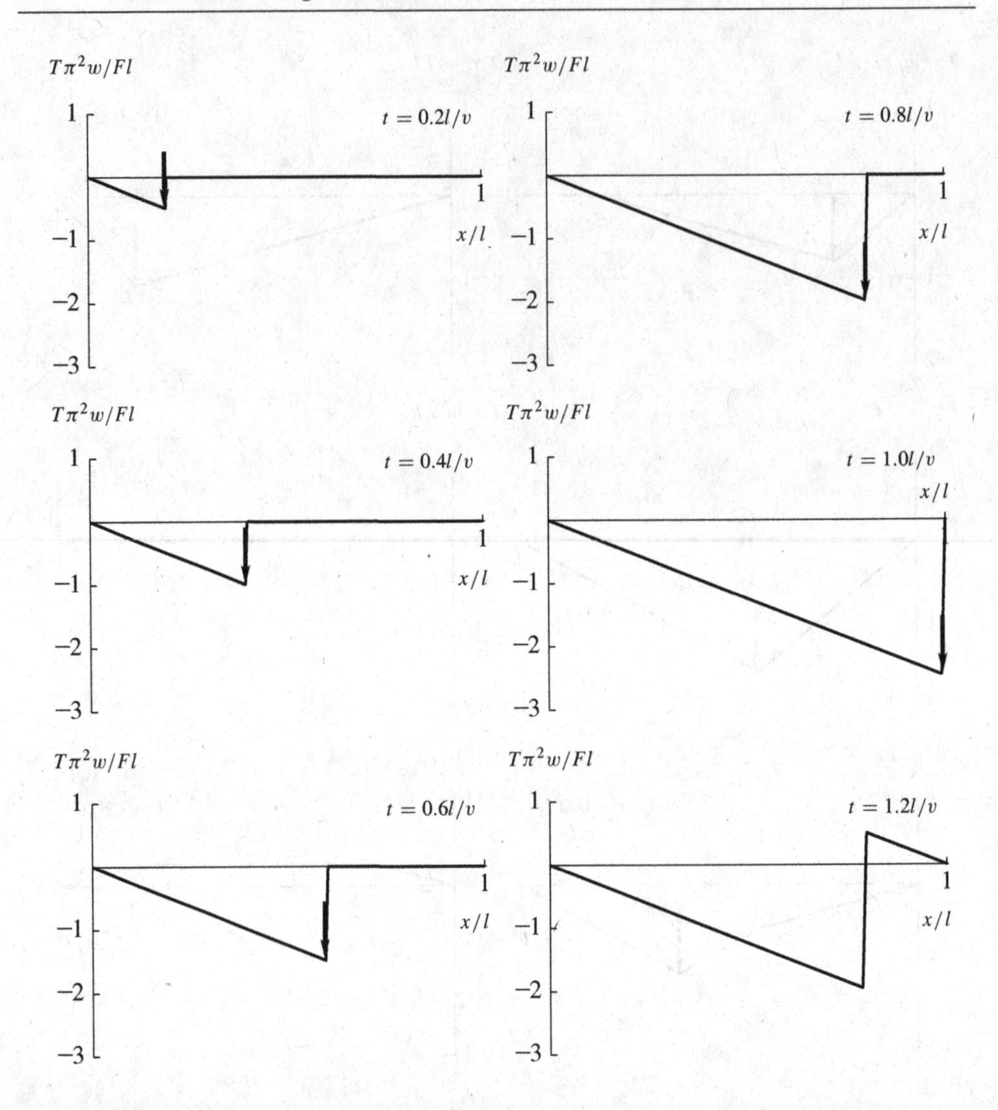

Figure 1.19 Shape of a string at selected times with a point force traveling at resonance speed

1.7 APPROXIMATE METHODS FOR CONTINUOUS SYSTEMS

In many problems of analysis of continuous systems, exact solutions are either not possible to obtain or become too cumbersome to use when a good estimate is all that is required. In certain cases, a given system may be close to an exactly solvable system, or a system with known solutions. In such situations, approximate methods provide sufficiently accurate results to serve the purpose. In the following, we discuss three approximate methods for analyzing continuous systems. Other approximate methods can be found in [5], [7], and [6] (see also Chapter 6).

1.7.1 Rayleigh method

The determination of the natural frequencies of vibration is of foremost importance in the analysis of vibratory systems. Rayleigh's method can be used to calculate or estimate the lowest (or fundamental) frequency of a self-adjoint (conservative) continuous system.

Consider the example of a bar of length l having a uniform cross-section of area A undergoing longitudinal vibrations. The total mechanical energy of the system comprising the kinetic and potential energies is given by

$$\mathcal{E} = \mathcal{T} + \mathcal{V} = \frac{1}{2}\int_0^l \rho A u_{,t}^2(x,t)\,\mathrm{d}x + \frac{1}{2}\int_0^l EAu_{,x}^2(x,t)\,\mathrm{d}x, \tag{1.171}$$

where ρ is the density of the material and E is Young's modulus. Since there is no dissipation, the total energy of the system is a constant. Assuming that the system is vibrating in one of its eigenmodes, we can write the solution as

$$u(x,t) = U(x)\cos\omega t, \tag{1.172}$$

where $U(x)$ is an eigenfunction of the system and ω the corresponding natural frequency. Substituting (1.172) in (1.171) yields on simplification

$$\mathcal{E} = \left[\frac{1}{2}\omega^2 \int_0^l \rho A U^2\,\mathrm{d}x\right]\sin^2\omega_k t + \left[\frac{1}{2}\int_0^l EAU'^2\,\mathrm{d}x\right]\cos^2\omega_k t. \tag{1.173}$$

Now, $\mathcal{E}$ given by (1.173) is a constant (i.e., independent of time) for a non-trivial solution if and only if the amplitudes of the kinetic and potential energy terms are equal, i.e.,

$$\frac{1}{2}\omega^2 \int_0^l \rho A U^2\,\mathrm{d}x = \frac{1}{2}\int_0^l EAU'^2\,\mathrm{d}x$$

$$\Rightarrow \omega^2 = \frac{\int_0^l EAU'^2(x)\,\mathrm{d}x}{\int_0^l \rho A U^2(x)\,\mathrm{d}x} := \mathcal{R}[U(x)]. \tag{1.174}$$

The ratio $\mathcal{R}[U(x)]$ defined in (1.174) is known as the *Rayleigh quotient*. If the eigenfunction $U_k(x)$ is known exactly, one can obtain the exact circular eigenfrequency from (1.174). However, if the eigenfunction is unknown, one can still use (1.174) to determine the fundamental circular frequency through the minimization problem

$$\omega_1^2 = \min_{\tilde{U}(x)\in\mathcal{U}} \mathcal{R}[\tilde{U}(x)] = \min_{\tilde{U}(x)\in\mathcal{U}} \frac{\int_0^l EA\tilde{U}'^2(x)\,\mathrm{d}x}{\int_0^l \rho A\tilde{U}^2(x)\,\mathrm{d}x}, \tag{1.175}$$

where the minimization is performed over the set $\mathcal{U}$ of all functions $\tilde{U}(x)$ that satisfy all the geometric boundary conditions of the problem, and are differentiable at least up to the highest order of space-derivative present in the energy integral. Such functions are known as

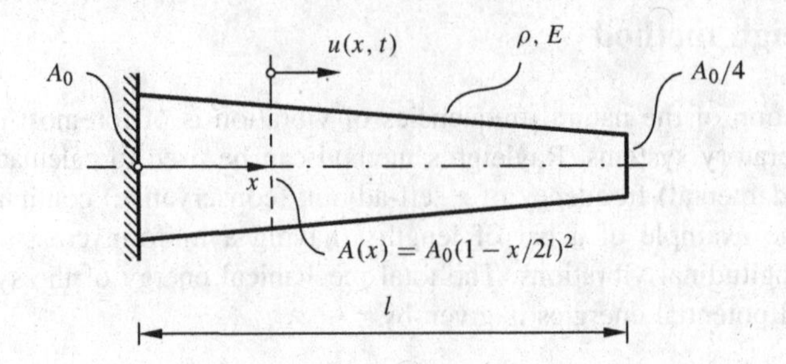

Figure 1.20 A tapered circular bar

admissible functions, and can be easily constructed using polynomials, trigonometric functions, or other elementary functions. The idea of using (1.174) (or (1.175)) for determining the fundamental frequency is due to Rayleigh, and is known as *Rayleigh's method*. The validity of (1.175) will be proved later in Section 1.7.2 in a more general context.

Assume the approximated fundamental mode-shape of longitudinal vibration of a fixed–free bar, as shown in Figure 1.20, in the form

$$F(x) = \left(\frac{x}{l}\right)^{\alpha}, \tag{1.176}$$

where α is a constant, and will be determined later. Computation of the numerator of the Rayleigh quotient in (1.174) yields

$$\begin{aligned} \int_0^l EAF'^2(x)\,\mathrm{d}x &= EA_0 \int_0^l \left(1 - \frac{x}{2l}\right)^2 \left(\frac{\alpha}{l}\right)^2 \left(\frac{x}{l}\right)^{2\alpha-2} \mathrm{d}x \\ &= \frac{EA_0}{l}\alpha^2 4 \frac{(2\alpha-1)2\alpha + 4(2\alpha+1)}{(2\alpha-1)2\alpha(2\alpha+1)}, \end{aligned} \tag{1.177}$$

where it is required that $\alpha > 1/2$ for the definite integral in the above to exist. Further, if $\alpha < 1$, then as $x \to 0$, $F'(x) \to \infty$. The denominator of the Rayleigh quotient yields

$$\begin{aligned} \int_0^l \rho AF^2(x)\,\mathrm{d}x &= \rho A_0 \int_0^l \left(1 - \frac{x}{2l}\right)^2 \left(\frac{x}{l}\right)^{2\alpha} \mathrm{d}x \\ &= \rho A_0 l \frac{(2\alpha+1)(2\alpha+2) + 4(2\alpha+3)}{4(2\alpha+1)(2\alpha+2)(2\alpha+3)}. \end{aligned} \tag{1.178}$$

Therefore, the Rayleigh quotient is obtained as

$$\mathcal{R}[F(x), \alpha] = \frac{E}{\rho l} \frac{\alpha(2\alpha^2 + 3\alpha + 2)(2\alpha^2 + 5\alpha + 3)}{(2\alpha^2 + 7\alpha + 7)(2\alpha - 1)}. \tag{1.179}$$

We can now minimize $\mathcal{R}[F(x), \alpha]$ with respect to α, which yields $\alpha \approx 0.93$, and the fundamental circular frequency as $\omega_1 \approx 2.08303c/l$. The exact solution was obtained in

Section 1.4.1.2 as $\omega_1^{\text{exact}} = 2.029c/l$. It may be noted that the obtained shape-function $F(x)$ after minimization cannot be used to determine the strain at the fixed end since $F'(0) = \infty$. However, it gives the frequency estimate within 3% of the exact value.

1.7.2 Rayleigh–Ritz method

Though Rayleigh's method can be used in principle to determine all the natural frequencies of a vibratory system (see Section 6.1.3), it is most convenient for determining the fundamental frequency. In the following, we discuss an extension of Rayleigh's idea using an expansion technique due to Ritz. This method, usually known as the Rayleigh–Ritz method, can be used to determine the natural frequencies of a continuous system.

We consider the expansion of the mode-shape in terms of N linearly independent admissible functions $U_i(x)$, $i = 1, 2, \ldots, N$ in the form

$$U(x) = \sum_{i=1}^{N} \alpha_i U_i(x), \tag{1.180}$$

where α_i are unknown constants which are to be chosen suitably to minimize the Rayleigh quotient. Substituting (1.180) in the Rayleigh quotient (1.174) leads to

$$\omega^2 = \frac{\sum_{i,j=1}^{N} \alpha_i \alpha_j m_{ij}}{\sum_{i,j=1}^{N} \alpha_i \alpha_j k_{ij}} = \frac{\boldsymbol{\alpha}^{\mathrm{T}} \mathbf{M} \boldsymbol{\alpha}}{\boldsymbol{\alpha}^{\mathrm{T}} \mathbf{K} \boldsymbol{\alpha}}, \tag{1.181}$$

where

$$m_{ij} = \int_0^l EAU_i'U_j'\,\mathrm{d}x \qquad \text{and} \qquad k_{ij} = \int_0^l \rho AU_iU_j\,\mathrm{d}x, \tag{1.182}$$

$\boldsymbol{\alpha}$ is the vector of the α_i, and T in the superscript indicates transposition. Now the minimization condition of the Rayleigh quotient can be written as

$$\begin{aligned}
&\frac{\partial}{\partial \alpha_p}\left(\frac{\boldsymbol{\alpha}^{\mathrm{T}} \mathbf{M} \boldsymbol{\alpha}}{\boldsymbol{\alpha}^{\mathrm{T}} \mathbf{K} \boldsymbol{\alpha}}\right) = 0, \qquad p = 1, 2, \ldots, N, \\
&\Rightarrow \boldsymbol{\alpha}^{\mathrm{T}} \mathbf{M} \boldsymbol{\alpha}\left(\frac{\partial \boldsymbol{\alpha}^{\mathrm{T}} \mathbf{K} \boldsymbol{\alpha}}{\partial \boldsymbol{\alpha}}\right) - \boldsymbol{\alpha}^T \mathbf{K} \boldsymbol{\alpha}\left(\frac{\partial \boldsymbol{\alpha}^{\mathrm{T}} \mathbf{M} \boldsymbol{\alpha}}{\partial \boldsymbol{\alpha}}\right) = \mathbf{0}, \\
&\Rightarrow 2\mathbf{K}\boldsymbol{\alpha} - 2\left(\frac{\boldsymbol{\alpha}^{\mathrm{T}} \mathbf{K} \boldsymbol{\alpha}}{\boldsymbol{\alpha}^{\mathrm{T}} \mathbf{M} \boldsymbol{\alpha}}\right)\mathbf{M}\boldsymbol{\alpha} = \mathbf{0}, \\
&\Rightarrow (\mathbf{K} - \omega^2 \mathbf{M})\boldsymbol{\alpha} = \mathbf{0} \qquad \text{(using (1.181))}.
\end{aligned} \tag{1.183}$$

Thus, it is observed that the extremization condition for Rayleigh's quotient (formed using a finite expansion) for a continuous system leads to the eigenvalue problem of a finite-dimensional system. It will be shown in Section 1.7.3 that this finite-dimensional system is actually the discretized version of the original continuous system. The eigenvalue problem (1.183) can now be solved to determine the first N approximate eigenvalues and eigenfunctions (from (1.180)) of the system. It must be mentioned, however, that the error is not uniform over all the eigenvalues. Convergence of the desired eigenvalues needs to be checked by increasing the number of terms in the expansion (1.180).

1.7.3 Ritz method

In this method, Ritz's idea of solution expansion is used in the variational formulation of system dynamics to obtain the discretized equations of motion of a continuous system.

Let us consider the example of a bar of varying cross-section in longitudinal vibration. The variational formulation of the problem can be stated in the form of Hamilton's principle as

$$\delta \int_{t_1}^{t_2} \frac{1}{2} \int_0^l \left(\rho A u_{,t}^2 - EA u_{,x}^2\right) \mathrm{d}x \, \mathrm{d}t = 0. \tag{1.184}$$

Consider an approximate solution of this system in the form

$$u(x,t) = \sum_{k=1}^{N} p_k(t) H_k(x) = \mathbf{H}^{\mathrm{T}} \mathbf{p}, \tag{1.185}$$

where $\mathbf{H} = [H_1(x), \ldots, H_N(x)]^{\mathrm{T}}$ is a vector of N linearly independent admissible functions and $\mathbf{p} = [p_1(t), \ldots, p_N(t)]^{\mathrm{T}}$ is a vector of the corresponding unknown coordinate functions. Substituting (1.185) in (1.184), we get

$$\delta \int_{t_1}^{t_2} \frac{1}{2} \int_0^l \left[\rho A \dot{\mathbf{p}}^{\mathrm{T}} \mathbf{H} \mathbf{H}^{\mathrm{T}} \dot{\mathbf{p}} - EA \mathbf{p}^{\mathrm{T}} \mathbf{H}' \mathbf{H}'^{\mathrm{T}} \mathbf{p}\right] \mathrm{d}x \, \mathrm{d}t = 0,$$

$$\Rightarrow \delta \int_{t_1}^{t_2} \frac{1}{2} \left[\dot{\mathbf{p}}^{\mathrm{T}} \mathbf{M} \dot{\mathbf{p}} - \mathbf{p}^{\mathrm{T}} \mathbf{K} \mathbf{p}\right] \mathrm{d}t = 0, \tag{1.186}$$

where

$$\mathbf{M} = \int_0^l \rho A \mathbf{H} \mathbf{H}^{\mathrm{T}} \, \mathrm{d}x, \quad \text{and} \quad \mathbf{K} = \int_0^l EA \mathbf{H}' \mathbf{H}'^{\mathrm{T}} \, \mathrm{d}x. \tag{1.187}$$

One can easily observe that (1.186) represents the variational formulation of dynamics of a discrete system with $\mathbf{p}$ as the vector of generalized coordinates. Therefore, one can directly use Lagrange's equations to obtain the discrete equations of motion

$$\mathbf{M}\ddot{\mathbf{p}} + \mathbf{K}\mathbf{p} = \mathbf{0}. \tag{1.188}$$

It may be observed from (1.187) that both **M** and **K** are symmetric. They are also positive definite since ρA and EA are positive functions, and the admissible functions chosen in the expansion (1.185) are linearly independent.

Consider the example of longitudinal vibration of the fixed–free tapered bar shown in Figure 1.20. We can choose the admissible functions as

$$H_j(x) = \frac{x}{l}\left(1 - \frac{x}{2l}\right)^{j-1}, \qquad j = 1, 2, \ldots, N, \tag{1.189}$$

since the geometric boundary conditions, $H_j(0) = 0$, are exactly satisfied. However, as can be checked, the natural boundary conditions, $H_j'(l) = 0$, are not satisfied. Considering only two admissible functions in the expansion (1.185), and following the steps discussed above, (1.188) takes the form

$$\rho A_0 l \begin{bmatrix} \frac{2}{15} & \frac{7}{80} \\ \frac{7}{80} & \frac{33}{560} \end{bmatrix} \begin{Bmatrix} \ddot{p}_1 \\ \ddot{p}_2 \end{Bmatrix} + \frac{EA_0}{l} \begin{bmatrix} \frac{7}{12} & \frac{17}{48} \\ \frac{17}{48} & \frac{31}{120} \end{bmatrix} \begin{Bmatrix} p_1 \\ p_2 \end{Bmatrix} = 0. \tag{1.190}$$

Assuming a modal solution $\mathbf{p}(t) = \mathbf{k}e^{i\omega t}$, (1.190) yields the eigenvalue problem

$$[-\omega^2\mathbf{M} + \mathbf{K}]\mathbf{k} = 0, \tag{1.191}$$

from which the characteristic equation is obtained as

$$\frac{81}{7}\omega^4 - 394\frac{c^2}{l^2}\omega^2 + 1455\frac{c^2}{l^2} = 0. \tag{1.192}$$

The first two approximate circular natural frequencies of longitudinal vibration are obtained from (1.192) as $\omega_1^R = 2.053c/l$ and $\omega_2^R = 5.462c/l$. The exact circular eigenfrequencies were obtained in Section 1.4.1.2 as $\omega_1^{\text{exact}} = 2.029c/l$, and $\omega_2^{\text{exact}} = 4.913c/l$. The eigenvectors are obtained from (1.191) as

$$\mathbf{k}_1 = \begin{Bmatrix} 1.0 \\ 1.475 \end{Bmatrix} \quad \text{and} \quad \mathbf{k}_2 = \begin{Bmatrix} 1.0 \\ -1.505 \end{Bmatrix}. \tag{1.193}$$

Using these eigenvectors, along with (1.189), in (1.185), we get the approximate eigenfunctions as

$$U_1(x) = \mathbf{H}^{\mathrm{T}}\mathbf{k}_1 = \frac{x}{l} + 1.457\frac{x}{l}\left(1 - \frac{x}{2l}\right) \tag{1.194}$$

and

$$U_2(x) = \mathbf{H}^{\mathrm{T}}\mathbf{k}_2 = \frac{x}{l} - 1.505\frac{x}{l}\left(1 - \frac{x}{2l}\right). \tag{1.195}$$

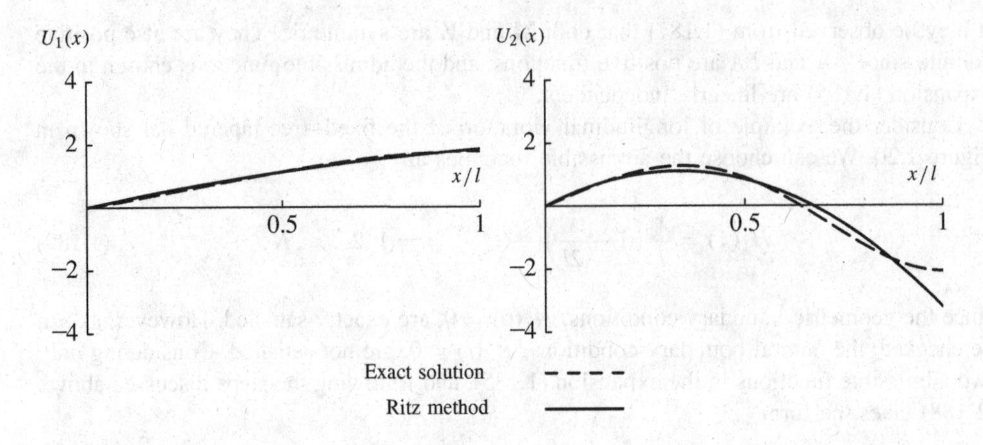

Figure 1.21 Comparison of first two mode-shapes from Ritz method and exact solution (mode-shapes normalized such that $\int_0^l \rho A U_i^2 \,\mathrm{d}x = \int_0^l \rho A \,\mathrm{d}x$)

The exact and approximate eigenfunctions are compared in Figure 1.21. It may be observed from the figure that the first mode-shape is determined reasonably accurately. However, the second mode-shape is in considerable error. For a better approximation of the second mode-shape, one must take more terms in the expansion (1.185). In general, the error in determination of the eigenfrequencies is less than that in the determination of the eigenfunctions. Further, the eigenfrequencies are always overestimated. In other words, we obtain an upper bound on the eigenfrequencies. This is expected since we are approximating an infinite degrees of freedom system by a finite degrees of freedom system, thereby increasing the stiffness of the system. The greatest advantage of the Ritz method is that only admissible functions are required to be constructed for the solution expansion.

When there are external forces, the variational principle (1.184) is modified to (see Appendix A)

$$\int_{t_1}^{t_2}\int_0^l \left[\delta\left(\frac{1}{2}\rho A u_{,t}^2 - \frac{1}{2}EAu_{,x}^2\right) + q(x,t)\delta u\right] \mathrm{d}x\,\mathrm{d}t = 0, \tag{1.196}$$

where $q(x,t)$ is the generalized force. Substituting the solution form (1.185) in (1.196) and following the procedure presented in this section, we obtain the discretized equation of motion with forcing as

$$\mathbf{M}\ddot{\mathbf{p}} + \mathbf{K}\mathbf{p} = \mathbf{f}(t),$$

where

$$\mathbf{f}(t) = \int_0^l q(x,t)\mathbf{H}(x)\mathrm{d}x.$$

1.7.4 Galerkin method

Consider the dynamics of a continuous system governed by the equation of motion

$$\mu(x)u_{,tt} + \mathcal{K}[u] = 0. \tag{1.197}$$

Let us construct an approximate solution of $u(x, t)$ in the form

$$u(x, t) = \sum_{k=1}^{N} p_k(t)P_k(x) = \mathbf{P}^{\mathrm{T}}\mathbf{p}, \tag{1.198}$$

where $\mathbf{p} = [p_1(t), \ldots, p_N(t)]^{\mathrm{T}}$, $\mathbf{P} = [P_1(x), \ldots, P_N(x)]^{\mathrm{T}}$, and $P_k(x)$ satisfy all the geometric and natural boundary conditions of the problem, and are differentiable at least up to the highest order of space-derivative in the differential equation of motion. Such functions are known as *comparison functions*. It is clear that the approximate solution (1.198) will also satisfy all the boundary conditions, but in general will not satisfy (1.197) identically. Thus, there remains a *residue* defined by

$$e(x, t) := \mu(x)\mathbf{P}^{\mathrm{T}}\ddot{\mathbf{p}} + \mathcal{K}[\mathbf{P}^{\mathrm{T}}]\mathbf{p}, \tag{1.199}$$

where $\mathcal{K}[\mathbf{P}^{\mathrm{T}}] = (\mathcal{K}[P_1(x)], \ldots, \mathcal{K}[P_N(x)])$. Since we are now searching for an approximate solution in a finite N-dimensional space, we can force the residue to have a zero projection on the chosen basis functions $P_j(x)$, $j = 1, 2, \ldots, N$, of this space. Therefore, we put

$$\langle e(x, t), P_j(x)\rangle := \int_0^l e(x, t)P_j(x)\,\mathrm{d}x = 0, \qquad j = 1, 2, \ldots, N. \tag{1.200}$$

Substituting the expression of the residue from (1.199) in (1.200), and writing in a compact form, we have

$$\mathbf{M}\ddot{\mathbf{p}} + \mathbf{K}\mathbf{p} = \mathbf{0}, \tag{1.201}$$

where the elements of the matrices **M** and **K** are obtained as

$$\mathbf{M} = \int_0^l \mu(x)\mathbf{P}\mathbf{P}^{\mathrm{T}}\,\mathrm{d}x \qquad \text{and} \qquad \mathbf{K} = \int_0^l \mathbf{P}\mathcal{K}[\mathbf{P}^{\mathrm{T}}]\,\mathrm{d}x. \tag{1.202}$$

It is evident that **M** is a symmetric matrix. If $\mathcal{K}[\cdot]$ is self-adjoint, **K** is also a symmetric matrix, as can be easily checked. The equations (1.188) (obtained from the Ritz method) and (1.201) look similar. However, they differ in the computation of the matrix **K**.

Galerkin's method can also be understood from the variational principle as follows. Consider once again the example of longitudinal vibration of a bar. The variational statement can be simplified and rewritten from (1.37) as

$$-\int_{t_1}^{t_2} EAu_{,x}\delta u\Big|_0^l\,\mathrm{d}t - \int_{t_1}^{t_2}\int_0^l [\rho Au_{,tt} - (EAu_{,x})_{,x}]\,\delta u\,\mathrm{d}x\,\mathrm{d}t = 0. \tag{1.203}$$

Let us assume the solution of $u(x,t)$ in the form (1.198). Since $\mathbf{P}$ satisfies all the boundary conditions, the assumed solution makes the boundary terms in (1.203) identically zero. The quantity in square brackets in the second term in (1.203) yields the residue in the form (1.199). Writing the variation of $u(x,t)$ from (1.198) as

$$\delta u(x,t) = \mathbf{P}^{\mathrm{T}}\delta\mathbf{p} = \delta\mathbf{p}^{\mathrm{T}}\mathbf{P}, \tag{1.204}$$

one can rewrite (1.203) in the form

$$\int_{t_1}^{t_2}\int_0^l \delta\mathbf{p}^{\mathrm{T}}\mathbf{P}e(x,t)\,\mathrm{d}x\,\mathrm{d}t = 0$$

or

$$\int_{t_1}^{t_2} \delta\mathbf{p}^T\left[\int_0^l \mathbf{P}e(x,t)\,\mathrm{d}x\right]\mathrm{d}t = 0. \tag{1.205}$$

Thus, the term in the square brackets in (1.205) must vanish identically for the above condition to hold for an arbitrary variation $\delta\mathbf{p}$. This again yields the condition (1.200), and the discretized equations (1.201). Since Galerkin's method works directly with the differential equation of motion, it offers certain advantages over the Ritz method. It is evident that one can in principle handle any kind of non-conservative and non-potential forces (non-self-adjoint problems) with Galerkin's method. However, generation of the comparison functions as required in Galerkin's method may be quite tedious for certain problems. Discretization of certain non-self-adjoint problems has been discussed in [6].

We consider the example of longitudinal vibration of the fixed–free tapered bar shown in Figure 1.20 once again. Two comparison functions are chosen as

$$P_k(x) = \left(1 - \frac{x}{l}\right)^{k+1} - 1, \qquad k = 1, 2. \tag{1.206}$$

It can be easily checked that these functions satisfy both the geometric and the natural boundary conditions of the problem. Following the steps discussed in this section, (1.201) takes the form

$$\rho A_0 l\begin{bmatrix} \frac{33}{140} & \frac{297}{1120} \\ \frac{297}{1120} & \frac{1517}{5040}\end{bmatrix}\begin{Bmatrix}\ddot{p}_1\\ \ddot{p}_2\end{Bmatrix} + \frac{EA_0}{l}\begin{bmatrix} \frac{31}{40} & \frac{49}{40} \\ \frac{49}{40} & \frac{213}{140}\end{bmatrix}\begin{Bmatrix}p_1\\ p_2\end{Bmatrix} = 0. \tag{1.207}$$

Assuming the solution in the form $\mathbf{p} = \mathbf{k}e^{i\omega t}$ the eigenvalue problem is solved, and the circular eigenfrequencies are obtained as $\omega_1^{\mathrm{G}} = 2.2029c/l$, and $\omega_2^{\mathrm{G}} = 5.258c/l$. As compared to the eigenfrequencies obtained from the Ritz method, these are closer to the exact eigenfrequencies (see Section 1.4.1.2). This is because, in the case of Galerkin's method, the eigenfunctions satisfy all the boundary conditions of the problem. The eigenvectors are obtained as

$$\mathbf{k}_1 = \begin{Bmatrix} 1.0 \\ -0.472 \end{Bmatrix} \quad \text{and} \quad \mathbf{k}_2 = \begin{Bmatrix} 1.0 \\ -0.898 \end{Bmatrix}, \tag{1.208}$$

and we get the approximate eigenfunctions as

$$U_1(x) = \frac{x}{l}\left(\frac{x}{l} - 2\right) + 0.472\frac{x}{l}\left(\frac{x^2}{l^2} - 3\frac{x}{l} + 3\right) \tag{1.209}$$

and

$$U_2(x) = \frac{x}{l}\left(\frac{x}{l} - 2\right) + 0.898\frac{x}{l}\left(\frac{x^2}{l^2} - 3\frac{x}{l} + 3\right). \tag{1.210}$$

A comparison of the exact and the approximate eigenfunctions obtained in (1.209)–(1.210) is shown in Figure 1.22. On comparing the results in Figure 1.21 and Figure 1.22 it can be observed that Galerkin's method yields better results than the Ritz method. However, it must be remembered that the eigenfunctions satisfy different conditions in the two methods. As mentioned before, the difficulty of Galerkin's method lies in the construction of the comparison functions.

In the presence of an external forcing $q(x, t)$ in (1.197), Galerkin's method yields

$$\mathbf{M}\ddot{\mathbf{p}} + \mathbf{K}\mathbf{p} = \mathbf{f}(t),$$

where **M** and **K** are defined by (1.202), and

$$\mathbf{f}(t) = \int_0^l q(x, t)\mathbf{P}(x)\mathrm{d}x.$$

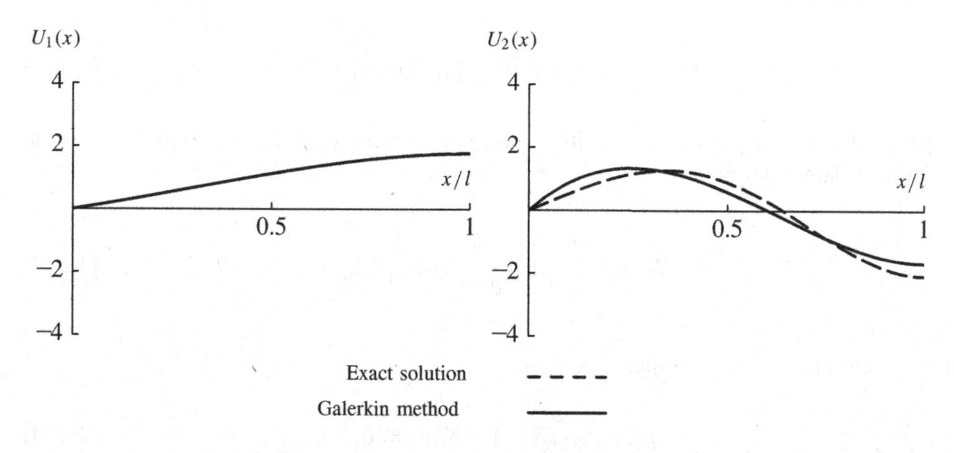

Figure 1.22 Comparison of first two mode-shapes from Galerkin method and exact solution (mode-shapes normalized such that $\int_0^l \rho A U_i^2 \,\mathrm{d}x = \int_0^l \rho A \,\mathrm{d}x$)

1.8 CONTINUOUS SYSTEMS WITH DAMPING

All vibratory systems experience energy dissipation, a phenomenon commonly known as *damping*. Damping forces may arise from external interactions of the system (external damping), or from within the system (internal damping). Damping from aerodynamic drag due to viscosity is the most common example of external damping, while internal damping occurs due to internal friction between the molecular layers as a result of differential straining. In these damping mechanisms, mechanical energy is converted irreversibly into thermal energy which flows out of the system. In Chapter 2, we will consider a damping mechanism in which energy is lost by a system through radiation.

Three damping models, namely viscous damping, Coulomb damping (or dry friction), and structural damping (or hysteretic damping) are usually used for engineering purposes. The viscous damping model, which is the most commonly used model, relates the damping forces with the time rate of change of the field variable, or its spatial derivatives. We will use only this model in our discussions below.

1.8.1 Systems with distributed damping

Consider the longitudinal oscillations of a uniform fixed–free bar. We assume that the internal damping in the material is such that the stresses are a linear function of both the strain and the strain rate. Thus, (1.14) is modified to

$$\sigma_x(x,t) = E\epsilon_x(x,t) + d_{\mathrm{I}}\epsilon_{x,t}(x,t) = Eu_{,x}(x,t) + d_{\mathrm{I}}u_{,xt}(x,t), \tag{1.211}$$

where $d_{\mathrm{I}} > 0$ is the coefficient of internal damping in the material. We also assume a distributed external damping force of the usual form $-d_{\mathrm{E}}u_{,t}(x,t)$, where $d_{\mathrm{E}} > 0$ is the coefficient of external damping. Then, proceeding similarly to what was done in Section 1.1.2, one obtains the equation of motion of the longitudinal dynamics of a bar with internal and external damping as

$$\rho A u_{,tt} - EAu_{,xx} - d_{\mathrm{I}}Au_{,xxt} + d_{\mathrm{E}}u_{,t} = 0, \tag{1.212}$$

instead of (1.15). The boundary conditions are not affected by these damping terms. One can define a damping operator

$$\mathcal{D}[\cdot] = \left(-d_{\mathrm{I}}A\frac{\mathrm{d}^2}{\mathrm{d}x^2} + d_{\mathrm{E}}\right)[\cdot], \tag{1.213}$$

and represent (1.212) in a compact form as

$$\rho A u_{,tt} + \mathcal{D}[u_{,t}] + \mathcal{K}[u] = 0, \tag{1.214}$$

where $\mathcal{K}[\cdot] = -EA[\cdot]_{,xx}$.

Multiplying both sides of (1.212) by $u_{,t}$ and integrating over the domain of the bar yields

$$\int_0^l (\rho A u_{,t} u_{,tt} - u_{,t} EA u_{,xx} - u_{,t} d_{\mathrm{I}} A u_{,xxt} + d_{\mathrm{E}} u_{,t}^2)\,\mathrm{d}x = 0$$

$$\Rightarrow \left[u_{,t} EA u_{,x} + u_{,t} d_{\mathrm{I}} A u_{,xt}\right]_0^l$$

$$+ \int_0^l \left[\left(\frac{1}{2}\rho A u_{,t}^2\right)_{,t} + u_{,xt} EA u_{,x} + d_{\mathrm{I}} A u_{,xt}^2 + d_{\mathrm{E}} u_{,t}^2\right] \mathrm{d}x = 0. \qquad (1.215)$$

Using the fixed–free boundary conditions, one can rewrite (1.215) as

$$\frac{\mathrm{d}}{\mathrm{d}t}\int_0^l \left(\frac{1}{2}\rho A u_{,t}^2 + \frac{1}{2} EA u_{,x}^2\right) \mathrm{d}x = -\int_0^l (d_{\mathrm{I}} A u_{,xt}^2 + d_{\mathrm{E}} u_{,t}^2)\,\mathrm{d}x. \qquad (1.216)$$

The integral on the left-hand side in (1.216) can be easily recognized to be the total mechanical energy of the bar. Since the right-hand side is always negative, (1.216) implies that the time rate of change of mechanical energy of the bar is always negative, i.e., mechanical energy monotonically decreases with time.

Consider now a system represented by

$$\mu(x) u_{,tt} + \mathcal{D}[u_{,t}] + \mathcal{K}[u] = 0. \qquad (1.217)$$

We explore the possibility of a solution of (1.217) in the form

$$u(x,t) = \sum_{k=1}^{\infty} p_k(t) U_k(x), \qquad (1.218)$$

where the shape-functions $U_k(x)$ are chosen to be the same as the eigenfunctions for the undamped case, i.e., they are solutions of the self-adjoint eigenvalue problem

$$-\lambda \mu(x) U + \mathcal{K}[U] = 0, \qquad (1.219)$$

with appropriate boundary conditions. We will assume that these eigenfunctions are orthonormal with respect to $\mu(x)$, i.e., $\langle \mu(x) U_j, U_k \rangle = \delta_{jk}$. Substituting (1.218) in (1.217) and taking the inner product with $U_j(x)$ yields

$$\ddot{p}_j + \sum_{k=1}^{\infty} d_{jk} \dot{p}_k + \lambda_j p_j = 0, \qquad j = 1, 2, \ldots, \infty \qquad (1.220)$$

where

$$d_{jk} = \langle \mu(x)(-d_I A U_{k,xx} + d_E U_k), U_j \rangle. \qquad (1.221)$$

It is evident that, in general, the damping matrix $\mathbf{D} = [d_{jk}]$ will not be diagonal. Therefore, all the coordinates p_j of the system are coupled through $\mathbf{D}$.

Consider the special situation when

$$\mathcal{D}[U_k(x)] = d_k \mu(x) U_k(x), \tag{1.222}$$

where d_k are constants. Then, it can be easily checked that the resulting damping matrix $\mathbf{D}$ is diagonal. It can be observed that (1.222) represents an eigenvalue problem for the damping operator similar to (1.219). It then follows that if the operators $\mathcal{D}[\cdot]$ and $\mathcal{K}[\cdot]$ have the same eigenfunctions, the resulting damping matrix $\mathbf{D}$ is diagonal. We can determine the condition for the two operators to have the same eigenfunctions as follows. From (1.222), one can write

$$\begin{aligned} \mathcal{K}[\mu^{-1}(x)\mathcal{D}[U_k(x)]] &= \mathcal{K}[d_k U_k(x)] \\ &= d_k \lambda_k U_k(x) \qquad \text{(using (1.219))}. \end{aligned} \tag{1.223}$$

Similarly, from (1.219), it follows that

$$\begin{aligned} \mathcal{D}[\mu^{-1}(x)\mathcal{K}[U_k(x)]] &= \mathcal{D}[\lambda_k U_k(x)] \\ &= \lambda_k d_k U_k(x) \qquad \text{(using (1.222))}. \end{aligned} \tag{1.224}$$

From (1.223) and (1.224), we can conclude that when $\mathcal{K}[\cdot]$ and $\mathcal{D}[\cdot]$ have the same eigenfunctions they satisfy

$$\begin{aligned} &\mathcal{K}[\mu^{-1}(x)\mathcal{D}[U_k]] - \mathcal{D}[\mu^{-1}(x)\mathcal{K}[U_k]] = 0, \qquad k = 1, 2, \ldots, \infty \\ &\Rightarrow (\mathcal{K}[\mu^{-1}(x)\mathcal{D}] - \mathcal{D}[\mu^{-1}(x)\mathcal{K}])[\cdot] = 0, \end{aligned} \tag{1.225}$$

i.e., the two operators commute with respect to $\mu^{-1}(x)$. The converse of this result can also be easily established. Let the two operators commute, i.e., (1.225) is satisfied. From (1.219), one can easily obtain

$$\begin{aligned} &-\lambda \mathcal{D}[U] + \mathcal{D}[\mu^{-1}(x)\mathcal{K}[U]] = 0 \\ &\Rightarrow -\lambda \mathcal{D}[U] + \mathcal{K}[\mu^{-1}(x)\mathcal{D}[U]] = 0 \qquad \text{(using (1.225))}, \end{aligned}$$

or

$$-\lambda \mu(x) V + \mathcal{K}[V] = 0, \tag{1.226}$$

where

$$V = \mu^{-1}(x)\mathcal{D}[U]. \tag{1.227}$$

It is evident that if V satisfies (1.226), in view of (1.219) it must be true that $V = \beta U$ for some constant factor β. Hence, from (1.227) we have

$$\mathcal{D}[U] = \beta\mu(x)U,$$

i.e., U must also be an eigenfunction of the damping operator $\mathcal{D}[\cdot]$. Therefore, (1.225) is the necessary and sufficient condition for $\mathcal{K}[\cdot]$ and $\mathcal{D}[\cdot]$ to have the same eigenfunctions, and hence for the damping matrix **D** to be diagonal. It is not difficult to show that the condition (1.225) implies that the operator $\mathcal{K}[\mu^{-1}(x)\mathcal{D}[\cdot]]$ is self-adjoint.

One clear advantage obtained if $\mathcal{D}[\cdot]$ satisfies (1.225) is that the discretized equations of motion are completely decoupled when the solution of the damped system is expanded in terms of the eigenfunctions of the undamped system. This decoupling allows us to solve the discretized equations in an easy manner. One special choice of the damping operator for which the commutation holds is

$$\mathcal{D}[\cdot] = \beta\mu(x) + \gamma\mathcal{K}[\cdot], \tag{1.228}$$

where β and γ are arbitrary constants. Such a damping is usually known as *classical damping* or *proportional damping*. The condition (1.228) is satisfied in the case of the damped bar described by (1.212). Therefore, the differential equation for the jth modal coordinate of the bar is given by

$$\ddot{p}_j + d_j\dot{p}_j + \lambda_j p_j = 0. \tag{1.229}$$

which can be easily solved for $p_j(t)$. Finally, the complete solution of the longitudinal vibration of the bar is obtained from (1.218).

1.8.2 Systems with discrete damping

In many practical situations, a continuous system may interact with discrete damping elements. For example, certain support points of a structure may provide substantially higher damping to the structure than its internal damping. In that case, the damping can be considered to be due to discrete dampers at such support points. Discrete damper elements are also routinely attached to structures for vibration control. Here we consider two specific cases, and discuss the effects of discrete damping.

Consider a uniform bar fixed at one end, and having an external damper at the other end, as shown in Figure 1.23. The equation of motion can be written as

$$u_{,tt} - c^2u_{,xx} = 0, \tag{1.230}$$

while the boundary conditions are

$$u(0,t) = 0, \quad \text{and} \quad EAu_{,x}(l,t) = -du_{,t}(l,t). \tag{1.231}$$

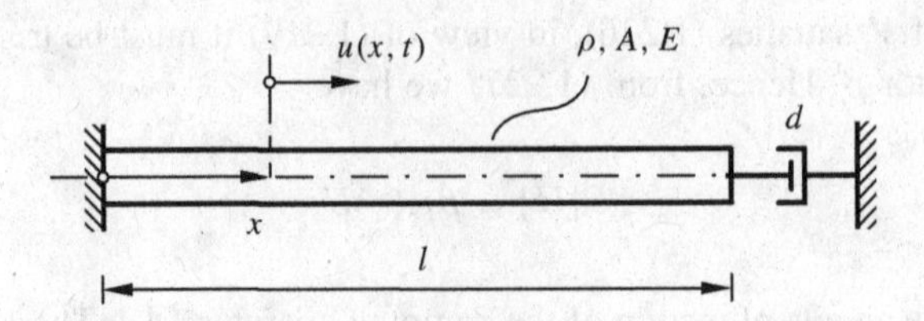

Figure 1.23 A uniform bar with boundary damping

Assuming a solution form

$$u(x,t) = U(x)e^{st}, \tag{1.232}$$

we obtain the eigenvalue problem

$$U'' - \frac{s^2}{c^2}U = 0, \tag{1.233}$$

with

$$U(0) = 0 \quad \text{and} \quad U'(l) = -\frac{sd}{EA}U(l). \tag{1.234}$$

Consider the general solution of (1.233) in the form

$$U(x) = Be^{sx/c} + Ce^{-sx/c}, \tag{1.235}$$

where B and C are constants of integration. Substituting this solution in the boundary conditions (1.234) yields on simplification

$$\begin{bmatrix} 1 & 1 \\ e^{\gamma}(1+a) & -e^{-\gamma}(1-a) \end{bmatrix} \begin{Bmatrix} B \\ C \end{Bmatrix} = 0, \tag{1.236}$$

where $\gamma = sl/c$, and $a = cd/EA$. The condition of non-triviality of the solution of (1.236) yields the characteristic equation as

$$e^{2\gamma} = \frac{a-1}{a+1}, \tag{1.237}$$

which can be solved for γ, and hence, the eigenvalues s of the system for $a \neq 1$. When $a = 1$, which occurs for the special value of boundary damping $d = EA/c$, it is observed from (1.237) that no eigenvalue exists. In this case, there is no solution of the assumed form (1.232). This will be discussed further in Chapter 2.

When $a \neq 1$, one can rewrite (1.237) using the definition $\gamma := \alpha + i\beta$ as

$$e^{2(\alpha+i\beta)} = \frac{a-1}{a+1},$$
$$\Rightarrow \quad \alpha = \frac{1}{2}\ln\left|\frac{a-1}{a+1}\right|$$

and

$$\beta_k = \begin{cases} (2k-1)\pi/2, & 0 \le a < 1 \\ k\pi, & a > 1 \end{cases} \qquad k = 1, 2, \ldots, \infty.$$

It can be easily checked that, when $d = 0$, this gives the eigenvalues of a fixed–free bar, while $d \to \infty$ yields the eigenvalues of a fixed–fixed bar. It is surprising to note that all the modes have the same decay rate, since α does not depend on k. Further, the transition in the imaginary part of the eigenvalues is discrete as a crosses unity. The locus of an eigenvalue with a as the parameter is depicted in Figure 1.24.

Consider next the case of a taut string with a discrete external damper, as shown in Figure 1.25. The equation of motion of the system can be written as

$$\rho A w_{,tt} + d w_{,t} \delta(x - x_d) - T w_{,xx} = 0, \tag{1.238}$$

where x_d is the location of the damper. Let us expand the solution in terms of the eigenfunctions of an undamped string as

$$w(x,t) = \sum_{k=1}^{\infty} p_k(t) \sin\frac{k\pi x}{l}. \tag{1.239}$$

Substituting this solution in (1.238) and taking the inner product with $\sin j\pi x/l$ yields the jth modal coordinate equation as

$$\ddot{p}_j + \sum_{k=1}^{\infty}\left(\frac{d}{\rho A}\sin\frac{k\pi x_d}{l}\sin\frac{j\pi x_d}{l}\right)\dot{p}_k + \frac{T}{\rho A}p_j = 0. \tag{1.240}$$

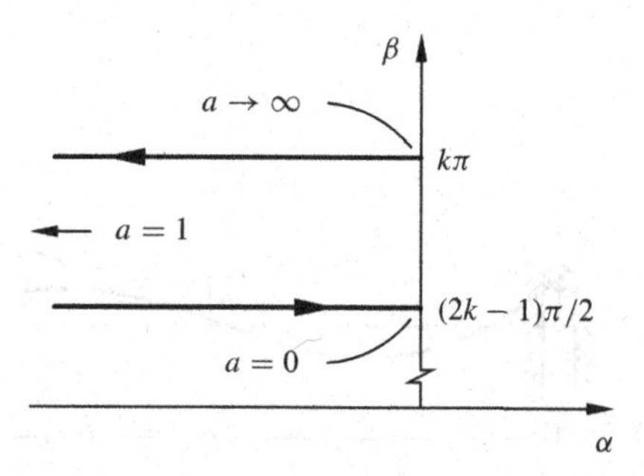

Figure 1.24 Locus of an eigenvalue with a as a parameter for a bar with boundary damping

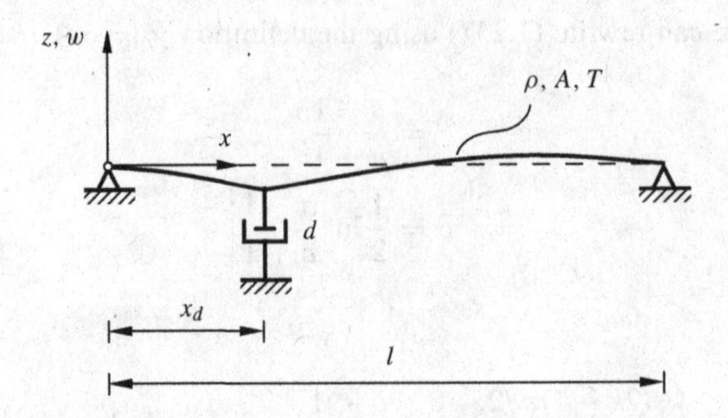

Figure 1.25 A string with discrete damping

It may be observed here that the damping matrix is positive semi-definite with rank one. Further, it couples all the modes of the undamped system. When x_d is chosen such that jx_d/l is never an integer for any j, it can be shown that all the modes are damped. In other words, the total mechanical energy of the string decreases monotonically in time. In this case, the damping is called *pervasive*. Such a damper location is most desirable when we want to damp any arbitrary string motion. In the case where jx_d/l is an integer for some j, the damping is not pervasive, and certain modes remain undamped since one of the nodes of such modes is at x_d. For example, if $x_d = l/3$, the 3rd, $6th, \ldots$ modes will remain undamped.

1.9 NON-HOMOGENEOUS BOUNDARY CONDITIONS

In all the preceding discussions, the boundary conditions were assumed to be homogeneous. However, there are situations where they are not. Non-homogeneity in boundary conditions occurs when either a motion or a force is prescribed at a boundary.

Consider a sliding–fixed string with a specified motion at the left boundary, as shown in Figure 1.26. The equation of motion and boundary conditions can be represented as

$$w_{,tt} - c^2 w_{,xx} = 0, \tag{1.241}$$

$$w(0,t) = h(t), \qquad \text{and} \qquad w(l,t) \equiv 0, \tag{1.242}$$

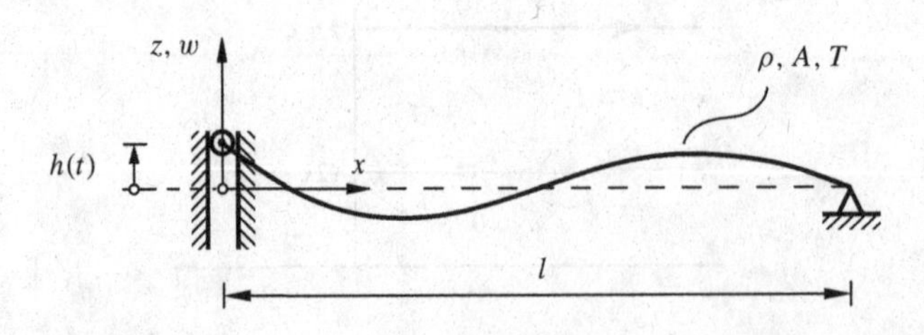

Figure 1.26 A string with a specified boundary motion

where $h(t)$ is an arbitrary function of time. For such non-homogeneous boundary conditions, the solution cannot be directly expanded in a series of eigenfunctions of a problem with homogeneous boundary conditions. However, the methods of integral transforms (such as Laplace transforms) may still be applicable. Alternatively, one may also convert a problem with non-homogeneous boundary conditions to an equivalent problem with homogeneous boundary conditions and an appropriate forcing in the equation of motion to take care of the boundary non-homogeneity. Once this is done, the modal expansion method becomes applicable. In the following, we shall pursue this method.

For the problem (1.241)–(1.242), let

$$w(x,t) = u(x,t) + h(t)\eta(x), \tag{1.243}$$

where $u(x,t)$ and $\eta(x)$ are unknown functions. Substituting this form in the boundary conditions (1.242), we have

$$w(0,t) = u(0,t) + h(t)\eta(0) = h(t) \quad \text{and} \quad w(l,t) = u(l,t) + h(t)\eta(l) = 0.$$

If we let

$$u(0,t) \equiv 0 \quad \text{and} \quad u(l,t) \equiv 0, \tag{1.244}$$

then $\eta(x)$ must be chosen such that $\eta(0) = 1$ and $\eta(l) = 0$. The simplest choice is then $\eta(x) = 1 - x/l$. Therefore, from (1.243),

$$w(x,t) = u(x,t) + h(t)\left(1 - \frac{x}{l}\right).$$

Substituting this in (1.241), one can write the equation of motion of the string using the field variable $u(x,t)$ as

$$u_{,tt} - c^2 u_{,xx} = -\left(1 - \frac{x}{l}\right)\ddot{h}(t),$$

along with the homogeneous boundary conditions (1.244). This transformed problem can be easily identified as a fixed–fixed string with distributed forcing, and can be solved using the modal expansion method.

1.10 DYNAMICS OF AXIALLY TRANSLATING STRINGS

Axially translating elastic continua are found in many situations of practical interest such as traveling threadlines in looms, rolling of rods in rolling mills, and traveling ropes in rope-ways. They exhibit very interesting dynamic characteristics which are not observed in non-translating continua. In this section, we will discuss the dynamics of a taut string translating along its length.

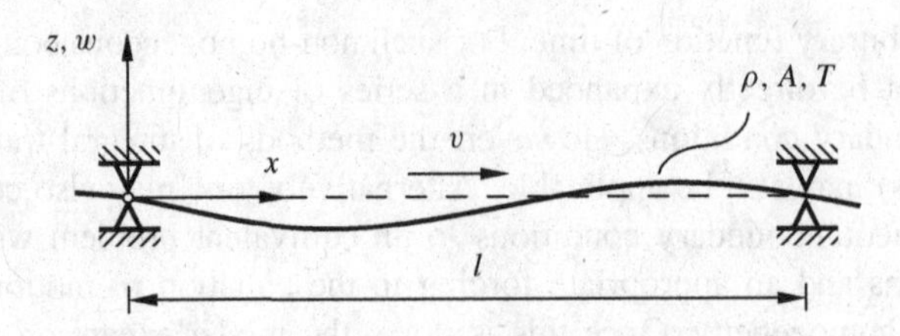

Figure 1.27 A translating string

1.10.1 Equation of motion

Consider a string under a tension T between two fixed supports, and translating along its length at a constant speed v, as shown in Figure 1.27. Let $w(x, t)$ denote the string displacement field variable which will be assumed to be small such that $w_{,x}(x, t) \ll 1$. Then, the Lagrangian of the system can be written as

$$\mathcal{L} = \frac{1}{2}\int_0^l \left[\rho A[(w_{,t} + vw_{,x})^2 + v^2] - Tw_{,x}^2\right] \mathrm{d}x,$$

where ρ is the density and A is the area of cross-section of the string. Following the procedure discussed in Appendix A, the equation of motion is obtained as

$$\rho A[w_{,tt} + 2vw_{,xt} + v^2 w_{,xx}] - Tw_{,xx} = 0$$

or

$$w_{,tt} + 2vw_{,xt} - (c^2 - v^2)w_{,xx} = 0 \tag{1.245}$$

where $c^2 = T/\rho A$, along with the boundary conditions

$$w(0, t) \equiv 0 \quad \text{and} \quad w(l, t) \equiv 0. \tag{1.246}$$

It may be mentioned here that the term $2vw_{,xt}$ in (1.245) is a result of Coriolis acceleration experienced by a string element moving at a speed v in a frame at the current location of the element and rotating at an angular speed $w_{,xt}$. This term is also known as the *gyroscopic term* for a reason that will become clear later. On the other hand, the term $v^2 w_{,xx}$ is due to the centripetal acceleration experienced by the element in the same rotating frame due to the tangential velocity v on a path of approximate curvature $w_{,xx}$.

1.10.2 Modal analysis and discretization

The next step is to study the free vibrations of a translating string. It can be easily verified that Bernoulli's solution procedure discussed in Section 1.3 cannot be applied in this case since the solution of (1.245) is non-separable. Let us assume a modal solution of the form

$$w(x, t) = \mathcal{R}[W(x)e^{i\omega t}], \tag{1.247}$$

where $W(x)$ is the eigenfunction, ω is the circular frequency, and $\mathscr{R}[\cdot]$ denotes the real part. Substituting this solution form in (1.245) yields on rearrangement, the eigenvalue problem for the traveling string as

$$-(c^2 - v^2)W'' + 2i\omega v W' - \omega^2 W = 0, \tag{1.248}$$

$$W(0) = 0, \qquad \text{and} \qquad W(l) = 0. \tag{1.249}$$

It may be noted that the differential operator in the eigenvalue problem is not self-adjoint. Further, it is evident from (1.248) that the eigenfunction $W(x)$ is complex. Substituting the solution form $W(x) = B\mathrm{e}^{ikx}$ in (1.248), one obtains

$$(c^2 - v^2)k^2 - 2\omega v k - \omega^2 = 0$$

$$\Rightarrow k = -\frac{\omega}{c+v} \qquad \text{or} \qquad k = \frac{\omega}{c-v}. \tag{1.250}$$

Using (1.250), the general solution of (1.248) can be written as

$$W(x) = D\mathrm{e}^{-i\omega x/(c+v)} + E\mathrm{e}^{i\omega x/(c-v)} \tag{1.251}$$

where D and E are arbitrary constants. Using the boundary conditions (1.249), one obtains

$$\begin{bmatrix} 1 & 1 \\ \mathrm{e}^{-i\omega l/(c+v)} & \mathrm{e}^{i\omega l/(c-v)} \end{bmatrix} \begin{Bmatrix} D \\ E \end{Bmatrix} = \mathbf{0}. \tag{1.252}$$

For non-trivial solution of $(D, E)^{\mathrm{T}}$, we must have

$$\mathrm{e}^{i\omega l[2c/(c^2-v^2)]} - 1 = 0, \tag{1.253}$$

which is the characteristic equation for the traveling string. Thus, the eigenvalues are obtained as

$$\omega_n = \frac{n\pi}{cl}(c^2 - v^2), \qquad n = 1, 2, \ldots, \infty. \tag{1.254}$$

The variation of the first three eigenvalues with speed of travel is shown in Figure 1.28. It is interesting to note that all the eigenvalues are zero when $v = c$, i.e., when the translation speed equals the wave speed in the string. Thus, when $v = c$, the string loses its stiffness completely and becomes neutrally (or marginally) stable. Hence, this speed is known as the *critical speed* of translation of the string. Next, the eigenfunctions are obtained using (1.251) and (1.252) as

$$W_n(x) = D_n \mathrm{e}^{in\pi v x/cl} \sin\frac{n\pi x}{l}, \tag{1.255}$$

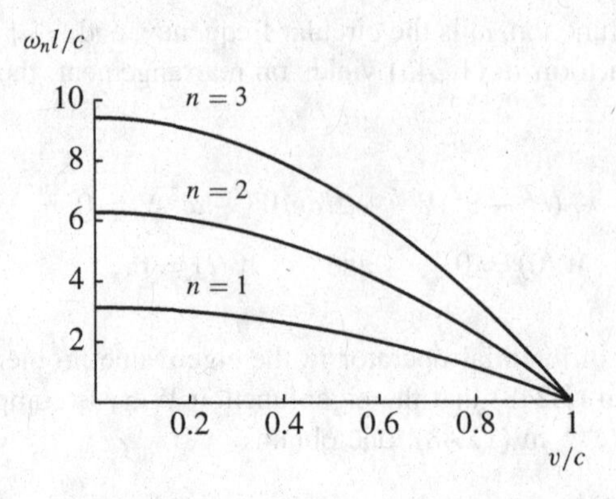

Figure 1.28 Variation of eigenvalues with speed of a translating string

where D_n is an arbitrary complex constant. Finally, the general solution of the string is obtained using (1.247) as

$$w(x,t) = \sum_{n=1}^{\infty} \left(B_n \cos\left[\frac{n\pi}{cl}\,(vx + (c^2 - v^2)t)\right] + C_n \sin\left[\frac{n\pi}{cl}\,(vx + (c^2 - v^2)t)\right]\right) \sin\frac{n\pi x}{l}, \tag{1.256}$$

where we have put $D_n = B_n - iC_n$, and B_n and C_n are arbitrary real constants which can be determined from the initial conditions. It may be noted that the solution (1.256) is non-separable in time and space. Since the eigenvalue problem of a traveling string is not self-adjoint, the determination of orthogonality relations (more appropriately *biorthogonality* relations) among the eigenfunctions is more involved (see [8]), and will not be pursued here. More discussions on non-self-adjoint eigenvalue problems can also be found in [7]. In certain problems of traveling strings, it may be convenient to use the method of Laplace transforms (see Exercise 1.15).

Consider a one-term complex solution of the string in terms of the eigenfunction (1.255) as

$$w(x,t) = z_n(t) W_n(x), \tag{1.257}$$

where $z_n(t) = z_{n1}(t) + i z_{n2}(t)$ is the nth complex modal coordinate. Substituting (1.257) in the equation of motion (1.245), multiplying by $W_n^*(x)$ (the complex conjugate of $W_n(x)$), and integrating over the domain of the string yields

$$\ddot{z}_n + i\frac{2n\pi v^2}{cl}\dot{z}_n - (c^4 - v^4)\frac{n^2\pi^2}{c^2 l^2} z_n = 0$$
$$\Rightarrow \ddot{\mathbf{z}} + \mathbf{G}\dot{\mathbf{z}} + \mathbf{K}\mathbf{z} = 0, \tag{1.258}$$

where $\mathbf{z} = (z_{n1}, z_{n2})^{\mathrm{T}}$,

$$\mathbf{G} = \begin{bmatrix} 0 & -2v^2 n\pi/cl \\ 2v^2 n\pi/cl & 0 \end{bmatrix},$$

and

$$\mathbf{K} = \begin{bmatrix} (c^4 - v^4)n^2\pi^2/c^2l^2 & 0 \\ 0 & (c^4 - v^4)n^2\pi^2/c^2l^2 \end{bmatrix}.$$

It is to be noted that $\mathbf{G}$ is skew-symmetric, i.e., $\mathbf{G}^{\mathrm{T}} = -\mathbf{G}$. The pair of ordinary differential equations in (1.258) represents a discrete *gyroscopic system* corresponding to the nth mode. It may be observed that the gyroscopic effect is due to the presence of the term $2vw_{,xt}$ in the equation of motion (1.245). An analysis of such discrete gyroscopic systems can be found in [9] (also see [7], and [10]).

1.10.3 Interaction with discrete elements

An ideal string, interacting with a discrete element or excited by a point force, can have a slope discontinuity at the interaction point due to its inability to transmit or resist moment. In the case of a traveling string, such a slope discontinuity at the interaction point brings in additional force terms due to abrupt change of momentum of the string in the transverse direction. As is well known in dynamics of mass flow systems (see [11]), a force term of the form

$$\mathbf{F}_{\mathrm{f}} = \dot{m}\mathbf{v}, \tag{1.259}$$

where $\dot{m}$ is the mass flow rate and $\mathbf{v}$ is the absolute velocity vector of the flow, appears in the equation of motion of such systems. In the following, we consider such a situation for a traveling string.

Let us consider a traveling string of length l interacting frictionlessly with a discrete spring at a location $x = a$, as shown in Figure 1.29. One can then write the equation of motion of the strings for the two regions separately as

$$\rho A w_{,tt} + 2\rho A v w_{,xt} - (T - \rho A v^2) w_{,xx} = 0, \qquad 0 \leq x < a, \tag{1.260}$$

and

$$\rho A w_{,tt} + 2\rho A v w_{,xt} - (T - \rho A v^2) w_{,xx} = 0, \qquad a < x \leq l, \tag{1.261}$$

where T is the tension in the string, which is the same in both the regions due to the assumption of no friction at the string–spring interface. The fixed end boundary conditions are

$$w(0, t) \equiv 0 \quad \text{and} \quad w(l, t) \equiv 0. \tag{1.262}$$

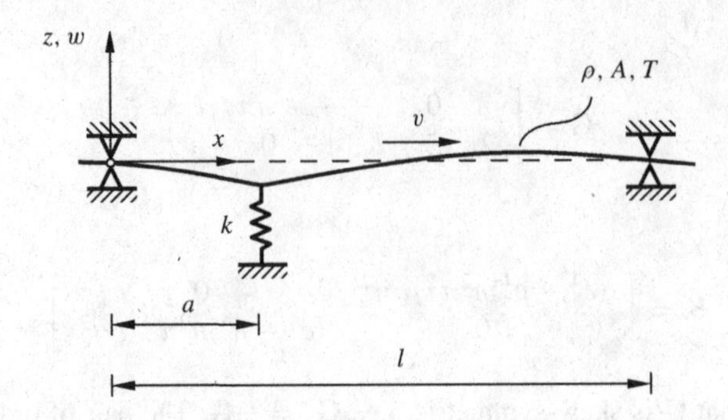

Figure 1.29 Traveling string interacting with a spring

At the point of attachment of the spring, we have the displacement condition

$$w(a^-, t) = w(a^+, t). \tag{1.263}$$

The transverse force condition at the string–spring interface consists of terms due to the tension in the two parts of the string, the extension of the spring, and the force due to mass flow. To determine the latter expression, we consider an infinitesimal control volume fixed to the spring at the interface point, and write down the net transverse force due to momentum flowing in from the region $x < a$ ($\dot{m}$ positive), and momentum flowing out into the region $x > a$ ($\dot{m}$ negative). With the help of (1.259), one can write

$$F_\mathrm{f} = \rho A v (w_{,t} + v w_{,x})\big|_{x=a^-} - \rho A v (w_{,t} + v w_{,x})\big|_{x=a^+}$$

$$\Rightarrow F_\mathrm{f} = \rho A v^2 [w_{,x}(a^-, t) - w_{,x}(a^+, t)] \qquad \text{(using (1.263))}, \tag{1.264}$$

where F_f is the net transverse force on the spring due to momentum flow. It may be mentioned that the force $-F_\mathrm{f}$ is responsible for changing the momentum of the string in the transverse direction. Now, one can write the force condition at the interaction point as

$$k w(a, t) = (\rho A v^2 - T)[w_{,x}(a^-, t) - w_{,x}(a^+, t)], \tag{1.265}$$

where k is the stiffness of the spring. The partial differential equations (1.260)–(1.261) along with the fixed–fixed boundary conditions, and matching conditions (1.263) and (1.265) completes the formulation of the problem of a traveling string interacting with a spring.

EXERCISES

1.1 Determine the eigenfrequencies and mode-shapes of transverse vibration of a taut string with a discrete mass, as shown in Figure 1.30. Discuss the cases when $m/\rho A l \to \infty$, and $m/\rho A l \to 0$.

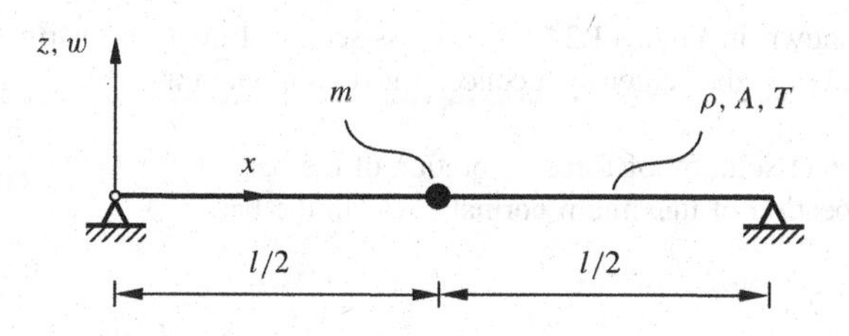

Figure 1.30 Exercise 1.1

1.2 A homogeneous bar is fixed at the left end, and sprung at the right end with a spring constant k, as shown in Figure 1.31.

(a) Using the variational principle, derive the equation of motion, and the boundary conditions of the system.

(b) For $k = EA/l$, determine the first two eigenfrequencies and the corresponding mode-shape-functions.

(c) Take $l = 2m, k = 5000$ N/m, $m = 5$ kg, and solve for two values of $m/\rho Al = 0.5$ and 2.

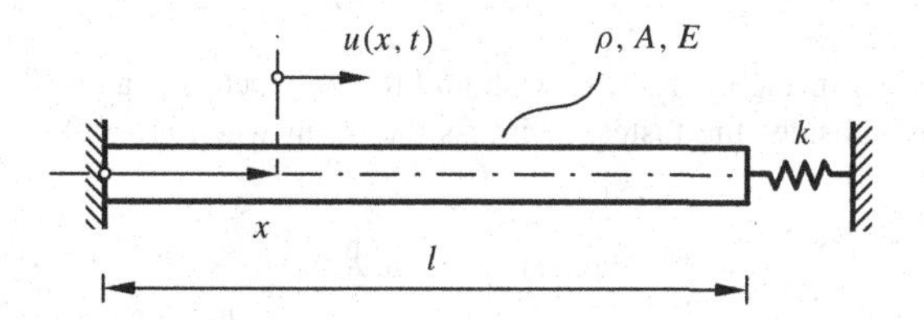

Figure 1.31 Exercise 1.2

1.3 Determine the eigenfrequencies and eigenfunctions for longitudinal vibrations of the system shown in Figure 1.32.

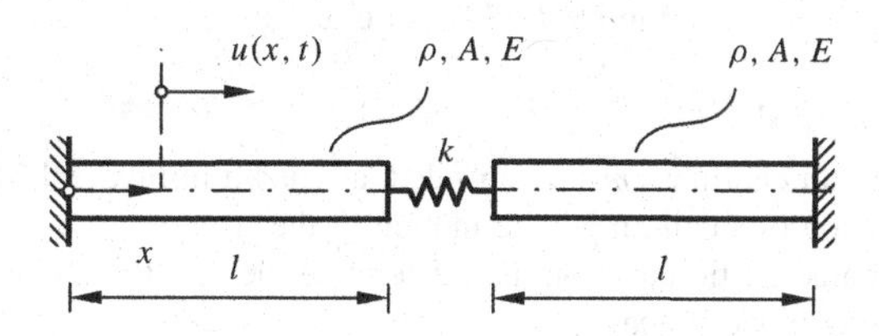

Figure 1.32 Exercise 1.3

1.4 A homogeneous uniform bar is kept under tension T with a string, as shown in Figure 1.33. If the string suddenly snaps, determine the response of the bar.

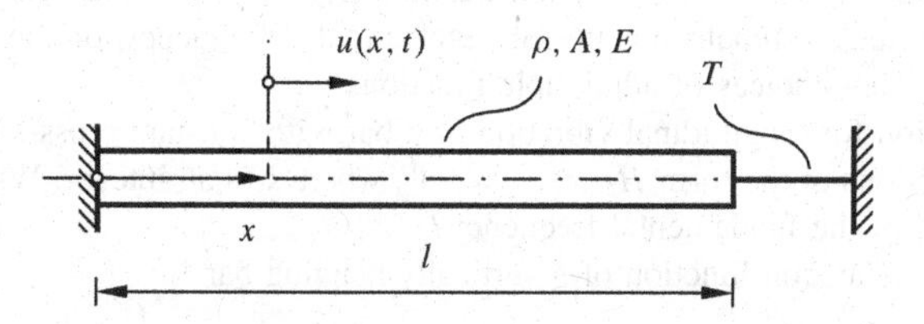

Figure 1.33 Exercise 1.4

1.5 The tapered bar shown in Figure 1.34 has a cross-sectional area that varies as $A(x) = A_0(1 - x/2l)^2$. The bar is excited at the center by a concentrated harmonic force $F(t) = F_0 \cos \Omega t$, as shown in the figure.

(a) Determine the exact solution of forced vibration of the bar.

(b) Determine the location of maximum normal stress in the bar.

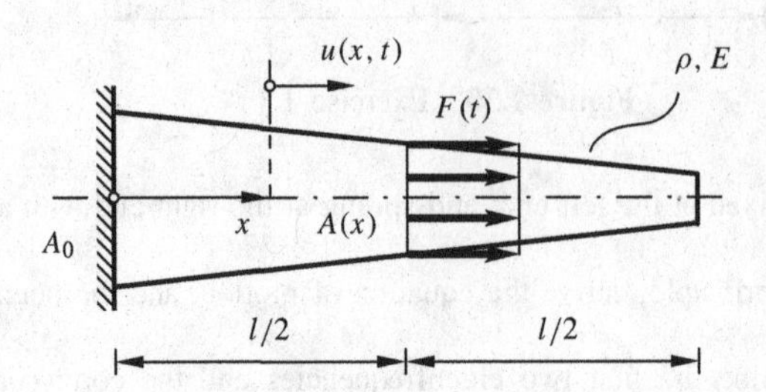

Figure 1.34 Exercise 1.5

1.6 For the systems shown in Figure 1.35, how should the parameters k and d be chosen so that the vibration of the mass m subsides the fastest? Discuss the result when $m \to 0$.

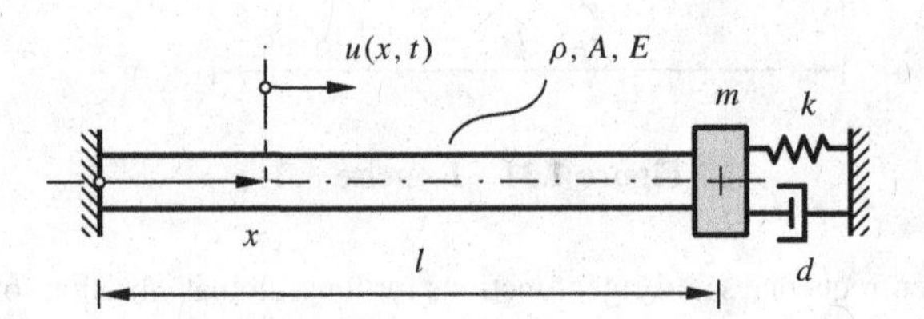

Figure 1.35 Exercise 1.6

1.7 A uniform string with an end-mass m is suspended, as shown in Figure 1.36.

(a) Determine the equation of small-amplitude motion of the string.

(b) Derive the exact characteristic equation, and determine the first three circular eigenfrequencies and the corresponding mode-shapes.

1.8 Using Galerkin's method, discretize the equation of motion of a hanging string. Use the comparison functions as $P_i(x) = x^i$, $i = 1, 2, \ldots, N$. For $N = 2$ determine the eigenfrequencies from the discretized system and compare with the exact solutions.

1.9 A homogeneous bar of circular cross-section with linearly varying radius is shown in Figure 1.37. Using Rayleigh's quotient, estimate the fundamental circular frequency of the bar in longitudinal vibration for the following choices of admissible functions:

(a) First eigenfunction for longitudinal vibration of a bar with constant cross-section.

(b) Admissible functions of the form $H_k(x) = (x/l)^k$, where k is an integer. What value of k yields the lowest value of the fundamental frequency?

(c) Using the static deflection function of a vertically hanging bar.

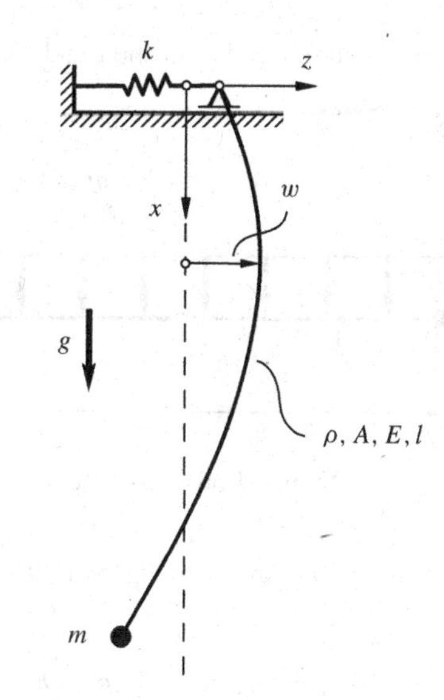

Figure 1.36 Exercise 1.7

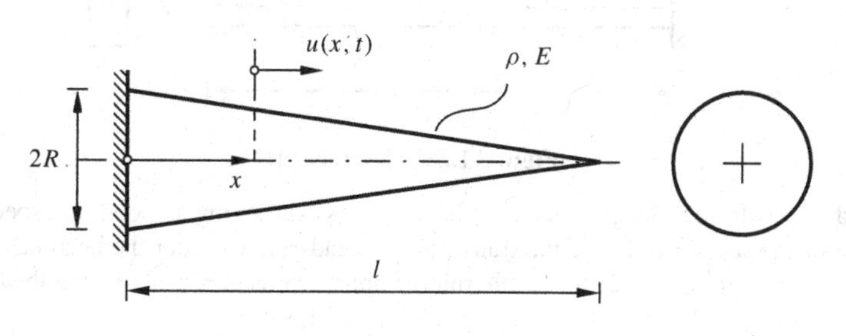

Figure 1.37 Exercise 1.9

1.10 Show that the initial value problem

$$\mu(x)w_{,tt} + \mathcal{K}[w] = 0, \qquad w(x,0) = w_0(x), \text{ and } w_{,t}(x,0) = v_0(x),$$

can be converted to the forced problem

$$\mu(x)w_{,tt} + \mathcal{K}[w] = \mu(x)w_0(x)\dot{\delta}(t) + \mu(x)v_0(x)\delta(t), \qquad w(x,0) = 0, \text{ and } w_{,t}(x,0) = 0.$$

1.11 A sliding–fixed string of length l is excited by a uniformly force $q(x,t) = Q_0 \cos \Omega t$, as shown in Figure 1.38. Determine the steady-state response of the string using: (a) Eigenfunction expansion method, and (b) Green's function method.

1.12 A bar of length l and varying cross-sectional area given by $A(x) = A_0(1 - x/2l)^2$ is fixed and $x = 0$ and free at $x = l$. The bar is harmonically forced by $F(t) = F_0 \sin \Omega t$ at $x = l/2$. Using the admissible functions $H_k(x) = (x/l)(1 - x/2l)^k$, $k = 1, 2, \ldots, N$, obtain the discretized equations of motion of the bar. With $N = 4$, determine the location and magnitude of the maximum response amplitude.

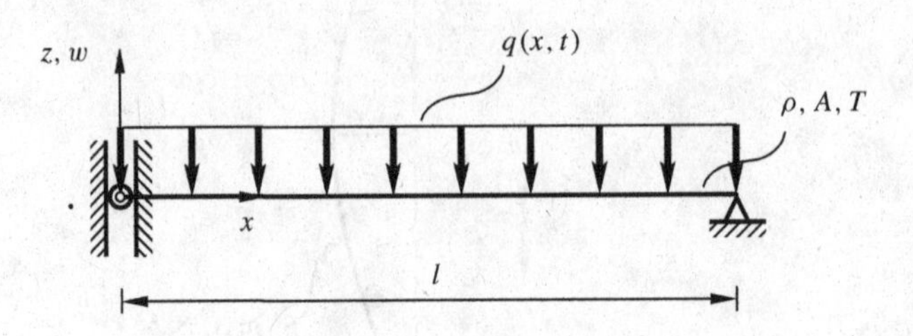

Figure 1.38 Exercise 1.11

1.13 Determine the response of the system shown in Figure 1.39 to a harmonic force $F(t) = B_0 \cos \Omega t$.

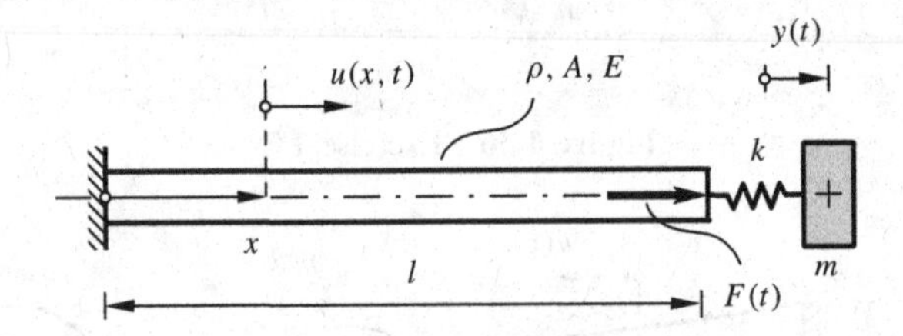

Figure 1.39 Exercise 1.13

1.14 A fixed–fixed string of length l carries a bead of mass m moving at a constant speed v. Determine the response of the string (a) during the transit of the bead, and (b) after the bead has left the span of the string. (One may refer to [12] for an alternative approach using methods described in Chapter 2.)

1.15 An axially translating string with an impulsive transverse point force at $x = \bar{x}$ is described by the equation of motion

$$\rho A[w_{,tt} + 2vw_{,xt} + v^2 w_{,xx}] - Tw_{,xx} = \delta(t - \tau)\delta(x - \bar{x}).$$

Show that the response (Green's function) of the string is given by

$$w(x, \bar{x}, t, \tau) = \mathcal{H}(t - \tau) \sum_{n=1}^{\infty} \frac{2}{n\pi\rho Ac} \sin\left[\frac{n\pi}{cl}\{(c^2 - v^2)(t - \tau) + v(x - \bar{x})\}\right] \sin\frac{n\pi\bar{x}}{l} \sin\frac{n\pi x}{l},$$

where $\mathcal{H}(\cdot)$ is the Heaviside step function and $c = \sqrt{T/\rho A}$. (*Hint*: Use Laplace transform, followed by a variable transformation $w(x, s) = e^{\alpha x}u(x, s)$, where $\alpha = vs/(c^2 - v^2)$, and s is the Laplace variable.)

1.16 A cable-car on a translating cable may be approximated by a constant point force W fixed to an axially translating string, where W is the weight of the car (see [8]). Using Green's function obtained in Exercise 1.15, determine the response of the string.

1.17 A traveling string is supported frictionlessly at the middle, as shown in Figure 1.40. If the middle support suddenly snaps leaving the string free, determine the subsequent motion of the string and plot its configurations. Assume the initial displacement a of the middle point in Figure 1.40 to be small so that the tension in the string does not change. (*Hint*: Use the idea of Exercise 1.10, and Green's function for the traveling string.)

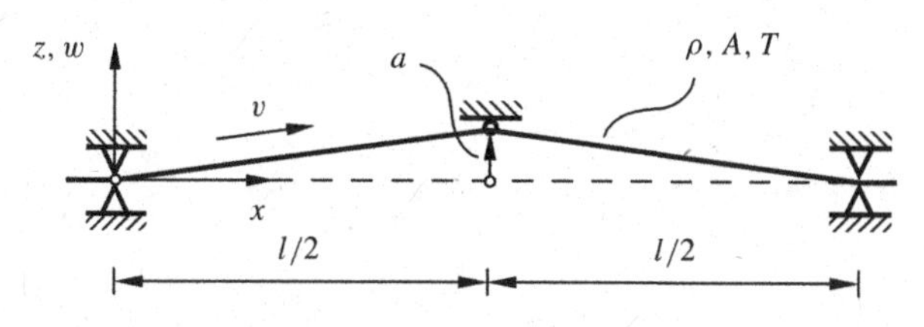

Figure 1.40 Exercise 1.17

REFERENCES

[1] Timoshenko, S.P., and Goodier, J.N., *Theory of Elasticity*, 3e, McGraw-Hill Book Co., Singapore, 1970.

[2] Kreyszig, E., *Advanced Engineering Mathematics*, 2e, Wiley Eastern Pvt. Ltd., New Delhi, 1969.

[3] Stakgold, I., *Boundary Value Problems of Mathematical Physics, Vol 1*, The Macmillan Co., New York, 1967.

[4] Butkovskiy, A.G., *Green's Functions and Transfer Functions Handbook*, Ellis Horwood Ltd., Chichester, UK, 1982.

[5] Hagedorn, P., *Technische Schwingungslehre: Lineare Schwingungen kontinuierlicher mechanischer Systeme*, Springer-Verlag, Berlin, 1989.

[6] Meirovitch, L., *Computational Methods in Structural Dynamics*, Sijthoff & Noordhoff, Alphen aan den Rijn, 1980.

[7] Meirovitch, L., and Hagedorn, P., A New Approach to the Modelling of Distributed Non-Self-Adjoint Systems, *J. of Sound and Vibration*, 178(2), 1994, pp. 227–241.

[8] Wickert, J.A., and Mote, Jr., C.D., Classical Vibration Analysis of Axially Moving Continua, *J. of Applied Mechanics, Trans. of ASME*, 57, 1990, pp. 738–744.

[9] Hagedorn, P., and Otterbein, S., *Technische Schwingungslehre*, Springer-Verlag, Berlin, 1982.

[10] Huseyin, K., *Vibrations and Stability of Multiple Parameter Systems*, Sijthoff & Noordhoff, Alphen aan den Rijn, 1978.

[11] Meriam, J.L., and Kraige, L.G., *Engineering Mechanics: Dynamics*, 4e, McGraw-Hill Book Co., Singapore, 1999.

[12] Smith, C. E., Motions of a Stretched String Carrying a Moving Mass Particle, *J. of Applied Mechanics, Trans. of ASME*, 31(1), 1964, pp. 29–37.

2

One-dimensional wave equation: d'Alembert's solution

The equation of motion of a taut string in transverse vibration or a uniform bar in longitudinal or torsional vibration is represented by the wave equation in one dimension. It is one of the fundamental equations appearing in vibrations of continuous systems. The general solution of the wave equation can be expressed in a special form, known as *d'Alembert's solution*, or the *traveling wave solution*. Such a solution is particularly suitable to represent the transient dynamics of the system, and has many applications in the study of continuous systems. Therefore, in this chapter, we will discuss the traveling wave solution of the wave equation and some of its applications.

2.1 D'ALEMBERT'S SOLUTION OF THE WAVE EQUATION

In the following, we will consider the wave equation for one-dimensional continua which may or may not be bounded. Thus, we may not have any boundary conditions. It may be recalled from the discussions in Chapter 1 that the concept of natural frequencies and modes of vibration is intimately related to the boundary conditions. On the other hand, in an infinite continuous system, we have only traveling waves which can be of any frequency. In a finite continuous medium, as long as a traveling wave does not encounter a boundary, the medium may be considered to be infinite. This consideration not only simplifies the situation but also provides important insights into the nature and properties of the system.

Consider the wave equation

$$w_{,tt} - c^2 w_{,xx} = 0, \tag{2.1}$$

where $w(x, t)$ is the displacement field variable (for example, the transverse displacement of a string) and c is the wave speed in the medium (say, a string). It is to be recognized that c is the speed with which the displacement field propagates through the medium, and is different from the velocity of the individual particles of the system. To make this distinction clear, one can think of a taut string in which the string particles oscillate transversely to the string with speed $w_{,t}(x, t)$, while the waves propagate along the string with speed c.

Vibrations and Waves in Continuous Mechanical Systems P. Hagedorn and A. DasGupta

One can rewrite the wave equation (2.1) as

$$\left(\frac{\partial}{\partial t}+c\frac{\partial}{\partial x}\right)\left(\frac{\partial}{\partial t}-c\frac{\partial}{\partial x}\right)w=\left(\frac{\partial}{\partial t}-c\frac{\partial}{\partial x}\right)\left(\frac{\partial}{\partial t}+c\frac{\partial}{\partial x}\right)w=0. \tag{2.2}$$

It is clear from (2.2) that we have a solution of the wave equation (2.1) whenever

$$\left(\frac{\partial}{\partial t}+c\frac{\partial}{\partial x}\right)w=0 \tag{2.3}$$

or

$$\left(\frac{\partial}{\partial t}-c\frac{\partial}{\partial x}\right)w=0. \tag{2.4}$$

Consider a function

$$w_+(x,t)=f(z)\big|_{z=x-ct}=f(x-ct), \tag{2.5}$$

where $f(z)$ is any function of z, as shown in Figure 2.1. Using the chain rule, we differentiate (2.5) with respect to the time coordinate t, and the space coordinate x, to obtain, respectively,

$$w_{+,t}=-cf'(x-ct) \tag{2.6}$$

and

$$w_{+,x}=f'(x-ct). \tag{2.7}$$

From (2.3), (2.6), and (2.7), we can easily conclude that $w_+(x,t)=f(x-ct)$ is a solution of (2.3). It can be easily shown that $f(x-ct)$ represents a waveform $f(z)$ traveling in the direction of the positive x-axis, as indicated in Figure 2.2, and will be referred to as the *positive-traveling wave*. Next, consider the function

$$w_-(x,t)=g(z)\big|_{z=x+ct}=g(x+ct). \tag{2.8}$$

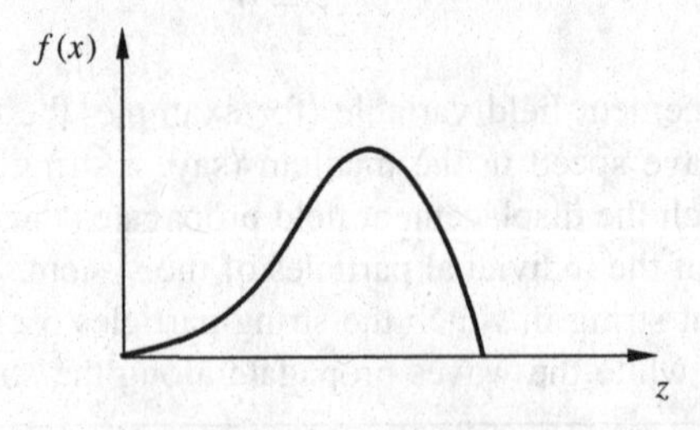

Figure 2.1 Representation of a pulse

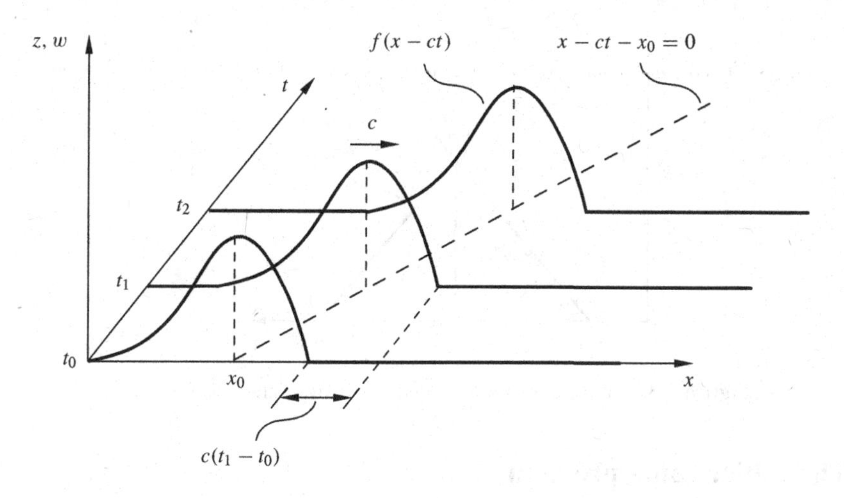

Figure 2.2 A traveling pulse in the x-t-plane

Following similar steps as in (2.6) and (2.7), one can easily observe that $w_-(x,t) = g(x+ct)$ is a solution of (2.4), and represents a waveform $g(z)$ traveling in the direction of the negative x-axis. This will be referred to as the *negative-traveling wave*. Therefore, the complete solution of the wave equation (2.1) is given by

$$w(x,t) = f(x-ct) + g(x+ct). \tag{2.9}$$

This solution is referred to as d'Alembert's solution (or the traveling wave solution) of the wave equation. It can be shown that any solution of (2.1) can be decomposed as in (2.9), i.e., into two waves traveling in opposite directions. Therefore, (2.9) represents the general solution of (2.1). As an example, consider the vibration of a fixed–fixed string vibrating in its kth mode with a frequency $\omega_k = k\pi c/l$. The motion of the string can be represented as

$$\begin{aligned} w(x,t) &= S \sin\frac{k\pi ct}{l} \sin\frac{k\pi x}{l} \\ &= \frac{S}{2}\cos\left[\frac{k\pi}{l}(x-ct)\right] - \frac{S}{2}\cos\left[\frac{k\pi}{l}(x+ct)\right], \end{aligned}$$

i.e., the standing wave can be represented by a superposition of two traveling waves.

It may be observed from Figure 2.2 that, on any line $x - ct - \xi_+ = 0$ on the x-t-plane, where $\xi_+ = x_0$ is an arbitrary point, the value of $w_+(x,t) = f(x-ct)$ is constant. These lines are known as the *characteristics* of the wave equation (see, for example, [1]). It is also easy to infer that the family of lines $x + ct - \xi_- = 0$, for arbitrary ξ_-, are also the characteristics of the wave equation since the solution $w_-(x,t) = g(x+ct)$ remains constant on them. Therefore, it follows that, at a point $(\overline{x}, \overline{t})$, given by intersection of the lines $x - ct - \xi_+ = 0$ and $x + ct - \xi_- = 0$, the solution is given by $w(x,t) = f(\xi_+) + g(\xi_-)$, as illustrated in Figure 2.3. The functions $f(\cdot)$ and $g(\cdot)$ can be determined from the initial conditions specified over the interval $[\xi_+, \xi_-]$ as discussed in the following section.

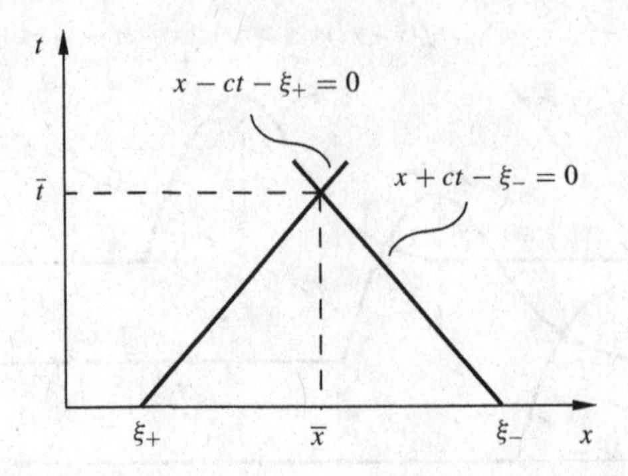

Figure 2.3 Solution of the wave equation from characteristics

2.1.1 The initial value problem

Let us now consider the solution of the initial value problem of (2.1) in terms of the decomposition (2.9). The task is to determine the functions $f(\cdot)$ and $g(\cdot)$. Let the initial conditions of (2.1) in an unbounded medium be specified as

$$w(x,0) = w_0(x) \quad \text{and} \quad w_{,t}(x,0) = v_0(x), \qquad -\infty < x < \infty. \tag{2.10}$$

Using (2.9) in (2.10) yields

$$f(x) + g(x) = w_0(x), \qquad -\infty < x < \infty, \tag{2.11}$$

and

$$-cf'(x) + cg'(x) = v_0(x), \qquad -\infty < x < \infty. \tag{2.12}$$

Integrating (2.12) with respect to x yields

$$\int_{x_0}^{x} [f(\xi) - g(\xi)]' \,\mathrm{d}\xi = [f(\xi) - g(\xi)]\big|_{x_0}^{x} = -\frac{1}{c}\int_{x_0}^{x} v_0(\xi)\,\mathrm{d}\xi$$

$$\Rightarrow f(x) - g(x) = -\frac{1}{c}\int_{x_0}^{x} v_0(\xi)\,\mathrm{d}\xi + f(x_0) - g(x_0), \tag{2.13}$$

where x_0 is any arbitrary point. From (2.11) and (2.13) one can solve for $f(x)$ and $g(x)$ as

$$f(x) = \frac{1}{2}\left[-\frac{1}{c}\int_{x_0}^{x} v_0(\xi)\,\mathrm{d}\xi + f(x_0) - g(x_0) + w_0(x)\right] \tag{2.14}$$

and

$$g(x) = \frac{1}{2}\left[\frac{1}{c}\int_{x_0}^{x} v_0(\xi)\,\mathrm{d}\xi - f(x_0) + g(x_0) + w_0(x)\right]. \tag{2.15}$$

Using (2.14) and (2.15) in (2.9) yields on simplification

$$
\begin{aligned}
w(x,t) &= f(x-ct) + g(x-ct) \\
&= \frac{1}{2}\left[w_0(x-ct) + w_0(x+ct) - \frac{1}{c}\int_{x_0}^{x-ct} v_0(\xi)\,\mathrm{d}\xi + \frac{1}{c}\int_{x_0}^{x+ct} v_0(\xi)\,\mathrm{d}\xi\right] \\
&= \frac{1}{2}\left[w_0(x-ct) + w_0(x+ct) + \frac{1}{c}\int_{x-ct}^{x+ct} v_0(\xi)\,\mathrm{d}\xi\right].
\end{aligned} \tag{2.16}
$$

This is d'Alembert's solution for the initial value problem of the wave equation. It may be noted that (2.16) is the exact solution of the initial value problem for a wave equation. The eigenfunction expansion method of solution discussed in Section 1.4.3 would require all the infinitely many terms in the expansion (1.108) to be summed up, in order to yield the solution (2.16).

It is evident from (2.16) that, in general, the solution at any point $(\bar{x}, \bar{t})$ in the x-t-plane depends upon the initial conditions specified over the interval $(\bar{x} - c\bar{t}, \bar{x} + c\bar{t})$, as shown in Figure 2.4. This interval is known as the *domain of dependence* (see, for example, [1]) for the solution at $(\bar{x}, \bar{t})$. Conversely, the solution at any (x, t) in the region B shown in Figure 2.5 is influenced by the initial condition at x_0. Hence, region B is known as the *range of influence* of the point x_0.

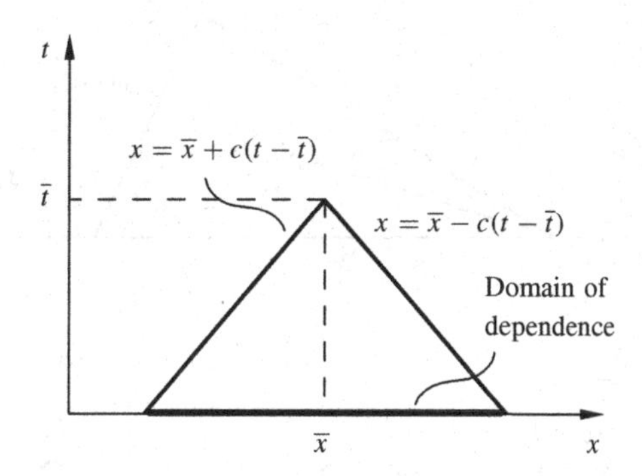

Figure 2.4 Domain of dependence for solution at $(\bar{x}, \bar{t})$

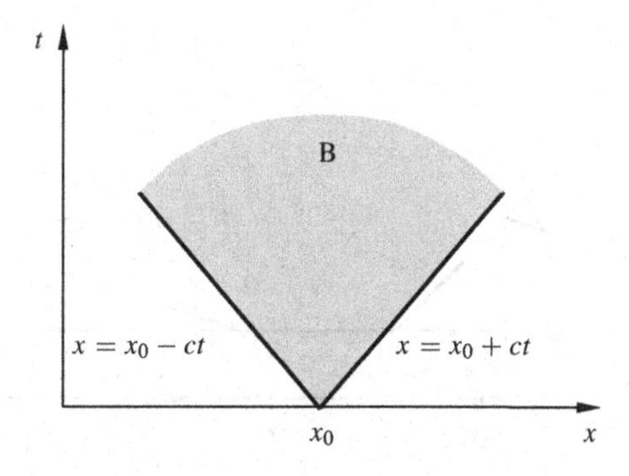

Figure 2.5 Range of influence of the point x_0

A particularly simple form of the solution (2.16) is obtained when the initial velocity of the system is identically zero, i.e., $v_0(x) \equiv 0$. Then, (2.14) and (2.15) reduce to $f(x) = g(x) = 1/2w_0(x)$, and the solution (2.16) simplifies to

$$\begin{aligned} w(x,t) &= \frac{1}{2}[w_0(x-ct) + w_0(x+ct)] \\ &= \frac{1}{2}w_0(z)\Big|_{z=x-ct} + \frac{1}{2}w_0(z)\Big|_{z=x+ct}. \end{aligned} \tag{2.17}$$

For a taut string, this solution is visualized in Figure 2.6. The initial shape of the string $w_0(z)$ is decomposed into two equal waveforms, as shown in the figure. One of the waveforms travels in the direction of the positive x-axis with a speed c, while the other travels in the opposite direction with the same speed, as illustrated in Figure 2.6.

Consider another initial condition for a taut string with $w_0(x) \equiv 0$ and $v_0(x) = v_0$ (constant) for $x_0 < x < x_0 + a$, as shown in Figure 2.7. Then, from (2.14)–(2.15), we have

$$-f(x) = g(x) = \frac{1}{2c}V_0(x), \tag{2.18}$$

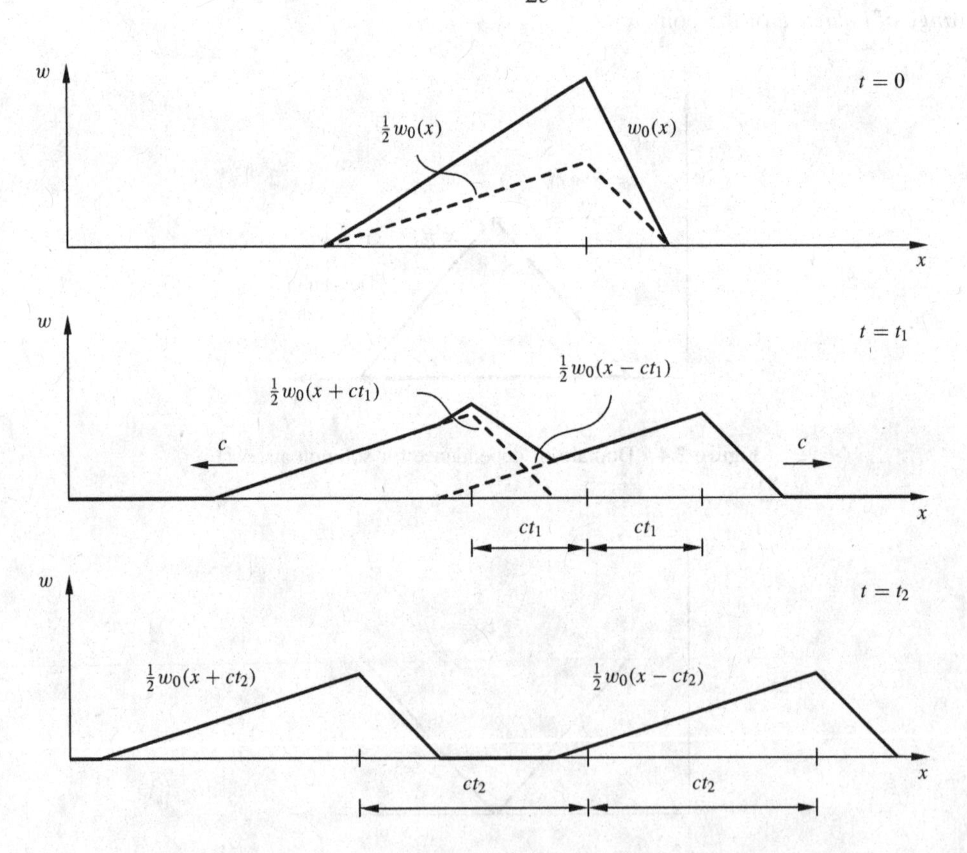

Figure 2.6 Traveling waves in an infinite string due to an initial displacement

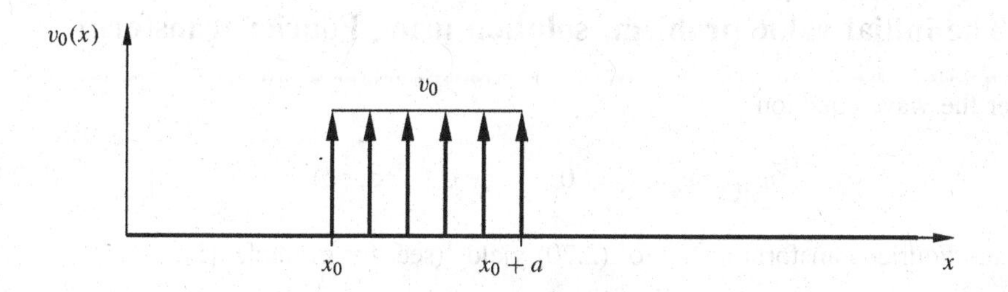

Figure 2.7 Initial velocity condition for an infinite string

where

$$V_0(x) = \int_{x_0}^{x} v_0(\xi)\,\mathrm{d}\xi.$$

The solution can now be written from (2.9) as

$$w(x, t) = \frac{1}{2c}[V_0(x + ct) - V_0(x - ct)]. \tag{2.19}$$

The shape of the string at a certain time instant $t = a/c$ is shown in Figure 2.8. It is interesting to note that as $t \to \infty$, $w(x, t) \to v_0 a/2c$ for all values of x, i.e., the string gets displaced by an amount $v_0 a/2c$.

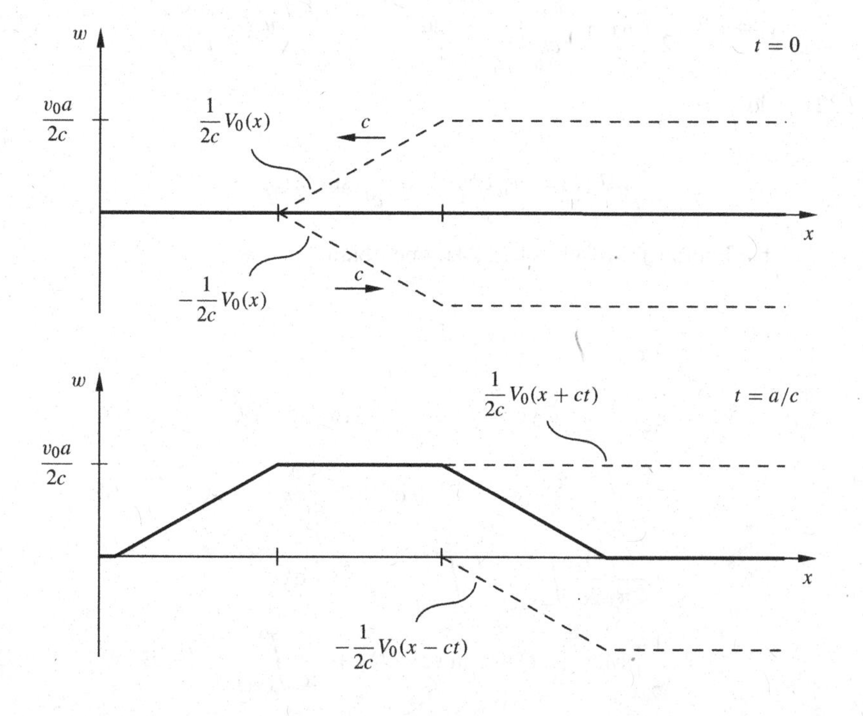

Figure 2.8 Traveling waves in an infinite string due to an initial velocity condition

2.1.2 The initial value problem: solution using Fourier transform

Consider the wave equation

$$w_{,tt} - c^2 w_{,xx} = 0, \qquad x \in (-\infty, \infty). \tag{2.20}$$

Taking the Fourier transform over x of (2.20) yields (see, for example, [2])

$$\ddot{\tilde{w}} + c^2 k^2 \tilde{w} = 0, \tag{2.21}$$

where

$$\tilde{w}(k,t) = \int_{-\infty}^{\infty} w(x,t) \mathrm{e}^{-ikx} \mathrm{d}x.$$

The general solution of (2.21) is given by

$$\tilde{w}(k,t) = a\mathrm{e}^{ikct} + b\mathrm{e}^{-ikct}, \tag{2.22}$$

where a and b are arbitrary constants. Using the initial conditions yields

$$a + b = \tilde{w}_0(k) \qquad \text{and} \qquad a - b = \frac{\tilde{v}_0}{ick},$$

$$\Rightarrow a = \frac{1}{2}\left(\tilde{w}_0 + \frac{\tilde{v}_0}{ick}\right) \qquad \text{and} \qquad b = \frac{1}{2}\left(\tilde{w}_0 - \frac{\tilde{v}_0}{ick}\right). \tag{2.23}$$

Thus, (2.22) yields

$$\tilde{w}(k,t) = \tilde{w}_0 \cos ckt + \frac{\tilde{v}_0}{ck} \sin ckt. \tag{2.24}$$

Taking the inverse Fourier transform of (2.24), one obtains

$$\begin{aligned}
w(x,t) &= \frac{1}{2\pi} \int_{-\infty}^{\infty} w(k,t) \mathrm{e}^{ikx} \mathrm{d}k \\
&= \frac{1}{2\pi} \int_{-\infty}^{\infty} \left[\tilde{w}_0 \cos ckt + \frac{\tilde{v}_0}{ck} \sin ckt\right] \mathrm{e}^{ikx} \mathrm{d}k \\
&= \frac{1}{2\pi} \int_{-\infty}^{\infty} \frac{\tilde{w}_0}{2} \left(\mathrm{e}^{ik(x+ct)} + \mathrm{e}^{ik(x-ct)}\right) \mathrm{d}k \\
&\quad + \frac{1}{2\pi} \frac{1}{2c} \int_{-\infty}^{\infty} \tilde{v}_0 \left(\int_{x-ct}^{x+ct} \mathrm{e}^{ik\xi} \mathrm{d}\xi\right) \mathrm{d}k \\
&= \frac{1}{2}\left[\tilde{w}_0(x+ct) + w_0(x-ct)\right] + \frac{1}{2c} \int_{x-ct}^{x+ct} v_0(\xi) \mathrm{d}\xi.
\end{aligned} \tag{2.25}$$

This is the same solution as given by (2.17).

2.2 HARMONIC WAVES AND WAVE IMPEDANCE

A special case of a traveling wave, known as a *harmonic wave*, can be expressed in complex notation as

$$w(x, t) = B\mathrm{e}^{i(kx-\omega t)}, \tag{2.26}$$

where B is a constant, $k = 2\pi/\lambda$ is the wave number, λ is the wavelength, and ω is the circular frequency of the wave. The actual wave is obtained by taking the real or imaginary part of (2.26). A discussion on harmonic waves is presented in detail in Appendix B. As noted in the Appendix, a harmonic traveling wave can be considered to be a Fourier component of a general traveling waveform, i.e., a general traveling waveform can be constructed from its harmonic traveling wave components. So it suffices to study the propagation of only harmonic waves in linear continuous systems. As can be easily checked, a positive-traveling harmonic wave (2.26) satisfies the wave equation (2.1) when

$$k = \frac{\omega}{c}. \tag{2.27}$$

Consider a semi-infinite taut string of density ρ under a tension T, as shown in Figure 2.9. If we want to set up a positive-traveling harmonic wave of the form (2.26), the end of the string has to be excited by a force

$$F(t) = -Tw_{,x}(0, t) = -iTkB\mathrm{e}^{-i\omega t}. \tag{2.28}$$

The corresponding velocity $v(t)$ of the end is given by

$$v(t) = w_{,t}(0, t) = -i\omega B\mathrm{e}^{-i\omega t}. \tag{2.29}$$

Analogous to the concept of impedance in electrical circuits, one can define *wave impedance* of the semi-infinite string as the ratio between the complex amplitudes of the complex harmonic force and the corresponding complex harmonic velocity. Thus, using (2.28) and (2.29), one can write

$$\mathcal{Z} := \frac{\mathscr{A}[F(t)]}{\mathscr{A}[v(t)]} = \frac{-iTkB}{-i\omega B} = \frac{Tk}{\omega} = \rho Ac, \qquad \text{(using (2.27))}$$

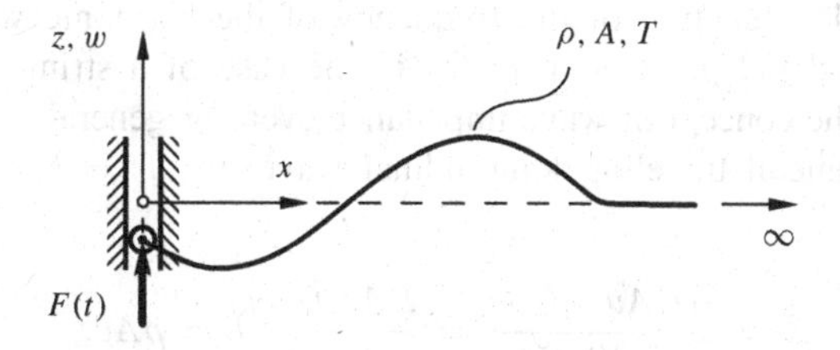

Figure 2.9 Harmonically forced semi-infinite string

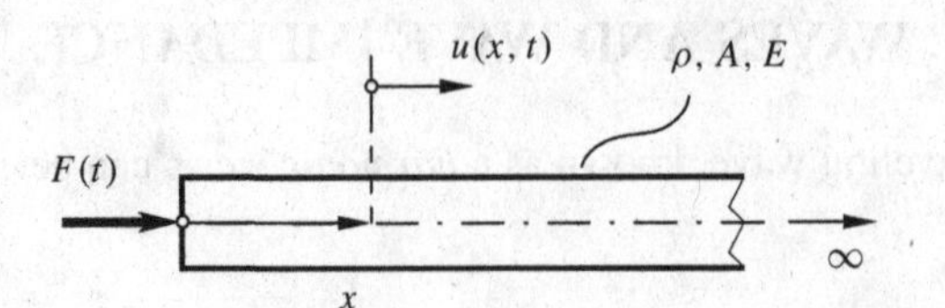

Figure 2.10 Harmonically forced semi-infinite bar

where $\mathscr{A}[\cdot]$ represents the complex amplitude, $\mathcal{Z}$ is the wave impedance, and $c = \sqrt{T/\rho A}$ is the speed of transverse waves in the string.

Consider next the case of a semi-infinite bar carrying a positive-traveling longitudinal harmonic wave $u(x,t) = Be^{i(kx-\omega t)}$, as shown in Figure 2.10. For such a motion, the force at the left boundary is given by $F(t) = -EAu_{,x}(0,t) = -iEAkBe^{i(kx-\omega t)}$, while the velocity is expressed as $v(t) = u_{,t}(0,t) = i\omega Be^{i(kx-\omega t)}$. One can then express the impedance of the bar as

$$\mathcal{Z} = \frac{\mathscr{A}[F(t)]}{\mathscr{A}[v(t)]} = \frac{EAk}{\omega} = \rho Ac, \tag{2.30}$$

where $c = \sqrt{E/\rho}$ is the speed of longitudinal waves in the bar. In the case of a bar, one can also define a *specific impedance* $\mathcal{Z}_S = \rho c$. It is to be noted that the specific impedance is defined as the ratio between the stress at the end of the bar and the corresponding end velocity.

The wave impedance (or the specific impedance) of a medium (for example, a string or bar) relates the velocity of the particles of the medium and the corresponding force (or stress, or pressure) required to produce that velocity. In other words, it provides a connection between the kinematics and the kinetics of the medium. This concept is very important in the study of various aspects of wave propagation, and will be discussed further in this chapter. It must be mentioned here that the concept of wave impedance is only defined in the context of complex harmonic waves in the medium. In a more general setting, the impedance is defined as the ratio of the Fourier transforms of the force to the corresponding velocity signals, i.e.,

$$\mathcal{Z}(\omega) := \frac{\mathscr{F}(\omega)}{\mathscr{V}(\omega)}, \tag{2.31}$$

where $\mathscr{F}(\omega)$ and $\mathscr{V}(\omega)$ are, respectively, the Fourier transforms of the force and velocity signals. The definition (2.31), however, results in the same expression of impedance as obtained using the complex harmonic signals/waves, since the Fourier integral is a superposition of complex harmonic signals/waves (see Appendix B). Thus, in general, the wave impedance is a complex function of the frequency of the harmonic wave. For problems in which $\mathcal{Z}$ is real and independent of ω (as in the case of a string or a bar discussed above), one may use the concept of wave impedance even for general traveling waveforms. For example, for a general traveling longitudinal wave $u(x,t) = f(x - ct)$ in a bar (as discussed above),

$$\mathcal{Z} = \frac{-EAu_{,x}(0,t)}{u_{,t}(0,t)} = \frac{EAf'(-ct)}{cf'(-ct)} = \rho Ac.$$

Table 2.1 Wave speeds and specific wave impedances of some common materials

Material	Wave speed, c (m/s) (Longitudinal waves)	Specific wave impedance, $\mathcal{Z}_S = \rho c$ (Ns/m^3)
Steel	5000	39×10^6
Aluminum	5100	13×10^6
Glass	5200	13.8×10^6
Concrete	4000	8×10^6
Water	1450	1.45×10^6
Cork	500	0.1×10^6
Rubber	40–150	$0.04–0.3 \times 10^6$
Air	340	410

It is evident that the difference in the transverse wave propagation in a taut uniform string, and longitudinal or torsional wave propagation of a uniform bar occurs in the definition of the wave speed c of the corresponding waves. In the case of a string, the wave speed depends on the tension T, which is an adjustable parameter. On the other hand, the longitudinal and torsional wave speeds $\sqrt{E/\rho}$ and $\sqrt{G/\rho}$, respectively, are material constants. For most engineering materials $E \approx 3G$, implying that the longitudinal wave speed is about $\sqrt{3}$ times the torsional wave speed. The longitudinal waves can be associated with normal stress or pressure (or dialatoric stress) waves, while the torsional waves are like shear (or deviatoric stress) waves in an elastic medium. Table 2.1 lists the longitudinal wave speed of some common materials along with their specific impedance $\mathcal{Z}_S$. The wave impedances of the semi-infinite or infinite media discussed in this section are all observed to be real. However, in general, the wave impedance can be complex. Such cases, and the implications of a complex wave impedance of a medium, will be discussed in the following sections.

2.3 ENERGETICS OF WAVE MOTION

Consider an infinite string of density ρ, area of cross-section A, and under tension T. The total mechanical energy density (energy per unit length), $\hat{\mathcal{E}}$, can be written in terms of the kinetic energy density $\hat{\mathcal{T}}$, and the potential energy density $\hat{\mathcal{V}}$ as

$$\hat{\mathcal{E}} = \hat{\mathcal{T}} + \hat{\mathcal{V}} = \frac{1}{2}\rho A w_{,t}^2 + \frac{1}{2} T w_{,x}^2. \tag{2.32}$$

Considering a positive-traveling wave $w(x,t) = f(x - ct)$ on the string, one obtains from (2.32)

$$\begin{aligned}
\hat{\mathcal{E}} &= \frac{1}{2}\rho A c^2 f'^2(x - ct) + \frac{1}{2} T f'^2(x - ct) \\
&= \frac{1}{2}\rho A c^2 f'^2(x - ct) + \frac{1}{2}\rho A c^2 f'^2(x - ct) \qquad (\text{using } c^2 = T/\rho A) \\
&= \rho A c^2 f'^2(x - ct) = \rho A w_{,t}^2 \qquad (2.33) \\
&= 2\hat{\mathcal{T}}. \qquad (2.34)
\end{aligned}$$

Using this in (2.32), one can easily obtain $\hat{\mathcal{T}} = \hat{\mathcal{V}}$. Thus, the total mechanical energy due to a positive-traveling wave is equally distributed in the form of kinetic and potential energy. It is evident that the same will hold for a negative-traveling wave as well. When both positive and negative-traveling waves exist simultaneously, i.e., $w(x,t) = f(x-ct) + g(x+ct)$, we have

$$\hat{\mathcal{T}} = \frac{1}{2}\rho A c^2 (f'^2 + g'^2 - 2f'g')$$

and

$$\hat{\mathcal{V}} = \frac{1}{2}T(f'^2 + g'^2 + 2f'g').$$

Therefore, in this case, the energy distribution in the form of kinetic and potential energy is unequal. The total energy density is obtained as

$$\hat{\mathcal{E}} = \rho A c^2 [f'^2(x-ct) + g'^2(x+ct)],$$

which is the sum of the energy densities due to the individual traveling waves.

Consider the forced semi-infinite string discussed in Section 2.2 once again. It may be noted from (2.28) and (2.29) that the force $F(t)$ applied at the end of a semi-infinite string, and the corresponding velocity $v(t)$ of the end are in phase. This implies that energy is continuously fed into the string by the force. This energy is carried forward by the positive-traveling harmonic wave in the string. However, the energy is transported through the string at a certain rate which depends on the impedance of the string and the frequency of the harmonic wave. In general, it is of considerable theoretical and practical interest to study the rate at which energy is transported through a continuous medium. In the following, we discuss a general procedure to study this energy propagation in any conservative continuous system with particular reference to a taut string.

Consider a section of the string between the coordinates $[x_1, x_2]$. The total mechanical energy for this section can be written as

$$\mathcal{E} = \frac{1}{2}\int_{x_1}^{x_2} \left(\rho A w_{,t}^2 + T w_{,x}^2\right) \mathrm{d}x.$$

Differentiating this expression with respect to time, one can write

$$\begin{aligned}\frac{\mathrm{d}\mathcal{E}}{\mathrm{d}t} &= \int_{x_1}^{x_2} \left(\rho A w_{,t} w_{,tt} + T w_{,x} w_{,xt}\right) \mathrm{d}x \\ &= T w_{,x} w_{,t}\Big|_{x_1}^{x_2} + \int_{x_1}^{x_2} \left(\rho A w_{,tt} - T w_{,xx}\right) w_{,t}\, \mathrm{d}x \\ &= \mathcal{P}(x_1, t) - \mathcal{P}(x_2, t),\end{aligned} \tag{2.35}$$

where we have used the equation of motion of the string, and introduced the definition

$$\mathcal{P}(x,t) := -T w_{,x} w_{,t}, \tag{2.36}$$

which represents the instantaneous energy transport per unit time, or the instantaneous *power* transported past the coordinate location x. For example, for a positive-traveling wave $w(x,t) = f(x-ct)$, the energy flux crossing a location x at time t is given by $Tcf'^2(x-ct)$, while for a negative-traveling wave $w(x,t) = g(x+ct)$ it is $-Tcg'^2(x+ct)$. Thus, the direction of power flow is associated with the expression (2.36), and hence $\mathcal{P}(x,t)$ is a signed scalar. Further, for a positive-traveling wave, one can write

$$\mathcal{P}(x,t) = Tcf'^2(x-ct) = \rho A c^3 f'^2(x-ct) = c\hat{\mathcal{E}}(x,t) \qquad \text{(using (2.33))}.$$

Thus, it can be easily inferred that the energy is transported at the same speed as the propagation of transverse waves in the string. It is evident that the expression (2.35) represents the energy balance equation for the section $[x_1, x_2]$ of the string. Similarly, in the case of longitudinal waves (such as in a bar in longitudinal vibration), one can easily obtain the instantaneous power crossing a location x as $\mathcal{P}(x,t) = -EAw_{,x}w_{,t}$. In this case, one can also define the instantaneous *intensity* $\mathcal{I}(x,t)$ of the longitudinal wave as the instantaneous power of the wave per unit area, i.e.,

$$\mathcal{I}(x,t) = \frac{\mathcal{P}(x,t)}{A} = \rho c^2 w_{,x}w_{,t},$$

where $c = \sqrt{E/\rho}$ is the speed of longitudinal waves in a bar.

All the above expressions of power evaluate the instantaneous power. However, the physically more relevant quantity is the *average power* over a cycle, which can be calculated for harmonic waves as follows. It may be recalled from the theory of vibrations of discrete systems that when a system is forced by $F(t) = F_0 \cos\omega t$, the corresponding velocity is obtained, in general, as $v(t) = v_0\cos(\omega t - \phi)$, where ϕ is an appropriate phase difference. The average power per cycle is then calculated as

$$\begin{aligned}\langle\mathcal{P}\rangle &= \frac{\omega}{2\pi}\int_0^{2\pi/\omega} F(t)v(t)\,\mathrm{d}t,\\ &= \frac{1}{2}F_0 v_0\cos\phi. \end{aligned} \tag{2.37}$$

In the complex notation, representing $\underline{F}(t) = F_0\mathrm{e}^{i\omega t}$ and $\underline{v}(t) = v_0\mathrm{e}^{i(\omega t-\phi)}$, one can rewrite (2.37) as

$$\langle\mathcal{P}\rangle = \frac{1}{2}\mathscr{R}[\underline{F}^*(t)\underline{v}(t)], \tag{2.38}$$

where $\underline{F}^*(t)$ is the complex conjugate of $\underline{F}(t)$, and $\mathscr{R}[\cdot]$ denotes the real part. It may be noted that, in this section, the quantities which are in general complex have been underlined to distinguish them from real quantities. The expression (2.38) is very convenient for calculating average power, and will be used here throughout.

Using the impedance relation $\underline{F} = \underline{\mathcal{Z}}\underline{v}$, where $\underline{\mathcal{Z}} = \mathcal{Z}_0\mathrm{e}^{i\phi}$ is a general complex impedance, one can rewrite (2.38) as

$$\begin{aligned}\langle\mathcal{P}\rangle &= \frac{1}{2}\mathscr{R}[\underline{F}^*(t)\underline{v}(t)] = \frac{1}{2}\mathscr{R}\left[\frac{\underline{F}^*\underline{F}}{\underline{\mathcal{Z}}}\right] = \frac{1}{2}\mathscr{R}\left[\frac{F_0^2}{\mathcal{Z}_0}\mathrm{e}^{-i\phi}\right]\\ &= \frac{1}{2}\frac{F_0^2}{\mathcal{Z}_0}\cos\phi. \end{aligned} \tag{2.39}$$

Hence, it is clear that when the impedance $\underline{\mathcal{Z}}$ is real (i.e., $\phi = n\pi$), power is either absorbed by or extracted from the medium (depending on the sign of $\langle \mathcal{P} \rangle$) over a cycle. On the other hand, when $\underline{\mathcal{Z}}$ is purely imaginary (i.e., $\phi = (2n+1)\pi/2$), no power can be absorbed over a cycle by the medium. In such a case, power absorbed by the medium over one part of a cycle is extracted back by the source over the remaining part of the cycle. As was observed in Section 2.2, the impedance of a uniform semi-infinite or infinite medium is always positive real. Therefore, power is always absorbed by the medium over a cycle. One can also rewrite (2.39) as

$$\begin{aligned}\langle \mathcal{P} \rangle &= \frac{1}{2}\mathscr{R}[\underline{F}^*(t)\underline{v}(t)] = \frac{1}{2}\mathscr{R}\left[\underline{\mathcal{Z}}^*\underline{v}^*\underline{v}\right] = \frac{1}{2}\mathscr{R}\left[\mathcal{Z}_0 v_0^2 e^{-i\phi}\right] \\ &= \frac{1}{2}\mathcal{Z}_0 v_0^2 \cos\phi. \end{aligned} \tag{2.40}$$

While the form (2.39) is useful for studying the average power absorbed by a medium from a source, the form (2.40) is convenient for determining the average power carried by a traveling wave.

Consider a mass–spring system connected to a semi-infinite string, as shown in Figure 2.11(a). Since the impedance of a semi-infinite string $\underline{\mathcal{Z}} = \rho c$ is purely real, energy is always absorbed by the string from the oscillator. Therefore, the oscillator loses energy through radiation into an infinite medium, and hence, the motion of the oscillator is damped. This example serves to illustrate an important phenomenon known as *radiation damping*. An example of such a process is the energy lost by a body (say, a vibrating machine body, or a bouncing ball) through acoustic radiation. Using the concept of impedance and calculating the force applied by the string on the mass, one can show that the string can be replaced by an equivalent damper of damping coefficient ρc, as shown in Figure 2.11(b). It can be easily checked that the energy lost by the oscillator through radiation and through viscous dissipation are the same. However, it is to be noted that the physical mechanisms of damping due to the semi-infinite string and the damper are completely different.

In the case of a general complex impedance $\underline{\mathcal{Z}} = \mathcal{Z}_R + i\mathcal{Z}_I$, the imaginary part contributes to an *added mass*, or an *added stiffness* depending on its sign. For example, one may write the impedance due to a mass m as $\underline{\mathcal{Z}}_m = i\omega m$, and due to a stiffness k as $\underline{\mathcal{Z}}_k = -ik/\omega$. Therefore, one can have either an added mass $m = \mathcal{Z}_I/\omega$ or an added stiffness $k = \mathcal{Z}_I\omega$.

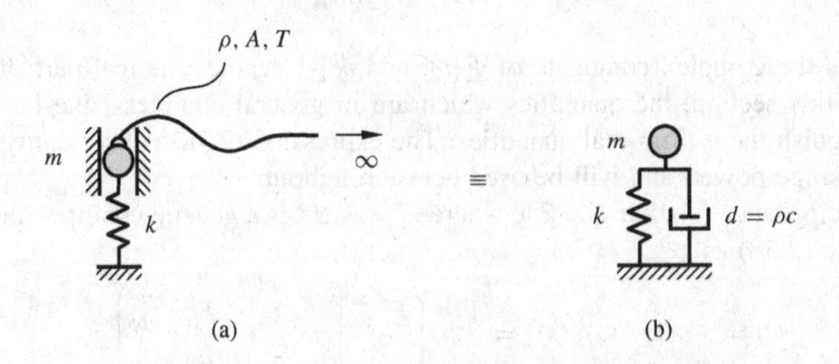

Figure 2.11 Equivalent damping due to a semi-infinite string

2.4 SCATTERING OF WAVES

The interaction of waves with a boundary or obstacles in a medium is generally termed *scattering*. In one dimension, scattering may result in wave reflection, transmission, or both. In this section, we will discuss the reflection and transmission of waves under various conditions.

2.4.1 Reflection at a boundary

In strings and bars, we usually have either a fixed boundary or a free boundary. These two cases are discussed below. Other cases of boundary conditions are discussed separately.

2.4.1.1 *Fixed boundary*

Consider a semi-infinite string, as shown in Figure 2.12, with a fixed boundary at $x = 0$. Let a positive-traveling harmonic wave of the form $B_{\mathrm{I}}\mathrm{e}^{i(kx-\omega t)}$ be incident on the boundary, where B_{I} is the amplitude of the wave and $k = \omega/c$. This results in a reflected negative-traveling harmonic wave of the form $B_{\mathrm{R}}\mathrm{e}^{i(-k'x-\omega' t)}$, where B_{R} is the amplitude, and k' and ω' are, respectively, the wave number and frequency of the reflected wave which satisfy the relation $k' = \omega'/c$. Then, one can write the net motion of the string in the form

$$\begin{aligned} w(x,t) &= B_{\mathrm{I}}\mathrm{e}^{i(kx-\omega t)} + B_{\mathrm{R}}\mathrm{e}^{i(-k'x-\omega' t)} \\ &= B_{\mathrm{I}}\mathrm{e}^{i(kx-\omega t)} + C_{\mathrm{R}}B_{\mathrm{I}}\mathrm{e}^{i(-k'x-\omega' t)}, \end{aligned} \tag{2.41}$$

where we write the amplitude of the reflected wave as a constant scaling of the incident wave amplitude by a factor C_{R} which is the *coefficient of reflection* of the boundary. It may be noted that C_{R} can be complex in general, implying that B_{R} can also be complex. On the other hand, B_{I} can be assumed to be real, since the phase of the incident wave can be taken as zero without loss of generality. Using (2.41) in the boundary condition $w(0,t) \equiv 0$, we obtain

$$B_{\mathrm{I}}\mathrm{e}^{i\omega t} + C_{\mathrm{R}}B_{\mathrm{I}}\mathrm{e}^{i\omega' t} = 0. \tag{2.42}$$

Since the coefficients in the above equation are independent of time, it can be identically satisfied if and only if $\omega = \omega'$. The wave number of the reflected wave is then given by $k' = \omega'/c = \omega/c = k$. Therefore, (2.42) can be written as

$$\begin{aligned} &(1 + C_{\mathrm{R}})B_{\mathrm{I}}\mathrm{e}^{i\omega t} = 0 \\ \Rightarrow\ &C_{\mathrm{R}} = -1 = \mathrm{e}^{i\pi}. \end{aligned} \tag{2.43}$$

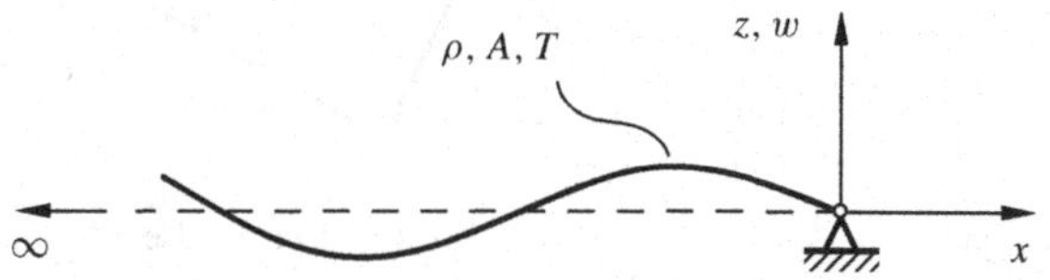

Figure 2.12 Semi-infinite string with a fixed boundary

Thus, the incident harmonic wave undergoes a phase change of π after reflection from a fixed boundary.

Let us now consider the energetics of the reflection process. The average power carried by the positive-traveling incident harmonic wave can be written using (2.40) as

$$\langle \mathcal{P}_{\mathrm{I}} \rangle = \frac{1}{2} \mathcal{R}[\mathcal{Z}^* v_{\mathrm{I}}^* v_{\mathrm{I}}]. \tag{2.44}$$

Using $v_{\mathrm{I}} = -i\omega B_{\mathrm{I}} e^{i(kx-\omega t)}$, and $\mathcal{Z} = \rho A c$ in (2.44), we obtain

$$\langle \mathcal{P}_{\mathrm{I}} \rangle = \frac{1}{2} \mathcal{R}[\rho A c \omega^2 B_{\mathrm{I}}^2] = \frac{1}{2} \rho A c \omega^2 B_{\mathrm{I}}^2. \tag{2.45}$$

Similarly, the wave energy carried by the reflected wave is given by

$$\langle \mathcal{P}_{\mathrm{R}} \rangle = \frac{1}{2} \rho A c \omega^2 |B_{\mathrm{R}}|^2 = \frac{1}{2} \rho A c \omega^2 |C_{\mathrm{R}}|^2 B_{\mathrm{I}}^2 = \frac{1}{2} \rho A c \omega^2 B_{\mathrm{I}}^2 \qquad \text{(using (2.43))}.$$

Therefore, the incident power is completely reflected without loss by a fixed boundary.

Let us now consider an arbitrarily shaped positive-traveling waveform $f(z)|_{z=x-ct}$ on a string, as shown in Figure 2.13(a). Let us denote the reflected waveform by $g(z)|_{x+ct}$. The motion of the string is then represented by

$$w(x, t) = f(x - ct) + g(x + ct). \tag{2.46}$$

Then applying the boundary condition $w(0, t) \equiv 0$ yields

$$\begin{aligned} & f(-ct) + g(ct) = 0 \\ \Rightarrow\ & g(z) = -f(-z) \\ \Rightarrow\ & g(x + ct) = -f(-x - ct). \end{aligned} \tag{2.47}$$

The function $g(z)$ is shown in Figure 2.13(b). The reflection process is depicted at certain time instants in Figure 2.14. As depicted in the figure, the reflection process may be

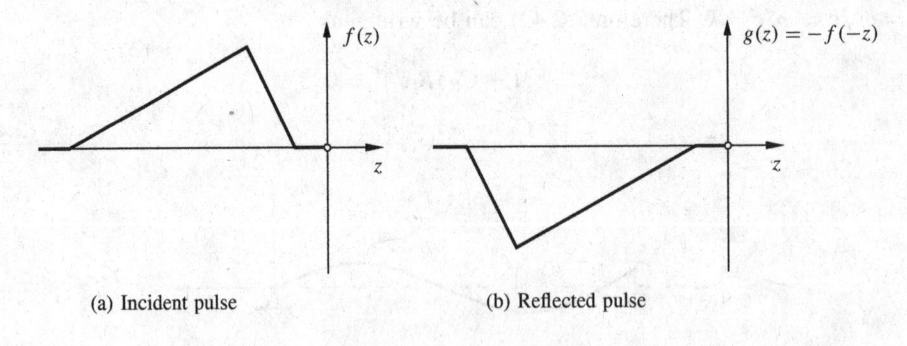

Figure 2.13 Incident and reflected pulse shapes

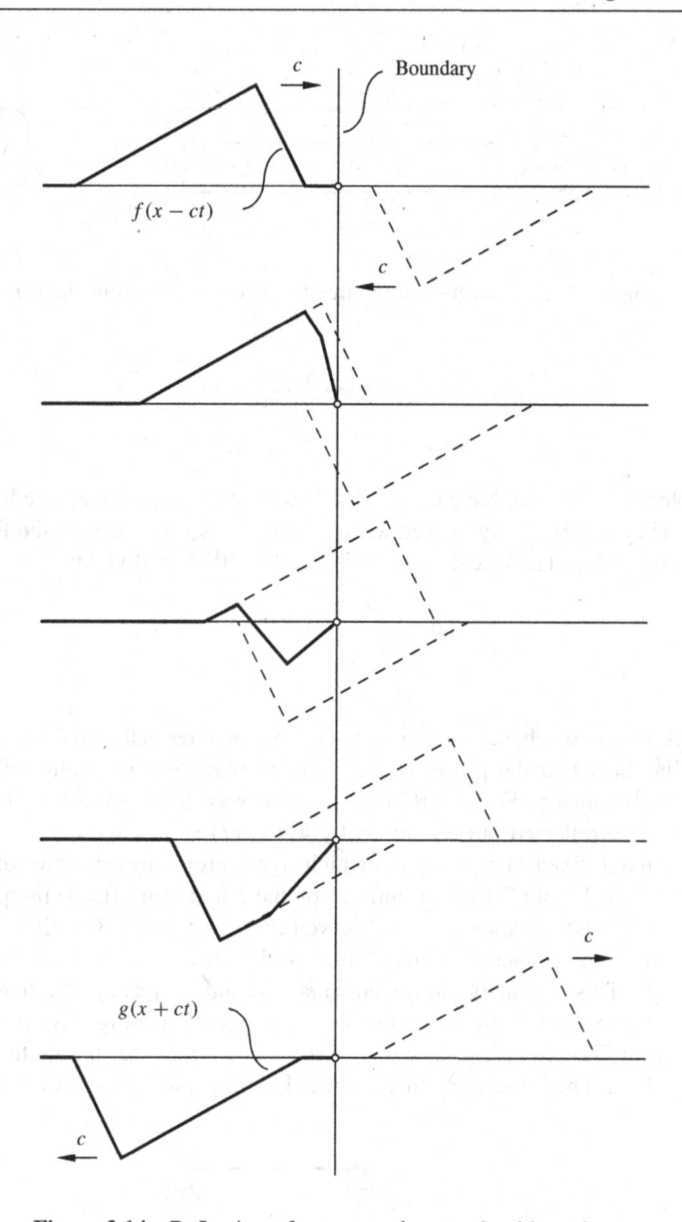

Figure 2.14 Reflection of a wave pulse at a fixed boundary

visualized as the superposition of the positive-traveling wave $f(x - ct)$ and the negative-traveling wave $g(x + ct) = -f(-x - ct)$.

2.4.1.2 Free boundary

Consider a semi-infinite bar with a free end, as shown in Figure 2.15. Let a positive-traveling harmonic wave be incident on the boundary, which results in a reflected negative-traveling harmonic wave. The complete wave field in the bar can be represented in complex

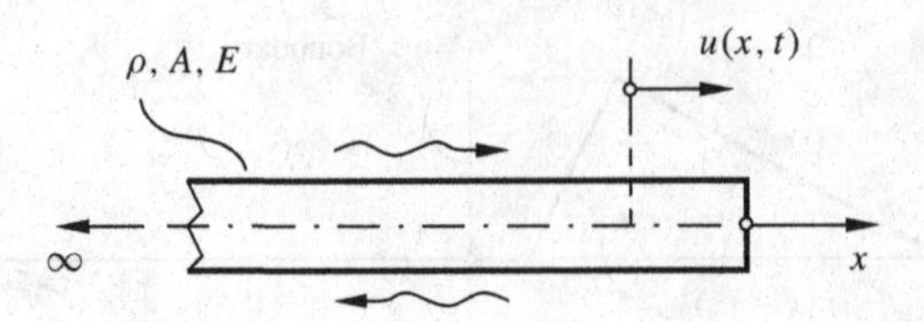

Figure 2.15 Reflection at the free boundary of a semi-infinite bar

notation as

$$u(x,t) = B_{\mathrm{I}}\mathrm{e}^{i(kx-\omega t)} + B_{\mathrm{R}}\mathrm{e}^{i(-kx-\omega t)}$$
$$= B_{\mathrm{I}}\mathrm{e}^{i(kx-\omega t)} + C_{\mathrm{R}}B_{\mathrm{I}}\mathrm{e}^{i(-kx-\omega t)}. \tag{2.48}$$

It may be noted here that we have assumed k and ω of the incident and reflected waves to be the same. They can be easily proved to be the same, as was done for the fixed boundary case in Section 2.4.1.1. The free end condition $EAu_{,x}(0,t) \equiv 0$ yields

$$ikB_{\mathrm{I}}\mathrm{e}^{-i\omega t} - ikC_{\mathrm{R}}B_{\mathrm{I}}\mathrm{e}^{-i\omega t} = 0$$
$$\Rightarrow C_{\mathrm{R}} = 1 = \mathrm{e}^{0i}. \tag{2.49}$$

Thus, there is no phase change in the harmonic wave after reflection from the free end. One can easily show that the power of the incident wave is completely reflected without loss from a free boundary. For an arbitrary positive waveform given by $f(x-ct)$, it can be checked that the reflected pulse is given by $g(x+ct) = f(-x-ct)$.

Consider a fixed–fixed taut string on which a waveform travels towards the negative x-axis, as shown in Figure 2.16. The motion of the waveform after subsequent reflection from the boundaries is also shown. It is observed that after a time $T = 2l/c$, the waveform repeats the motion again. Therefore, the motion is $2l/c$-periodic, and the circular frequency is $2\pi/T = \pi c/l$. This is exactly the fundamental circular frequency of a fixed–fixed string as determined in Section 1.3. Similarly, the motion of a left-traveling wave in a sliding–fixed string is shown in Figure 2.17. It is observed from this figure that the motion repeats after a period $T = 4l/c$. Thus, the fundamental circular frequency of a sliding–fixed string is given by $\pi c/2l$.

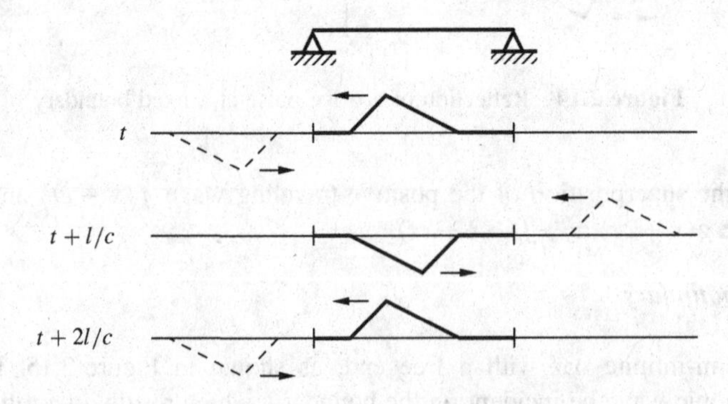

Figure 2.16 Reflection of waves in a taut string with fixed boundaries

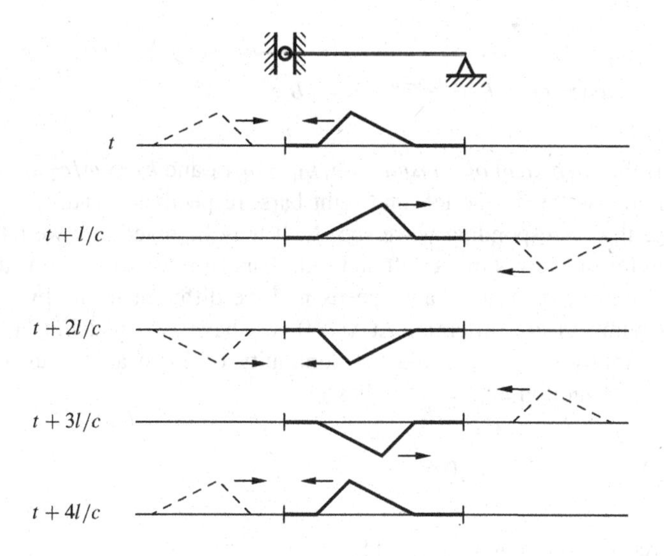

Figure 2.17 Reflection of waves in a taut string with a sliding boundary

2.4.2 Scattering at a finite impedance

From the concept of impedance discussed in Section 2.2, it can be easily shown that a fixed boundary has infinite impedance, while a free boundary has zero impedance. In this section, we consider scattering from boundaries or obstacles of finite impedance.

Consider two semi-infinite bars of two different materials connected at $x = 0$, as shown in Figure 2.18. The left and right bars have densities ρ_1 and ρ_2, respectively, and the corresponding Young's moduli are E_1 and E_2. Let a positive-traveling longitudinal wave from the left impinge on the boundary. This causes a partial transmission and a partial reflection of the wave as indicated in Figure 2.18. Using complex notation, one can write the motion of the bar on the left and right, respectively, as

$$u_1(x,t) = B_{\mathrm{I}} e^{i(k_1 x - \omega t)} + C_{\mathrm{R}} B_{\mathrm{I}} e^{i(-k_1 x - \omega t)} \tag{2.50}$$

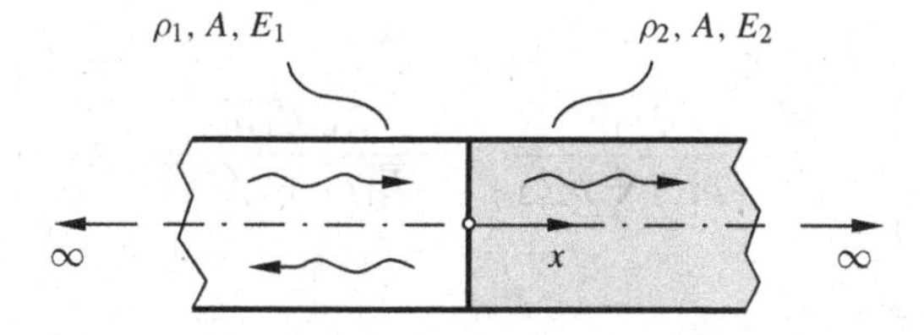

Figure 2.18 Scattering at the boundary between two semi-infinite bars of different materials

and

$$u_2(x,t) = B_{\rm T}{\rm e}^{i(k_2x-\omega t)} = C_{\rm T}B_{\rm I}{\rm e}^{i(k_2x-\omega t)}, \tag{2.51}$$

where $C_{\rm T}$ is known as the *coefficient of transmission*, $k_1 = \omega/c_1$ and $k_2 = \omega/c_2$ are the wave numbers of the harmonic waves in the left and right bars, respectively, and $c_1 = \sqrt{E_1/\rho_1}$ and $c_2 = \sqrt{E_2/\rho_2}$ are the corresponding wave speeds. It may be noted that we have taken the same frequency ω for the waves in the left and right bars from considerations discussed in Section 2.4.1.1. However, since the wave speeds may be different in the two bars, the corresponding wave numbers are different. At $x = 0$, we have two matching conditions which ensure the continuity of motion, and the continuity of force at the junction. The condition of continuity of motion at $x = 0$ implies

$$u_1(0,t) = u_2(0,t). \tag{2.52}$$

The continuity of force condition at $x = 0$ yields

$$E_1Au_{1,x}(0,t) = -E_2Au_{2,x}(0,t) = -\mathcal{Z}_2u_{2,t}(0,t), \tag{2.53}$$

where $\mathcal{Z}_2 = \rho_2Ac_2$ is the impedance of the semi-infinite right bar. Using (2.50)–(2.51) in (2.52), we obtain

$$\begin{aligned} B_{\rm I}{\rm e}^{-i\omega t} + C_{\rm R}B_{\rm I}{\rm e}^{-i\omega t} &= C_{\rm T}B_{\rm I}{\rm e}^{-i\omega t} \\ \Rightarrow 1 + C_{\rm R} &= C_{\rm T}. \end{aligned} \tag{2.54}$$

Similarly, from (2.53), we have

$$\begin{aligned} E_1A(ik_1B_{\rm I}{\rm e}^{-i\omega t} - ik_1C_{\rm R}B_{\rm I}{\rm e}^{-i\omega t}) &= i\omega\rho_2c_2AC_{\rm T}B_{\rm I}{\rm e}^{-i\omega t} \\ \Rightarrow -1 + C_{\rm R} &= -\frac{\omega}{k_1E_1}\rho_2c_2C_{\rm T} = -\frac{\rho_2c_2}{\rho_1c_1}C_{\rm T}. \end{aligned} \tag{2.55}$$

From (2.54) and (2.55), we can solve for the transmission and reflection coefficients, respectively, as

$$C_{\rm T} = \frac{2\rho_1c_1}{\rho_1c_1 + \rho_2c_2} = \frac{2\sqrt{E_1\rho_1}}{\sqrt{E_1\rho_1} + \sqrt{E_1\rho_2}} \tag{2.56}$$

and

$$C_{\rm R} = \frac{\rho_1c_1 - \rho_2c_2}{\rho_1c_1 + \rho_2c_2} = \frac{\sqrt{E_1\rho_1} - \sqrt{E_2\rho_2}}{\sqrt{E_1\rho_1} + \sqrt{E_2\rho_2}}. \tag{2.57}$$

Let us now look at the energetics of this scattering process. The average power carried by the incident, reflected, and transmitted harmonic waves are given by, respectively,

$$\langle \mathcal{P}_{\mathrm{I}} \rangle = \frac{1}{2}\rho_1 A c_1 \omega^2 B_{\mathrm{I}}^2,$$

$$\langle \mathcal{P}_{\mathrm{R}} \rangle = \frac{1}{2}\rho_1 A c_1 \omega^2 |C_{\mathrm{R}}|^2 B_{\mathrm{I}}^2,$$

and

$$\langle \mathcal{P}_{\mathrm{T}} \rangle = \frac{1}{2}\rho_2 A c_2 \omega^2 |C_{\mathrm{T}}|^2 B_{\mathrm{I}}^2.$$

The coefficients of power reflection C_{PR}, and power transmission C_{PT} can be defined as, respectively,

$$C_{\mathrm{PR}} := \frac{\langle \mathcal{P}_{\mathrm{R}} \rangle}{\langle \mathcal{P}_{\mathrm{I}} \rangle} = |C_{\mathrm{R}}|^2 \quad \text{and} \quad C_{\mathrm{PT}} := \frac{\langle \mathcal{P}_{\mathrm{T}} \rangle}{\langle \mathcal{P}_{\mathrm{I}} \rangle} = \frac{\rho_2 c_2}{\rho_1 c_1}|C_{\mathrm{T}}|^2.$$

One can easily check that

$$\langle \mathcal{P}_{\mathrm{R}} \rangle + \langle \mathcal{P}_{\mathrm{T}} \rangle = \left(|C_{\mathrm{R}}|^2 + \frac{\rho_2 c_2}{\rho_1 c_1}|C_{\mathrm{T}}|^2\right)\langle \mathcal{P}_{\mathrm{I}} \rangle = \langle \mathcal{P}_{\mathrm{I}} \rangle \qquad \text{(using (2.56) and (2.57))}.$$

Thus, the incident average power is partially reflected and partially transmitted at the junction of the bars, and the sum of the scattered average power equals the incident average power.

The scattering process in the above example of two bars is also phenomenologically similar to the process of *radiation damping* discussed previously in Section 2.3. The left bar (the radiating body) loses the energy it transmits to the right bar. In certain situations, radiation damping may be a desirable mode of energy dissipation. From the above calculations, it is clear that a fraction of the energy is reflected back into the left bar. One way to cut down this reflection (and hence maximize the energy radiation), is to introduce a film of appropriate thickness of a third material between the two bars (see Exercise 2.5). Also, in certain other applications such as ultrasonic testing, a fluid medium is used for maximizing the transmission of ultrasonic waves from the ultrasonic generator into the specimen.

Next, consider the example of a semi-infinite bar with a damper, as shown in Figure 2.19. Let a positive-traveling harmonic wave in the bar be incident on the boundary at $x = 0$. The incident and reflected waves in the bar can be represented in complex notation as

$$u(x,t) = B_{\mathrm{I}} e^{i(kx-\omega t)} + C_{\mathrm{R}} B_{\mathrm{I}} e^{i(-kx-\omega t)}. \tag{2.58}$$

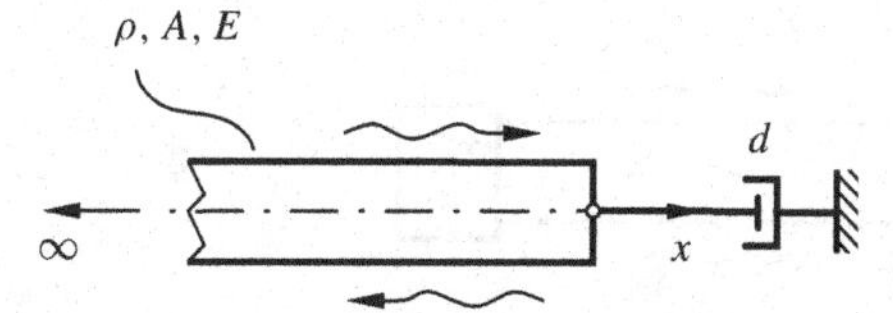

Figure 2.19 Reflection at the boundary of a semi-infinite bar with a damper

Here, we have only a dynamic boundary condition of the form $EAu_{,x}(0,t) = -du_{,t}(0,t)$. Substituting the solution form (2.58) in the boundary condition yields

$$kEA(1 - C_R) = d\omega(1 + C_R) \quad \Rightarrow \quad C_R = \frac{\rho Ac - d}{\rho Ac + d}, \tag{2.59}$$

where we have used $k = \omega/c$, and $c = \sqrt{E/\rho}$.

An interesting situation occurs when

$$d = \rho Ac \tag{2.60}$$

in (2.59), implying $C_R = 0$. Thus, there is no reflected wave, and the total incident energy is lost. The reason for this phenomenon becomes clear if we compare the special value of the damper impedance d in (2.60) with the impedance of a semi-infinite bar given in (2.30). Thus, when the damper impedance d equals the impedance of a semi-infinite bar, to a positive-traveling wave, the damper 'appears' like a semi-infinite bar. Hence, there is no reflection from the common boundary of the bar and the damper. While solving the eigenvalue problem of a fixed–damped bar in Section 1.8.2, it was observed that no eigenfrequency exists for the special choice of boundary damping d. It is now clear that this occurs because of the absence of any reflected wave from the damper–bar interface.

2.4.2.1 *Equivalent impedance*

Consider a finite bar with a general complex boundary impedance $\mathcal{Z}_B$, as shown in Figure 2.20. In many cases, it is convenient to replace the complex system by a single equivalent impedance $\mathcal{Z}_E$ as shown in the figure. In the following, we will determine the equivalent impedance of the bar, and study its scattering properties.

Consider a harmonic force $F(t) = F_0 e^{-i\Omega t}$ at the left end of the bar. One can then write the motion of the bar as

$$u(x,t) = B_1 e^{i(kx - \Omega t)} + B_2 e^{i(-kx - \Omega t)}, \tag{2.61}$$

where B_1 and B_2 represent the amplitudes of the corresponding harmonic waves, and $k = \Omega/c$ is the wave number. The force condition at the left end can be written as

$$F(t) = -EAu_{,x}(0,t) \quad \Rightarrow \quad F_0 = -ikEA[B_1 - B_2]. \tag{2.62}$$

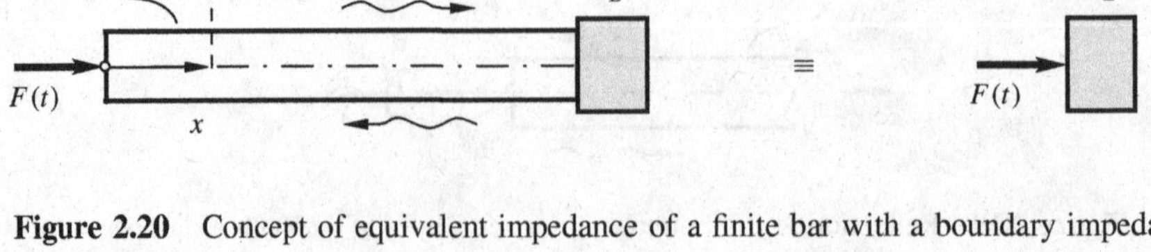

Figure 2.20 Concept of equivalent impedance of a finite bar with a boundary impedance

Defining an equivalent impedance of the bar as $\mathcal{Z}_E$, we can also write

$$F(t) = \mathcal{Z}_E u_{,t} \quad \Rightarrow \quad F_0 = -i\Omega\mathcal{Z}_E[B_1 + B_2]. \tag{2.63}$$

At the right end of the bar, we have the dynamic boundary condition

$$EAu_{,x}(l,t) = -\mathcal{Z}_B u_{,t}(l,t)$$

or

$$ikEA(B_1 e^{ikl} - B_2 e^{-ikl}) = i\Omega\mathcal{Z}_B(B_1 e^{ikl} + B_2 e^{-ikl})$$

$$\Rightarrow B_2 = re^{2ikl} B_1, \tag{2.64}$$

where

$$r = \frac{kEA - \Omega\mathcal{Z}_B}{kEA + \Omega\mathcal{Z}_B} = \frac{\rho Ac - \mathcal{Z}_B}{\rho Ac + \mathcal{Z}_B} \tag{2.65}$$

is a complex scalar. Eliminating F_0 from (2.62) and (2.63), and subsequently using (2.64), we obtain the equivalent impedance as

$$\mathcal{Z}_E = \rho Ac \frac{1 - re^{2ikl}}{1 + re^{2ikl}}. \tag{2.66}$$

Solving for B_1 and B_2 from (2.62) and (2.64), we obtain

$$B_1 = -\frac{F_0}{i\Omega\rho Ac(1 - re^{2ikl})} \tag{2.67}$$

and

$$B_2 = -\frac{F_0 re^{2ikl}}{i\Omega\rho Ac(1 - re^{2ikl})}. \tag{2.68}$$

Let us now consider some special cases of the impedance $\mathcal{Z}_B$. In the case of a free right end of the bar, $\mathcal{Z}_B = 0$ (i.e., $r = 1$), and we have from (2.67)

$$B_1 = -\frac{F_0 e^{-ikl}}{2\Omega\rho Ac \sin kl}. \tag{2.69}$$

When $k_n l = \Omega_n l/c = n\pi, n = 1, 2, \ldots, \infty$, the denominator of (2.69) vanishes. This is then the resonance condition, and Ω_n are the circular eigenfrequencies of the free–free bar. For a fixed-end, $\mathcal{Z}_B = \infty$ (i.e., $r = -1$), and we obtain from (2.67)

$$B_1 = -\frac{F_0 e^{-ikl}}{2i\Omega\rho Ac \cos kl}. \tag{2.70}$$

In this case, we can obtain the eigenfrequencies of a free–fixed bar from the resonance condition $k_n l = \Omega_n l/c = (2n-1)\pi/2$, $n = 1, 2, \ldots, \infty$. For $\mathcal{Z}_B = \rho Ac$, we have $r = 0$, and (2.68) yields $B_2 = 0$. This implies that there is no reflection from the right end, i.e., the right end behaves like a perfect absorber.

Next, we study the scattering properties of the bar with boundary impedance. Using (2.64) in (2.61), and representing the complex scalar $r = |r|e^{i\gamma}$, we can write the motion of the bar as

$$u(x,t) = (1 + |r|e^{i[2k(l-x)+\gamma]})B_1 e^{i(kx-\Omega t)} = p(x)B_1 e^{i(kx-\Omega t)}, \tag{2.71}$$

where $p(x) = 1 + |r|e^{i[2k(l-x)+\gamma]}$. It is evident from (2.71) that the amplitude of motion of the bar is decided by the amplitude factor $p(x)$, which is geometrically visualized in Figure 2.21. One can conclude from the figure that the magnitude of the amplitude factor lies between $1 + |r|$ and $1 - |r|$, and is nowhere zero if $|r| \neq 1$. Thus, nodes can form only when $|r| = 1$, which implies from (2.65) that $\mathscr{R}[\mathcal{Z}_B] = 0$, i.e., $\mathcal{Z}_B$ is purely imaginary. As discussed in Section 2.3, when an impedance is purely imaginary, it cannot absorb energy over a cycle. The coordinate of a node x_N can be obtained easily from the condition

$$p(x) = 1 + e^{i[2k(l-x_N)+\gamma]} = 0$$
$$\Rightarrow 2k(l - x_N) + \gamma = (2n-1)\pi,$$

where n is an integer. It is to be noted, however, that a node can form within the length of the bar only for a range of values of γ and n. In particular, a node is formed at the right end of the bar when $\gamma = (2n-1)\pi$.

The normal stress in the bar can be evaluated from the expression

$$\sigma(x,t) = Eu_{,x}(x,t) = ikE\left(1 - |r|e^{i[2k(l-x)+\gamma]}\right) B_1 e^{i(kx-\Omega t)}. \tag{2.72}$$

From (2.71) and (2.72), it can be easily observed that the points of minimum amplitude correspond to maximum stress, and vice versa. The velocity at any point of the bar can be obtained as

$$v(x,t) = u_{,t}(x,t) = -i\Omega\left(1 - |r|e^{i[2k(l-x)+\gamma]}\right) B_1 e^{i(kx-\Omega t)}. \tag{2.73}$$

Using (2.72) and (2.73), one can determine the equivalent impedance at any location x as

$$\mathcal{Z}_L(x) := \frac{-A\sigma(x,t)}{u_{,t}(x,t)} = \mathcal{Z}\frac{1 - |r|e^{i[2k(l-x)+\gamma]}}{1 + |r|e^{i[2k(l-x)+\gamma]}}, \tag{2.74}$$

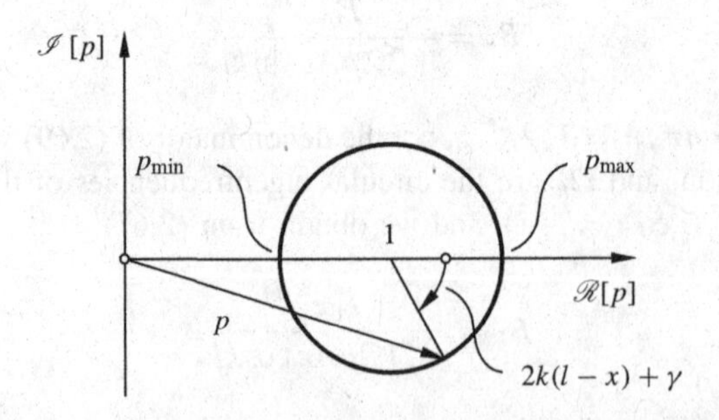

Figure 2.21 Graphical representation of the amplitude factor $p(x)$

where $\mathcal{Z} = EA/c = \rho Ac$ is the impedance of a semi-infinite bar. It can be easily checked that $\mathcal{Z}_L(0) = \mathcal{Z}_{\mathrm{E}}$ (as defined in (2.66)), and $\mathcal{Z}_L(l) = \mathcal{Z}_{\mathrm{B}}$, the complex boundary impedance.

2.5 APPLICATIONS OF THE WAVE SOLUTION

In this section, we consider the application of the traveling wave solution in studying the transient dynamics of some physical systems.

2.5.1 Impulsive start of a bar

Consider a free–free bar of length l set into motion by an impulsive force $F(x,t) = F_0\delta(t)\delta(x)$, as shown in Figure 2.22. The subsequent motion of the bar can be determined as follows. Using the impulse–momentum equation, one can write

$$\int_0^{0^+} \rho A u_{,tt}\,\mathrm{d}t = \int_0^{0^+} F_0\delta(t)\delta(x)\,\mathrm{d}t$$

$$\Rightarrow u_{,t}(x,0^+) = v_0(x) = \frac{F_0}{\rho A}\delta(x). \tag{2.75}$$

Thus, at $t = 0^+$, the left end of the bar is imparted a velocity $F_0/\rho A$, while the rest of the bar is at rest. One may, alternatively, start with the boundary condition

$$-EAu_{,x}(0,t) = F_0\delta(t). \tag{2.76}$$

Approaching the problem with the boundary condition (2.76) is left as an exercise (see Exercise 2.11).

Assuming a displacement field $u(x,t) = f_1(x - ct)$ in (2.75), one can write

$$u_{,t}(x,0) = -cf_1'(x) = v_0 = \frac{F_0}{\rho A}\delta(x)$$

$$\Rightarrow f_1(x) = \frac{F_0}{\rho Ac}[1 - \mathcal{H}(x)]$$

$$\Rightarrow u(x,t) = f_1(x - ct) = \frac{F_0}{\rho Ac}[1 - \mathcal{H}(x - ct)], \qquad 0 < t < l/c, \tag{2.77}$$

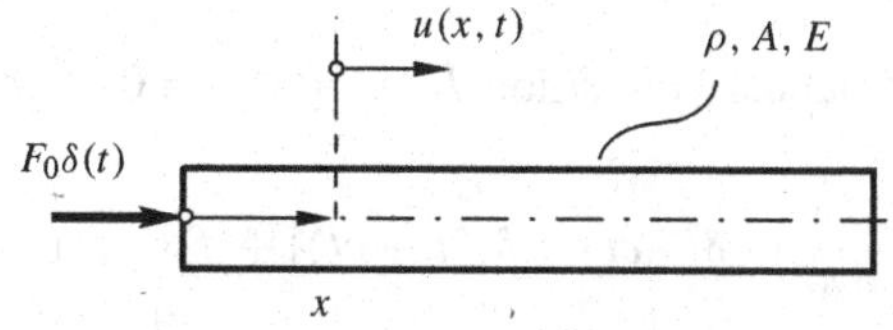

Figure 2.22 A free–free bar started with an impulse

where the constant of integration has been found from the initial zero displacement condition, and $\mathcal{H}(\cdot)$ is the *Heaviside step function* defined as

$$\mathcal{H}(x) = \begin{cases} 0, & x < 0 \\ 1, & x \geq 0 \end{cases}.$$

When the wave reaches the right end at $t = l/c$, a reflected wave, denoted by $g_1(x + ct)$, is created. The motion of the bar can be represented as

$$u(x,t) = \frac{F_0}{\rho A c}[1 - \mathcal{H}(x - ct)] + g_1(x + ct), \qquad l/c \leq t < 2l/c. \tag{2.78}$$

Applying the right-end boundary condition $EAu_{,x}(l, t) \equiv 0$ yields

$$-\frac{F_0}{\rho A c}\delta(l - ct) + g_1'(l + ct) = 0$$

$$\Rightarrow g_1'(l + ct) = \frac{F_0}{\rho A c}\delta(l - ct).$$

Defining $z = l + ct$, and writing $l - ct = 2l - z$ in the above, we have

$$g_1'(z) = \frac{F_0}{\rho A c}\delta(2l - z)]$$

$$\Rightarrow g_1(z) = -\frac{F_0}{\rho A c}\mathcal{H}(2l - z) + C_{\mathrm{I}}$$

$$\Rightarrow g_1(x + ct) = -\frac{F_0}{\rho A c}\mathcal{H}(2l - x - ct) + C_{\mathrm{I}},$$

where C_{I} is a constant of integration. From the condition $g_1(x + ct)|_{t=l/c} \equiv 0$, we have $C_{\mathrm{I}} = F_0/\rho A c$. Therefore, the complete solution after one reflection is given by

$$u(x,t) = \frac{F_0}{\rho A c}[2 - \mathcal{H}(x - ct) - \mathcal{H}(2l - x - ct)], \qquad l/c \leq t < 2l/c. \tag{2.79}$$

At $t = 2l/c$, the wave reaches the left end of the bar, and reflects again to produce a reflected wave denoted by, say $f_2(x - ct)$. Writing the motion of the bar as

$$u(x,t) = \frac{F_0}{\rho A c}[2 - \mathcal{H}(x - ct) - \mathcal{H}(2l - x - ct)] + f_2(x - ct), \qquad t > 2l/c, \tag{2.80}$$

and using the force–free boundary condition $EAu_{,x}(0, t) \equiv 0$ for $t \geq 2l/c$ yields

$$\frac{F_0}{\rho A c}[-\delta(-ct) + \delta(2l - ct)] + f_2'(-ct) = 0$$

$$\Rightarrow f_2'(-ct) = -\frac{F_0}{\rho A c}\delta(2l - ct).$$

Defining $z = -ct$, we have

$$f_2'(z) = -\frac{F_0}{\rho Ac}\delta(2l + z)$$

$$\Rightarrow f_2(z) = -\frac{F_0}{\rho Ac}\mathcal{H}(2l + z) + C_\text{I},$$

$$\Rightarrow f_2(x - ct) = -\frac{F_0}{\rho Ac}\mathcal{H}(2l + x - ct) + C_\text{I}. \tag{2.81}$$

Using the condition $f_2(x - ct)|_{t=2l/c} \equiv 0$ yields $C_\text{I} = F_0/\rho Ac$. Therefore, the complete solution after the second reflection is given by

$$u(x, t) = \frac{F_0}{\rho Ac}[3 - \mathcal{H}(x - ct) - \mathcal{H}(2l - x - ct) - \mathcal{H}(2l + x - ct)], \qquad 2l/c \leq t < 3l/c.$$

This solution procedure can be continued indefinitely. It may be observed that the bar continuously moves to the right. However, the ends of the bar do not move simultaneously.

The propagation of the displacement wave in the bar is shown in Figure 2.23. The motions of the left end and right end are shown in Figure 2.24 It is seen that the whole bar shifts a distance $F_0/\rho Ac$ in time l/c. The average velocity of the bar is, therefore, $F_0/\rho Al = F_0/m$, where m is the mass of the bar. This is also the velocity acquired by a rigid bar of mass m when acted upon by an impulsive force of strength F_0. Thus, the average motion of an elastic bar is identical to that of a rigid bar.

2.5.2 Step-forcing of a bar with boundary damping

Consider a finite bar with a damper on one end, and free at the other, as shown in Figure 2.25. At time $t = 0$, a constant force of magnitude F_0 is suddenly applied at the free end. The problem is to determine the subsequent motion of the bar.

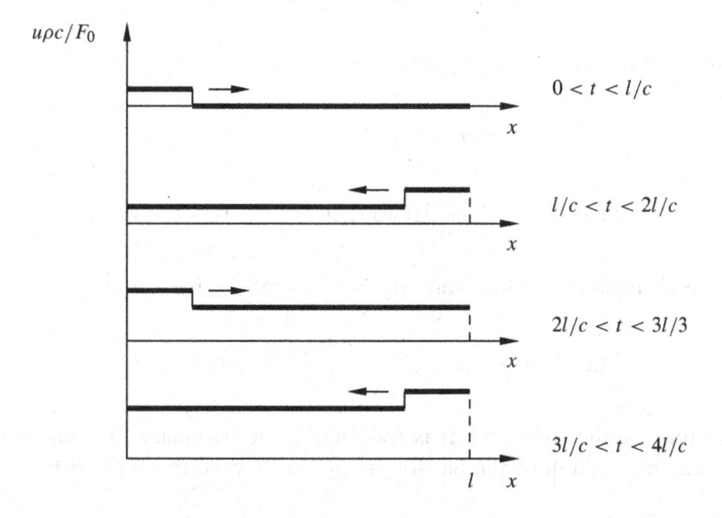

Figure 2.23 Propagation of displacement wave in a free–free bar started with an impulsive force

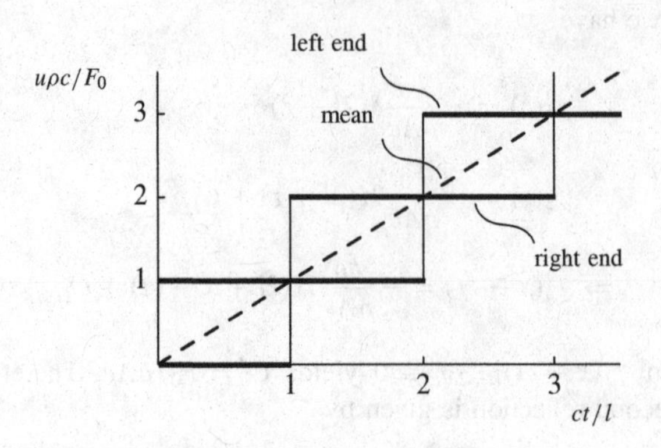

Figure 2.24 Motion of the ends of a free–free bar started with an impulse

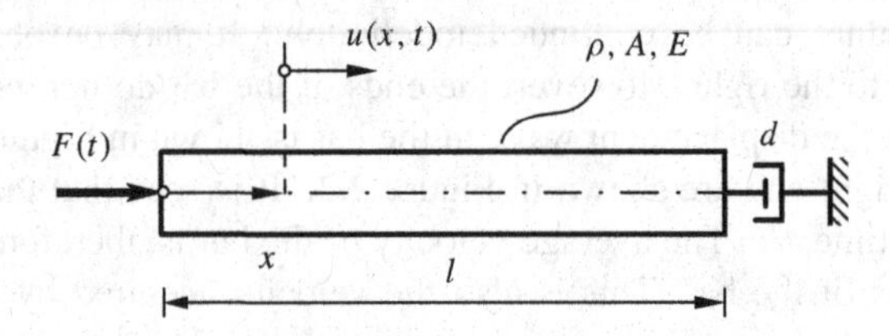

Figure 2.25 A finite bar with a discrete boundary damping

Let the displacement of the bar be represented as $u(x,t) = f_1(x - ct)$, $0 \leq t < l/c$, which is a positive-traveling wave. The boundary condition at the left end can be written as

$$-EAu_{,x}(0,t) = F_0\mathcal{H}(t)$$

$$\Rightarrow f_1'(-ct) = -\frac{F_0}{EA}\mathcal{H}(t) \tag{2.82}$$

Using the definition $z := -ct$ in (2.82) yields

$$f_1'(z) = -\frac{F_0}{EA}\mathcal{H}(-z/c) \tag{2.83}$$

$$\Rightarrow f_1'(x - ct) = -\frac{F_0}{EA}\mathcal{H}(-x/c + t), \qquad 0 \leq t < l/c, \tag{2.84}$$

where $\mathcal{H}(\cdot)$ is the Heaviside step function. The velocity of the bar can be represented as

$$u_{,t}(x,t) = -cf_1'(x - ct) = \frac{F_0 c}{EA}\mathcal{H}(-x/c + t). \tag{2.85}$$

When the wave reaches the right end, it is reflected at the boundary to create a negative-traveling wave. Let the motion of the bar for $l/c \leq t < 2l/c$ be represented by

$$u(x,t) = f_1(x - ct) + g_1(x + ct). \tag{2.86}$$

The boundary condition at the right end can be written as

$$EAu_{,x}(l,t) = -du_{,t}(l,t), \tag{2.87}$$

where d is the impedance of the damper. Using the solution form (2.86) in the boundary condition yields

$$EA[f_1'(l-ct) + g_1'(l+ct)] = -d[-cf_1'(l-ct) + cg_1'(l+ct)] \tag{2.88}$$

or

$$g_1'(l+ct) = -\left(\frac{EA-dc}{EA+dc}\right) f_1'(l-ct)$$

$$\Rightarrow g_1'(z) = -r\mathcal{H}(z/c - 2l/c) \qquad \text{(using (2.83))} \tag{2.89}$$

$$\Rightarrow g_1'(x+ct) = r\frac{F_0}{EA}\mathcal{H}(x/c + t - 2l/c), \tag{2.90}$$

where

$$r = \frac{EA-dc}{EA+dc} = \frac{\rho Ac - d}{\rho Ac + d}. \tag{2.91}$$

The velocity of the bar can then be represented as

$$u_{,t}(x,t) = \frac{F_0}{\rho Ac}[\mathcal{H}(-x/c+t) + r\mathcal{H}(x/c+t-2l/c)]$$

$$= \frac{F_0}{\rho Ac}[1 + r\mathcal{H}(x/c+t-2l/c)], \qquad l/c \le t < 2l/c. \tag{2.92}$$

For $t \ge 2l/c$, the wave reflects from the left boundary, and the solution can be written as

$$u(x,t) = f_1(x-ct) + g_1(x+ct) + f_2(x-ct). \tag{2.93}$$

Applying the boundary condition of the left end again yields

$$-EA[f_1'(-ct) + g_1'(ct) + f_2'(-ct)] = F_0\mathcal{H}(t)$$

$$\Rightarrow g_1'(ct) + f_2'(-ct) = 0 \qquad \text{(using (2.82))}$$

$$\Rightarrow f_2'(z) = -g_1'(-z) \tag{2.94}$$

$$\Rightarrow f_2'(x-ct) = -g_1'(-x+ct). \tag{2.95}$$

It can be easily inferred from (2.94) that, at the left boundary, all subsequent reflections can be computed using the recursive relation

$$f_{n+1}'(z) = -g_n'(-z), \tag{2.96}$$

where $g_1'(z)$ is given by (2.89). Considering the reflection from the right boundary for $t \geq 4l/c$, we can write the solution as

$$u(x,t) = f_1(x-ct) + g_1(x+ct) + f_2(x-ct) + g_2(x+ct). \tag{2.97}$$

Again using the boundary condition for the right end yields

$$\begin{aligned}
&EA[f_1'(l-ct) + g_1'(l+ct) + f_2'(l-ct) + g_2'(l+ct)] = \\
&-d[-cf_1'(l-ct) + cg_1'(l+ct) - cf_2'(l-ct) + cg_2(l+ct)] \\
&\Rightarrow g_2'(l+ct) = -rf_2'(l-ct) \qquad \text{(using (2.88) and (2.91))}, \\
&\Rightarrow g_2'(z) = -rf_2'(-z+2l) \\
&\Rightarrow g_2'(x+ct) = -rf_2'(-x-ct+2l).
\end{aligned} \tag{2.98}$$

It follows from (2.98) that all subsequent reflections from the right boundary can be computed from the recursion

$$g_n'(z) = rf_n'(-z+2l), \tag{2.99}$$

where $f_1'(z)$ is given by (2.83). Therefore, as $t \to \infty$, one can write the stress in the bar as

$$\begin{aligned}
\sigma_\infty &= \lim_{t\to\infty} Eu_{,x} = \lim_{n\to\infty} E[f_1' + g_1' + f_2' + g_2' + f_3' + g_3' + \ldots + f_n' + g_n'] \\
&= \lim_{n\to\infty} E[f_1' - rf_1' + rf_1' - r^2 f_1' + r^2 f_1' + r^3 f_1' + \ldots \\
&\quad + r^{(n-1)} f_1' - r^n f_1'] \\
&= \lim_{n\to\infty} E[f_1' - r^n f_1'] \\
&= Ef_1' \qquad (\text{if } |r| < 1) \\
&= -\frac{F_0}{A}.
\end{aligned} \tag{2.100}$$

Similarly, in the limit $t \to \infty$, the velocity can be determined as

$$\begin{aligned}
v_\infty &= \lim_{n\to\infty} [-cf_1' + cg_1' - cf_2' + cg_2' - cf_3' + cg_3' + \ldots \\
&\quad -cf_n' + cg_n'] \\
&= \lim_{n\to\infty} c[-f_1' - rf_1' - rf_1' - r^2 f_1' - r^2 f_1' - r^3 f_1' - \ldots \\
&\quad -r^{(n-1)} f_1' - r^n f_1'] \\
&= \lim_{n\to\infty} -cf_1'[1 + 2r + 2r^2 + 2r^3 + \ldots + 2r^{(n-1)} + r^n] \\
&= \lim_{n\to\infty} \frac{F_0 c}{EA}\left(2\frac{1-r^{(n+1)}}{1-r} - 1 - r^n\right) \qquad (2.101) \\
&= \frac{F_0 c}{EA}\left(\frac{1+r}{1-r}\right) \qquad (\text{if } |r| < 1) \\
&= \frac{F_0}{d} \qquad \text{(using (2.91))}.
\end{aligned} \tag{2.102}$$

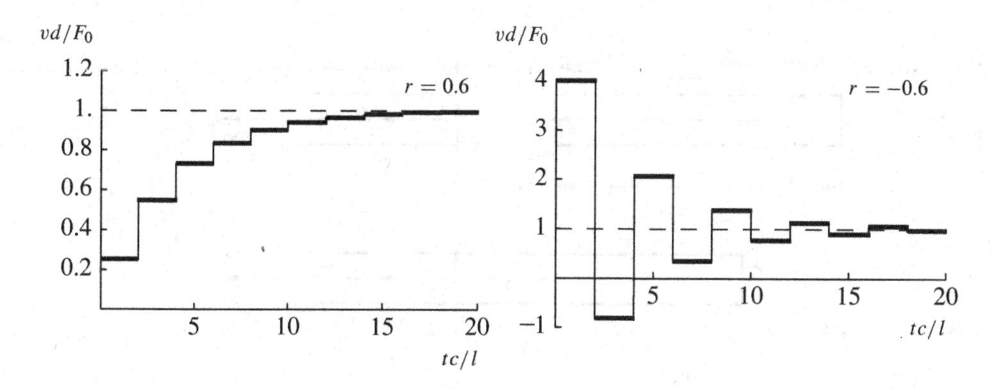

Figure 2.26 Velocity of the left end of a step-forced bar for two values of boundary damping

The velocity profiles are shown in Figure 2.26 for two values of r. It is observed that for $r > 0$, the convergence is monotonous, while for $r < 0$, the convergence is oscillatory. To show this, we rewrite (2.101) as

$$\begin{aligned} v_n &= \frac{F_0}{d}(1 - r^n), \\ &= \frac{F_0}{d}(1 - \mathrm{e}^{n \ln r}). \end{aligned} \tag{2.103}$$

For $r \in (0, 1)$, the exponent is always negative, indicating $v_\infty \to F_0/d$. For $r \in (-1, 0)$, one can rewrite (2.103) as

$$v_n = \frac{F_0}{d}(1 - \mathrm{e}^{n \ln |r|} \mathrm{e}^{in\pi}) = \frac{F_0}{d}[1 - (-1)^n \mathrm{e}^{n \ln |r|}].$$

This solution is clearly oscillatory, and converges in the limit $n \to \infty$ to $v_\infty = F_0/d$. A special case occurs for $dc = EA$ (i.e., $r = 0$), as already seen in Section 1.8.2. In this case the final velocity $v = F_0/d = F_0/A\sqrt{E\rho}$ is reached after time $t = l/c$.

2.5.3 Axial collision of bars

Consider a homogeneous uniform bar of length l, moving axially with a velocity v_0, colliding at $t = 0$ with a semi-infinite bar at rest, as shown in Figure 2.27. Let the density, Young's modulus, and area of cross-section of the moving bar be, respectively, ρ, E, and A, while the (real) impedance of the semi-infinite bar is denoted by $\mathcal{Z}$. The motion of the bar just after collision can be written as

$$u(x, t) = v_0 t + g(x + ct), \tag{2.104}$$

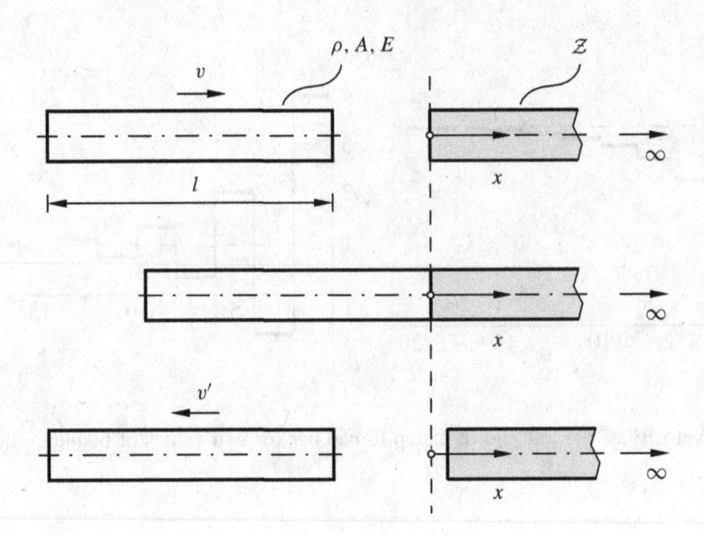

Figure 2.27 Collision of a bar of finite length with a semi-infinite bar

where $g(x+ct)$ is a left-traveling wave created as a result of the collision. The boundary condition at the right end can be represented as

$$EAu_{,x}(0,t) = -\mathcal{Z}u_{,t}(0,t).$$

Using the solution (2.104) in the boundary condition yields

$$\begin{aligned} EAg'(ct) &= -\mathcal{Z}[v_0 + cg'(ct)] \\ \Rightarrow g'(ct) &= \frac{-\mathcal{Z}v_0}{EA + \mathcal{Z}c} \\ \Rightarrow g'(z) &= \frac{-\mathcal{Z}v_0}{EA + \mathcal{Z}c}\mathcal{H}(z) \\ \Rightarrow g'(x+ct) &= \frac{-\mathcal{Z}v_0}{EA + \mathcal{Z}c}\mathcal{H}(x+ct), \end{aligned} \tag{2.105}$$

where the Heaviside step function $\mathcal{H}(\cdot)$ is introduced to enforce the causality condition $g'(z) \equiv 0$ for $z \leq 0$. At $t = l/c$, the left-traveling wave reflects from the left boundary, and the subsequent motion can be represented by

$$u(x,t) = v_0 t + g(x+ct) + f(x-ct). \tag{2.106}$$

The force–free boundary condition at the left boundary can be represented as

$$EAu_{,x}(-l,t) = 0.$$

Using the solution form (2.106) in this boundary condition yields

$$\begin{aligned} g'(-l+ct) + f'(-l-ct) &= 0 \\ \Rightarrow f'(z) &= -g'(-z-2l) \end{aligned}$$

$$\Rightarrow f'(x-ct) = -g'(-x+ct-2l)$$
$$\Rightarrow f'(x-ct) = \frac{\mathcal{Z}v_0}{EA+\mathcal{Z}c}\mathcal{H}(-x+ct-2l). \tag{2.107}$$

The velocity of the bar after the reflection from the left end can be written as

$$\begin{aligned} u_{,t}(x,t) &= v_0 + cg'(x+ct) - cf'(x-ct) \\ &= v_0 - \frac{\mathcal{Z}v_0c}{EA+\mathcal{Z}c}\mathcal{H}(x+ct) - \frac{\mathcal{Z}v_0c}{EA+\mathcal{Z}c}\mathcal{H}(-x+ct-2l). \end{aligned} \tag{2.108}$$

Thus, at $t = 2l/c$ the velocity is given by

$$\begin{aligned} u_{,t}(x,2l/c) = v &= v_0 - \frac{\mathcal{Z}v_0c}{EA+\mathcal{Z}c} - \frac{\mathcal{Z}v_0c}{EA+\mathcal{Z}c} \\ &= \frac{EA-\mathcal{Z}c}{EA+\mathcal{Z}c}v_0 = \frac{\rho Ac-\mathcal{Z}}{\rho Ac+\mathcal{Z}}v_0 = rv_0, \end{aligned} \tag{2.109}$$

where $r = (\rho Ac - \mathcal{Z})/(\rho Ac + \mathcal{Z})$. The stress in the bar at $t = 2l/c$ can be calculated as

$$\sigma(x,2l/c) = Eu_{,x}(x,2l/c) = E[g'(x+ct) + f'(x-ct)]\big|_{t=2l/c} \equiv 0. \tag{2.110}$$

Thus, the finite bar is completely stress-free at $t = 2l/c$.

Since the contact between the two bars is unilateral, there can be various interesting situations just after $t = 2l/c$, depending on the value of $\mathcal{Z}$. When $\mathcal{Z} = \infty$, i.e. the boundary is rigid, the final velocity of the finite bar after collision is obtained from (2.109) as $v_c = -v_0$. Thus, the bar rebounds elastically, and there is no loss of energy. In the case of $\mathcal{Z}$ being a finite real impedance, there can be three cases, namely, $\mathcal{Z} > \rho Ac$, $\mathcal{Z} = \rho Ac$, and $\mathcal{Z} < \rho Ac$. In the first case (i.e., $\mathcal{Z} > \rho Ac$), there is a rebound of the bar, although with a velocity magnitude less than v_0, which can be computed from (2.109). The fraction of the energy lost by the bar can be computed as

$$\frac{\Delta\mathcal{E}}{\mathcal{E}} = \frac{\frac{1}{2}\rho l(rv_0)^2 - \frac{1}{2}\rho lv_0^2}{\frac{1}{2}\rho lv_0^2} = r^2 - 1. \tag{2.111}$$

Since $|r| < 1$ for a finite $\mathcal{Z} > \rho Ac$, energy is always lost by the colliding bar. In the second case $r = 0$, and the finite bar comes to a complete stop. Thus, it loses its energy completely to the semi-infinite bar. Since the contact force at the interface vanishes, the end of the semi-infinite bar also stops. However, the two bars remain in contact. In the case $\mathcal{Z} < \rho Ac$, there is a velocity discontinuity at $t = 2l/c$. At this time, the colliding bar has a velocity $v_c = rv_0 > 0$, while the semi-infinite bar end is moving with $v_s = EAv_0/(EA + \mathcal{Z}c)$, as can be easily obtained from (2.108). As soon as the velocity discontinuity occurs, the interface force vanishes, and the end of the semi-infinite bar also stops. However, the finite bar is still moving in the positive direction with v_c. Thus, the problem now reduces to that of a finite bar traveling with an initial velocity v_c, and colliding with a stationary semi-infinite bar. It is not difficult to conclude that, at the second collision at $t = 4l/c$, the velocity of the finite bar is obtained as $v_c^{(2)} = r^2 v_0$. Generalizing, we have at $t = 2nl/c$,

$$v_c^{(n)} = r^n v_0 = e^{n \ln r} v_0$$

$$\Rightarrow v_c(t) = e^{\frac{tc}{2l} \ln r} v_0. \tag{2.112}$$

Since $0 < r < 1$, the velocity of the colliding bar decays exponentially, and as $t \to \infty$, it loses its energy completely.

2.5.4 String on a compliant foundation

Let us consider next an example of wave propagation in a dispersive medium. Consider a stretched infinite string on a compliant foundation which is modeled as a distributed stiffness, as shown in Figure 2.28. The equations of motion can be easily written as

$$\rho A w_{,tt} - T w_{,xx} + \kappa w = 0, \tag{2.113}$$

where κ is the stiffness per unit length of the foundation. Assuming a traveling wave solution $w(x, t) = Be^{i(kx - \omega t)}$ for (2.113), we obtain the dispersion relation

$$-\omega^2 \rho A + k^2 T + \kappa = 0, \tag{2.114}$$

as shown in Figure 2.29. The phase velocity of the wave is then obtained as (see Appendix B)

$$c_P = \frac{\omega}{k} = \sqrt{c^2 + \frac{\kappa}{\rho A k^2}}, \tag{2.115}$$

where $c = \sqrt{T/\rho A}$ is the wave speed in a normal taut string. The group velocity is given by

$$c_G = \frac{d\omega}{dk} = \frac{Tk}{\rho A \omega} = \frac{c^2}{c_P}. \tag{2.116}$$

Thus, c_P and c_G are both functions of the wave number k, and are shown in Figure 2.30. It may be noted from (2.116) that $c_P c_G = c^2$. This relation also holds for the wave equation.

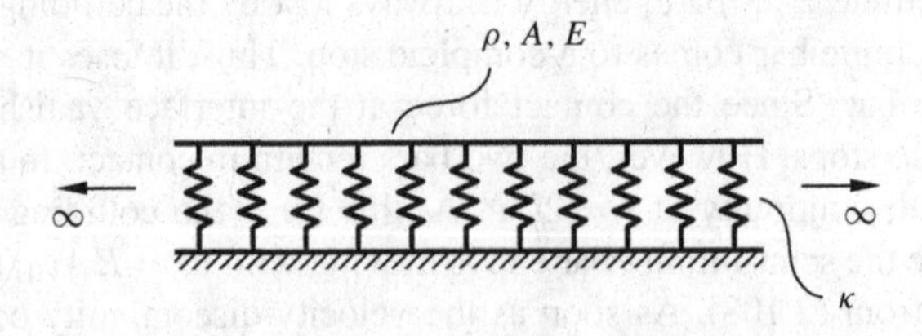

Figure 2.28 An infinite stretched string on a compliant foundation

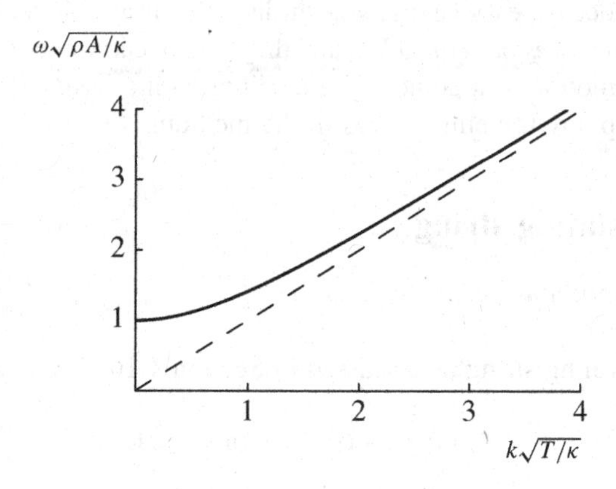

Figure 2.29 Dispersion relation of an infinite stretched string on a compliant foundation

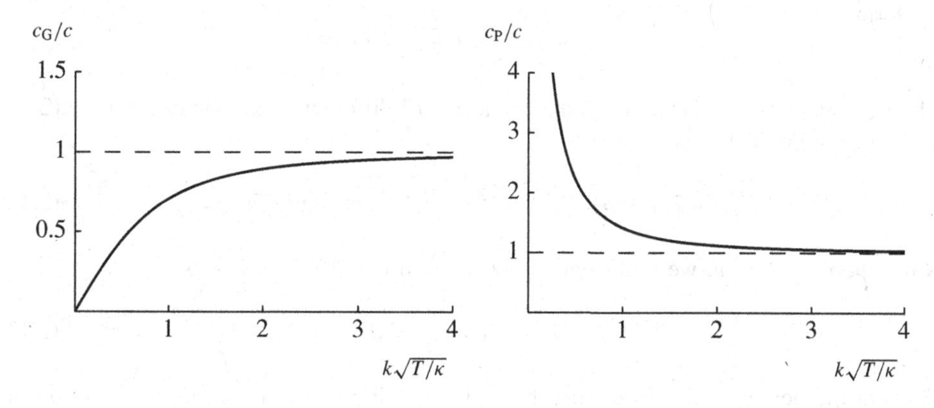

Figure 2.30 Variation of phase velocity (c_P) and group velocity (c_G) with wave number for a string on a compliant foundation

Further, as $k \to \infty$, both c_P and c_G tend to the taut string wave speed c. For a given frequency ω, the solution of k is obtained from (2.114) as

$$k = \pm\frac{\omega}{c}\sqrt{1 - \frac{\kappa}{\rho A\omega^2}}.$$

It is interesting to observe that, for values of $\omega < \omega_c := \sqrt{\kappa/\rho A}$, where ω_c is known as the *cut-off frequency*, the wave number becomes complex, say $k = \pm i\tilde{k}$. The wave solution is then of the form

$$w(x,t) = B_+ e^{-\tilde{k}x - i\omega t} + B_- e^{\tilde{k}x - i\omega t}, \tag{2.117}$$

where the first term comes from the positive-traveling wave component, and the second term corresponds to the negative-traveling wave component. It is evident from (2.117) that the solution amplitude decays exponentially with distance in any direction. Such solutions are known as *evanescent waves* or *near-fields*, and they do not carry energy (see, for example, [3]). Thus, a harmonic motion of a point of the medium with a frequency below the cut-off frequency will not setup any traveling waves in the medium.

2.5.5 Axially translating string

2.5.5.1 D'Alembert's solution

Consider an axially traveling string as discussed in Section 1.10. The equation of motion is given by

$$w_{,tt} + 2vw_{,xt} - (c^2 - v^2)w_{,xx} = 0, \tag{2.118}$$

where v is the constant travel speed of the string. Assuming the string to be of infinite length, we do not have any boundary conditions. Consider a coordinate transformation of the form

$$\xi = t, \qquad \eta = x - \beta t, \tag{2.119}$$

where β is a constant. Then, using the chain rule of differentiation, one can rewrite (2.118) in the new coordinates as

$$\tilde{w}_{,\xi\xi} + 2(v - \beta)\tilde{w}_{,\eta\xi} + [\beta^2 - 2v\beta - (c^2 - v^2)]\tilde{w}_{,\eta\eta} = 0. \tag{2.120}$$

Now, choosing $\beta = v$, we immediately obtain from (2.120)

$$\tilde{w}_{,\xi\xi} - c^2\tilde{w}_{,\eta\eta} = 0. \tag{2.121}$$

Thus, in the new coordinates defined by (2.119) with $\beta = v$, the equation of motion of a traveling string transforms to a wave equation. Now, one can write the general solution of (2.121) as

$$\begin{aligned} \tilde{w}(\xi, \eta) &= f(\eta - c\xi) + g(\eta + c\xi) \\ \Rightarrow w(x, t) &= f[x - (c + v)t] + g[x + (c - v)t]. \end{aligned} \tag{2.122}$$

The solution (2.122) is a superposition of two traveling waves, one with speed $c + v$ in the direction of travel of the string (the forward traveling wave), and the other in the opposite direction with speed $c - v$ (the backward traveling wave). It may be observed that when $v = c$, the backward traveling wave becomes stationary in the fixed frame. Thus, there are no backward propagating waves for travel speeds beyond the critical speed of the string.

Now consider the initial value problem for a traveling string with initial configuration $w(x, 0) = w_0(x)$ and initial velocity $w_{,t}(x, 0) = v_0(x)$. Then, using (2.122), one can write

$$f(x) + g(x) = w_0(x)$$

and

$$-(c + v)f'(x) + (c - v)g'(x) = v_0(x).$$

Proceeding similarly to what was done in Section 2.1.1, one can obtain the final solution of the initial value problem for a traveling string as

$$w(x,t) = \frac{1}{2c}\Bigg[(c-v)w_0(x-(c+v)t) + (c+v)w_0(x+(c-v)t)$$
$$+ \int_{x-(c+v)t}^{x+(c-v)t} v_0(\xi)\,\mathrm{d}\xi\Bigg]. \tag{2.123}$$

As an example, consider a string translating with speed v in the positive x-axis direction. Assume an initial velocity distribution $v_0(x) = V_0\delta(x)$ for the initially undisturbed string. Then, the subsequent motion of the string is obtained from (2.123) as

$$w(x,t) = \frac{V_0}{2c}[\mathcal{H}(x+(c-v)t) - \mathcal{H}(x-(c+v)t)],$$

where $\mathcal{H}(\cdot)$ is the Heaviside step function. This solution is shown is Figure 2.31 for three possible velocity regimes, namely $v < c$, $v = c$, and $v > c$. These three velocity regimes show distinctive behavior. For $v < c$, the disturbance spreads in the whole string, while for $v = c$, only the downstream side of $x = 0$ is disturbed. On the other hand, for $v > c$, the disturbance is completely convected away downstream.

2.5.5.2 *Energetics of waves*

Consider the total mechanical energy of a section of a traveling string between the coordinates $[x_1, x_2]$ given by

$$\mathcal{E} = \frac{1}{2}\int_{x_1}^{x_2} [\rho A(w_{,t} + vw_{,x})^2 + Tw_{,x}^2]\,\mathrm{d}x. \tag{2.124}$$

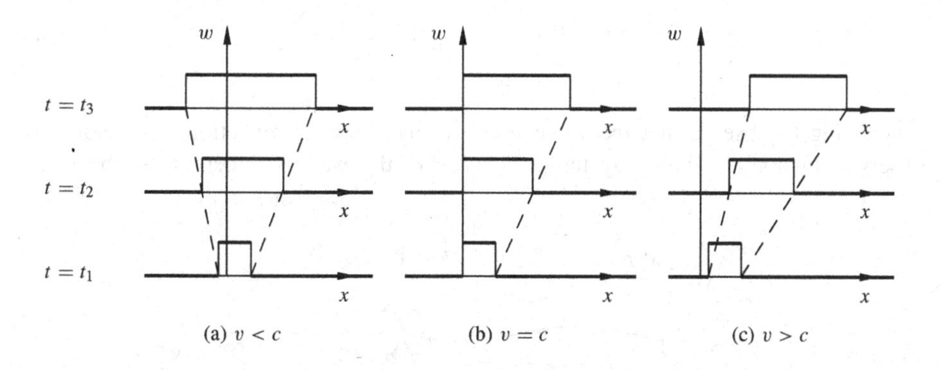

Figure 2.31 Solution of the initial value problem for an axially translating string for three travel speed regimes

Recalling that $\mathrm{d}(\cdot)/\mathrm{d}t = \partial(\cdot)/\partial t + v\partial(\cdot)/\partial x$, one can write

$$\begin{aligned}\frac{\mathrm{d}\mathcal{E}}{\mathrm{d}t} &= \int_{x_1}^{x_2} [\rho A(w_{,t} + vw_{,x})(w_{,tt} + 2vw_{,xt} + v^2 w_{,xx}) \\ &\quad + Tw_{,x}(w_{,xt} + vw_{,xx})]\,\mathrm{d}x \\ &= Tw_{,x}(w_{,t} + vw_{,x})\Big|_{x_1}^{x_2} \quad \text{(using (2.118))}. \end{aligned} \tag{2.125}$$

Similarly as in Section 2.3, one can define the instantaneous power flowing past a coordinate location x as

$$\mathcal{P}(x,t) := -Tw_{,x}(w_{,t} + vw_{,x}). \tag{2.126}$$

Using this expression, one can now calculate the energy flux of traveling waves on a translating string. It may be recalled from the discussion in Section 2.3 that the sign of $\mathcal{P}(x,t)$ gives the direction of flow of power.

2.5.5.3 Reflection of waves

Consider a traveling string passing through a rigid support point at $x = 0$, as shown in Figure 2.32. Let us take a positive-traveling harmonic wave on the string to the left of $x = 0$, and study the reflection process at the support point. The total wave field in the region $x < 0$ is given by

$$w(x,t) = A\mathrm{e}^{ik[x-(c+v)t]} + B\mathrm{e}^{ik'[-x-(c-v)t]}, \tag{2.127}$$

where k and k' are, respectively, the wave numbers of the incident and reflected waves. The support condition $w(0,t) \equiv 0$ leads to

$$\begin{aligned} &A\mathrm{e}^{-ik(c+v)t} + B\mathrm{e}^{-ik'(c-v)t} = 0 \\ &\Rightarrow k' = \left(\frac{c+v}{c-v}\right)k, \quad \text{and} \quad B = -A. \end{aligned} \tag{2.128}$$

It is interesting to observe that the wave number changes after reflection, a phenomenon not observed in non-translating systems. For $v < c$, the wave number of the backward

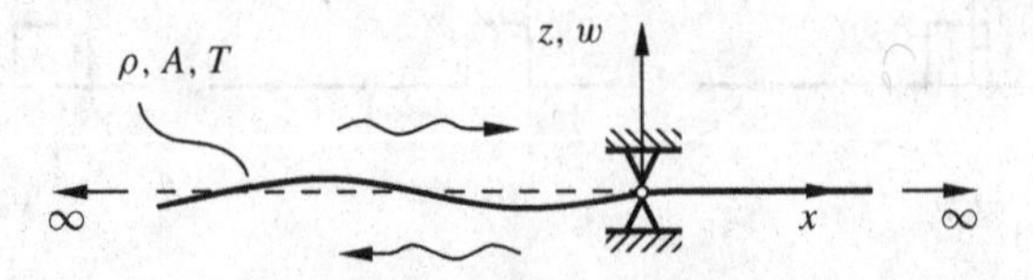

Figure 2.32 Reflection of waves in a traveling string from a fixed support

wave increases. Travel speeds $v \geq c$ lead to a complex phenomenon due to accumulation of the backward wave. Such a situation will require non-linear analysis, and will not be discussed here.

Next, we analyze the energetics of the reflection process. Taking the real (or imaginary) part of the incident and reflected waves, one can use (2.126) to calculate the instantaneous power of the individual waves. The average incident and reflected power can then be obtained as, respectively,

$$\langle \mathcal{P}_{\mathrm{I}} \rangle = \frac{1}{2} k^2 T c A^2 \quad \text{and} \quad \langle \mathcal{P}_{\mathrm{R}} \rangle = -\frac{1}{2} \left(\frac{c+v}{c-v} \right)^2 k^2 T c A^2. \tag{2.129}$$

It is evident that the magnitude of the reflected power is higher than the incident power. This observation leads to the conclusion that the wave reflection process in traveling strings from a fixed support is a non-conservative process. The reflected wave draws the extra energy from the axial motion of the string.

The above conclusions can be used to understand the dynamics of a translating string between two supports (see Figure 1.27). When a small disturbance reflects from the right support, the reflected (backward) wave gains some extra energy. For $v < c$, the backward wave can travel up to the left support and lose the extra energy on reflection at the left support (see Exercise 2.15). Hence, for sub-critical travel speeds, the energy balance of the string between the supports is maintained. However, for $v \geq c$, the reflected (backward) wave cannot propagate, and hence, the extra energy cannot be shed. This may result in a continuous build-up of energy in the string, leading to instability.

EXERCISES

2.1 Show by differentiations that the solution

$$w(x,t) = \frac{1}{2} \left[w_0(x-ct) + w_0(x+ct) + \frac{1}{c} \int_{x-ct}^{x+ct} v_0(\xi) \mathrm{d}\xi \right]$$

satisfies the wave equation $w_{,tt} - c^2 w_{,xx} = 0$, and the initial conditions $w(x,0) = w_0(x)$ and $w_{,t}(x,0) = v_0(x)$.

2.2 A semi-infinite taut string of density ρ, and cross-sectional area A, and under a tension T, as shown in Figure 2.9, is harmonically forced by an actuator at the sliding end $x = 0$. If the actuator can deliver an average power of $\overline{P}$, determine the amplitude of harmonic waves that can be produced in the string in terms of the properties of the string and the frequency.

2.3 A semi-infinite string is connected at the boundary $x = 0$ to a spring–mass system as shown in Figure 2.33. At $t = 0$, a waveform given by

$$f(\xi) = \begin{cases} A\left(1 - \cos \frac{2\pi\xi}{l}\right), & \xi \in [0, l] \\ 0, & \xi \geq l \end{cases}$$

is incident on the boundary. Taking the mass m and stiffness k, determine the reflected wave in the case of no resonance, and when there is resonance. How is the resonance condition identified?

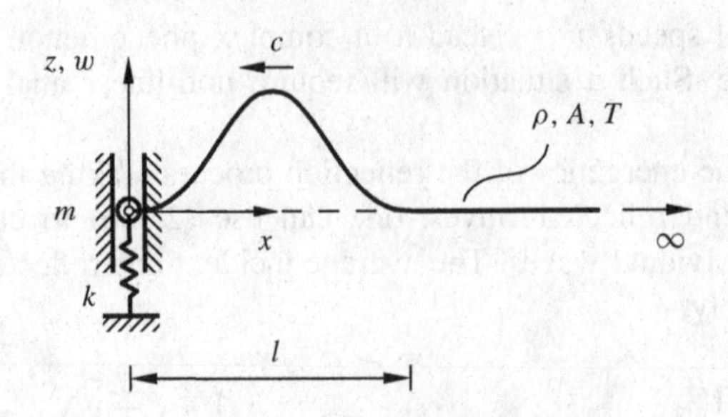

Figure 2.33 Exercise 2.3

2.4 A mass–spring system is connected to a string of length l, density ρ, and tension T, as shown in Figure 2.34. The other end of the string is attached to a damper of damping coefficient d.

(a) Determine the impedance of the string–damper system as seen by the mass-spring system.

(b) If the mass is excited by a given harmonic force, determine the added mass and damping on the mass–spring system due to the string–damper arrangement. For values of $d/\rho c = 0.5$, 1.0 and 1.5, plot the added mass and damping as a function of the circular frequency Ω. Explain the high value of added damping at certain excitation frequencies when $d/\rho c \ll 1$.

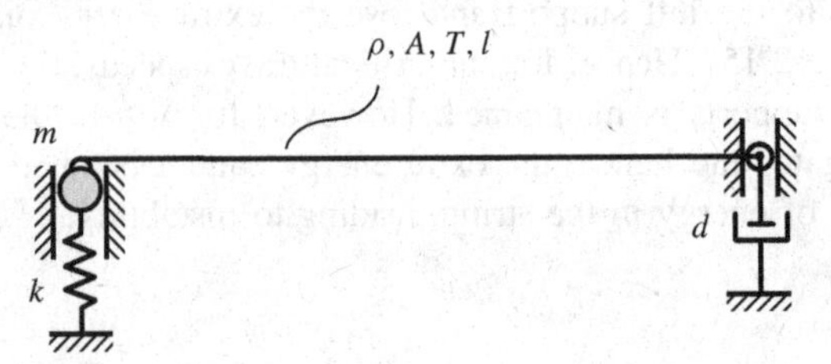

Figure 2.34 Exercise 2.4

2.5 Two semi-infinite bars of the same diameter but different materials are joined by a film of a third material as shown in Figure 2.35. A positive-traveling harmonic wave of circular frequency ω in the left bar is incident at the junction $x = 0$. Determine the thickness of the film required to ensure that no wave is reflected back into the left bar from the junction. Physically interpret the no-reflection condition. Show that, if this condition is achieved, the incident energy is completely transmitted to the right bar. For a particular film thickness, plot the fraction of the incident energy reflected into the left bar as a function of the frequency of the incident harmonic wave.

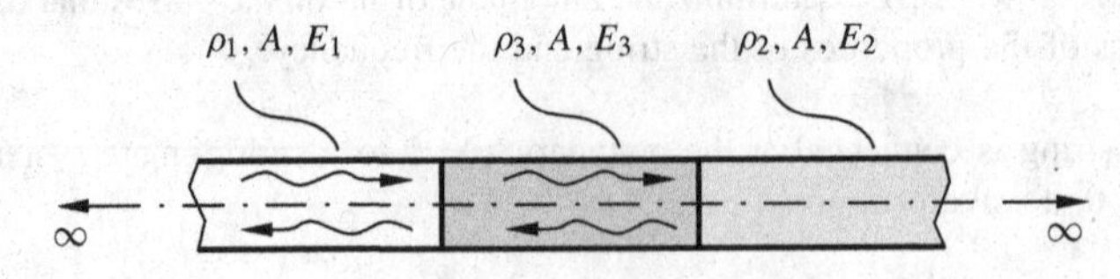

Figure 2.35 Exercise 2.5

2.6 Two semi-infinite bars of different materials and diameters are joined as shown in Figure 2.36. A positive-traveling longitudinal wave $f_1(x - c_1 t)$ in the left bar is incident on the junction at $x = 0$,

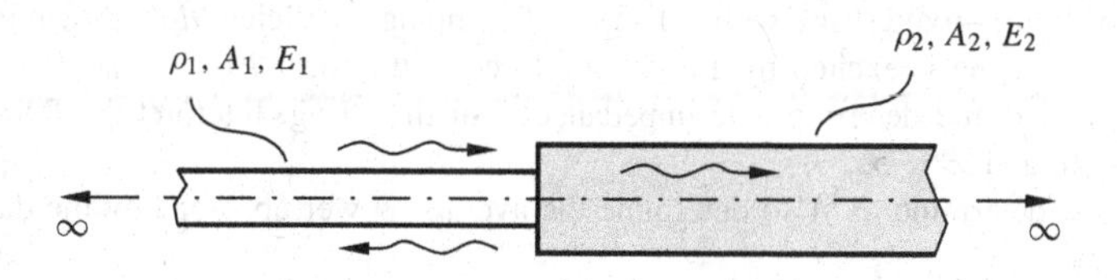

Figure 2.36 Exercise 2.6

and is partially reflected in the form $f_R(x + c_1t)$ and partially transmitted to the right bar in the form $f_T(x - c_2t)$. Determine the waveforms $f_R(\xi)$ and $f_T(\xi)$ in terms of $f_I(\xi)$.

2.7 Two identically tensioned cables (modeled as strings) are connected by a spacer–damper as shown in Figure 2.37.

(a) In the first case, consider that positive-traveling harmonic waves of the same phase are incident from the left of the spacer–damper in both cables. Determine the values of m, k, and d so that maximum vibration energy is dissipated through the dampers.

(b) In the second case, consider that positive-traveling harmonic waves of opposite phase are incident from the left of the spacer–damper in both cables. Solve for the values of m, k, and d for maximum dissipation of vibration energy.

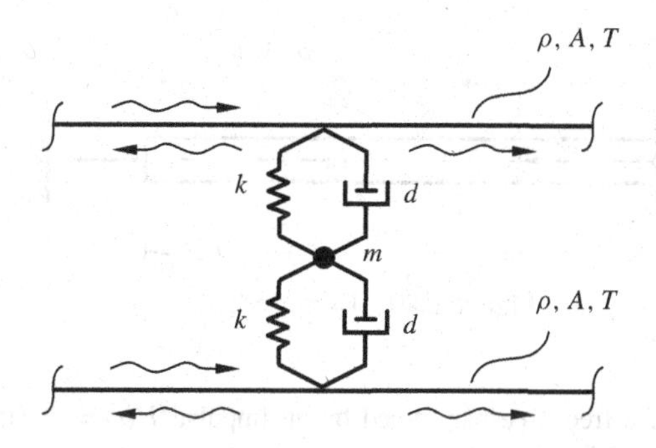

Figure 2.37 Exercise 2.7

2.8 A Stockbridge damper, used for damping the vibrations of high-voltage conductors (modeled as strings), is attached as shown in two cases (a) and (b) in Figure 2.38. For both cases shown in the figure, determine the optimal impedance $\mathcal{Z}$ of the Stockbridge damper as a function of the point of attachment l of the damper, and the string parameters so that a negative-traveling harmonic wave from the right is not reflected back.

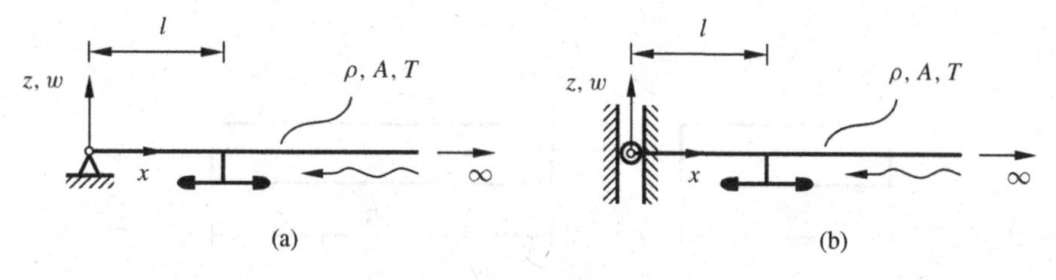

Figure 2.38 Exercise 2.8

2.9 Consider the sliding–fixed string with a damper of damping coefficient d as shown in Figure 2.39. The free end of the string is excited by a harmonic force of the form $F(t) = F_0 e^{-i\Omega t}$.

(a) First, take $d = 0$ and determine the impedance $\mathcal{Z}$ of the string. Interpret the frequencies Ω for which $\mathcal{Z} = 0$, and $\mathcal{Z} = \infty$.

(b) When $d \neq 0$, determine $\mathcal{Z}$. Also determine the average power absorbed by the damper.

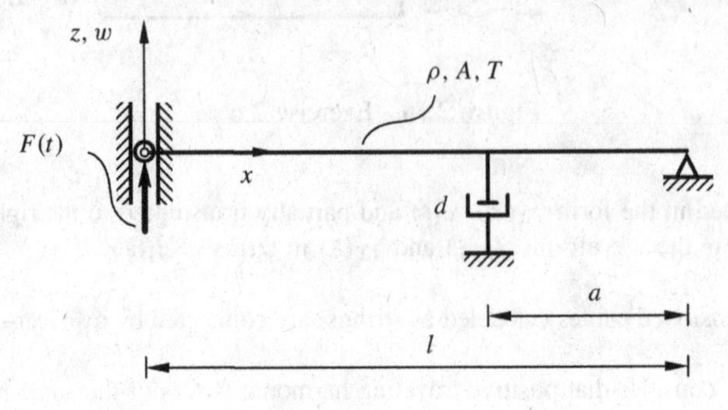

Figure 2.39 Exercise 2.9

2.10 A homogeneous uniform bar is kept under tension by a string as shown in Figure 2.40. If the string suddenly snaps, determine the transient motion and stress in the bar in terms of the propagating waves.

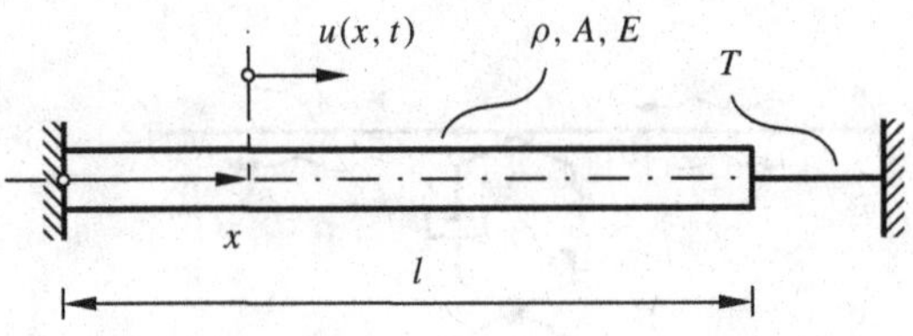

Figure 2.40 Exercise 2.10

2.11 Rework the motion of a free–free bar forced by an impulse $F(t) = F_0\delta(t)$ at one end starting with the boundary condition (2.76).

2.12 A homogeneous uniform bar of length l_1, moving axially with a velocity v_0 to the right, hits a similar stationary bar of length $l_2 > l_1$ as shown in Figure 2.41.

(a) At the time instants $t_k = kl_1/2c$, where $k = 0, 1, 2, 3, 4$, determine the velocity and normal stress in both the bars.

(b) Show that for time $t > 2l_1/c$, the bar of length l_1 becomes stationary. What is the velocity of the center of mass of both the bars taken together.

(c) What is the coefficient of restitution e between the two bars?

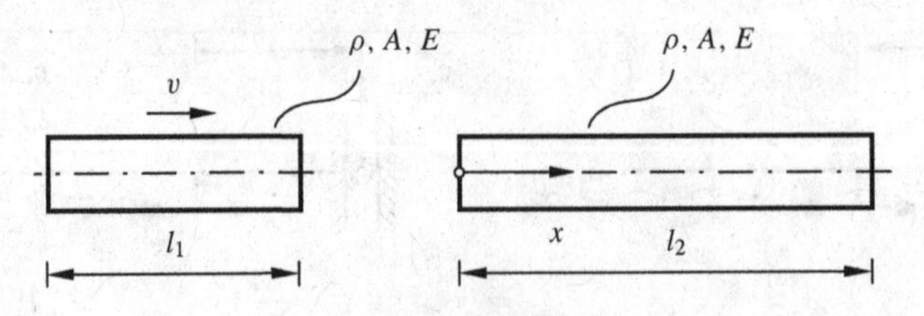

Figure 2.41 Exercise 2.12

2.13 A uniform homogeneous bar of length l, density ρ, and section-modulus EA is acted upon by an axial force of the form $F(t) = F_0[\mathcal{H}(t) - \mathcal{H}(t - \tau)]$, where τ is a constant, as shown in Figure 2.42. Determine the motion of the bar.

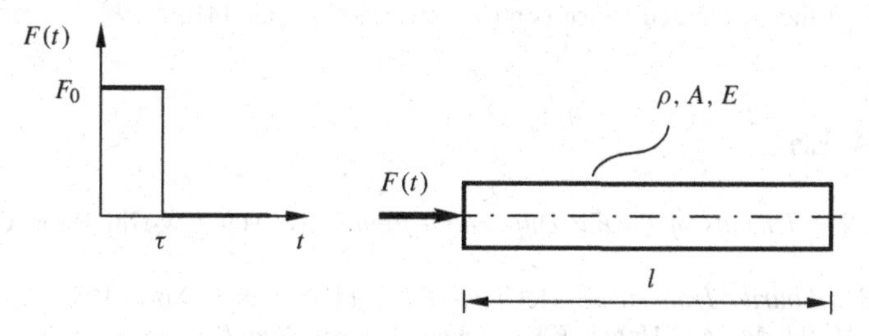

Figure 2.42 Exercise 2.13

2.14 Using d'Alembert's solution (2.122), determine the traveling wave solution of the initial value problem for a traveling string with $w(x, 0) \equiv 0$, and $w_{,t}(x, 0) = V_0[\mathcal{H}(x) - \mathcal{H}(x - a)]$, where $\mathcal{H}(\cdot)$ is the Heaviside step function, and a is a positive constant.

2.15 Consider a string traveling between two supports at a speed v, as shown in Figure 1.27. A triangular waveform

$$y = \begin{cases} ax, & 0 \le x \le a \\ 0, & a < x < 0 \end{cases}$$

is incident from the left on the right support. Determine the reflected (backward-traveling) waveform. Assuming $v < c$, where c is the wave speed, determine the reflected wave when the backward-traveling wave is reflected at the left boundary.

2.16 A traveling string of infinite extent is supported frictionlessly by a spring of stiffness k, as shown in Figure 2.43. A positive-traveling harmonic wave is incident from the left. Determine the transmitted and reflected waves in the string. Also determine the power carried by the transmitted and reflected waves.

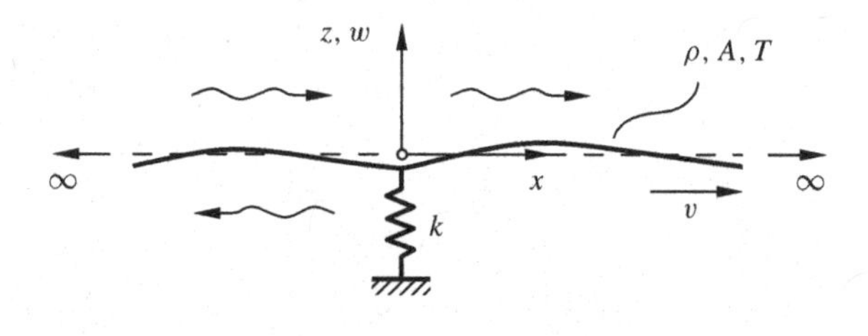

Figure 2.43 Exercise 2.16

2.17 Using the expression (2.126), show that the rate of change of total mechanical energy of a traveling string between two fixed supports at $x = 0$ and $x = l$ is given by $\mathrm{d}\mathcal{E}/\mathrm{d}t = Tv[w_{,x}^2(l, t) - w_{,x}^2(0, t)]$. Using the solution for the nth mode from (1.256), determine $\mathrm{d}\mathcal{E}/\mathrm{d}t$ for the nth mode.

2.18 For an axially translating string with a sliding boundary at $x = l$, show that the rate of change of total mechanical energy is given by $\mathrm{d}\mathcal{E}/\mathrm{d}t = Tv[w_{,x}^2(l,t) - w_{,x}^2(0,t)] + Tw_{,x}(l,t)\dot{h}(t)$, where $h(t)$ is the prescribed motion of the sliding end. Investigate how one can choose an appropriate $h(t)$ so as to dissipate energy of the string, and hence control its vibrations (see [4]).

REFERENCES

[1] Sneddon, I.N., *Elements of Partial Differential Equations*, McGraw-Hill Book Co., Singapore, 1957.
[2] Sneddon, I.N., *Fourier Transforms*, McGraw-Hill Book Co., New York, 1951.
[3] Cremer, L., Heckl, M., and Ungar, E.E., *Structure-borne Sound*, Springer-Verlag, Berlin, 1973.
[4] Lee, S. Y., and Mote, Jr., C.D., Vibration Control of an Axially Moving String by Boundary Control, *J. of Dynamic Systems, Measurement, and Control, Trans. of ASME*, 118, 1996, pp. 66–74.

3

Vibrations of beams

In the previous chapters, we have modeled a string as a one-dimensional elastic continuum that does not transmit or resist bending moment. However, such an assumption is seldom satisfied by any elastic continuum. In this chapter, we will consider planar transverse vibrations of one-dimensional elastic continua known as *beams*, which transmit or resist not only bending moment but also shear. Some simple beam models will be discussed, and their solutions under various conditions will be studied.

3.1 EQUATION OF MOTION

In this section, we discuss a beam theory which considers only the effect of bending moment on the dynamics of the beam. Thus, it is required that the shear forces be small so that the shear deformation of the beam is negligible. Another way of stating this assumption is to say that the beam is almost infinitely stiff in shear. The effect of shear deformation on the transverse dynamics of a beam will be considered separately later in this chapter.

3.1.1 The Newtonian formulation

Consider a straight beam undergoing a planar deflection in uni-axial bending as represented schematically in Figure 3.1. The simplest of all beam theories starts with the assumption that planar cross-sections of the undeformed beam remain planar even after the beam undergoes a deformation, as illustrated in the figure. From elementary theory of elasticity (see, for example, [1]), it is known that when the beam is deflected, certain hypothetical longitudinal lines or *fibers* are elongated, while others are compressed. There, however, exist fibers which are neither elongated nor compressed, but are merely deflected. Such a fiber is called the *neutral fiber*, and is shown in Figure 3.1. It is assumed that cross-sections orthogonal to the neutral fiber before deformation are also orthogonal to the neutral fiber in the deformed beam. These assumptions are referred to as the *Euler–Bernoulli hypotheses*. The Euler–Bernoulli hypotheses hold good as long as the ratio of the height of the beam h to the radius of curvature $\rho(x, t)$ of the neutral fiber after deformation is much smaller than unity. In terms of the forces, the assumptions remain meaningful for small bending moment gradient (i.e., for small shear) along the length of the beam. With this assumption, and referring to Figure 3.2,

Vibrations and Waves in Continuous Mechanical Systems P. Hagedorn and A. DasGupta

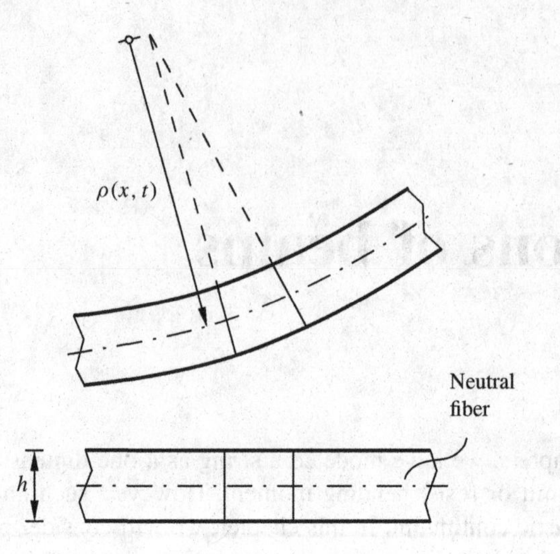

Figure 3.1 Schematic representation of a beam under planar deflection

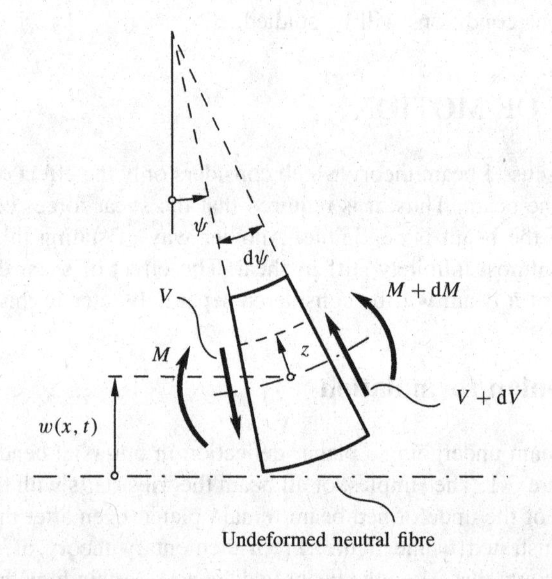

Figure 3.2 Infinitesimal element of a deflected beam

the strain–displacement relation at any height z measured from the plane of the neutral fibers, can be written from the theory of elasticity as

$$\begin{aligned}\epsilon_x(x, z, t) &= \frac{(\rho(x, t) - z)\,\mathrm{d}\psi - \rho(x, t)\,\mathrm{d}\psi}{\rho(x, t)\,\mathrm{d}\psi} = -\frac{z}{\rho(x, t)} \\ &= -\frac{zw_{,xx}(x, t)}{[1 + w_{,x}^2(x, t)]^{3/2}} \\ &\approx -zw_{,xx}(x, t) \qquad \text{(assuming } w_{,x} \ll 1\text{)}, \end{aligned} \tag{3.1}$$

where $w(x, t)$ is the transverse deflection field. Next, the constitutive relation for a linearly elastic material can be written from Hooke's law as

$$\sigma_x(x, z, t) = E\epsilon_x(x, z, t) = -Ezw_{,xx}(x, t), \tag{3.2}$$

where E is Young's modulus. The bending moment at any section can then be written as

$$\begin{aligned} M(x, t) &= -\int_{-h/2}^{h/2} z\sigma_x(x, z, t)\,\mathrm{d}A \\ &= \int_{-h/2}^{h/2} Ew_{,xx}(x, t)z^2\,\mathrm{d}A \\ &= EI(x)w_{,xx}(x, t), \end{aligned} \tag{3.3}$$

where $I(x)$ is the second moment of area of cross-section of the beam about the *neutral axis*. The neutral axis is the line of intersection of the plane of the neutral fibers, and the plane of the cross-section of the beam. Now, the equation of translational dynamics of an infinitesimal element can be written as

$$(\rho A(x)\,\mathrm{d}x)w_{,tt} = p(x, t)\mathrm{d}x + (V + \mathrm{d}V)\cos(\psi + \mathrm{d}\psi) - V\cos\psi$$

or

$$\rho A w_{,tt} = p(x, t) + V_{,x}, \tag{3.4}$$

where $p(x, t)$ is the external transverse force density, V is the shear force at any cross-section, and it is assumed that $\cos\psi \approx 1$. The rotational dynamics of the infinitesimal element is represented by

$$(\rho I(x)\,\mathrm{d}x)\psi_{,tt} = (M + \mathrm{d}M) - M + (V + \mathrm{d}V)\frac{\mathrm{d}x}{2} + V\frac{\mathrm{d}x}{2}$$

or

$$\rho I(x)\psi_{,tt} = M_{,x} + V. \tag{3.5}$$

Using the relation $\tan\psi = w_{,x}$, one can write

$$\psi_{,t} = \frac{w_{,xt}}{(1 + w_{,x}^2)} \approx w_{,xt} \tag{3.6}$$

and

$$\psi_{,tt} = \frac{w_{,xtt}}{(1 + w_{,x}^2)} - \frac{2w_{,x}w_{,xt}^2}{(1 + w_{,x}^2)^2} \approx w_{,xtt}, \tag{3.7}$$

where all non-linear terms have been dropped. Using (3.3) and (3.7) in (3.5), and subsequently eliminating V between (3.5) and (3.4) yields on simplification

$$\rho A w_{,tt} + [EIw_{,xx}]_{,xx} - [\rho I w_{,xtt}]_{,x} = p(x,t). \tag{3.8}$$

This equation of motion is known as the *Rayleigh beam model*. The term $(EIw_{,xx})_{,xx}$ is usually referred to as the *flexure term*, where EI is called the *flexural stiffness*, and $(\rho I w_{,xtt})_{,x}$ is known as the rotary inertia term. When the rotary inertia term is neglected, we obtain

$$\rho A w_{,tt} + [EIw_{,xx}]_{,xx} = p(x,t), \tag{3.9}$$

which is referred to as the *Euler–Bernoulli beam model*. It is observed that the equations of motion (3.8) or (3.9) are fourth-order partial differential equations in space, and second-order in time. Thus, we require four boundary conditions and two initial conditions. The boundary conditions are discussed in a later section.

3.1.2 The variational formulation

The variational principle (see Appendix A) provides an alternative convenient approach for obtaining the equation of motion and the boundary conditions for beams. The total kinetic energy $\mathcal{T}$ due to translation and rotation of an infinitesimal beam element can be written as

$$\begin{aligned}\mathcal{T} &= \frac{1}{2}\int_0^l \left[\rho A w_{,t}^2 + \rho I \psi_{,t}^2\right] \mathrm{d}x \\ &= \frac{1}{2}\int_0^l \left[\rho A w_{,t}^2 + \rho I w_{,xt}^2\right] \mathrm{d}x \quad \text{(using (3.6))}.\end{aligned} \tag{3.10}$$

The potential energy $\mathcal{V}$ can be written from the theory of elasticity as

$$\begin{aligned}\mathcal{V} &= \frac{1}{2}\int_0^l \int_A \sigma_x \epsilon_x \, \mathrm{d}A \, \mathrm{d}x \\ &= \frac{1}{2}\int_0^l \int_A E w_{,xx}^2 z^2 \, \mathrm{d}A \, \mathrm{d}x \qquad \text{(using (3.1) and (3.2))} \\ &= \frac{1}{2}\int_0^l EI w_{,xx}^2 \, \mathrm{d}x.\end{aligned} \tag{3.11}$$

The Lagrangian is given by $\mathcal{L} = \mathcal{T} - \mathcal{V}$, and the variational formulation yields

$$\delta \int_{t_1}^{t_2} \mathcal{L} \, \mathrm{d}t = 0$$

or

$$\delta \int_{t_1}^{t_2} \frac{1}{2} \int_0^l \left[\rho A w_{,t}^2 + \rho I w_{,xt}^2 - E I w_{,xx}^2\right] \mathrm{d}x\mathrm{d}t = 0. \tag{3.12}$$

Following the procedure discussed in Appendix A, we have from (3.12)

$$\int_{t_1}^{t_2} \int_0^l [\rho A w_{,t}\, \delta w_{,t} + \rho I w_{,xt}\, \delta w_{,xt} - E I w_{,xx}\, \delta w_{,xx}] \mathrm{d}x\mathrm{d}t = 0$$

or

$$-\int_{t_1}^{t_2} E I w_{,xx}\, \delta w_{,x}\Big|_0^l \mathrm{d}t - \int_{t_1}^{t_2} [(E I w_{,xx})_{,x} - \rho I w_{,xtt}]\, \delta w\Big|_0^l \mathrm{d}t$$
$$+ \int_{t_1}^{t_2} \int_0^l [-\rho A w_{,tt} + (\rho I w_{,xtt})_{,x} - (E I w_{,xx})_{,xx}]\, \delta w \,\mathrm{d}x\mathrm{d}t = 0, \tag{3.13}$$

where we have used the fact that the variation of the field variable and its spatial derivatives at the initial and final times is zero, i.e., $\delta w|_{t_i} = \delta w_{,x}|_{t_i} \equiv 0$ for $i = 0, 1$. The condition (3.13) must hold for arbitrary variations δw. This yields, from the last integral in (3.13), the equation of motion

$$\rho A w_{,tt} + [E I w_{,xx}]_{,xx} - [\rho I w_{,xtt}]_{,x} = 0. \tag{3.14}$$

Similarly, boundary conditions are obtained from the first and second integrals in (3.13). For example, one possible set of boundary conditions is given by

$$E I w_{,xx}(0, t) \equiv 0 \quad \text{or} \quad w_{,x}(0, t) \equiv 0, \tag{3.15}$$

$$E I w_{,xx}(l, t) \equiv 0 \quad \text{or} \quad w_{,x}(l, t) \equiv 0, \tag{3.16}$$

$$[(E I w_{,xx})_{,x} - \rho I w_{,xtt}](0, t) \equiv 0 \quad \text{or} \quad w(0, t) \equiv 0, \tag{3.17}$$

and

$$[(E I w_{,xx})_{,x} - \rho I w_{,xtt}](l, t) \equiv 0 \quad \text{or} \quad w(l, t) \equiv 0. \tag{3.18}$$

The first condition in (3.15) and (3.16) implies zero moment at the ends, while the first condition in (3.17) and (3.18) implies zero shear force. The second equation in each of the above conditions is a geometric boundary condition which implies either a zero displacement or a zero slope.

In other kinds of boundary conditions, linear combinations of the boundary terms in (3.13) may be set equal to zero. For example, for the beam shown in Figure 3.3(a), the geometric boundary conditions are given by $w(0, t) \equiv 0$, $w(l, t) \equiv 0$, and $w_{,x}(0, t) = w_{,x}(l, t)$. In this case, the difference of the boundary term evaluated at $x = l$ and $x = 0$ under the first integral in (3.13) must vanish. This leads to the fourth (natural) boundary condition $E I w_{,xx}(0, t) = E I w_{,xx}(l, t)$ (since $\delta w_{,x}(0, t) = \delta w_{,x}(l, t)$). In Figure 3.3(b), the boundary conditions are $w(0, t) \equiv 0$, $E I w_{,xx}(l, t) \equiv 0$, and $a w_{,x}(0, t) = w(l, t)$. The natural boundary condition in this case also can be obtained easily from the boundary terms in (3.13).

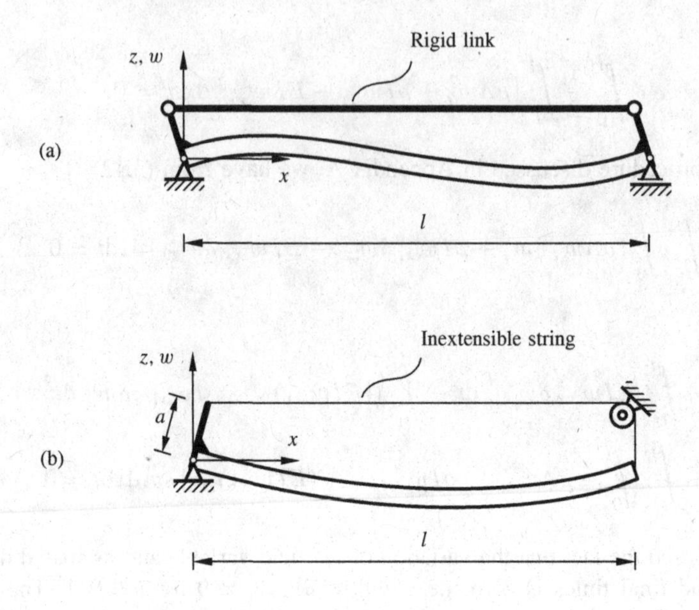

Figure 3.3 Special boundary conditions for beams

3.1.3 Various boundary conditions for a beam

Some of the above boundary conditions are realized in various combinations in beams depending on the support, as illustrated in Figure 3.4. The simplest support conditions can be any combination of pinned, clamped, free, and sliding, as illustrated in Figure 3.4. When

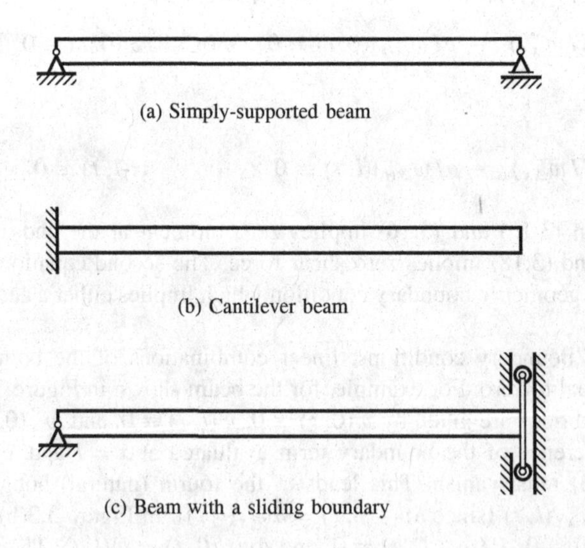

Figure 3.4 Various boundary conditions for a beam

the end is pinned without friction, there is zero transverse displacement (geometric boundary condition) and zero moment (dynamic boundary condition) at that end. Thus, we have in Figure 3.4(a)

$$w(0,t) \equiv 0, \quad w(l,t) \equiv 0, \quad EIw_{,xx}(0,t) \equiv 0, \quad \text{and} \quad EIw_{,xx}(l,t) \equiv 0. \tag{3.19}$$

At a clamped end, as shown in Figure 3.4(b), the displacement and slope of the beam are zero (both are geometric boundary conditions). Therefore, we have

$$w(0,t) \equiv 0 \qquad \text{and} \qquad w_{,x}(0,t) \equiv 0. \tag{3.20}$$

At a free boundary, it is evident that the moment and the shear force vanish (both dynamic boundary conditions). Hence, one can write

$$EIw_{,xx}(l,t) \equiv 0 \qquad \text{and} \qquad \rho I w_{,xtt}(l,t) - [EIw_{,xx}]_{,x}(l,t) \equiv 0. \tag{3.21}$$

A sliding boundary is characterized by zero slope and zero shear. Thus, the mathematical conditions for the right boundary of the beam shown in Figure 3.4(c) are

$$w_{,x}(l,t) \equiv 0 \qquad \text{and} \qquad EIw_{,xxx}(l,t) \equiv 0. \tag{3.22}$$

When there are external forces over the beam, or at the boundaries, appropriate forcing terms can be added, respectively, to (3.14), or in the moment and shear boundary conditions.

For example, consider a uniform cantilever beam with discrete damping and a stiffened free-end, as shown in Figure 3.5. The equation of motion and the boundary conditions can be written as

$$\rho A w_{,tt} + [EIw_{,xx}]_{,xx} - [\rho I w_{,xtt}]_{,x} + d\,\delta(x-a)w_{,t} = 0,$$

$$w(0,t) = 0, \qquad w_{,x}(0,t) = 0, \qquad EIw_{,xx}(l,t) = -k_M w_{,x}(l,t),$$

and

$$EIw_{,xxx}(l,t) - \rho I w_{,xtt}(l,t) = -k_S w(l,t).$$

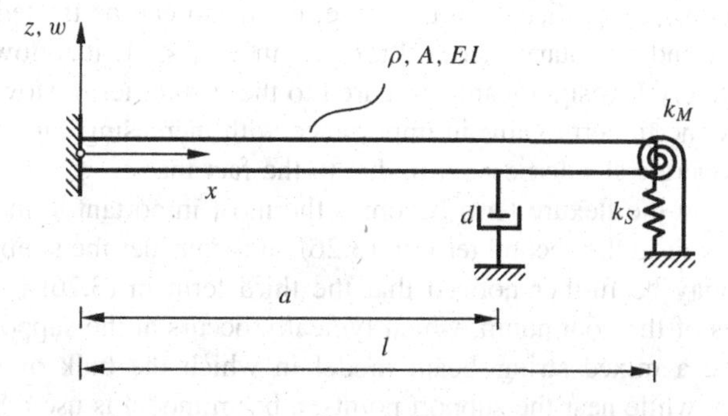

Figure 3.5 Cantilever beam with discrete damping and stiffened free-end

3.1.4 Taut string and tensioned beam

Before proceeding further, let us compare a taut string and a tensioned beam. Consider a uniform beam under axial tension. One can easily show that the equation of motion of this beam is obtained as

$$\rho A w_{,tt} - T w_{,xx} + EI w_{,xxxx} - \rho I w_{,ttxx} = 0, \tag{3.23}$$

where T is the tension in the beam. Consider a non-dimensionalization scheme

$$\overline{w} = \frac{w}{r_g}, \qquad \overline{x} = \frac{x}{l}, \qquad \text{and} \qquad \overline{t} = \frac{tc}{l} = \frac{t}{l}\sqrt{\frac{T}{\rho A}},$$

where $r_g := \sqrt{I/A}$ is the radius of gyration of the cross-section about the neutral axis of the beam. Using these non-dimensionalized variables, (3.23) can be written as

$$\overline{w}_{,\overline{t}\overline{t}} - \overline{w}_{,\overline{x}\overline{x}} + \frac{EI}{Tl^2}\overline{w}_{,\overline{x}\overline{x}\overline{x}\overline{x}} - \frac{I}{Al^2}\overline{w}_{,\overline{t}\overline{t}\overline{x}\overline{x}} = 0. \tag{3.24}$$

Thus, the non-dimensional quantity EI/Tl^2 decides the relative importance of the flexure term $w_{,xxxx}$, while I/Al^2 reflects the relative importance of the rotary inertia term $w_{,ttxx}$. One can also write these non-dimensional numbers as

$$\frac{EI}{Tl^2} = \frac{EI/\rho A}{Tl^2/\rho A} = \frac{1}{T/EA}\frac{1}{l^2 A/I} = \frac{1}{\epsilon_x}\frac{1}{s_r^2} \qquad \text{and} \qquad \frac{I}{Al^2} = \frac{1}{s_r^2}, \tag{3.25}$$

where ϵ_x is the longitudinal strain in the x-axis direction due to pre-tension, and $s_r := l/r_g$ is defined as the slenderness ratio. Then, one can rewrite (3.24) as

$$\overline{w}_{,\overline{t}\overline{t}} - \overline{w}_{,\overline{x}\overline{x}} + \frac{1}{\epsilon_x}\frac{1}{s_r^2}\overline{w}_{,\overline{x}\overline{x}\overline{x}\overline{x}} - \frac{1}{s_r^2}\overline{w}_{,\overline{t}\overline{t}\overline{x}\overline{x}} = 0. \tag{3.26}$$

It is clear from (3.25) that when the beam is very slender (i.e., $s_r \gg 1$), the third and fourth terms in (3.26) become insignificant. In that case, the beam can be treated as a string with no flexural stiffness and no rotary inertia. Further, since $\epsilon_x \ll 1$, it follows that the rotary inertia term is relatively less significant compared to the flexure term. However, as we shall see later, the rotary inertia term gains in importance with increasing curvature of the beam. In the case of a moderate slenderness ratio, due to the fact that $\epsilon_x \ll 1$ (i.e., $T \ll EA$), we have $1/\epsilon_x s_r^2 \gg 1$, and the flexure term becomes the most important term in the dynamics. In that case, we may drop the second term in (3.26), and consider the simple beam equation (3.8), or (3.9). It may be further noticed that the third term in (3.26) becomes important for large curvatures of the continuum, which typically occurs at the support points. In such cases, one may use a mixed string–beam model in which the bulk of the continuum is modeled as a string, while near the support points, a beam model is used. Such analysis can be found in [2].

3.2 FREE VIBRATION PROBLEM

The free vibration problem is essentially the determination of the eigenfrequencies and the corresponding eigenfunctions of the system. Hence, we begin here with the modal analysis of the beam models derived above. The solution of the initial value problem can be formulated using the eigenfunctions obtained from the modal analysis.

3.2.1 Modal analysis

3.2.1.1 The eigenvalue problem

Consider a Rayleigh beam described by the equation of motion

$$\rho A w_{,tt} + (EIw_{,xx})_{,xx} - (\rho I w_{,xtt})_{,x} = 0. \quad (3.27)$$

Assume a modal solution of (3.27) in the form

$$w(x,t) = W(x)e^{i\omega t}, \quad (3.28)$$

where ω is the circular eigenfrequency and $W(x)$ is the eigenfunction. The actual real solution is obtained by taking the real part (though we could equally well take the imaginary part) of the complex expression in (3.28). Substituting the modal solution in the field equation (3.27) yields on rearrangement

$$-\omega^2[\rho A W - (\rho I W')'] + (EIW'')'' = 0, \quad (3.29)$$

which together with the boundary conditions represents the eigenvalue problem for a Rayleigh beam. One may consider (3.29) as a general eigenvalue problem of the form

$$-\omega^2\mathcal{M}[W] + \mathcal{K}[W] = 0, \quad (3.30)$$

where

$$\mathcal{M}[\cdot] = \left[\rho A - \frac{\mathrm{d}}{\mathrm{d}x}\left(\rho I \frac{\mathrm{d}}{\mathrm{d}x}\right)\right][\cdot] \quad \text{and} \quad \mathcal{K}[\cdot] = \frac{\mathrm{d}^2}{\mathrm{d}x^2}\left(EI\frac{\mathrm{d}^2}{\mathrm{d}x^2}\right)[\cdot]. \quad (3.31)$$

In the case of an Euler–Bernoulli beam described by

$$\rho A w_{,tt} + (EIw_{,xx})_{,xx} = 0, \quad (3.32)$$

substituting the solution form (3.28) leads to

$$-\omega^2\rho A W + (EIW'')'' = 0. \quad (3.33)$$

It is evident that (3.33) is a special case of (3.30) with

$$\mathcal{M}[\cdot] = \rho A[\cdot] \quad \text{and} \quad \mathcal{K}[\cdot] = \frac{\mathrm{d}^2}{\mathrm{d}x}\left(EI\frac{\mathrm{d}^2}{\mathrm{d}x^2}\right)[\cdot].$$

The general solution of (3.30) (and hence (3.33)) cannot be obtained in closed form for arbitrary $EI(x)$ and/or $\rho A(x)$. Therefore, we will solve the eigenvalue problem for uniform beams only. Before proceeding further to solve the eigenvalue problem, let us first discuss the orthogonality property of eigenfunctions of (3.30).

3.2.1.2 Orthogonality relations

Consider the eigenvalue problem of the Rayleigh beam described by (3.30) together with the boundary conditions, as discussed in Section 3.1.3. For two different modes j and k, one can write (3.30) as

$$-\omega_j^2 \mathcal{M}[W_j] + \mathcal{K}[W_j] = 0 \tag{3.34}$$

and

$$-\omega_k^2 \mathcal{M}[W_k] + \mathcal{K}[W_k] = 0, \tag{3.35}$$

where $\mathcal{M}[\cdot]$ and $\mathcal{K}[\cdot]$ are given by (3.31). Multiplying (3.34) by W_k, (3.34) by W_j, subtracting one equation from the other, and integrating the result over the length of the beam gives

$$\begin{aligned}&[((EIW_j'')' - \omega_j^2 \rho I W_j')W_k - ((EIW_k'')' - \omega_j^2 \rho I W_k')W_j]\big|_0^l \\ &\quad + [EIW_k'' W_j' - EIW_j'' W_k']\big|_0^l + (\omega_j^2 - \omega_k^2)\int_0^l [\rho A W_k - (\rho I W_k')']W_j \, \mathrm{d}x = 0.\end{aligned} \tag{3.36}$$

Using the boundary conditions defined by (3.15)–(3.18), it can be easily checked that the boundary terms in (3.36) disappear. Hence, we immediately obtain the orthogonality relation from (3.36) as

$$\int_0^l [\rho A W_k - (\rho I W_k')']W_j \, \mathrm{d}x = 0, \qquad j \neq k, \tag{3.37}$$

or

$$\int_0^l \mathcal{M}[W_k] W_j \, \mathrm{d}x = 0, \qquad j \neq k. \tag{3.38}$$

In the case of an Euler–Bernoulli beam, (3.37) simplifies further to

$$\int_0^l \rho A W_k W_j \, \mathrm{d}x = 0, \qquad j \neq k. \tag{3.39}$$

One may normalize the eigenfunctions with respect to an inner product such that

$$\int_0^l \mathcal{M}[W_k] W_j \, \mathrm{d}x = \delta_{jk}, \tag{3.40}$$

where δ_{jk} represents the Kronecker delta function. The eigenfunctions so normalized form an orthonormal basis. As a consequence of this orthonormality, from (3.30) and (3.38), one can easily write

$$\int_0^l \mathcal{K}[W_k]W_j \mathrm{d}x = \omega_k^2 \delta_{jk}.$$

3.2.1.3 Modal analysis of uniform beams

Consider the eigenvalue problem of a uniform Rayleigh beam described by

$$-\omega^2[\rho AW - \rho IW''] + EIW'''' = 0, \tag{3.41}$$

along with the corresponding boundary conditions. Substituting in (3.41) a solution of the form

$$W(x) = B\mathrm{e}^{\tilde{\beta}x}, \tag{3.42}$$

where B and $\tilde{\beta}$ are constants, one can write

$$EI\tilde{\beta}^4 - \omega^2\rho I\tilde{\beta}^2 - \omega^2\rho A = 0$$
$$\Rightarrow \tilde{\beta}^2 = \frac{1}{2EI}\left[\omega^2\rho I \pm \sqrt{\omega^4\rho^2 I^2 + 4\omega^2 EI\rho A}\right]. \tag{3.43}$$

It is easily observed that the bracketed term in (3.43) will take both a positive and a negative value. Therefore, $\tilde{\beta}$ has four solutions given as $\tilde{\beta} = \pm\beta_1, \pm i\beta_2$, where

$$\beta_1 = \frac{1}{\sqrt{2EI}}\left[\omega^2\rho I + \sqrt{\omega^4\rho^2 I^2 + 4\omega^2 EI\rho A}\right]^{1/2} \tag{3.44}$$

and

$$\beta_2 = \frac{1}{\sqrt{2EI}}\left[-\omega^2\rho I + \sqrt{\omega^4\rho^2 I^2 + 4\omega^2 EI\rho A}\right]^{1/2}. \tag{3.45}$$

Thus, the general (complex) solution of (3.29) is obtained as

$$W(x) = A_1\mathrm{e}^{\beta_1 x} + A_2\mathrm{e}^{-\beta_1 x} + A_3\mathrm{e}^{i\beta_2 x} + A_4\mathrm{e}^{-i\beta_2 x}, \tag{3.46}$$

where A_i, $i = 1, \ldots, 4$, are (complex) constants. Alternatively, the solution may also be expressed in the real form as

$$W(x) = B_1\cosh\beta_1 x + B_2\sinh\beta_1 x + B_3\cos\beta_2 x + B_4\sin\beta_2 x, \tag{3.47}$$

where B_i, $i = 1, \ldots, 4$ are real constants to be obtained from the boundary conditions.

Next, we consider the case of a uniform Euler–Bernoulli beam. Substituting the solution (3.42) in (3.32) yields

$$-\omega^2 \rho A W + EIW'''' = 0, \tag{3.48}$$

where ρA and EI are constants, we obtain

$$-\rho A\omega^2 + EI\tilde{\beta}^4 = 0$$
$$\Rightarrow \tilde{\beta}^2 = \sqrt{\frac{\omega^2 \rho A}{EI}}. \tag{3.49}$$

Therefore, we have the four solutions $\tilde{\beta} = \pm\beta, \pm i\beta$, where

$$\beta = (\omega^2 \rho A/EI)^{1/4}. \tag{3.50}$$

Now, one can write the general (complex) solution (for $\omega \neq 0$) of (3.33) as

$$W(x) = A_1 e^{\beta x} + A_2 e^{-\beta x} + A_3 e^{i\beta x} + A_4 e^{-i\beta x}, \tag{3.51}$$

where A_i, $i = 1, \ldots, 4$ are (complex) constants, or in the real form as

$$W(x) = B_1 \cosh \beta x + B_2 \sinh \beta x + B_3 \cos \beta x + B_4 \sin \beta x, \tag{3.52}$$

where B_i, $i = 1, \ldots, 4$, are real constants of integration which are determined by the boundary conditions of the problem.

In the following, we consider beams with some typical support conditions, and determine their eigenfrequencies and eigenfunctions.

(a) Uniform simply-supported beam

Consider a simply-supported (pinned–pinned) uniform Rayleigh beam. The boundary conditions for the corresponding eigenvalue problem (3.30) are

$$W(0) = 0, \quad W''(0) = 0, \quad W(l) = 0, \quad \text{and} \quad W''(l) = 0. \tag{3.53}$$

Using the first two conditions from (3.53) in (3.47) yields $B_1 = B_3 = 0$. The last two boundary conditions in (3.53) yield

$$B_2 \sinh \beta_1 l + B_4 \sin \beta_2 l = 0 \tag{3.54}$$

and

$$B_2 \sinh \beta_1 l - B_4 \sin \beta_2 l = 0. \tag{3.55}$$

For non-trivial solutions of (B_2, B_4) from (3.54)–(3.55), one must have

$$\sinh\beta_1 l \, \sin\beta_2 l = 0$$
$$\Rightarrow \sin\beta_2 l = 0 \qquad (\text{since } \sinh\beta_1 l \neq 0 \text{ for any } \beta_1 l \neq 0), \tag{3.56}$$

which is the characteristic equation for the problem. The solutions of the characteristic equation are obtained as

$$\beta_2 = \frac{n\pi}{l}, \qquad n = 1, 2, \ldots, \infty. \tag{3.57}$$

Substituting this expression of β_2 in (3.45) and solving for ω yields the circular natural frequencies of a simply-supported uniform Rayleigh beam as

$$\omega_n^{\mathrm{R}} = \frac{n^2\pi^2}{l^2} \frac{1}{\left[1 + n^2\pi^2 \dfrac{I}{l^2 A}\right]^{1/2}} \sqrt{\frac{EI}{\rho A}}, \qquad n = 1, 2, \ldots, \infty. \tag{3.58}$$

Taking $n \gg 1$ such that $1 + n^2\pi^2 I/l^2 A \approx n^2\pi^2 I/l^2 A$, one obtains from (3.58) the approximation $\omega_n^{\mathrm{R}} \approx (n\pi/l)\sqrt{E/\rho}$. As can be easily checked, these are the circular eigenfrequencies of longitudinal vibrations of a fixed–fixed bar.

The final step of modal analysis is to determine the eigenfunctions. From (3.54) and (3.55), and the characteristic equation (3.56), one can easily conclude that $B_2 = 0$. Substituting this in (3.47), along with $B_1 = B_3 = 0$ and (3.57), the eigenfunctions of a simply-supported uniform Rayleigh beam can be written as

$$W_n(x) = B \sin\frac{n\pi x}{l}, \qquad n = 1, 2, \ldots, \infty, \tag{3.59}$$

where B is an arbitrary constant. These eigenfunctions are clearly orthogonal, and can be normalized to make them orthonormal.

In the case of a simply-supported uniform Euler–Bernoulli beam, we have the same expression for β_n given by (3.57), as one can easily check. Therefore, the circular natural frequencies of an Euler–Bernoulli beam are obtained by substituting the expression of β from (3.57) in (3.50), and solving for ω_n. This yields

$$\omega_n^{\mathrm{EB}} = \frac{n^2\pi^2}{l^2}\sqrt{\frac{EI}{\rho A}}, \qquad n = 1, 2, \ldots, \infty. \tag{3.60}$$

It may be observed by comparing (3.58) and (3.60) that, in the case of a very slender beam (i.e., $s_{\mathrm{r}} = l^2 A/I \gg 1$), the natural frequencies of the lower modes given by the Rayleigh beam model tend to be the same as those obtained from the Euler–Bernoulli beam model (since $1 + n^2\pi^2 I/l^2 A \approx 1$). Thus, for lower modes of very slender beams, the effect of rotary inertia is insignificant. It can be checked that the eigenfunctions for the simply-supported uniform Rayleigh and Euler–Bernoulli beams are the same.

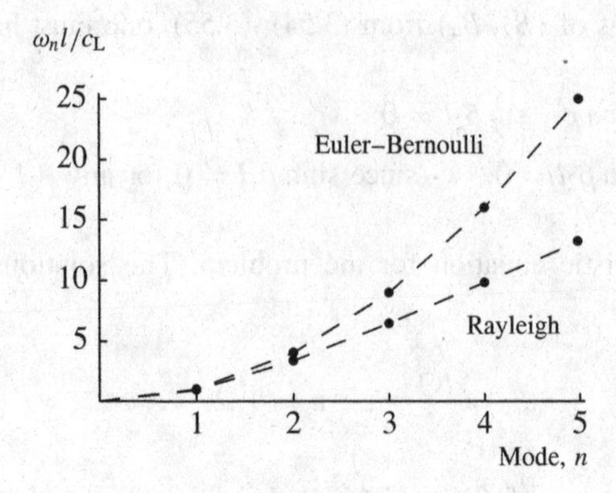

Figure 3.6 Comparison of natural frequencies of a simply-supported Rayleigh beam and an Euler–Bernoulli beam for a fixed slenderness ratio $s_r = 10$

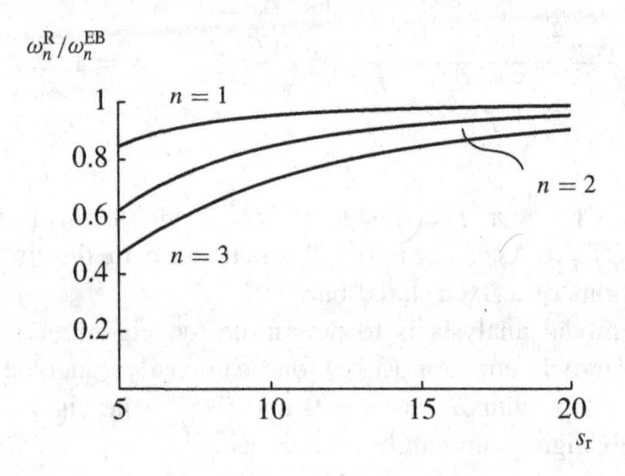

Figure 3.7 Variation of ω_n^R/ω_n^{EB} with slenderness ratio s_r for the first three natural frequencies

The two non-dimensional natural frequencies $\omega_n^R l/c_L$ and $\omega_n^{EB} l/c_L$, where $c_L = \sqrt{E/\rho}$, are compared for first few modes in Figure 3.6. It is observed that for lower modes, the two frequencies tend to match. However, divergence is observed at higher modes. This is primarily due to the effect of rotary inertia in the Rayleigh beam. It is easy to conclude from the eigenfunctions (3.59) that, for higher modes, the curvature of the beam increases, thereby increasing the influence of rotary inertia on the dynamics of the beam.

The ratio ω_n^R/ω_n^{EB} as a function of the slenderness ratio s_r is plotted in Figure 3.7 for the first three modes. At low slenderness ratios, the frequency ratio is widely different for different modes. However, as the beam gets slender, the two frequencies tend to agree, as can be observed from the figure. Further, for the higher modes, the effect of rotary inertia becomes more pronounced at low slenderness ratios.

(b) Uniform cantilever beam

Here we consider a uniform Euler–Bernoulli cantilever beam for which the boundary conditions are given by

$$W(0) = 0, \quad W'(0) = 0, \quad W''(l) = 0, \quad \text{and} \quad W'''(l) = 0. \tag{3.61}$$

Substituting the solution form (3.52) in these boundary conditions yields

$$B_1 + B_3 = 0, \tag{3.62}$$

$$B_2 + B_4 = 0, \tag{3.63}$$

$$B_1 \cosh \beta l + B_2 \sinh \beta l - B_3 \cos \beta l - B_4 \sin \beta l = 0, \tag{3.64}$$

and

$$B_1 \sinh \beta l + B_2 \cosh \beta l + B_3 \sin \beta l - B_4 \cos \beta l = 0. \tag{3.65}$$

For a non-trivial solution of the $(B_1, \ldots, B_4)$, we must have

$$\begin{vmatrix} 1 & 0 & 1 & 0 \\ 0 & 1 & 0 & 1 \\ \cosh \beta l & \sinh \beta l & -\cos \beta l & -\sin \beta l \\ \sinh \beta l & \cosh \beta l & \sin \beta l & -\cos \beta l \end{vmatrix} = 0$$

$$\Rightarrow \cos \beta l \cosh \beta l + 1 = 0, \tag{3.66}$$

which is the characteristic equation of a cantilever Euler–Bernoulli beam. The solutions of the characteristic equation (3.66) are visualized graphically by circles in Figure 3.8. It can be observed that the function $1/\cosh z$ converges to zero rapidly, and the characteristic equation (3.66) essentially reduces to $\cos \beta l = 0$ for higher modes. The analytical solution can be expressed in the form

$$\beta_n = \left(\omega_n \sqrt{\frac{\rho A}{EI}} \right)^{1/2} = \left(\frac{2n-1}{2}\pi + e_n \right) \frac{1}{l} \tag{3.67}$$

$$\Rightarrow \omega_n = \left(\frac{2n-1}{2}\pi + e_n \right)^2 \frac{1}{l^2} \sqrt{\frac{EI}{\rho A}}, \qquad n = 1, 2, \ldots, \infty, \tag{3.68}$$

where e_n are small correction terms, and obtained as $e_1 = 0.3042$, $e_2 = -0.018$, $e_3 = 0.001, \ldots$. The corrections in the higher modes tend to zero rapidly, and can be neglected.

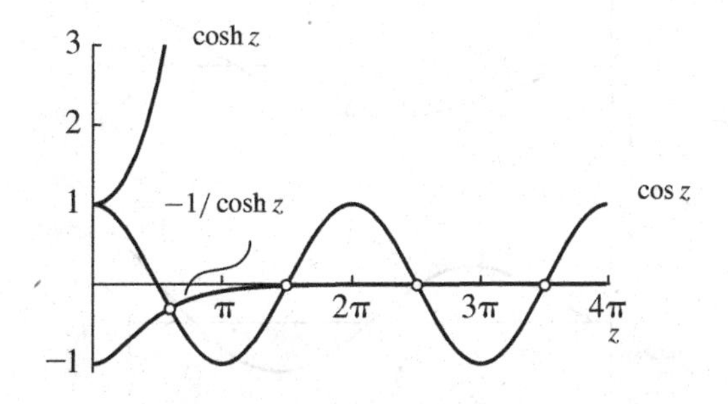

Figure 3.8 Graphical representation of the solutions of the characteristic equation of a cantilever beam

For values of β given by (3.67), a non-trivial solution of $(B_1, \ldots, B_4)$ can be determined from (3.62)–(3.65) by considering any three of the four equations. For example, using (3.62) and (3.63) we can eliminate B_3 and B_4 from (3.64) to obtain

$$B_1 = -\frac{\sinh \beta_n l + \sin \beta_n l}{\cosh \beta_n l + \cos \beta_n l} B_2 := \alpha_n B_2. \tag{3.69}$$

Therefore, taking $B_2 = 1$, one possible solution is given by

$$B_1 = \alpha_n, \qquad B_2 = 1, \qquad B_3 = -\alpha_n, \qquad \text{and} \qquad B_4 = -1, \tag{3.70}$$

which yields the nth eigenfunctions as

$$W_n(x) = \sinh \beta_n x - \sin \beta_n x - \left[\frac{\sinh \beta_n l + \sin \beta_n l}{\cosh \beta_n l + \cos \beta_n l}\right] (\cosh \beta_n x - \cos \beta_n x). \tag{3.71}$$

The first three eigenfunctions are shown in Figure 3.9. These eigenfunctions satisfy the orthogonality condition (3.39) as can be checked.

(c) Uniform free–free beam

Consider a free–free Euler–Bernoulli beam. The boundary conditions in this case are zero moment and shear force at both ends of the beam. This implies

$$W''(0) = 0, \quad W'''(0) = 0, \quad W''(l) = 0, \quad \text{and} \quad W'''(l) = 0. \tag{3.72}$$

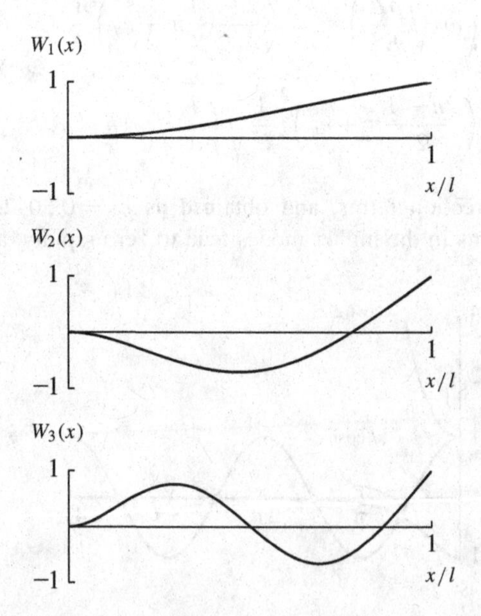

Figure 3.9 First three eigenfunctions of a cantilever beam

Using these boundary conditions in the shape-function (3.52), we have

$$B_1 - B_3 = 0, \tag{3.73}$$

$$B_2 - B_4 = 0, \tag{3.74}$$

$$B_1 \cosh \beta l + B_2 \sinh \beta l - B_3 \cos \beta l - B_4 \sin \beta l = 0, \tag{3.75}$$

and

$$B_1 \sinh \beta l + B_2 \cosh \beta l + B_3 \sin \beta l - B_4 \cos \beta l = 0. \tag{3.76}$$

A non-trivial solution of the B_is is obtained if and only if

$$\begin{vmatrix} 1 & 0 & -1 & 0 \\ 0 & 1 & 0 & -1 \\ \cosh \beta l & \sinh \beta l & -\cos \beta l & -\sin \beta l \\ \sinh \beta l & \cosh \beta l & \sin \beta l & -\cos \beta l \end{vmatrix} = 0$$

$$\Rightarrow \cos \beta l \cosh \beta l - 1 = 0. \tag{3.77}$$

The geometric visualization of the solution of the characteristic equation (3.77) is done by circles in Figure 3.10. It can be observed that $\beta l = 0$ is a solution to (3.77). However, for $\beta = 0$, (3.51) is no longer the form of solution to the differential equation. This case, therefore, has to be considered separately. For higher modes, since $\cosh z$ is an exponentially divergent function, the characteristic equation can be approximated by $\cos \beta l = 0$. The solution of (3.77) can be represented in the form

$$\beta_n = \omega_n \sqrt{\frac{\rho A}{EI}} = \left(\frac{2n+1}{2} \pi + e_n \right) \frac{1}{l} \tag{3.78}$$

$$\Rightarrow \omega_n = \left(\frac{2n+1}{2} \pi + e_n \right)^2 \frac{1}{l^2} \sqrt{\frac{EI}{\rho A}}, \qquad n = 1, 2, \ldots, \infty, \tag{3.79}$$

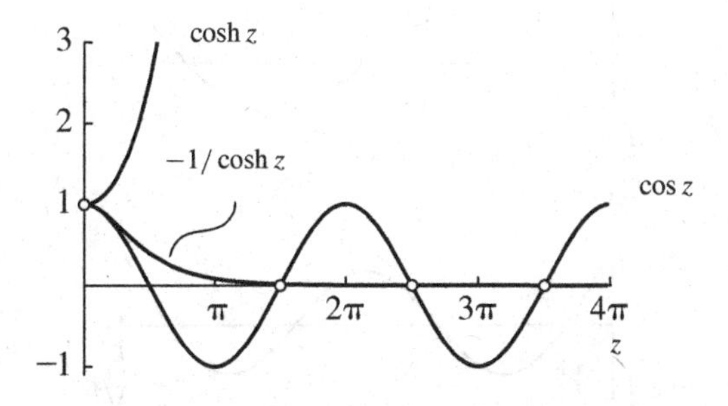

Figure 3.10 Graphical representation of the solutions of the characteristic equation of a free–free beam

where e_n are small correction terms. For example, $e_1 = 0.01766$, $e_2 = -0.00078, \ldots$. The corrections for higher modes are negligibly small, and can be dropped for all practical purposes.

The eigenfunctions for the non-zero eigenfrequencies can be determined from (3.73)–(3.76). It can be easily checked that solving for B_1 from (3.73)–(3.75) yields

$$B_1 = \frac{-\sinh \beta_n l + \sin \beta_n l}{\cosh \beta_n l - \cos \beta_n l} B_2 := \alpha_n B_2. \tag{3.80}$$

Therefore, taking $B_2 = 1$, a possible solution is given by

$$B_1 = \alpha_n, \qquad B_2 = 1, \qquad B_3 = \alpha_n, \qquad \text{and} \qquad B_4 = 1, \tag{3.81}$$

which yields the nth eigenfunctions as

$$W_n(x) = \sinh \beta_n x + \sin \beta_n x + \left[\frac{-\sinh \beta_n l + \sin \beta_n l}{\cosh \beta_n l - \cos \beta_n l} \right] (\cosh \beta_n x + \cos \beta_n x). \tag{3.82}$$

The first three eigenfunctions of the free–free beam are shown in Figure 3.11. Once again it can be checked that these eigenfunctions are orthogonal.

For the case $\beta = 0$ (i.e., $\omega_n = 0$), (3.33) implies that

$$W'''' = 0$$
$$\Rightarrow W(x) = B_1 + B_2 x + B_3 x^2 + B_4 x^3. \tag{3.83}$$

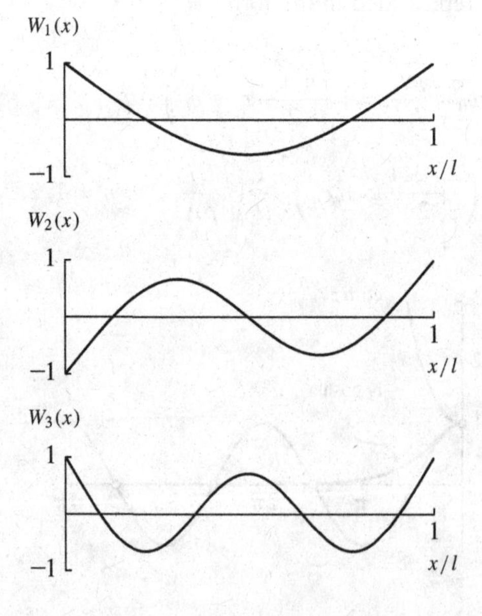

Figure 3.11 First three eigenfunctions of a free–free beam

This solution corresponds to the rigid-body motion, and consists of translation and rotation of the beam. Using (3.72), we can easily obtain $B_3 = 0$ and $B_4 = 0$. Then, the solution for the rigid-body motion for small time can be written as

$$w_0(x,t) = (K + Lt) + \left[\frac{x}{l} - \frac{1}{2}\right](G + Ht), \tag{3.84}$$

where K, L, G, and H are appropriate constants to be determined from the initial conditions. The complete solution of the free–free beam can now be written as

$$w(x,t) = (K + Lt) + \left[\frac{x}{l} - \frac{1}{2}\right](G + Ht) + \sum_{n=1}^{\infty}(C_n \cos\omega_n t + S_n \sin\omega_n t)W_n(x), \tag{3.85}$$

where $W_n(x)$ are given by (3.82).

3.2.1.4 Approximate methods

In the case of an arbitrary geometry of the beam, or in the presence of discrete elements, the exact modal analysis becomes at least difficult, and usually even impossible. In such situations, the approximate methods such as the Ritz and the Galerkin methods are useful. These methods have already been discussed in Chapter 1, and will be studied in some more detail in Chapter 6.

In both methods, we approximate the solutions of the variational problem (3.12) in the form

$$w(x,t) = \sum_{j=1}^{N} p_j(t)\psi_j(x) = \mathbf{\Psi}^{\mathrm{T}}\mathbf{p}, \tag{3.86}$$

where $p_j(t)$ are time functions to be determined, and the $\psi_j(x)$ are suitably chosen shape-functions. In the Ritz method, for convergence, the shape-functions must satisfy all the geometric boundary conditions of the problem and be differentiable at least up to the highest order of the space-derivative in the Lagrangian (*admissible functions*). Substituting (3.86) in (3.12), and following the procedure detailed in Section 1.7.3, we obtain the discretized equations of motion as

$$\mathbf{M}\ddot{\mathbf{p}} + \mathbf{K}\mathbf{p} = \mathbf{0}, \tag{3.87}$$

where

$$\mathbf{M} = \int_0^l [\rho A \mathbf{\Psi}\mathbf{\Psi}^{\mathrm{T}} + \rho I \mathbf{\Psi}'(\mathbf{\Psi}')^{\mathrm{T}}]\,\mathrm{d}x \quad \text{and} \quad \mathbf{K} = \int_0^l EI\mathbf{\Psi}''(\mathbf{\Psi}'')^{\mathrm{T}}\,\mathrm{d}x. \tag{3.88}$$

The approximate eigenfrequencies and eigenfunctions can now be obtained as discussed in Section 1.7.3.

In the case of the Galerkin method, for convergence, $\psi_j(x)$ must satisfy all the boundary conditions of the problem, and must be differentiable at least up to the highest derivative in the equation of motion (*comparison functions*). Substituting the solution form in the equation of motion, and using the procedure of Section 1.7.4, we obtain the discretized equations of motion (3.87). However, the definitions in this case are

$$\mathbf{M} = \int_0^l [\rho A \mathbf{\Psi} \mathbf{\Psi}^{\mathrm{T}} - \mathbf{\Psi}(\rho I (\mathbf{\Psi}')^{\mathrm{T}})']\,\mathrm{d}x \quad \text{and} \quad \mathbf{K} = \int_0^l \mathbf{\Psi}[(EI\mathbf{\Psi}'')^{\mathrm{T}}]''\,\mathrm{d}x. \tag{3.89}$$

3.2.2 The initial value problem

The initial value problem for a beam is specified in terms of the initial position and velocity conditions as $w(x,0) = w_0(x)$ and $w_{,t}(x,0) = v_0(x)$. The solution of this problem can be conveniently represented as

$$w(x,t) = \sum_{j=1}^{\infty} (C_j \cos \omega_j t + S_j \sin \omega_j t) W_j(x), \tag{3.90}$$

where $W_j(x)$ are the eigenfunctions of the beam, and C_j and S_j are unknown constants which are to be determined from the initial conditions. Here we have assumed that there are no rigid body motions of the type (3.84); they could of course easily be taken into account.

Consider a simply-supported beam of uniform cross-section with a concentrated force at the center, as shown in Figure 3.12. If the force is suddenly removed, we have an initial value problem with an initial deflected shape $w_0(x)$ and zero initial velocity. The initial value problem can then be defined as

$$EIw_{,xxxx}(x,t) + \rho A w_{,tt} = 0, \qquad w(0,t) \equiv 0, \ \text{and} \ w(l,t) \equiv 0, \tag{3.91}$$

with the initial conditions $w(x,0) = w_0(x)$ and $w_{,t}(x,0) = v_0(x) \equiv 0$. The initial deflected shape can be determined from the statics boundary value problem

$$EIw_{,xxxx}(x,0) = -F\delta(x - l/2), \qquad w(0,0) = 0, \ \text{and} \ w(l,0) = 0. \tag{3.92}$$

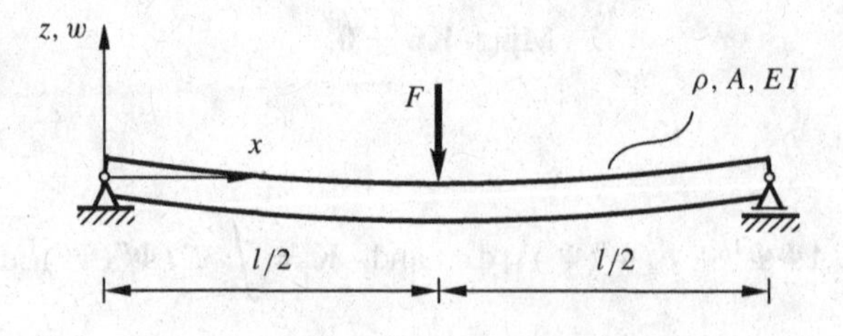

Figure 3.12 Beam with an initial constant point force

From the solution of the eigenvalue problem of a simply-supported beam, it is already known that the eigenfunctions are given by $W_n(x) = \sin n\pi x/l$. Therefore, the solution of the initial value problem (3.91) can be sought using the expansion

$$w(x,t) = \sum_{n=1}^{\infty}(C_n \cos\omega_n t + S_n \sin\omega_n t)\sin\frac{n\pi x}{l}, \tag{3.93}$$

where C_n and S_n are unknown constants to be determined from the initial conditions

$$w(x,0) = w_0(x) = \sum_{n=1}^{\infty} C_n \sin\frac{n\pi x}{l} \quad \text{and} \quad w_{,t}(x,0) = v_0(x) \equiv 0. \tag{3.94}$$

The initial shape $w_0(x)$ should satisfy (3.92). Therefore, substituting the first condition from (3.94) in (3.92), and taking the inner product of both sides with $\sin m\pi x/l$ yields on simplification

$$C_m = \begin{cases} \dfrac{2Fl^3}{m^4\pi^4 EI}(-1)^{(m-1)/2}, & m = 1, 3, 5, \ldots, \infty, \\ 0, & m = 2, 4, 6, \ldots, \infty. \end{cases}$$

Using the initial condition on the velocity $w_{,t}(x,0) \equiv 0$ one can easily obtain

$$S_m = 0, \qquad m = 1, 2, \ldots, \infty.$$

Thus, the solution of the initial value problem is of the form

$$w(x,t) = \sum_{n=1,3,5\ldots}^{\infty} \frac{2Fl^3}{n^4\pi^4 EI}(-1)^{(n-1)/2}\cos\omega_n t \sin\frac{n\pi x}{l}.$$

3.3 FORCED VIBRATION ANALYSIS

The general forced vibration problem for an Euler–Bernoulli beam can be represented as

$$\rho A w_{,tt} + (EIw_{,xx})_{,xx} = q(x,t), \tag{3.95}$$

along with the corresponding boundary conditions, where $q(x,t)$ is a general forcing function. In the following, we will discuss some solution methods for (3.95).

3.3.1 Eigenfunction expansion method

The solution of (3.95) can be written as

$$w(x,t) = w_{\mathrm{H}}(x,t) + w_{\mathrm{P}}(x,t), \tag{3.96}$$

where $w_{\mathrm{H}}(x,t)$ and $w_{\mathrm{P}}(x,t)$ are, respectively, the general solution to the homogeneous problem (i.e., $q(x,t)=0$) and a particular solution to the inhomogeneous problem. The homogeneous solution has already been discussed in the previous section, and is of the form (3.90). Consider a particular solution in the form of the eigenfunction expansion

$$w_{\mathrm{P}}(x,t) = \sum_{j=1}^{\infty} p_j(t) W_j(x), \tag{3.97}$$

where $W_j(x)$ are the eigenfunctions and $p_j(t)$ are the corresponding modal coordinates. This expression will then automatically fulfill the boundary conditions. Using (3.97) in (3.96), and substituting (3.96) in (3.95) yields

$$\sum_{j=1}^{\infty} \rho A \ddot{p}_j W_j + (EIW_j'')'' p_j = q(x,t)$$

or

$$\sum_{j=1}^{\infty} \rho A[\ddot{p}_j + \omega_j^2 p_j] W_j = q(x,t), \qquad \text{(using (3.48))}. \tag{3.98}$$

Taking the inner product with $W_k(x)$ on both sides, and using the orthonormality condition (3.40), we get

$$\ddot{p}_k + \omega_k^2 p_k = f_k(t), \qquad k = 1, 2, \ldots, \infty. \tag{3.99}$$

where

$$f_k(t) = \int_0^l q(x,t) W_k(x)\, \mathrm{d}x.$$

Thus, (3.99) represents the modal dynamics of the forced Euler–Bernoulli beam. These equations can be solved using standard techniques such as Green's function method or the Laplace transform method. The complete solution is then obtained from (3.96) as

$$w(x,t) = \sum_{j=1}^{\infty} (C_j \cos\omega_j t + S_j \sin\omega_j t) W_j(x) + \sum_{j=1}^{\infty} p_j(t) W_j(x), \tag{3.100}$$

where C_j and S_j are the constants of integration to be determined from the initial conditions.

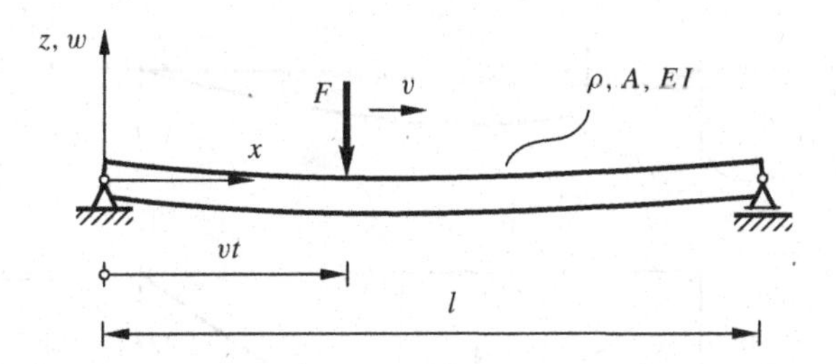

Figure 3.13 Beam with a traveling constant point force

Consider the case of a traveling force on a uniform Euler–Bernoulli beam, as shown in Figure 3.13. In this case, $q(x,t) = F\delta(x - vt)$, where v is the speed of travel of the constant force of magnitude F. Therefore, from (3.99), we have the equations of modal dynamics as

$$\ddot{p}_k + \omega_k^2 p_k = \frac{F}{\rho A} \sin \frac{k\pi vt}{l}, \qquad 0 \le t \le l/c, \qquad k = 1, 2, \ldots, \infty. \tag{3.101}$$

A particular solution of (3.101) is given by

$$p_k(t) = \frac{2Fl^3}{\pi^4 EI} \frac{\sin(k\pi vt/l)}{k^2(k^2 - \rho A l^2 v^2/\pi^2 EI)}, \qquad k = 1, 2, \ldots, \infty. \tag{3.102}$$

Using the initial conditions $w(x, 0) \equiv 0$ and $w_{,t}(x, 0) \equiv 0$ one obtains the complete solution for $0 \le t \le l/c$ as

$$w(x,t) = \frac{2Fl^3}{\pi^4 EI} \sum_{j=1}^{\infty} \frac{1}{j^2(j^2 - \rho A l^2 v^2/\pi^2 EI)} (\sin \frac{j\pi vt}{l} - \frac{j\pi v}{l\omega_j} \sin \omega_j t) \sin \frac{j\pi x}{l}. \tag{3.103}$$

The shapes of the beam at certain selected time instants are shown in Figure 3.14 for $v/l = \omega_1/4\pi$, and in Figure 3.15 for $v/l = \omega_1\pi/4$.

3.3.2 Approximate methods

The approximate methods of Ritz and Galerkin can also be used for studying the forced motion of beams. Expressing the solution of (3.95) as (3.86), the discretized equations of motion are obtained in the form

$$\mathbf{M}\ddot{\mathbf{p}} + \mathbf{K}\mathbf{p} = \mathbf{f}(t),$$

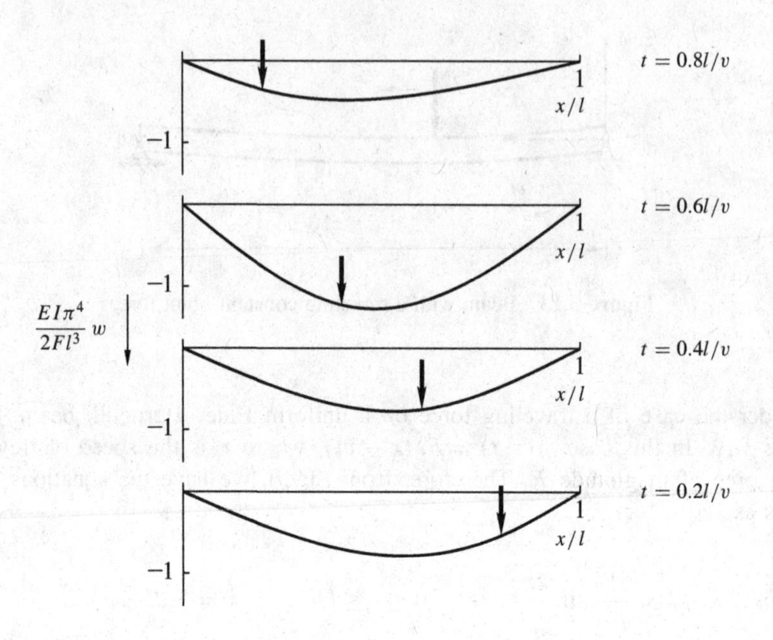

Figure 3.14 Response of a beam with constant force with with $v/l = \omega_1/4\pi$

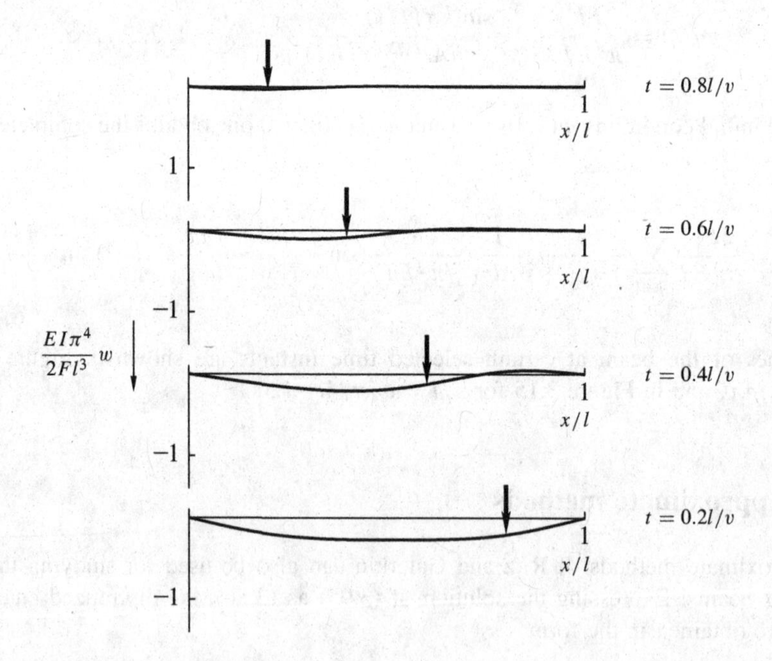

Figure 3.15 Response of a beam with constant force with with $v/l = \omega_1\pi/4$

where **M** and **K** are as defined in (3.88) (for the Ritz method) or (3.89) (for Galerkin's method), and

$$\mathbf{f}(t) = \int_0^l \boldsymbol{\Psi}(x) q(x,t)\,\mathrm{d}x, \tag{3.104}$$

where $\boldsymbol{\Psi}(x)$ is the vector of admissible functions (for the Ritz method), or comparison functions (for the Galerkin method).

3.4 NON-HOMOGENEOUS BOUNDARY CONDITIONS

As discussed before in Chapter 1, in the presence of non-homogeneous boundary conditions, we cannot use the expansion theorem for studying the dynamics of a continuous system. Moreover, generating the comparison functions for the Galerkin method also becomes difficult. Here, we use the approach discussed in Section 1.9 to convert a non-homogeneous boundary condition to a homogeneous one, along with an appropriate forcing in the equation of motion.

Consider a simply-supported Euler–Bernoulli beam with a specified time-varying moment $M(t)$ at one end, as shown in Figure 3.16. The equation of motion and the boundary conditions are given by

$$\rho A w_{,tt} + EI w_{,xxxx} = 0, \tag{3.105}$$

$$w(0,t) \equiv 0, \quad w_{,xx}(0,t) = \frac{M(t)}{EI}, \quad w(l,t) \equiv 0, \quad \text{and} \quad w_{,xx}(l,t) \equiv 0. \tag{3.106}$$

Let us rewrite the field variable $w(x,t)$ as

$$w(x,t) = u(x,t) + \eta(x)\frac{M(t)}{EI}, \tag{3.107}$$

where $u(x,t)$ is a new field variable and $\eta(x)$ is an unknown function. Substituting (3.107) in the equation of motion (3.105) yields

$$\rho A u_{,tt} + EI u_{,xxxx} = -\eta(x)\frac{\ddot{M}(t)}{EI} - \eta''''(x)\frac{M(t)}{EI}. \tag{3.108}$$

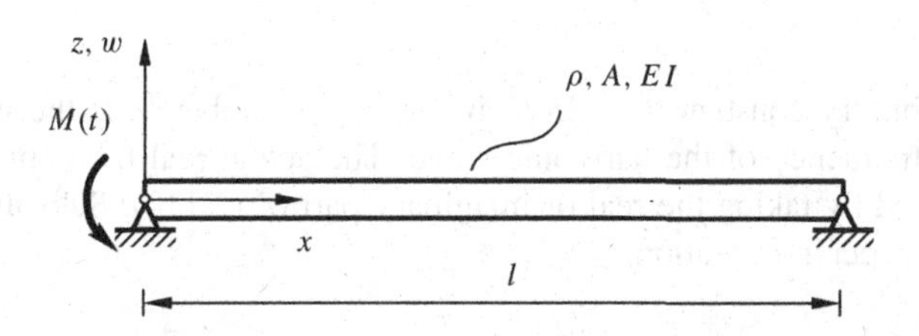

Figure 3.16 Beam with non-homogeneous boundary condition

Next, substituting (3.107) in the boundary conditions (3.106), one obtains

$$u(0,t)+\eta(0)\frac{M(t)}{EI}=0,\ \ u_{,xx}(0,t)+\eta''(0)\frac{M(t)}{EI}=\frac{M(t)}{EI},$$
$$u(l,t)+\eta(l)\frac{M(t)}{EI}=0,\ \ u_{,xx}(0,t)+\eta''(l)\frac{M(t)}{EI}=0.$$

We assume homogeneous boundary conditions for (3.108), i.e., $u(0,t)\equiv 0$, $u_{,xx}(0,t)\equiv 0$, $u(l,t)\equiv 0$, and $u_{,xx}(l,t)\equiv 0$. Then, it is evident from the above that the function $\eta(x)$ must satisfy the conditions

$$\eta(0)=0,\qquad \eta''(0)=1,\qquad \eta(l)=0,\qquad \text{and}\qquad \eta''(l)=0. \tag{3.109}$$

Let us assume $\eta(x)=a_0+a_1x+a_2x+a_3x^3$. Substituting this form of $\eta(x)$ in (3.109), one can easily obtain

$$\eta(x)=\frac{lx}{6}\left(-\frac{x^2}{l^2}+3\frac{x}{l}-2\right).$$

This determines the right-hand side of the transformed equation of motion (3.108), which now represents a simply-supported beam with forcing, and has homogeneous boundary conditions. The transformed problem can be solved easily for $u(x,t)$, and the solution of the original problem (3.105)–(3.106) is then obtained from (3.107). It must be mentioned that $\eta(x)$ is not a unique function. However, the solution of the original problem can be correctly determined by the above procedure.

3.5 DISPERSION RELATION AND FLEXURAL WAVES IN A UNIFORM BEAM

Consider a harmonic traveling wave solution for the Rayleigh beam (3.14) in the complex notation

$$w(x,t)=De^{i(kx-\omega t)}, \tag{3.110}$$

where D is an arbitrary constant, $k=2\pi/\lambda$ is the wave number, λ is the wavelength, and ω is the circular frequency of the harmonic wave. The actual real harmonic wave solution of (3.14) is obtained by taking the real or imaginary part of (3.110). Substituting (3.110) in (3.14) yields the dispersion relation

$$EIk^4-\rho I\omega^2k^2-\rho A\omega^2=0. \tag{3.111}$$

As discussed in Section 3.2.1, the four solutions of k are obtained as $k_1 = \pm i\beta_1$, and $k_2 = \pm\beta_2$, where β_1 and β_2 are defined by (3.44) and (3.45), respectively. The complete wave solution is then obtained as

$$w(x,t) = B_1 e^{\beta_1 x - i\omega t} + B_2 e^{-\beta_1 x - i\omega t} + B_3 e^{i(\beta_2 x - \omega t)} + B_4 e^{i(-\beta_2 x - \omega t)}. \tag{3.112}$$

The first two terms correspond to spatially decaying waves in the negative and positive directions, respectively, and are known as *evanescent waves* or *near-fields*. The last two terms in (3.112) are the harmonic traveling waves in the positive and negative directions, respectively. The harmonic traveling waves can propagate without distortion, and carry energy. The evanescent waves fall off exponentially with distance, and hence cannot transport energy.

The speed of propagation of a harmonic wave is given by its *phase velocity* (see Appendix B). Using the definition in Appendix B, the phase velocity of a harmonic wave in a Rayleigh beam can be obtained from (3.111) as

$$c_{\mathrm{P}}^{\mathrm{R}} = \frac{\omega}{k} = \left[\frac{EIk^2}{\rho I k^2 + \rho A}\right]^{1/2} = \frac{c_{\mathrm{L}}\tilde{k}}{(\tilde{k}^2 + 1)^{1/2}}, \tag{3.113}$$

where $c_{\mathrm{L}} = \sqrt{E/\rho}$ is the wave speed of longitudinal waves in a bar and $\tilde{k} = k\sqrt{I/A}$ is the non-dimensional wave number. For an Euler–Bernoulli beam (i.e., with no rotary inertia term), the phase velocity is obtained as

$$c_{\mathrm{P}}^{\mathrm{EB}} = \left[\frac{EIk^2}{\rho A}\right]^{1/2} = c_{\mathrm{L}}\tilde{k}. \tag{3.114}$$

The variation of the non-dimensional phase velocities $c_{\mathrm{P}}^{\mathrm{R}}/c_{\mathrm{L}}$ and $c_{\mathrm{P}}^{\mathrm{EB}}/c_{\mathrm{L}}$ with the non-dimensional wave number $\tilde{k}$ are compared in Figure 3.17. It is observed from the figure that these phase velocities match for low wave numbers. However, the phase velocity of the Euler–Bernoulli beam increases indefinitely with increasing wave numbers. This is unrealistic, and a drawback of the Euler–Bernoulli beam theory. It can be easily concluded

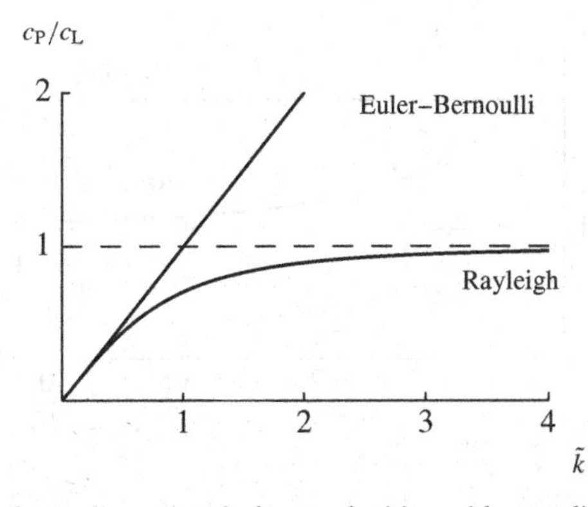

Figure 3.17 Variation of non-dimensional phase velocities with non-dimensional wave number for Euler–Bernoulli and Rayleigh beams

that the presence of rotary inertia in the Rayleigh beam is responsible for limiting the phase velocity at high wave numbers (i.e., small wavelengths).

When a wave pulse consisting of harmonic waves of different wave numbers travels in a beam, the speed of the pulse is determined by the *group velocity* of waves in the medium (see Appendix B). Following the definition in Appendix B, one can easily compute the group velocity of a wave group in a Rayleigh beam from the dispersion relation (3.111) as

$$c_G = c_P\left[1 + \frac{1}{\tilde{k}^2+1}\right]. \tag{3.115}$$

It can be observed from (3.115) that $c_G > c_P$. For small wave numbers (large wavelengths), i.e., $\tilde{k} \to 0$, we have $c_G \approx 2c_P$, while $\tilde{k} \to \infty$ yields $c_G \approx c_P$. In the case of the Euler–Bernoulli beam it can be easily checked that $c_G = 2c_P$ for all values of $\tilde{k}$. The variations of the ratio of group and phase velocities with the wave number for Rayleigh and Euler–Bernoulli beams are compared in Figure 3.18.

3.5.1 Energy transport

The total mechanical energy contained in a finite interval $[x_1, x_2]$ of the beam is given by

$$\begin{aligned}\mathcal{E} = \mathcal{T} + \mathcal{V} &= \int_{x_1}^{x_2}\left(\frac{1}{2}\rho A w_{,t}^2 + \frac{1}{2}EI w_{,xx}^2 + \frac{1}{2}\rho I w_{,xt}^2\right)\mathrm{d}x,\\ &= \int_{x_1}^{x_2} \hat{\mathcal{E}}\,\mathrm{d}x,\end{aligned} \tag{3.116}$$

where $\hat{\mathcal{E}}$ is the mechanical energy density. This leads to

$$\frac{\mathrm{d}\mathcal{E}}{\mathrm{d}t} = \int_{x_1}^{x2}[\rho A w_{,t} w_{,tt} + EI w_{,xx} w_{,xxt} + \rho I w_{,xt} w_{,xtt}]\,\mathrm{d}x. \tag{3.117}$$

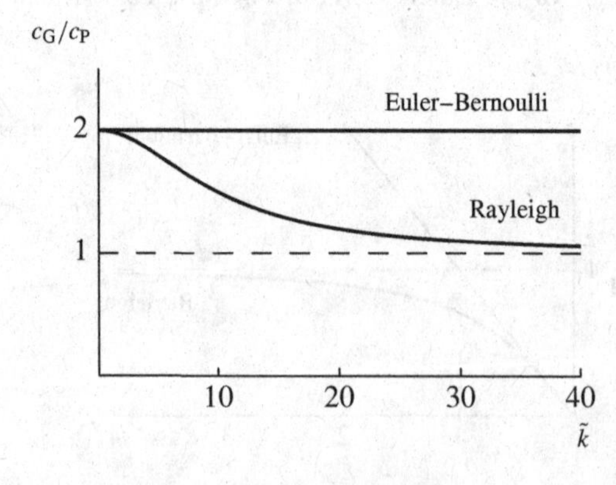

Figure 3.18 Variation of ratio of group and phase velocities with non-dimensional wave number for Euler–Bernoulli and Rayleigh beams

Integrating by parts, one can rewrite the above expression as

$$\begin{aligned}\frac{\mathrm{d}\mathcal{E}}{\mathrm{d}t} &= [EIw_{,xx}w_{,xt} - EIw_{,xxx}w_{,t} + \rho Iw_{,xtt}w_{,t}]\Big|_{x_1}^{x_2} \\ &\quad + \int_{x_1}^{x_2} [\rho Aw_{,tt} + EIw_{,xxxx} - \rho Iw_{,xxtt}]w_{,t}\,\mathrm{d}x \\ &= [EIw_{,xx}w_{,xt} - EIw_{,xxx}w_{,t} + \rho Iw_{,xtt}w_{,t}]\Big|_{x_1}^{x_2} \\ &= \mathcal{P}(x_1, t) - \mathcal{P}(x_2, t),\end{aligned}$$

where we have used the equation of motion of the Rayleigh beam, and defined

$$\mathcal{P}(x, t) := -EIw_{,xx}w_{,xt} + EIw_{,xxx}w_{,t} - \rho Iw_{,xtt}w_{,t}, \tag{3.118}$$

representing the instantaneous power flowing in the positive x-axis direction past a coordinate location x. For a harmonic wave of frequency ω traveling through the beam, the average power flowing past any point of the beam can be represented by

$$\langle\mathcal{P}\rangle = \frac{\omega}{2\pi}\int_0^{2\pi/\omega} [-EIw_{,xx}w_{,xt} + EIw_{,xxx}w_{,t} - \rho Iw_{,xtt}w_{,t}]\,\mathrm{d}t. \tag{3.119}$$

Consider a positive-traveling harmonic wave in the beam represented by

$$w(x, t) = \sin(kx - \omega t). \tag{3.120}$$

Substituting (3.120) in (3.119), the average power flowing across any point is obtained as

$$\begin{aligned}\langle\mathcal{P}\rangle &= \frac{\omega}{2\pi}\int_0^{2\pi/\omega} [EI\omega k^3 - \rho I\omega^3 k\cos^2(kx - \omega t)]\,\mathrm{d}t \\ &= \frac{\omega}{2\pi}\int_0^{2\pi/\omega} \left[\frac{\omega^3}{k}(\rho Ik^2 + \rho A) + \rho I\omega^3 k\cos^2(kx - \omega t)\right]\mathrm{d}t \quad \text{(using (3.111))} \\ &= \frac{\rho A\omega^3}{k}\left[1 + \frac{1}{2}r_g^2k^2\right].\end{aligned} \tag{3.121}$$

The mechanical energy density $\hat{\mathcal{E}}$ of a Rayleigh beam was defined through (3.116). When a harmonic wave (3.120) travels through the beam, one can define an average mechanical energy density as

$$\begin{aligned}\langle\hat{\mathcal{E}}\rangle &= \frac{\omega}{2\pi}\int_0^{2\pi/\omega} \hat{\mathcal{E}}\,\mathrm{d}t \\ &= \rho A\omega^2\frac{\omega}{2\pi}\int_0^{2\pi/\omega}\left[\frac{1}{2} + r_g^2k^2\sin^2(kx - \omega t)\right]\mathrm{d}t \qquad \text{(using (3.111))} \\ &= \rho A\omega^2\left[\frac{1}{2} + \frac{1}{2}r_g^2k^2\right].\end{aligned} \tag{3.122}$$

Comparing (3.121) and (3.122), one can write

$$\langle \mathcal{P} \rangle = \frac{\omega}{k}\left[1 + \frac{1}{r_g^2 k^2 + 1}\right]\langle \hat{\mathcal{E}} \rangle$$
$$= c_{\mathrm{G}}\langle \hat{\mathcal{E}} \rangle \qquad \text{(using (3.115))}. \tag{3.123}$$

This shows that energy propagates at the group velocity of harmonic flexural waves in a Rayleigh beam. It is not difficult to show that the same holds for an Euler–Bernoulli beam.

3.5.2 Scattering of flexural waves

Consider a positive-traveling harmonic wave incident on the boundary of a semi-infinite Euler–Bernoulli beam, as shown in Figure 3.19. As discussed at the beginning of Section 3.5, there can exist two kinds of waves in a beam, namely the harmonic traveling waves and the evanescent waves (or near-fields). Therefore, the reflected wave can consist of both a negative-traveling harmonic wave and an evanescent wave. The complete wave field in the beam can be written as

$$w(x,t) = A\mathrm{e}^{i(k_1 x - \omega t)} + B\mathrm{e}^{i(-k_1 x - \omega t)} + C\mathrm{e}^{k_2 x - i\omega t}, \tag{3.124}$$

where k_1 is the wave number of the harmonic waves, which may be different from the wave number k_2 of the evanescent wave. For a given frequency ω, the values of k_1 and k_2 are computed from the dispersion relation (3.111). It is to be noted that the positive solution of k_2 is to be taken to ensure that the solution is finite (and hence realistic) as $x \to -\infty$. The boundary conditions at $x = 0$ are given by

$$w_{,xx}(0,t) \equiv -\alpha_{\mathrm{M}} w_{,x}(0,t) \qquad \text{and} \qquad w_{,xxx}(0,t) \equiv \alpha_{\mathrm{L}} w(0,t), \tag{3.125}$$

where $\alpha_{\mathrm{M}} = k_{\mathrm{M}}/EI$, k_{M} is the angular stiffness, $\alpha_{\mathrm{L}} = k_{\mathrm{L}}/EI$, and k_{L} is the linear stiffness. Substituting (3.124) in the boundary conditions above yields, respectively,

$$(ik\alpha_{\mathrm{M}} - k_1^2)A - (ik\alpha_{\mathrm{M}} + k_1^2)B + (\alpha_{\mathrm{M}} k_2 + k_2^2)C = 0$$

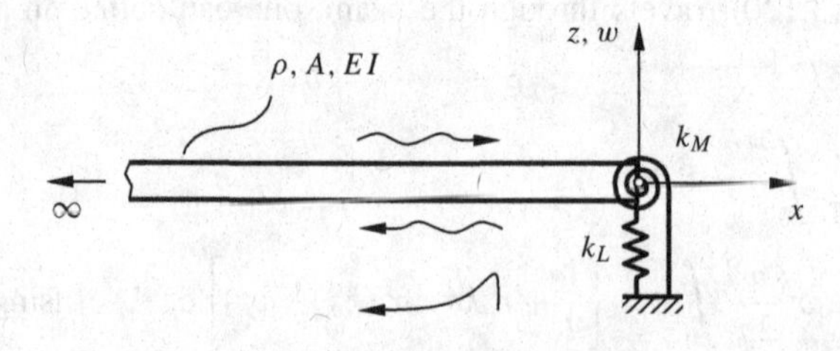

Figure 3.19 Boundary scattering in a semi-infinite beam

and

$$-(\alpha_L + ik^3)A - (-\alpha_L + ik^3)B + (-\alpha_L + k_2^3)C = 0.$$

One can now define the reflection coefficients C_{RH} and C_{RE} for the harmonic wave and the evanescent wave, respectively, as

$$C_{RH} := \frac{B}{A} = \frac{(-\alpha_L + k_2^3)(ik\alpha_M - k^2) + (\alpha_M k_2 + k_2^2)(\alpha_L + ik_1^3)}{(-\alpha_L + k_2^3)(ik\alpha_M + k_1^2) + (\alpha_M k_2 + k_2^2)(-\alpha_L + ik_1^3)} \quad (3.126)$$

and

$$C_{RE} := \frac{C}{A} = \frac{1}{\alpha_L - k_2^3}\left[-\alpha_L\left(\frac{B}{A} + 1\right) + ik^3\left(\frac{B}{A} - 1\right)\right]. \quad (3.127)$$

Various special cases can now be considered as follows:

1. **Pinned end:** When $\alpha_M = 0$ and $\alpha_L = \infty$, we have a pinned boundary. The reflection coefficients can be obtained from (3.126) and (3.127) as

$$C_{RH} = -1 \quad \text{and} \quad C_{RE} = 0.$$

2. **Clamped end:** A clamped boundary condition is realized when $\alpha_M = \infty$ and $\alpha_L = \infty$. In this case, the reflection coefficients are obtained as

$$C_{RH} = \frac{-k_2 + ik_1}{k_2 + ik_1} \quad \text{and} \quad C_{RE} = -\frac{2ik_1}{ik_1 + k_2}.$$

3. **Sliding end:** When $\alpha_M = \infty$ and $\alpha_L = 0$, we obtain the reflection coefficients for a sliding boundary, as shown in Figure 3.4. The reflection coefficients are given by

$$C_{RH} = 1 \quad \text{and} \quad C_{RE} = 0.$$

4. **Free end:** The values $\alpha_M = 0$ and $\alpha_L = 0$ correspond to a free boundary for which we have

$$C_{RH} = \frac{-k_2 + ik_1}{k_2 + ik_1} \quad \text{and} \quad C_{RE} = \frac{k_1^2}{k_2^2}\frac{2ik_1}{k_2 + ik_1}.$$

It is interesting to observe that there is no reflected evanescent wave for either a pinned or a sliding boundary.

3.6 THE TIMOSHENKO BEAM

In the previous sections, we have neglected the effect of shear deformation on the dynamics of a beam. However, just like rotary inertia, shearing of a beam cross-section has significant effects at high wave numbers (i.e., for large curvatures). A beam model taking into consideration the shear deformation effect is usually known as the Timoshenko beam model.[1] In this section, we will discuss the modeling, and some elementary aspects of the Timoshenko (or Bresse) beam.

3.6.1 Equations of motion

3.6.1.1 The Newtonian formulation

Consider the beam element subjected to bending and shear deformations, as shown in Figure 3.20. Let the angular deformation field variable due to pure moment be denoted by $\psi(x, t)$, and the deformation field variable due to shear be denoted by $\theta(x, t)$. Thus, it is evident that the Timoshenko beam requires two field variables for complete description of its configuration. From Figure 3.20, one can easily write

$$w_{,x}(x, t) = \psi(x, t) + \theta(x, t). \tag{3.128}$$

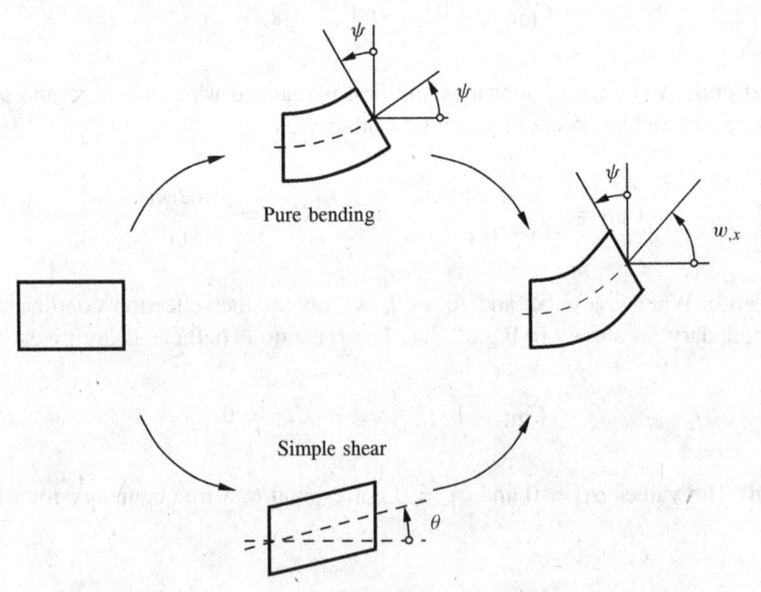

Figure 3.20 Deformation components of a Timoshenko beam element

[1] Such a beam model was proposed earlier by J.A. Bresse in 1859 in his book *Cours de mécanique Applique* (see [3]), which was largely unknown when Timoshenko proposed his model in 1922.

where $w(x,t)$ represents the displacement field of a neutral fiber of the beam. The relation (3.128) allows us to choose $w(x,t)$ as one of the field variables. In the following, we will take $w(x,t)$ and $\psi(x,t)$ as the two field variables.

Since the longitudinal strain ϵ_x in the beam is produced only from bending, one can write

$$\epsilon_x = \frac{(R-z)\,\mathrm{d}\psi - R\,\mathrm{d}\psi}{\mathrm{d}x} = -z\psi_{,x}. \tag{3.129}$$

Using Hooke's law, the longitudinal stress is obtained as $\sigma_x = -Ez\psi_x$, and the bending moment can be expressed as

$$M = -\int_{-h/2}^{h/2} \sigma_x z\,\mathrm{d}A = EI\psi_x. \tag{3.130}$$

Similarly, the net shear force acting at a section can be written as

$$V = GA_s\theta = GA_s(w_{,x} - \psi),$$

where $A_s = A/\kappa$, A is the area of cross-section of the beam, and κ is known as the *shear correction factor*. The shear correction is introduced to take care of the non-uniformity in the shear force across the section. For a rectangular section $\kappa \approx 1.20$, for a circular section $\kappa \approx 1.11$, while for an I-section $\kappa \approx 2$–2.4. Writing the Newton's second law for the transverse motion yields

$$(\rho A\,\mathrm{d}x)w_{,tt} = V(x+\mathrm{d}x) - V(x)$$
$$\Rightarrow \rho A w_{,tt} = V_{,x}$$

or

$$\rho A w_{,tt} = [GA_s(w_{,x} - \psi)]_{,x}. \tag{3.131}$$

The equation of rotational dynamics can be written as

$$(\rho I\,\mathrm{d}x)\psi_{,tt} = V\frac{\mathrm{d}x}{2} + (V + \mathrm{d}V)\frac{\mathrm{d}x}{2} + (M + \mathrm{d}M) - M$$
$$\Rightarrow \rho I\psi_{,tt} = V + \frac{\mathrm{d}M}{\mathrm{d}x}$$

or

$$\rho I\psi_{,tt} = GA_s(w_{,x} - \psi) + [EI\psi_{,x}]_{,x}. \tag{3.132}$$

The two differential equations (3.131) and (3.132) in $w(x,t)$ and $\psi(x,t)$ represent the dynamics of a Timoshenko beam. The boundary conditions of a Timoshenko beam are derived in the next section.

In the case of a uniform beam, we have the simplification

$$\rho A w_{,tt} = G A_s (w_{,xx} - \psi_{,x}) \tag{3.133}$$

and

$$\rho I \psi_{,tt} = G A_s (w_{,x} - \psi) + E I \psi_{,xx}. \tag{3.134}$$

Differentiating (3.134) once with respect to x we have

$$\rho I \psi_{,xtt} = G A_s (w_{,xx} - \psi_{,x}) + E I \psi_{,xxx}. \tag{3.135}$$

Solving for $\psi_{,x}$ from (3.133) and substituting in (3.135) yields on simplification

$$\frac{\rho I}{G A_s} w_{,tttt} + w_{,tt} - \left(\frac{I}{A} + \frac{E I}{G A_s} \right) w_{,ttxx} + \frac{E I}{\rho A} w_{,xxxx} = 0. \tag{3.136}$$

This is a fourth-order differential equation in space and time, and represents the transverse dynamics of a uniform Timoshenko beam.

3.6.1.2 The variational formulation

The total kinetic energy density of a beam element consists of the translational and rotational kinetic energy densities, and is given by

$$\hat{\mathcal{T}} = \frac{1}{2} \rho A w_{,t}^2 + \frac{1}{2} \rho I \psi_{,t}^2. \tag{3.137}$$

From theory of elasticity, the potential energy density can be written as

$$\begin{aligned} \hat{\mathcal{V}} &= \frac{1}{2} E I \psi_{,x}^2 + \frac{1}{2} G A_s \theta^2 \\ &= \frac{1}{2} E I \psi_{,x}^2 + \frac{1}{2} G A_s (w_{,x} - \psi)^2. \end{aligned} \tag{3.138}$$

Hamilton's variational principle for the dynamics of the beam yields

$$\begin{aligned} &\delta \int_{t_1}^{t_2} \int_0^l (\hat{\mathcal{T}} - \hat{\mathcal{V}}) \,\mathrm{d}x \,\mathrm{d}t = 0 \\ \Rightarrow &\int_{t_1}^{t_2} \int_0^l \Big[\rho A w_{,t} \, \delta w_{,t} + \rho I \psi_{,t} \, \delta \psi_{,t} \\ &\quad - E I \psi_{,x} \, \delta \psi_{,x} - G A_s (w_{,x} - \psi)(\delta w_{,x} - \delta \psi) \Big] \,\mathrm{d}x \,\mathrm{d}t = 0. \end{aligned} \tag{3.139}$$

Integrating by parts and rearranging, we have

$$\int_0^l \left[\rho A w_{,t}\,\delta w + \rho I \psi_{,t}\,\delta\psi\right]\Bigg|_{t_1}^{t_2} \mathrm{d}x$$
$$+\int_{t_1}^{t_2}\left[-EI\psi_{,x}\,\delta\psi - GA_{\mathrm{s}}(w_{,x}-\psi)\,\delta w\right]\Bigg|_0^l \mathrm{d}t$$
$$+\int_{t_1}^{t_2}\int_0^l \Big[(-\rho A w_{,tt} + [GA_{\mathrm{s}}(w_{,x}-\psi)]_{,x})\,\delta w$$
$$+(-\rho I\psi_{,tt} + (EI\psi_{,x})_{,x} + GA_{\mathrm{s}}(w_{,x}-\psi))\,\delta\psi\Big]\,\mathrm{d}x\,\mathrm{d}t = 0. \quad (3.140)$$

The first integral above vanishes from the statement of the variational principle. The integrand in the double integral yields the two equations of motion (3.131) and (3.132). The boundary conditions are obtained from the integrand of the second integral in (3.140), for example, as

$$[EI\psi_{,x}](0,t) \equiv 0 \quad \text{or} \quad \psi(0,t) \equiv 0,$$
$$[EI\psi_{,x}](l,t) \equiv 0 \quad \text{or} \quad \psi(l,t) \equiv 0,$$
$$[GA_s(w_{,x}-\psi)](0,t) \equiv 0 \quad \text{or} \quad w(0,t) \equiv 0,$$

and

$$[GA_s(w_{,x}-\psi)](l,t) \equiv 0 \quad \text{or} \quad w(l,t) \equiv 0.$$

3.6.2 Harmonic waves and dispersion relation

Consider an infinite uniform Timoshenko beam as described by (3.136). It can be easily checked that one can rewrite (3.136) as

$$\left(c_{\mathrm{L}}^2\frac{\partial^2}{\partial x^2} - \frac{\partial^2}{\partial t^2}\right)\left(\frac{c_{\mathrm{S}}^2}{\kappa}\frac{\partial^2}{\partial x^2} - \frac{\partial^2}{\partial t^2}\right) w + \frac{c_{\mathrm{S}}^2}{\kappa r_{\mathrm{g}}^2}\frac{\partial^2 w}{\partial t^2} = 0, \quad (3.141)$$

where $c_{\mathrm{L}} = \sqrt{E/\rho}$ is the longitudinal wave speed, $c_{\mathrm{S}} = \sqrt{G/\rho}$ is the shear wave speed, and $r_{\mathrm{g}} = \sqrt{I/A}$ is the radius of gyration. Assume a harmonic traveling wave solution of the form

$$w(x,t) = D\mathrm{e}^{i(kx-\omega t)}.$$

Substituting this solution in (3.141) yields, on rearrangement, the dispersion relation

$$(-c_{\mathrm{L}}^2 k^2 + \omega^2)\left(-\frac{c_{\mathrm{S}}^2}{\kappa}k^2 + \omega^2\right) - \frac{c_{\mathrm{S}}^2}{\kappa r_{\mathrm{g}}^2}\omega^2 = 0 \quad (3.142)$$

or

$$(\tilde{\omega}^2 - \tilde{k}^2)(\tilde{\omega}^2 - \gamma^2\tilde{k}^2) - \gamma^2\tilde{\omega}^2 = 0, \tag{3.143}$$

where $\tilde{\omega} = \omega r_g/c_L$ and $\tilde{k} = r_g k$ are non-dimensional frequency and wave number, respectively, and $\gamma^2 = c_S^2/\kappa c_L^2$ is a non-dimensional parameter.

Solving for $\tilde{\omega}$ from (3.143), we obtain

$$\tilde{\omega}_1(\tilde{k}) = \pm\frac{1}{\sqrt{2}}\left[\gamma^2 + (1+\gamma^2)\tilde{k}^2 + \sqrt{[(1-\gamma)^2\tilde{k}^2 + \gamma^2][(1+\gamma)^2\tilde{k}^2 + \gamma^2]}\right]^{1/2}, \tag{3.144}$$

and

$$\tilde{\omega}_2(\tilde{k}) = \pm\frac{1}{\sqrt{2}}\left[\gamma^2 + (1+\gamma^2)\tilde{k}^2 - \sqrt{[(1-\gamma)^2\tilde{k}^2 + \gamma^2][(1+\gamma)^2\tilde{k}^2 + \gamma^2]}\right]^{1/2}. \tag{3.145}$$

Thus, there are two branches of the dispersion relation of a Timoshenko beam, which are discussed further later on. The variations of $\tilde{\omega}_1(\tilde{k})$ and $\tilde{\omega}_2(\tilde{k})$ with $\tilde{k}$ for $\gamma^2 = 1/3$ are shown in Figure 3.21. It is observed from the figure that, below a certain *cut-off frequency* given by $\tilde{\omega}_c = \gamma$, there is only one real value of $\tilde{k}$, implying that there is only one propagating mode. The non-dimensional phase velocity $c_P/c_L = \tilde{\omega}/\tilde{k}$, and the non-dimensional group velocity $c_G/c_L = \mathrm{d}\tilde{\omega}/\mathrm{d}\tilde{k}$ can be easily obtained from (3.144)–(3.145), and are plotted in Figure 3.22 for $\gamma^2 = 1/3$.

A comparison of the Euler–Bernoulli, Rayleigh and Timoshenko beam theories in terms of the non-dimensional phase speeds is shown in Figure 3.23. Some significant observations can be made in this figure. For $\tilde{k} \to 0$, the non-dimensional phase speed for the Euler–Bernoulli and the Rayleigh beam theories match with the lower branch of the Timoshenko beam. It can be shown that, for small wave numbers, the lower branch corresponds essentially to flexural mode of vibration of the beam, while the upper branch

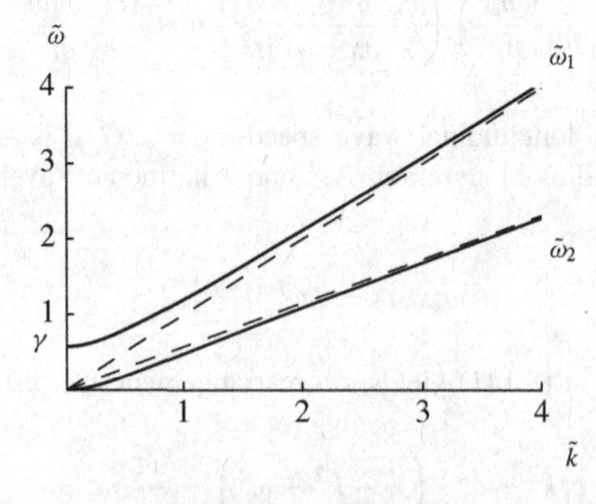

Figure 3.21 Non-dimensional dispersion relation of a Timoshenko beam

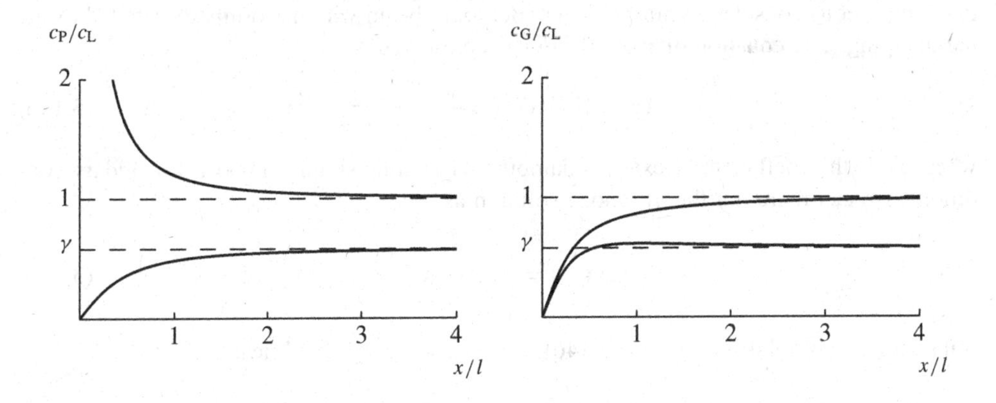

Figure 3.22 Non-dimensional phase velocity (c_P/c_L), and non-dimensional group velocity (c_G/c_L) of the two wave modes in a Timoshenko beam

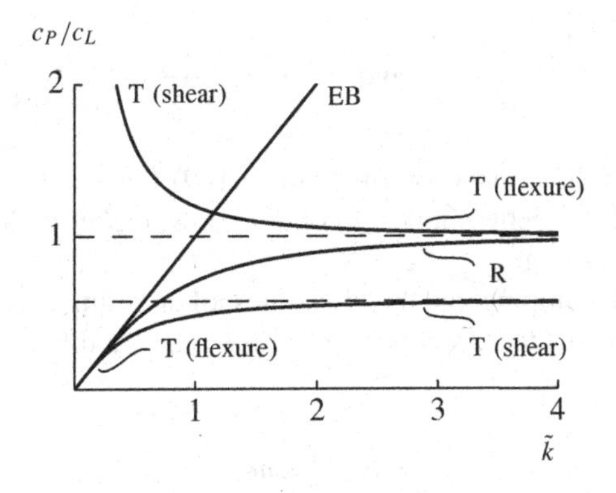

Figure 3.23 Comparison of non-dimensional phase speeds of Euler–Bernoulli (EB), Rayleigh (R) and Timoshenko (T) beams

corresponds essentially to the shear mode of vibration (see [4]). As $\tilde{k} \to \infty$, the branch of the Rayleigh beam and the upper branch of the Timoshenko beam tend to match. Both these branches correspond essentially to flexural mode of vibration of the beam at high values of $\tilde{k}$ as observed in Figure 3.23. Another interesting observation in Figure 3.23 is that, for $\tilde{k} \to 0$, the non-dimensional phase speed of the shear mode tends to infinity. This implies that the beam tends to become infinitely stiff in shear. Hence, for small wave numbers, the shearing of the beam can be neglected, and the Euler–Bernoulli or Rayleigh beam models are adequate for analysis.

3.7 DAMPED VIBRATION OF BEAMS

The motion of a beam can be damped by either external or internal damping. The external damping can be distributed or discrete, while the internal damping is usually a distributed damping. Let us consider a uniform Euler–Bernoulli beam with uniformly distributed external damping. The equation of motion can be represented as

$$\rho A w_{,tt} + EI w_{,xxxx} + d_{\mathrm{E}} w_{,t} = 0, \tag{3.146}$$

where d_{E} is the coefficient of external damping. Assuming simply-supported boundary conditions, we can represent the nth-mode solution as

$$w_n(x,t) = p_n(t) \sin \frac{n\pi x}{l}. \tag{3.147}$$

Substituting this solution form in (3.146), one can write on simplification

$$\ddot{p}_n + 2\zeta_n \omega_n \dot{p}_n + \omega_n^2 p_n = 0, \tag{3.148}$$

where ω_n is the natural frequency of the undamped beam, and ζ_n is the damping factor, which are given by, respectively,

$$\omega_n^2 = \frac{n^4 \pi^4}{l^4} \frac{EI}{\rho A} \quad \text{and} \quad \zeta_n = \frac{l^2}{n^2 \pi^2} \frac{d_{\mathrm{E}}}{2\sqrt{EI\rho A}}. \tag{3.149}$$

It can be concluded from the expression of ζ_n in (3.149) that, with external damping, the damping factor diminishes at higher modes. In other words, higher modes cannot be damped effectively by external damping.

The effect of internal damping on the dynamics of a beam may be incorporated by including a viscous term in the expression of stress. The modified Hooke's law is then written as

$$\sigma = E\epsilon + E\eta\dot{\epsilon}, \tag{3.150}$$

where η is known as the *loss factor*. Using this expression, and following the steps discussed in Section 3.1.1, one can easily obtain the equation of motion of an internally damped Euler–Bernoulli beam as

$$\rho A w_{,tt} + (\eta EI w_{,txx})_{,xx} + (EI w_{,xx})_{,xx} = 0. \tag{3.151}$$

In the case of a uniform beam with simply-supported boundary conditions, substitution of the solution form (3.147) into the equation of motion yields on simplification the temporal dynamics in the form (3.148). However, the expression of ζ in this case is

$$\zeta_n = \frac{n^2 \pi^2}{l^2} \frac{\eta}{2} \sqrt{\frac{EI}{\rho A}}.$$

It is interesting to observe here that the damping factor increases quadratically with the mode number. Thus, the suppression of higher modes of vibration in a beam is primarily due to internal damping, if a material law of the type (3.150) is assumed.

3.8 SPECIAL PROBLEMS IN VIBRATIONS OF BEAMS

In this section, we will discuss a few special problems to illustrate some interesting phenomena associated with vibrations of beams.

3.8.1 Influence of axial force on dynamic stability

The presence of axial force can greatly influence the dynamics of a beam. In this section, we consider two types of axial forces resulting in two distinctly different types of dynamic behavior.

3.8.1.1 Simply-supported beam with constant axial force

Consider a simply-supported uniform Euler–Bernoulli beam with a constant compressive axial force F applied at the end, as shown in Figure 3.24. The equation of motion is

$$\rho A w_{,tt} + F w_{,xx} + EI w_{,xxxx} = 0. \tag{3.152}$$

Consider a modal solution

$$w(x,t) = \sum_{j=1}^{\infty} p_j(t) \sin \frac{j\pi x}{l}.$$

Substituting this solution in (3.152) and taking the inner product with the kth-mode-shape yields on simplification

$$\rho A \ddot{p}_k + \frac{k^2\pi^2}{l^2}\left(EI\frac{k^2\pi^2}{l^2} - F\right) p_k = 0. \tag{3.153}$$

Setting $p_k(t) = Ce^{st}$, it is straightforward to show that $s = \pm i\omega_k$, where

$$\omega_k = \sqrt{\frac{k^2\pi^2(EIk^2\pi^2/l^2 - F)/l^2}{\rho A}} \tag{3.154}$$

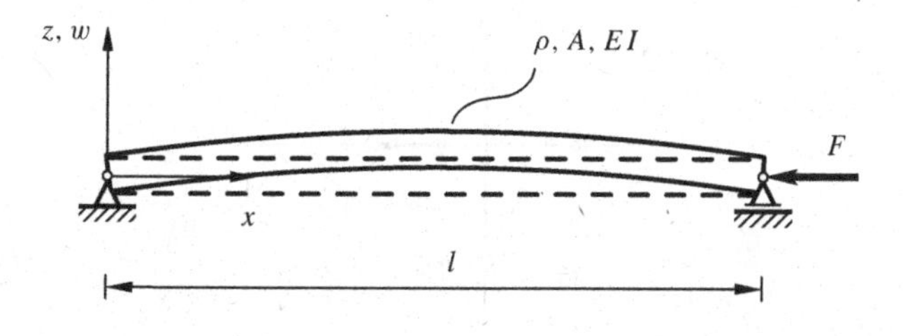

Figure 3.24 A simply-supported beam with a constant axial force

is the kth circular eigenfrequency. One can rewrite (3.154) in non-dimensional form as

$$\tilde{\omega}_k = \sqrt{k^2\pi^2(k^2\pi^2 - S)},$$

where $\tilde{\omega}_k = \omega_k\sqrt{\rho A l^4/EI}$ and $S = Fl^2/EI$. For $k = 1$, the variation of the first eigenfrequency with axial force can be visualized from Figure 3.25. It is clear from (3.154) that if the axial force F crosses the critical value $F_1^c = \pi^2 EI/l^2$, the first eigenfrequency becomes imaginary (i.e., s in the solution $p_k(t) = Ce^{st}$ is real), implying the existence of a divergent solution of (3.153). This critical value of axial force F_1^c can be easily recognized to be the first Euler buckling load of a simply-supported uniform Euler–Bernoulli beam. Thus, the undeformed equilibrium configuration of the beam becomes linearly unstable, and the beam suffers a *divergence instability*. This kind of instability is characterized by the eigenvalues s going through zero before becoming real as shown in Figure 3.26.

3.8.1.2 Cantilever beam with follower force

Consider a constant tip-force on a cantilever Euler–Bernoulli beam always acting along the neutral fiber, as shown in Figure 3.27. Such a force is usually named a *follower force*. It it

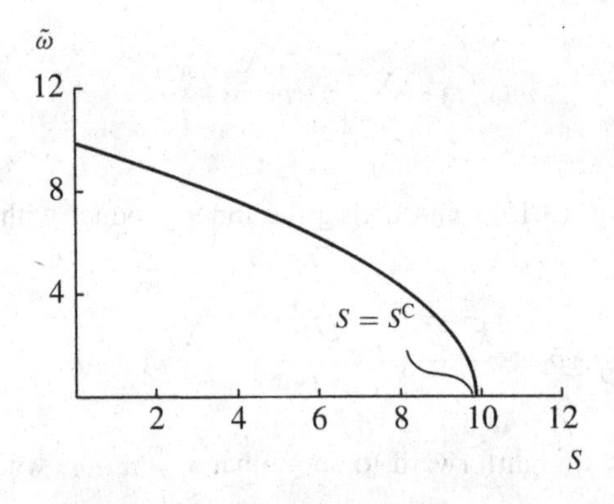

Figure 3.25 Variation of the first non-dimensional circular eigenfrequency $\tilde{\omega}$ of a simply-supported beam with non-dimensional axial force S

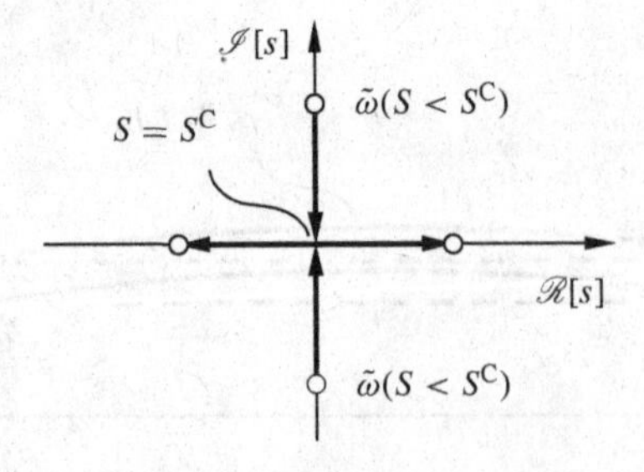

Figure 3.26 Loci of eigenfrequencies of a simply-supported beam with variation of axial force

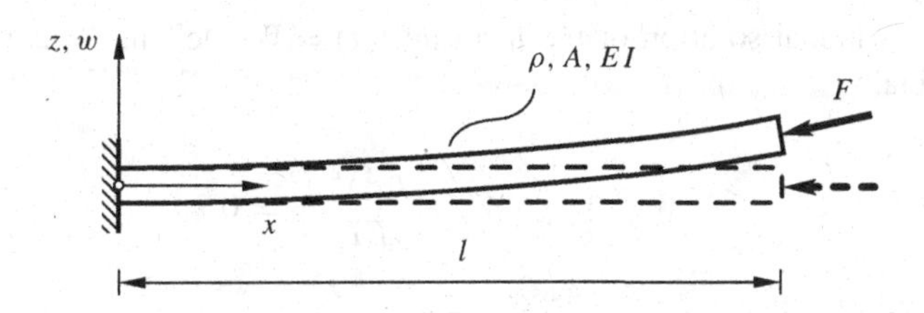

Figure 3.27 A cantilever beam with a follower force

important to observe here that the force has a component in the transverse direction given by $-Fw_{,x}(l,t)$, for $w_{,x}(l,t) \ll 1$. This transverse force component, which now has to be introduced in the dynamics, cannot be derived from a potential energy function. Hence, the work integral depends on the path of integration, implying thereby that the force is non-conservative.

The equation of motion in this case may be obtained using the *extended Hamilton's principle* (see Appendix A)

$$\int_{t_1}^{t_2} (\delta\mathcal{L} + \overline{\delta\mathcal{W}})\,\mathrm{d}t = 0, \tag{3.155}$$

where

$$\mathcal{L} = \frac{1}{2}\int_0^l (\rho A w_{,t}^2 - EIw_{,xx}^2 + Fw_{,x}^2)\,\mathrm{d}x \tag{3.156}$$

and

$$\overline{\delta\mathcal{W}} = -Fw_{,x}(l,t)\delta w(l,t). \tag{3.157}$$

Substituting (3.156) and (3.157) in (3.155), and following the steps discussed in Appendix A, one obtains

$$\begin{aligned}&\int_{t_1}^{t_2} \left[\left(-EIw_{,xx}\delta w_{,x} + (EIw_{,xxx} + Fw_{,x})\delta w\right)\Big|_0^l - Fw_{,x}(l,t)\delta w(l,t)\right]\mathrm{d}t\\ &\quad - \int_{t_1}^{t_2}\int_0^l (\rho A w_{,tt} + EIw_{,xxxx} + Fw_{,xx})\,\delta w\,\mathrm{d}x\,\mathrm{d}t.\end{aligned} \tag{3.158}$$

Thus, the equation of motion and boundary conditions for the problem are given by

$$\rho A w_{,tt} + EIw_{,xxxx} + Fw_{,xx} = 0, \tag{3.159}$$

$$w(0,t) \equiv 0, \quad w_{,x}(0,t) \equiv 0, \quad w_{,xx}(l,t) \equiv 0, \quad \text{and} \quad w_{,xxx}(l,t) \equiv 0. \tag{3.160}$$

Substituting a modal solution of the form $w(x,t) = W(x)e^{st}$ in the equation of motion (3.159), we obtain

$$W'''' + \frac{F}{EI}W'' + \frac{\rho A s^2}{EI}W = 0. \tag{3.161}$$

Writing the solution of (3.161) in the form $W(x) = e^{\beta x}$, we get

$$\beta^4 + \frac{F}{EI}\beta^2 + \frac{\rho A s^2}{EI} = 0. \tag{3.162}$$

Using the definitions

$$S := \frac{Fl^2}{EI} \quad \text{and} \quad \tilde{\omega}^2 := -\frac{\rho A s^2 l^4}{EI}, \tag{3.163}$$

one can rewrite (3.162) as

$$l^4\beta^4 + Sl^2\beta^2 - \tilde{\omega}^2 = 0. \tag{3.164}$$

Therefore, the general solution of $W(x)$ can be represented as

$$W(x) = B_1 \cosh \beta_1 x + B_2 \sinh \beta_1 x + B_3 \cos \beta_2 x + B_4 \sin \beta_2 x, \tag{3.165}$$

where β_1 and β_2 are obtained from (3.164) as

$$\beta_1 = \frac{1}{\sqrt{2}l}[-S + \sqrt{S^2 + 4\tilde{\omega}^2}]^{1/2}$$

and

$$\beta_2 = \frac{1}{\sqrt{2}l}[S + \sqrt{S^2 + 4\tilde{\omega}^2}]^{1/2}.$$

Substituting the solution form (3.165) in the boundary conditions (3.160) yields

$$B_1 + B_3 = 0, \tag{3.166}$$

$$B_2 + B_4 = 0, \tag{3.167}$$

$$B_1 \cosh \beta_1 l + B_2 \sinh \beta_1 l - B_3 \cos \beta_2 l - B_4 \sin \beta_2 l = 0, \tag{3.168}$$

and

$$B_1 \sinh \beta_1 l + B_2 \cosh \beta_1 l + B_3 \sin \beta_2 l - B_4 \cos \beta_2 l = 0. \tag{3.169}$$

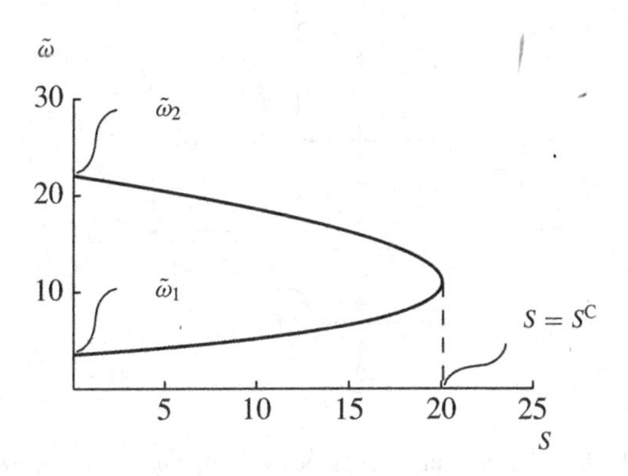

Figure 3.28 Variation of the first two non-dimensional eigenfrequencies of a cantilever beam with the non-dimensional follower force S

The above set of linear homogeneous equations can be compactly represented as $[\mathbf{A}(\tilde{\omega}, S)]$ $\mathbf{b} = \mathbf{0}$, where $\mathbf{b} = (B_1, B_2, B_3, B_4)^T$. The condition of non-triviality of a solution of $\mathbf{b}$ yields the characteristic equation $\det[\mathbf{A}(\tilde{\omega}, S)] = 0$, which can be solved for the non-dimensional eigenfrequencies $\tilde{\omega}$ of the beam as a function of the non-dimensional axial force S.

Solving the characteristic equation numerically, the first two non-dimensional eigenfrequencies of the beam as a function of the non-dimensional axial force are obtained, as shown in Figure 3.28. It may be mentioned here that, for small values of S, the eigenfrequencies $\tilde{\omega}$ are purely real (i.e., s calculated from (3.163) is purely imaginary), implying a harmonically oscillating solution. As the axial force is increased, the two eigenfrequencies come close, and they coalesce at a critical value of axial force given by $S_1^C \approx 20.05$, i.e., $F^c \approx 20.05EI/l^2$. Beyond this value of the axial force, we obtain two pairs of complex conjugate roots of $\tilde{\omega}$ that are opposite in sign. This implies that s calculated from (3.163) also occur as two pairs of complex conjugate roots with opposite signs (i.e., s, s^*, $-s$, and $-s^*$). The locus of s as a function of the axial force S is depicted in Figure 3.29. When an eigenvalue s becomes complex with a positive real part, the solution form $w(x, t) = W(x)e^{st}$ implies that the beam oscillates with the amplitude increasing exponentially with time. This phenomenon is known as *flutter instability*. Therefore, the critical value of the follower force is obtained when the two frequencies coalesce.

3.8.2 Beam with eccentric mass distribution

Consider a uniform beam of circular cross-section of radius r with an eccentric mass m, as shown in Figure 3.30. It is evident that if the beam vibrates transversely in the z-axis direction, the eccentric mass will excite the torsional mode of the beam, and vice versa. This important phenomenon, known as *flexure–torsion coupling*, is observed in beams having cross-sections with less than two axes of symmetry (for example, in channel sections, and L-beams). In addition, cross-sections of such beams also warp due to torsion. However, we have taken a beam of circular cross-section to keep the complexity due to torsional warping of the section out of our analysis.

Let the density of the beam be ρ, area of cross-section A, length l, and flexural rigidity EI. Considering flexural and torsional vibrations, the field variables are taken as $w(x, t)$

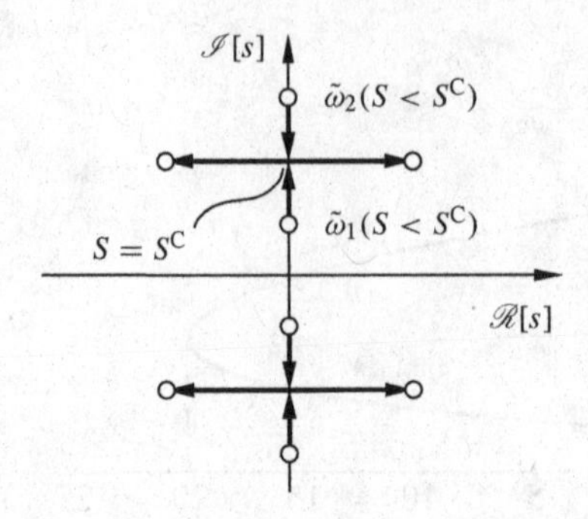

Figure 3.29 Loci of natural frequencies of a beam with a follower force

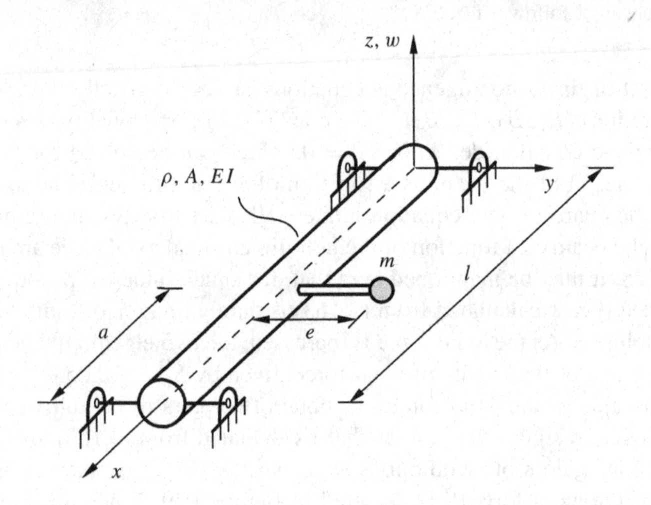

Figure 3.30 Beam of circular cross-section with eccentric mass

for flexural vibrations along the z-axis, and $\phi(x, t)$ for torsional vibrations about the x-axis. The Lagrangian of the system can be written as

$$\mathcal{L} = \frac{1}{2}\int_0^l [\mu(x)(w_{,t} + e\phi_{,t})^2 + \rho I_\mathrm{p}\phi_{,t}^2 + \rho A w_{,t}^2 - EIw_{,xx}^2 - GI_\mathrm{p}\phi_{,x}^2]\,\mathrm{d}x, \tag{3.170}$$

where $\mu(x) = m\delta(x - a)$ represents the eccentric mass distribution. It may be noted that if the line joining the center of the beam and the eccentric mass makes a non-zero angle with the y-axis, we must also consider the coupling of transverse vibration along the y-axis direction. Using the variational principle (see Appendix A), the equations of motion can be easily obtained as

$$(\rho A + \mu(x))w_{,tt} + \mu(x)e\phi_{,tt} + EIw_{,xxxx} = 0 \tag{3.171}$$

and

$$\mu(x)ew_{,tt} + (\rho I_p + \mu(x)e^2)\phi_{,tt} - GI_p\phi_{,xx} = 0. \tag{3.172}$$

The boundary conditions are obtained, for example, as

$$[EIw_{,xx}](0,t) \equiv 0, \quad \text{or} \quad w_{,x}(0,t) \equiv 0,$$

$$[EIw_{,xx}](l,t) \equiv 0, \quad \text{or} \quad w_{,x}(l,t) \equiv 0,$$

$$[EIw_{,xxx}](0,t) \equiv 0, \quad \text{or} \quad w(0,t) \equiv 0,$$

$$[EIw_{,xxx}](l,t) \equiv 0, \quad \text{or} \quad w(l,t) \equiv 0,$$

$$[GI_p\phi_{,x}](0,t) \equiv 0, \quad \text{or} \quad \phi(0,t) \equiv 0,$$

and

$$[GI_p\phi_{,x}](l,t) \equiv 0, \quad \text{or} \quad \phi(l,t) \equiv 0.$$

Let us now consider for simplicity, a uniform eccentric mass distribution, i.e., $\mu(x) = \mu$ is a constant. One can non-dimensionalize (3.171)–(3.172) as

$$(1 + m_r)\tilde{w}_{,\tilde{t}\tilde{t}} + m_r e_r \phi_{,\tilde{t}\tilde{t}} + \frac{1}{s_r^2}\tilde{w}_{,\tilde{x}\tilde{x}\tilde{x}\tilde{x}} = 0 \tag{3.173}$$

and

$$m_r e_r \tilde{w}_{,\tilde{t}\tilde{t}} + \left(\frac{1}{2} + m_r e_r^2\right)\phi_{,\tilde{t}\tilde{t}} - \frac{\gamma^2}{2}\phi_{,\tilde{x}\tilde{x}} = 0, \tag{3.174}$$

where $\tilde{x} = x/l$, $\tilde{t} = c_L t/l$, $\tilde{w} = w/r$, $c_L = \sqrt{E/\rho}$ is the speed of longitudinal waves, $m_r = \mu/\rho A$ is the mass ratio, $e_r = e/r$ is the eccentricity ratio, $s_r = l/\sqrt{I/A}$ is the slenderness ratio, and $\gamma^2 = G/E$. The boundary conditions for the beam are assumed to be pinned–pinned which do not allow torsional motion, as shown in Figure 3.30. Consider an approximate solution of (3.171)–(3.172) in the form

$$\tilde{w}(\tilde{x},\tilde{t}) = \sum_j a_{1j}(\tilde{t}) \sin j\pi\tilde{x} \quad \text{and} \quad \phi(\tilde{x},\tilde{t}) = \sum_j a_{2j}(\tilde{t}) \sin j\pi\tilde{x}, \tag{3.175}$$

where we have used the eigenfunctions of transverse vibration of a pinned–pinned beam and torsional vibration of a fixed–fixed circular bar, and $a_{1j}(\tilde{t})$ and $a_{2j}(\tilde{t})$ are the modal coordinates. Substituting the solution forms given by (3.175) in (3.173)–(3.174), and taking the inner product of the resulting equations with the respective eigenfunctions leads to decoupled sets of differential equations (on account of orthogonality) for each of the modal coordinate vector $\mathbf{a}_j = (a_{1j}, a_{2j})^{\mathrm{T}}$. These are of the form

$$\mathbf{M}\,\ddot{\mathbf{a}}_j + \mathbf{K}_j\mathbf{a}_j = \mathbf{0}, \tag{3.176}$$

where

$$\mathbf{M} = \begin{bmatrix} (1+m_{\mathrm{r}}) & m_{\mathrm{r}}e_{\mathrm{r}} \\ m_{\mathrm{r}}e_{\mathrm{r}} & \left(\dfrac{1}{2} + m_{\mathrm{r}}e_{\mathrm{r}}^2\right) \end{bmatrix} \quad \text{and} \quad \mathbf{K}_j = \begin{bmatrix} \dfrac{j^4\pi^4}{s_{\mathrm{r}}^2} & 0 \\ 0 & \dfrac{\gamma^2}{2}j^2\pi^2 \end{bmatrix}.$$

It is evident from the structure of $\mathbf{M}$ in (3.176) that when $m_{\mathrm{r}}e_{\mathrm{r}} = 0$, the flexural and torsional modes decouple, and we obtain the corresponding natural frequencies.

In order to see the effect of variation of m_{r} on the two frequencies, we consider a numerical example with $s_{\mathrm{r}} = 5$, $e_{\mathrm{r}} = 0.05$, and $\gamma^2 = 1/3$. For the first mode (i.e., $j = 1$ in (3.176)), substituting the solution form

$$\mathbf{a}_1(\tilde{t}) = \begin{Bmatrix} C_1 \\ D_1 \end{Bmatrix} \mathrm{e}^{i\omega_1\tilde{t}}$$

in (3.176), we obtain a pair of non-dimensional eigenfrequencies ω_1^{U} and ω_1^{L} as functions of m_{r}, as shown in Figure 3.31. The two frequency branches correspond to the flexural and torsional modes. In order to identify the corresponding modes, we must look at the corresponding eigenvector $(C_1, D_1)^{\mathrm{T}}$. The components of the normalized eigenvector $(C_1^{\mathrm{U}}, D_1^{\mathrm{U}})^{\mathrm{T}}$ corresponding to ω_1^{U} (see Figure 3.31) is shown in Figure 3.32. It is observed from the eigenvector components that, as $m_{\mathrm{r}} \to 0$, $C_1^{\mathrm{U}} \to 1$, and $D_1^{\mathrm{U}} \to 0$. Hence, the mode consists of primarily flexural vibration of the beam. However, as $m_{\mathrm{r}} \to \infty$, the mode is primarily torsional vibration of the beam. Around $m_{\mathrm{r}} \approx 0.17$, the flexural and torsional vibration modes get mixed. A similar observation can be drawn from the eigenvector corresponding to the lower branch in Figure 3.31.

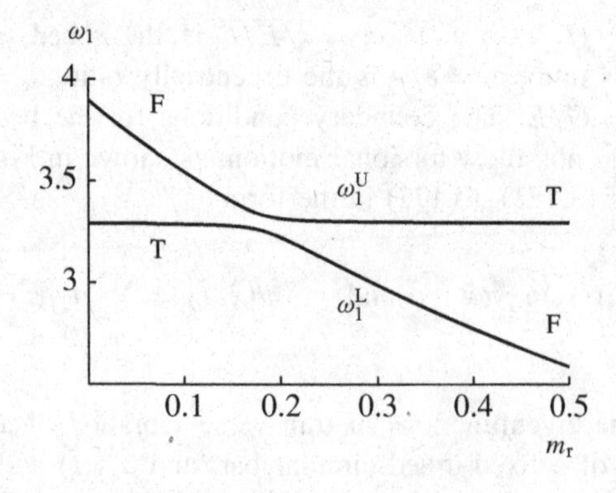

Figure 3.31 Variation of flexural (F) and torsional (T) mode eigenfrequencies with mass ratio

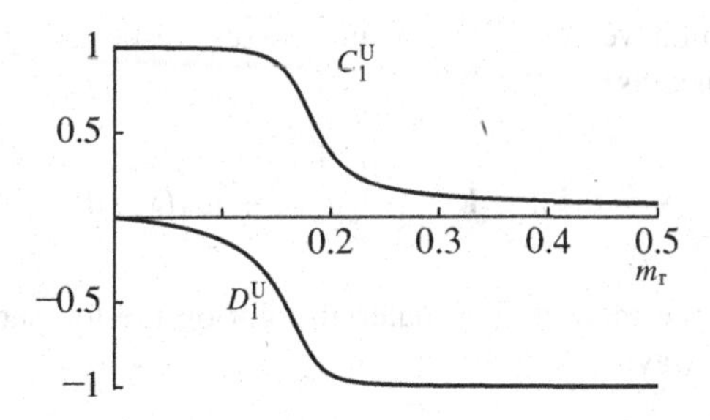

Figure 3.32 Variation of the eigenvector components corresponding to ω_{11}^U with mass ratio

3.8.3 Problems involving the motion of material points of a vibrating beam

In all preceding discussions on beams, we have considered only the motion of the neutral fiber of the beam represented by the field variable $w(x, t)$. It was assumed that the points on the neutral fiber move only in the transverse direction. However, material points elsewhere in the beam execute a combination of transverse and axial motion, due to rotation of the cross-section. This has interesting consequences as discussed in this section.

Consider a section of a uniform beam of rectangular cross-section having a thickness h, as shown in Figure 3.33. We assume the Euler–Bernoulli hypothesis to be valid, and take $w_{,x}(x,t) \ll 1$. As the beam deforms, the position of any point on the top surface of the beam can be written in the inertial frame as (see figure)

$$\mathbf{p} = p_x\hat{\mathbf{i}} + p_z\hat{\mathbf{k}} = \left[x - \frac{h}{2}w_{,x}(x,t)\right]\hat{\mathbf{i}} + \left[\frac{h}{2} + w(x,t)\right]\hat{\mathbf{k}}, \tag{3.177}$$

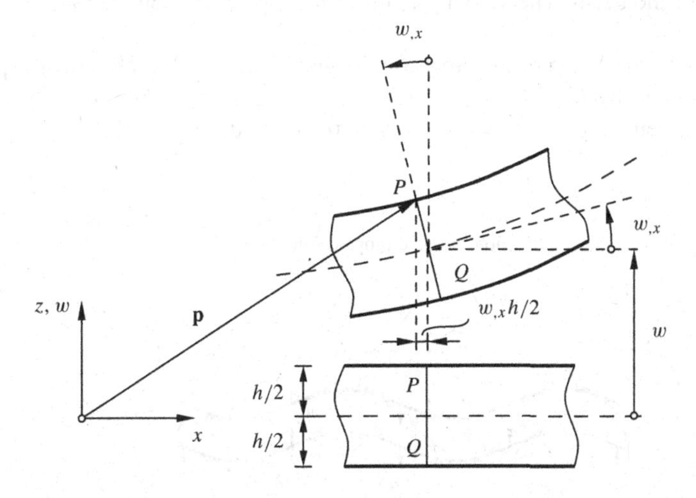

Figure 3.33 Displacement of a material point due to deformation of a beam

where $\hat{\mathbf{i}}$ and $\hat{\mathbf{k}}$ are the unit vectors along x-and z-axes, respectively. The velocity of any point can then be obtained as

$$\mathbf{v} = v_x\hat{\mathbf{i}} + v_z\hat{\mathbf{k}} = -\frac{h}{2}w_{,xt}\hat{\mathbf{i}} + w_{,t}(x,t)\hat{\mathbf{k}}. \tag{3.178}$$

Let us first consider the motion of a material point on the top surface of the beam due to a harmonic traveling wave

$$w(x,t) = B\sin(kx - \omega t),$$

where B is the amplitude, k is the wave number, and ω is the circular frequency of the wave. Substituting this wave solution in (3.177) and (3.178) yields

$$p_x = x - \frac{h}{2}Bk\cos(kx - \omega t), \qquad p_z = \frac{h}{2} + B\sin(kx - \omega t)$$

and

$$v_x = -\frac{h}{2}Bk\omega\sin(kx - \omega t), \qquad v_z = -B\omega\cos(kx - \omega t).$$

Thus, the particles move on an elliptical path given by

$$\frac{(p_x - x)^2}{(Bkh/2)^2} + \frac{(p_z - h/2)^2}{B^2} = 1.$$

Similarly, one can also determine the displacement and velocity fields of material points on the lower surface of the beam. The velocity fields at the top and bottom surfaces are shown in Figure 3.34.

Consider the points at the top of the crests on the upper surface. The phase corresponding to these points is given by $kx - \omega t = (2n-1)\pi/2$, $n = 1, 2, \ldots$. Therefore, the velocity of these points is given by $(v_x, v_y) = (-Bk\omega h/2, 0)$. Consider a rigid block lying on top

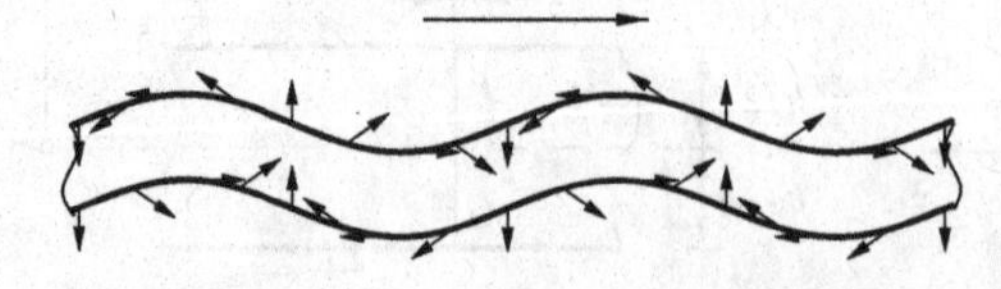

Figure 3.34 Velocity field of material points on top and bottom surfaces of a beam for a harmonic traveling wave

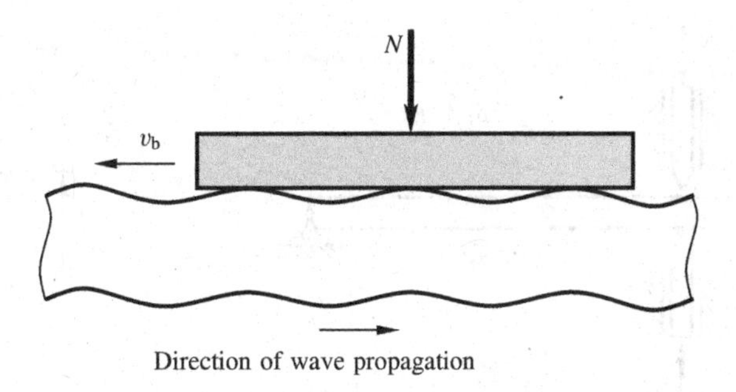

Figure 3.35 Generation of linear motion using traveling waves in a beam

of the beam, as shown in Figure 3.35. Assume, for simplicity, that the block contacts the points only at the top of the crests as shown, and the frictional force between the block and the beam is large enough to prevent slip at the contact points. Then it follows that the block will move at a horizontal speed $v_b = -Bk\omega h/2\hat{\mathbf{i}}$. The negative sign of v_b indicates that the block will move in the direction opposite to that of the wave propagation. Since v_b is frequency dependent, the higher the frequency, the higher is the travel speed of the block. The maximum force that the block can generate (i.e., when it starts slipping) is given by μN, where μ is the coefficient of friction and N is the total normal force between the block and the beam. The above principle is used practically in ultrasonic linear actuators which operate at very high frequencies. One of the key requirements in this case is the generation of traveling waves in finite systems. This can be achieved by using an array of actuators along the beam which are harmonically driven with appropriate phase difference. Another way to achieve this is by impedance matching, as can be found in Exercise 3.17.

Now consider a modal solution for a simply-supported beam of length l of the form $w(x, t) = B_n \sin \omega_n t \sin n\pi x/l$. Substituting this standing wave solution in (3.177) and (3.178) yields

$$p_x = -\frac{h}{2}\frac{n\pi}{l} B_n \sin \omega_n t \cos \frac{n\pi x}{l} \qquad p_z = B_n \sin \omega_n t \sin \frac{n\pi x}{l},$$

and

$$v_x = -\frac{h}{2}\frac{n\pi}{l} \omega_n B_n \cos \omega_n t \cos \frac{n\pi x}{l}, \qquad v_z = \omega_n B_n \cos \omega_n t \sin \frac{n\pi x}{l}.$$

It is evident in this case that a particle on the top of the beam at a location x moves in a straight line of slope $p_z/p_x = (2l/n\pi h) \tan n\pi x/l$.

The small axial motion of material points on the top and bottom surfaces can also give rise to interesting effects. Let us consider a simply-supported uniform beam with two massless pressure pads held against the beam by a constant force N, as shown in Figure 3.36. We will assume for simplicity that there is no sticking between the pads and the beam at any time. When the motion of the beam is such that $w_{,xt}(a, t) > 0$, it is evident that the friction forces on the top and bottom surfaces will produce a net opposing moment about the neutral

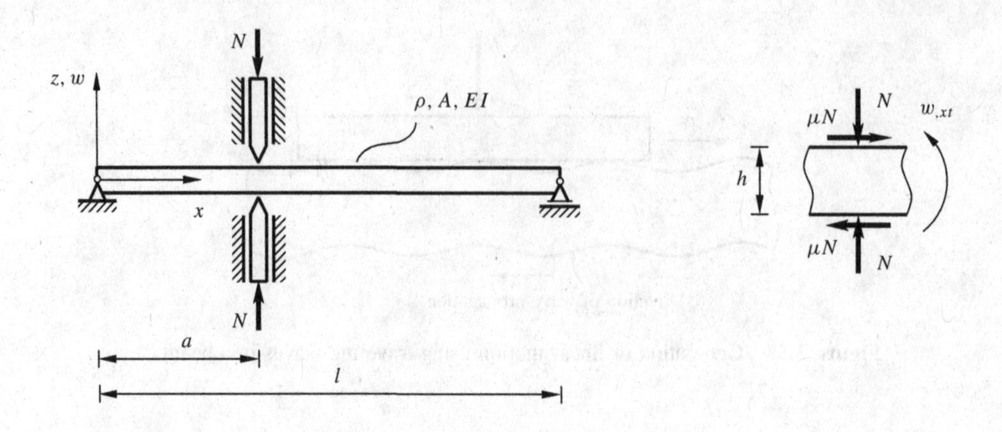

Figure 3.36 Beam with friction pads

axis point at $x = a$. This moment can be expressed as

$$M_f(t) = -\mu N h \,\mathrm{sgn}[w_{,xt}(a,t)], \tag{3.179}$$

where μ is the coefficient of friction and $\mathrm{sgn}(\cdot)$ is the signum function. With this moment, the equation of motion can be written as

$$\rho A w_{,tt} + EI w_{,xxxx} = -M_f(t)\delta'(x-a) = \mu N h \delta'(x-a)\,\mathrm{sgn}[w_{,xt}(a,t)]. \tag{3.180}$$

Multiplying both sides by $w_{,t}(x,t)$ and integrating over the length of the beam, one can write

$$\int_0^l \left[\frac{\mathrm{d}}{\mathrm{d}t}\left(\frac{1}{2}\rho A w_{,t}^2\right) + EI w_{,xxxx} w_{,t}\right]\mathrm{d}x = \int_0^l \mu N h w_{,t} \delta'(x-a)\,\mathrm{sgn}[w_{,xt}(a,t)]\,\mathrm{d}x.$$

Integrating by parts twice the second term on the left-hand side, and using the boundary conditions, we obtain

$$\frac{\mathrm{d}}{\mathrm{d}t}\int_0^l \left[\frac{1}{2}\rho A w_{,t}^2 + \frac{1}{2}EI w_{,xx}^2\right]\mathrm{d}x = -\mu N h w_{,xt}(a,t)\,\mathrm{sgn}[w_{,xt}(a,t)]$$

$$\Rightarrow \frac{\mathrm{d}\mathcal{E}}{\mathrm{d}t} = -\mu N h |w_{,xt}(a,t)|. \tag{3.181}$$

Since the right-hand side in (3.181) is always negative, the total mechanical energy of the beam always decreases. Thus, a stationary load in friction contact with a beam can damp the vibrations of the beam.

3.8.4 Dynamics of rotating shafts

Rotating shafts or rotors are very commonly found in machines. The dynamics of rotors is very complex, and depends on their construction, support conditions and speed. In the following, we discuss a simple model of a rotor assumed to be a torsionally stiff beam rotating about its axis. Shear deformation and rotational inertia effects are neglected.

Consider a simply-supported uniform asymmetric rotor of rectangular cross-section rotating about its axis at a constant angular speed Ω, as shown in Figure 3.37. Consider an inertial frame x–y, a body fixed frame ξ'–η', and a frame ξ–η rotating with the shaft but located at the undeformed axis of the rotor, as depicted in Figure 3.38. We represent the dynamics of the shaft using the rotating frame field variables $[u_1(x,t), u_2(x,t)]$, as indicated in Figure 3.38. One can easily express the position vector $[w_1(x,t), w_2(x,t)]$ in the inertial frame x–y as

$$\left\{ \begin{array}{c} w_1 \\ w_2 \end{array} \right\} = \left[\begin{array}{cc} \cos \Omega t & -\sin \Omega t \\ \sin \Omega t & \cos \Omega t \end{array} \right] \left\{ \begin{array}{c} u_1 \\ u_2 \end{array} \right\}. \tag{3.182}$$

The absolute kinetic energy of the shaft is

$$\begin{aligned} T &= \frac{1}{2}\int_0^l \left[\rho A(w_{1,t}^2 + w_{2,t}^2) + J\Omega^2\right] \mathrm{d}x \\ &= \frac{1}{2}\int_0^l \left[\rho A[u_{1,t}^2 + u_{2,t}^2 + 2\Omega(u_{2,t}u_1 - u_{1,t}u_2) \right. \\ &\qquad \left. + \Omega^2(u_1^2 + u_2^2)] + J\Omega^2\right] \mathrm{d}x, \end{aligned} \tag{3.183}$$

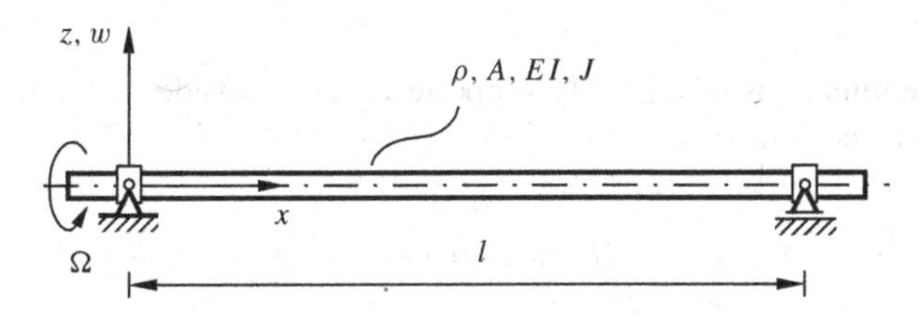

Figure 3.37 A simply-supported flexible rotor

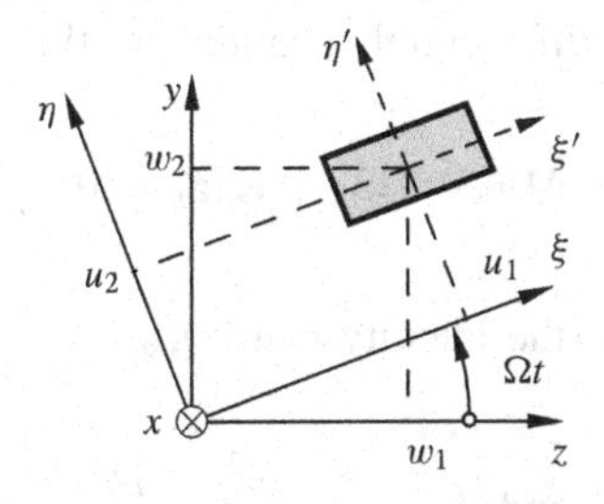

Figure 3.38 Coordinate frames for a rotor

where J is the polar moment of inertia of the cross-sectional area, l is the length, ρ is the density, and A is the area of cross-section of the rotor. The potential energy is

$$\mathcal{V} = \frac{1}{2}\int_0^l [EI_1 u_{1,xx}^2 + EI_2 u_{2,xx}^2]\,\mathrm{d}x, \tag{3.184}$$

where E is Young's modulus and I_1 and I_2 are the second moments of the area about the η'- and ξ'-axes, respectively. Using the variational formulation

$$\delta \int_{t_1}^{t_2} (\mathcal{T} - \mathcal{V})\,\mathrm{d}t = 0,$$

one obtains the equations of motion of the rotor as

$$\rho A u_{1,tt} - 2\rho A\Omega u_{2,t} + EI_1 u_{1,xxxx} - \rho A\Omega^2 u_1 = 0 \tag{3.185}$$

and

$$\rho A u_{2,tt} + 2\rho A\Omega u_{1,t} + EI_2 u_{2,xxxx} - \rho A\Omega^2 u_2 = 0, \tag{3.186}$$

along with the boundary conditions for the simply-supported ends as

$$u_j(0,t) \equiv 0, \quad u_{j,xx}(0,t) \equiv 0, \quad u_j(l,t) \equiv 0, \quad \text{and} \quad u_{j,xx}(l,t) \equiv 0, \tag{3.187}$$

where $j = 1, 2$.

Using the eigenfunctions of a simply-supported beam, consider a solution expansion for (3.185)–(3.186) of the form

$$u_1(x,t) = \sum_n a_{1n}(t)\sin n\pi x/l \quad \text{and} \quad u_2(x,t) = \sum_n a_{2n}(t)\sin n\pi x/l, \tag{3.188}$$

where $a_{1n}(t)$ and $a_{2n}(t)$ are the modal coordinates. Substituting (3.188) in (3.185)–(3.186), taking the inner product with the modal functions, and using the orthogonality relations, one can easily obtain the set of differential equations for the nth modal coordinate vector as

$$\mathbf{M}\ddot{\mathbf{a}}_n + \mathbf{G}\dot{\mathbf{a}}_n + \mathbf{K}_n\mathbf{a}_n = 0, \tag{3.189}$$

where $\mathbf{a}_n = (a_{1n}, a_{2n})^{\mathrm{T}}$, $\mathbf{M} = \mathbf{I}$ (the identity matrix),

$$\mathbf{G} = \begin{bmatrix} 0 & -2\Omega \\ 2\Omega & 0 \end{bmatrix}, \quad \text{and} \quad \mathbf{K}_n = \begin{bmatrix} (\omega_{1n}^2 - \Omega^2) & 0 \\ 0 & (\omega_{2n}^2 - \Omega^2) \end{bmatrix}, \tag{3.190}$$

with

$$\omega_{jn} = \frac{n^2\pi^2}{l^2}\sqrt{\frac{EI_j}{\rho A}}, \qquad j = 1, 2.$$

The stability of the discrete gyroscopic system (3.189) can be easily analyzed by taking a modal solution $\mathbf{a}_n(t) = \mathbf{A}e^{st}$ and obtaining the characteristic equation

$$s^4 + (\omega_{1n}^2 + \omega_{2n}^2 + 2\Omega^2)s^2 + (\omega_{1n}^2 - \Omega^2)(\omega_{2n}^2 - \Omega^2) = 0. \tag{3.191}$$

The eigenvalues are then obtained as

$$s = \pm\frac{1}{\sqrt{2}}\Bigg[- (\omega_{1n}^2 + \omega_{2n}^2 + 2\Omega^2) \pm\sqrt{(\omega_{1n}^2 + \omega_{2n}^2 + 2\Omega^2)^2 - 4(\omega_{1n}^2 - \Omega^2)(\omega_{2n}^2 - \Omega^2)}\,\Bigg]^{1/2}.$$

It can be easily checked that the system has purely imaginary eigenvalues whenever $(\omega_{1n}^2 - \Omega^2)(\omega_{2n}^2 - \Omega^2) > 0$. When this condition is not satisfied, the system has a pair of real roots of equal magnitude but opposite signs signifying clearly a divergence instability. A repeated root occurs when either

$$(\omega_{1n}^2 - \Omega^2)(\omega_{2n}^2 - \Omega^2) = 0 \tag{3.192}$$

or

$$(\omega_{1n}^2 + \omega_{2n}^2 + 2\Omega^2)^2 - 4(\omega_{1n}^2 - \Omega^2)(\omega_{2n}^2 - \Omega^2) = 0. \tag{3.193}$$

The condition (3.192) yields the divergence speed boundaries of the rotor. If Ω is such that the left-hand side of (3.193) is negative, we have two pairs of complex conjugate roots which are opposite in signs (i.e., $s, s^*, -s, -s^*$). This indicates flutter instability, and hence, condition (3.193) yields the flutter speed boundaries. In the case of a symmetric rotor, we have $I_1 = I_2$, and hence $\omega_{n1} = \omega_{n2}$. Then, as can be easily verified, the eigenvalues are given by $s = \pm i(\omega_n \pm \Omega)$. Thus, the system is neutrally stable whenever the rotation speed $\Omega = \omega_n$, and oscillatory otherwise.

3.8.5 Dynamics of axially translating beams

Axially translating beams form an important class of problems with many applications. Dynamics of band saws, belt drives, and rolling of beams and rods may all be modeled as translating beams with different boundary conditions. In the following, we consider an elementary model of a translating beam passing through two frictionless guides.

Consider a beam translating axially along its length at a constant speed v, as shown in Figure 3.39. The kinetic energy of the system is

$$\mathcal{T} = \frac{1}{2}\int_0^l \rho A(w_{,t} + vw_{,x})^2\,\mathrm{d}x, \tag{3.194}$$

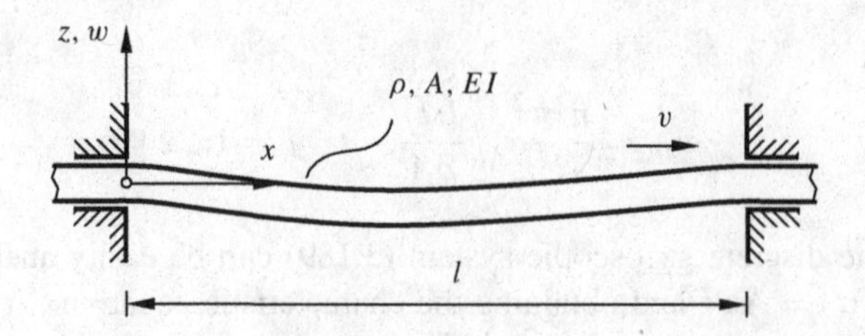

Figure 3.39 A beam translating along its axis

where ρ is the density, A the cross-sectional area, and $w(x, t)$ is the transverse deflection field variable. The potential energy of the beam is

$$\mathcal{V} = \frac{1}{2}\int_0^l EI w_{,xx}^2 \, \mathrm{d}x, \tag{3.195}$$

where EI is the flexural stiffness of the beam. Using the variational formulation, one can write

$$\delta \int_{t_1}^{t_2} (\mathcal{T} - \mathcal{V}) \, \mathrm{d}t = 0$$

$$\Rightarrow \rho A(w_{,tt} + 2vw_{,xt} + v^2 w_{,xx}) + EI w_{,xxxx} = 0, \tag{3.196}$$

with the boundary conditions

$$w(0, t) \equiv 0, \quad w_{,x}(0, t) \equiv 0, \quad w(l, t) \equiv 0, \quad \text{and} \quad w_{,x}(l, t) \equiv 0, \tag{3.197}$$

for the guided ends, as shown in Figure 3.39.

Before proceeding further, we non-dimensionalize the equation of motion (3.196) using the definitions

$$\tilde{x} = x/l, \qquad \tilde{w} = w/l, \qquad \tilde{t} = t\frac{\pi^2}{l^2}\sqrt{\frac{EI}{\rho A}}, \qquad \text{and} \qquad \tilde{v} = v\frac{l}{\pi^2}\sqrt{\frac{\rho A}{EI}},$$

to obtain the non-dimensional equation of motion as

$$\tilde{w}_{,\tilde{t}\tilde{t}} + 2\tilde{v}\tilde{w}_{,\tilde{x}\tilde{t}} + \tilde{v}^2\tilde{w}_{,\tilde{x}\tilde{x}} + \frac{1}{\pi^4}\tilde{w}_{,\tilde{x}\tilde{x}\tilde{x}\tilde{x}} = 0. \tag{3.198}$$

For notational convenience, we will drop the tilde symbols from (3.198) in the following analysis.

Let us construct a solution of (3.198) in the form

$$w(x, t) = \sum_k p_k(t)\phi_k(x), \tag{3.199}$$

where $p_k(t)$ are the modal coordinates, and $\phi_k(x)$ are the eigenfunctions of the non-translating beam (i.e., for $v = 0$) with clamped–clamped boundary conditions (i.e., $\phi_k(0) = 0$, $\phi_k'(0) = 0$, $\phi_k(l) = 0$, and $\phi_k'(l) = 0$). Substituting (3.199) in (3.196) and using the orthogonality property, one can write the discretized equations of motion of the traveling beam as

$$\mathbf{M}\ddot{\mathbf{p}} + \mathbf{G}\dot{\mathbf{p}} + \mathbf{K}\mathbf{p} = 0, \tag{3.200}$$

where $\mathbf{p} = (p_1, p_2, \ldots, p_n)^{\mathrm{T}}$,

$$M_{jk} = \int_0^1 \rho A \phi_j \phi_k \,\mathrm{d}x, \qquad G_{jk} = \int_0^1 2v\phi_j \phi_k' \,\mathrm{d}x,$$

and

$$K_{jk} = \int_0^1 \left[v^2 \phi_j \phi_k'' + \frac{1}{\pi^4} \phi_j \phi_k'''' \right] \mathrm{d}x.$$

It can be easily verified through integration by parts and using the clamped–clamped boundary conditions that one can rewrite

$$G_{jk} = \int_0^1 v(\phi_j \phi_k' - \phi_k \phi_j') \,\mathrm{d}x$$

and

$$K_{jk} = \int_0^1 \left[-v^2 \phi_j' \phi_k' + \frac{1}{\pi^4} \phi_j'' \phi_k'' \right] \mathrm{d}x.$$

Thus, $\mathbf{G}$ is a skew-symmetric matrix and $\mathbf{K}$ is a symmetric matrix.

The eigenfunctions of the clamped–clamped static beam are obtained as

$$\phi_j(x) = \cosh \beta_j x - \cos \beta_j x - \left[\frac{\cosh \beta_j - \cos \beta_j}{\sinh \beta_j - \sin \beta_j} \right] (\sinh \beta_j x - \sin \beta_j x), \tag{3.201}$$

where β_j are obtained by solving the characteristic equation of a clamped–clamped beam given by $\cosh \beta \, \cos \beta = 1$. It may be easily verified that the eigenfunctions (3.201) satisfy the orthogonality property

$$\int_0^1 \phi_j(x) \phi_k(x) \,\mathrm{d}x = 0, \qquad j \neq k.$$

Substituting the solution form (3.199) with the first two eigenfunctions of the form (3.201), one obtains, using the orthogonality property, the discretized equations of motion (3.200), where $\mathbf{M} = \mathbf{I}$ (identity matrix),

$$\mathbf{G} = \begin{bmatrix} 0 & -\gamma v \\ \gamma v & 0 \end{bmatrix}, \qquad \mathbf{K} = \begin{bmatrix} \omega_1^2 - \mu_1 v^2 & 0 \\ 0 & \omega_2^2 - \mu_2 v^2 \end{bmatrix},$$

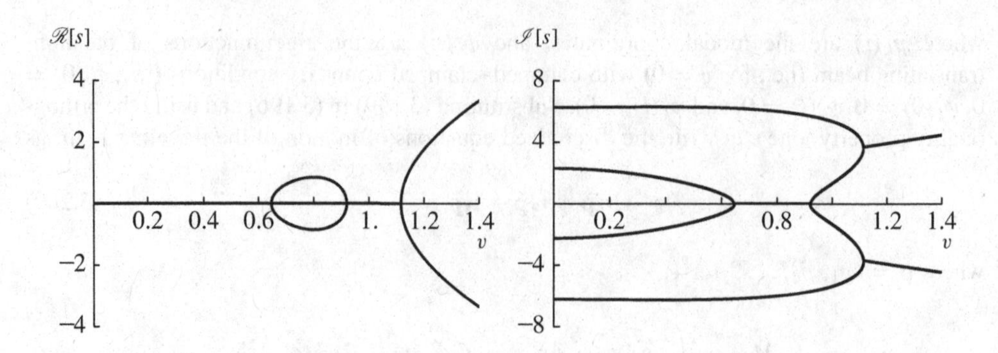

Figure 3.40 Variation of the real and imaginary parts of the eigenvalues s of a translating beam with travel speed v

$\gamma = 3.342$, $\omega_1 = 5.139$, $\omega_2 = 39.047$, $\mu_1 = 12.303$, and $\mu_2 = 46.050$. Using the modal solution $\mathbf{z} = \mathbf{Z}e^{st}$ in (3.200), one obtains the characteristic equation

$$s^4 + [\omega_1^2 + \omega_2^2 - v^2(\mu_1 + \mu_2 - \gamma^2)]s^2 + (\omega_1^2 - \mu_1 v^2)(\omega_2^2 - \mu_2 v^2) = 0, \qquad (3.202)$$

from which the eigenvalues of the discretized system can be calculated. As expected, the eigenvalues are functions of the speed of travel v. The variations of the real and imaginary parts of the eigenvalues shown as functions of v are shown in Figure 3.40. It may be observed that for values of v such that $\omega_1/\sqrt{\mu_1} = 0.646 < v < \omega_2/\sqrt{\mu_2} = 0.921$, there is a pair of real eigenvalues with opposite signs. The presence of a positive real eigenvalue indicates a divergence instability. For values of v such that $[\omega_1^2 + \omega_2^2 - v^2(\mu_1 + \mu_2 - \gamma^2)]^2 - 4(\omega_1^2 - \mu_1 v^2)(\omega_2^2 - \mu_2 v^2) < 0$ (which occurs for $v > 1.115$), there are two pairs of complex conjugate eigenvalues which are of opposite signs (i.e., $s, s^*, -s, -s^*$). This is characteristic of flutter instability. Outside these two velocity regions discussed above, the system has purely imaginary eigenvalues, implying oscillatory solutions.

3.8.6 Dynamics of fluid-conveying pipes

A fluid conveying pipe exhibits very interesting and varied dynamical behavior depending on the support conditions of the pipe, and variation of the flow velocity of the fluid (see [5]). In the following discussion, we consider two cases of boundary conditions. The fluid is assumed to be inviscid and incompressible, and the pipe is assumed to retain its geometrical properties under small transverse deflections.

3.8.6.1 Simply-supported pipe

First, we consider the simplest case, that of a pinned–pinned uniform pipe conveying an incompressible inviscid fluid, as shown in Figure 3.41. We treat the pipe as a (hollow) beam of density ρ, area of the annular cross-section A, flexural stiffness EI, and length l.

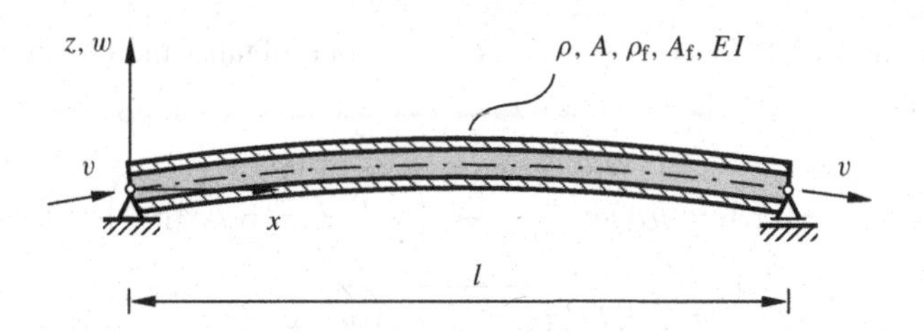

Figure 3.41 A uniform pipe conveying a fluid

Let the constant velocity of the fluid be v, the density of the fluid be ρ_f, and the area of cross-section of the fluid inside the pipe be A_f. The kinetic energy of the pipe and the fluid can be written as

$$\mathcal{T} = \frac{1}{2}\int_0^l \left[\rho A w_{,t}^2 + \rho_\mathrm{f} A_\mathrm{f}(w_{,t} + v w_{,x})^2\right] \mathrm{d}x. \tag{3.203}$$

The potential energy, which is only due to the pipe, is

$$\mathcal{V} = \frac{1}{2}\int_0^l EI w_{,xx}^2 \,\mathrm{d}x. \tag{3.204}$$

The equation of motion is then obtained from Hamilton's principle

$$\delta \int_{t_1}^{t_2} (\mathcal{T} - \mathcal{V})\,\mathrm{d}t = 0$$

$$\Rightarrow (\rho_\mathrm{f} A_\mathrm{f} + \rho A) w_{,tt} + 2\rho_\mathrm{f} A_\mathrm{f} v w_{,xt} + \rho_\mathrm{f} A_\mathrm{f} v^2 w_{,xx} + EI w_{,xxxx} = 0, \tag{3.205}$$

where we have used the boundary conditions

$$w(0,t) \equiv 0, \quad EI w_{,xx}(0,t) \equiv 0, \quad w(l,t) \equiv 0, \quad \text{and} \quad EI w_{,xx}(l,t) \equiv 0. \tag{3.206}$$

Consider a solution of the form

$$w(x,t) = \sum_j b_j(t) \sin\frac{j\pi x}{l},$$

where we take the eigenfunctions of a pinned–pinned beam as comparison functions for the problem. Using Galerkin's method, one easily obtains the decoupled differential equation for the jth modal coordinate as

$$(\rho_\mathrm{f} A_\mathrm{f} + \rho A)\ddot{b}_j + \frac{j^2\pi^2}{l^2}\left(EI\frac{j^2\pi^2}{l^2} - \rho_\mathrm{f} A_\mathrm{f} v^2\right) b_j = 0. \tag{3.207}$$

Writing the solution of (3.207) as $b_j(t) = B_j e^{i\omega_j t}$, one obtains the characteristic equation for the jth mode as

$$-(\rho_f A_f + \rho A)\omega_j^2 + \frac{j^2\pi^2}{l^2}\left(EI\frac{j^2\pi^2}{l^2} - \rho_f A_f v^2\right) = 0$$

$$\Rightarrow \omega_j = \frac{j\pi}{l}\sqrt{\frac{EIj^2\pi^2/l^2 - \rho_f A_f v^2}{\rho_f A_f + \rho A}}.$$

This clearly shows that $\omega_j = 0$ for a critical flow velocity of the fluid given by

$$v_j^c = \frac{j\pi}{l}\sqrt{\frac{EI}{\rho_f A_f}}.$$

For flow velocities above this critical value, ω_j will become imaginary, and the system will exhibit divergence instability for the jth mode. Thus, a pinned–pinned pipe conveying an incompressible fluid will become unstable at flow speeds $v > v_1^c = \pi/l\sqrt{EI/\rho_f A_f}$.

3.8.6.2 *Cantilever pipe*

Next, consider a cantilever pipe conveying fluid, as shown in Figure 3.42. In this case, it is important to realize that the fluid absolute velocity vector at the exit may not be tangential to the axis of the pipe at the free end. Considering the pipe to be the control volume, and based on the discussions in Section 1.10.3, one can then write (up to first order) the force experienced by the pipe at the free end due to momentum efflux as $\mathbf{F}_f = \dot{m}\mathbf{v} = -\rho_f A_f v[v\hat{\mathbf{i}} + (w_{,t} + vw_{,x})\hat{\mathbf{k}}]|_{x=l}$, where $\hat{\mathbf{i}}$ and $\hat{\mathbf{k}}$ are the unit vectors along the x- and z-axes, respectively. Thus, the force in the direction transverse to the pipe is given by

$$F_z = -\rho_f A_f v(w_{,t} + vw_{,x})\big|_{x=l}. \tag{3.208}$$

With the external force (3.208), the extended Hamilton's principle now reads

$$\int_{t_1}^{t_2} \left[\delta\mathcal{L} + F_z\delta w\big|_{x=l}\right] \mathrm{d}t = 0,$$

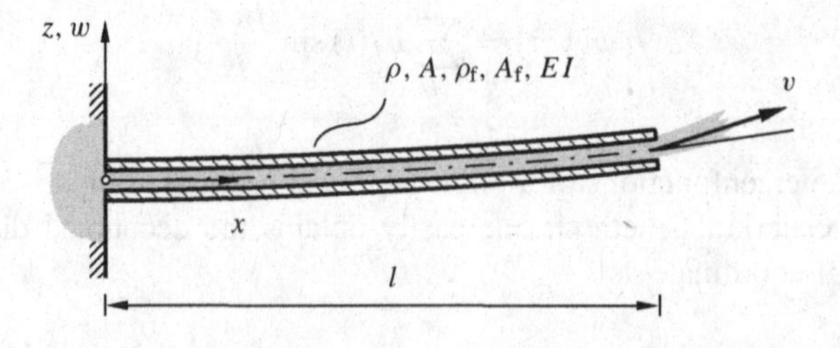

Figure 3.42 A cantilever pipe conveying fluid

where $\mathcal{L} = \mathcal{T} - \mathcal{V}$ is defined using (3.203)–(3.204). Carrying out the variation again leads to the equation of motion (3.205), along with the boundary conditions

$$w(0,t) \equiv 0, \quad w_{,x}(0,t) \equiv 0, \quad EIw_{,xx}(l,t) \equiv 0, \quad \text{and} \quad EIw_{,xxx}(l,t) \equiv 0.$$

Using the eigenfunctions of a cantilever beam, one can expand the solution of (3.205) in the form

$$w(x,t) = \sum_j p_j(t) W_j(x),$$

and obtain the discretized equations for the system, which are now of the form

$$\mathbf{M}\ddot{\mathbf{p}} + (\mathbf{D} + \mathbf{G})\dot{\mathbf{p}} + (\mathbf{K} + \mathbf{N})\mathbf{p} = \mathbf{0},$$

where **M**, **D** and **K** are symmetric matrices, and **G** and **N** are skew-symmetric matrices. Here, the matrix elements are given by

$$M_{jk} = (\rho_f A_f + \rho A) \int_0^l W_j W_k \, dx, \qquad D_{jk} = \rho_f A_f v W_j W_k \big|_l,$$

$$G_{jk} = \rho_f A_f v \int_0^l (W_j W_k' - W_k W_j') \, dx,$$

$$K_{jk} = \rho_f A_f v^2 \left[\frac{1}{2} (W_j W_k' + W_k W_j') \big|_l - \int_0^l W_j' W_k' \, dx \right] + EI \int_0^l W_j'' W_k'' \, dx,$$

$$N_{jk} = \frac{1}{2} \rho_f A_f v^2 [W_j W_k' - W_k W_j'] \big|_l .$$

EXERCISES

3.1 Determine the eigenfrequencies and eigenfunctions of the uniform Euler–Bernoulli beams shown in Figure 3.43.

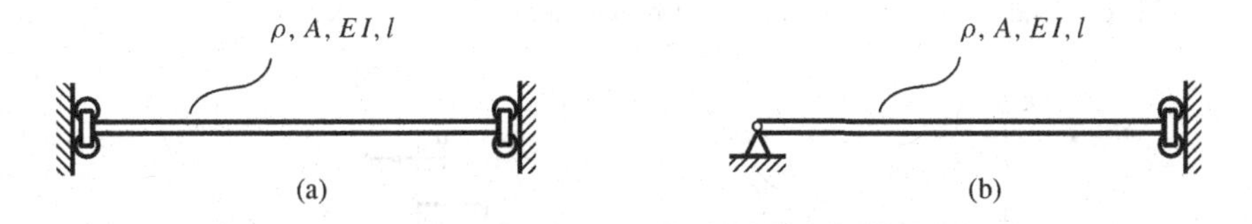

Figure 3.43 Exercise 3.1

3.2 A uniform Euler–Bernoulli beam of length $a + b$, flexural stiffness EI and linear density ρA is supported, as shown in Figure 3.44. Determine the characteristic equation, and calculate the eigenfrequencies for the special values of $b = 0$ and $b = a$. Plot the variation of the first few eigenvalues for variation of b in the range $(0, a)$. Using Rayleigh's method, determine the upper bounds on the lowest eigenfrequency in each case. Use the simplest possible polynomial shape-functions for the purpose.

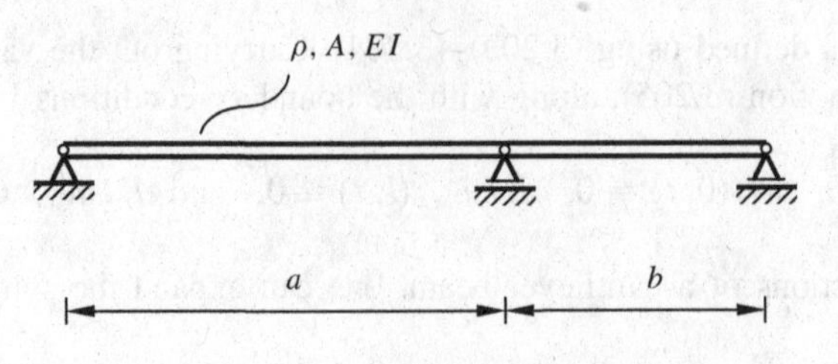

Figure 3.44 Exercise 3.2

3.3 The frame shown in Figure 3.45 is made by welding two homogeneous Euler–Bernoulli beams infinitely stiff in tension. Write down the boundary conditions, and the matching conditions at the weld for the system. Derive the characteristic equation for the frame, and determine the eigenfrequencies and mode-shapes of free vibration.

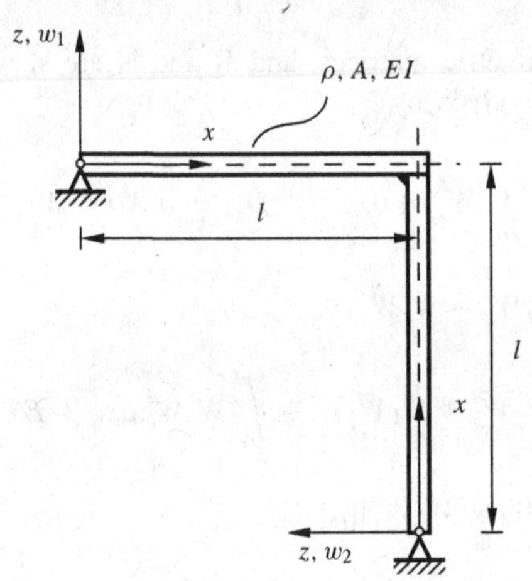

Figure 3.45 Exercise 3.3

3.4 A uniform homogeneous beam is clamped on one end, and carries a body of mass m and moment of inertia I_{cm} on the other end, as shown in Figure 3.46. Determine the characteristic equation, and the eigenfrequencies and the mode-shapes of the free vibrations.

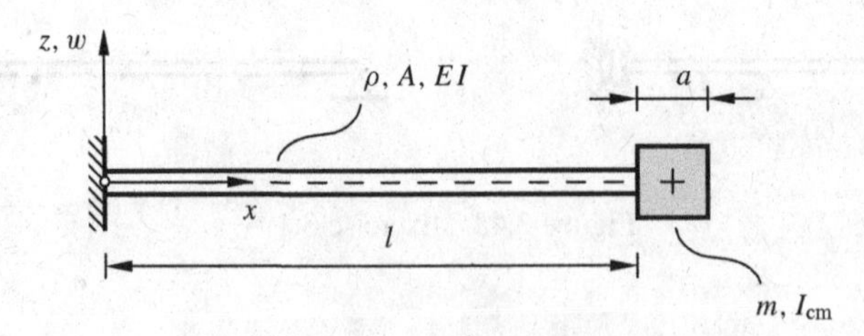

Figure 3.46 Exercise 3.4

3.5 A simply-supported uniform beam is loaded by a constant distributed force $q(x, t) = Q_0$, as shown in Figure 3.47. Determine the motion of the beam when, at $t = 0$, the force is removed suddenly.

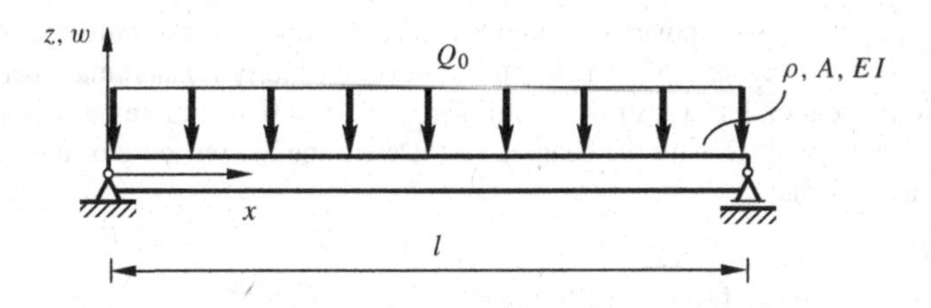

Figure 3.47 Exercise 3.5

3.6 Determine the steady-state response of a uniform cantilever beam with end-mass when excited by a force $F(t) = F_0 \cos \Omega t$, as shown in Figure 3.48.

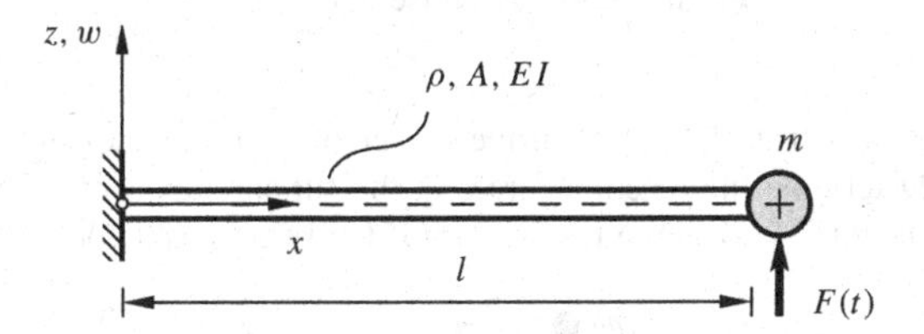

Figure 3.48 Exercise 3.6

3.7 A uniform homogeneous Timoshenko beam of length l having flexural rigidity EI, shear rigidity GA_s, linear density ρA, and rotary inertia ρI is simply-supported. Determine the eigenfrequencies and eigenfunctions for flexural vibrations of the beam.

3.8 A frame shown in Figure 3.49 is made of three uniform beams each having flexural rigidity EI and mass per unit length ρA, and is assumed to be infinitely stiff in tension. A constant force P acts at the center of the horizontal beam as shown. Determine the first eigenfrequency of the frame $\omega_1(P)$ as a function of the force P. Re-calculate the first eigenfrequency if the frame is pinned at the support B.

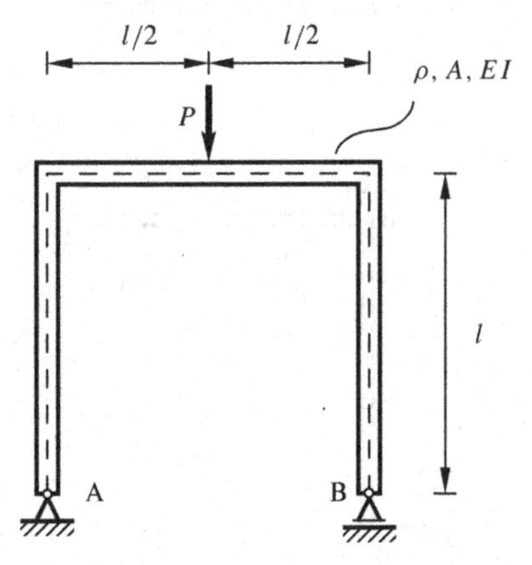

Figure 3.49 Exercise 3.8

3.9 An automobile of mass m moves with uniform acceleration a over a uniform beam on a flexible foundation, as shown in Figure 3.50. The beam has flexural rigidity EI and mass per unit length ρA, and the flexible foundation has a distributed stiffness κ. At $t = 0$, the automobile passes over the left support point with a speed v_0, with the beam at rest. Determine the response of the beam, treating the automobile as a point mass.

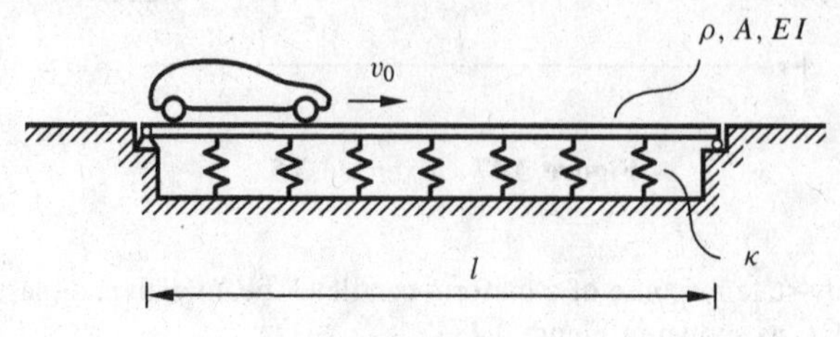

Figure 3.50 Exercise 3.9

3.10 A particle of mass m falls from height h on the center of a simply-supported uniform beam, as shown in Figure 3.51. Determine the reaction forces at the supports as a function of time. When does the particle leave the beam? How do the results vary if the beam has a constant modal damping $D = 0.04$?

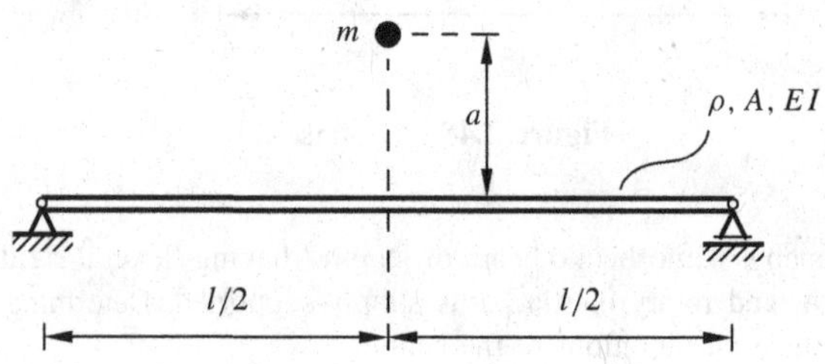

Figure 3.51 Exercise 3.10

3.11 A uniform beam, pinned at one end, is released from rest from a horizontal position, as shown in Figure 3.52. When the beam reaches a vertical position, the free end A hits a rigid edge. Determine the reaction forces at the ends A and B as functions of time.

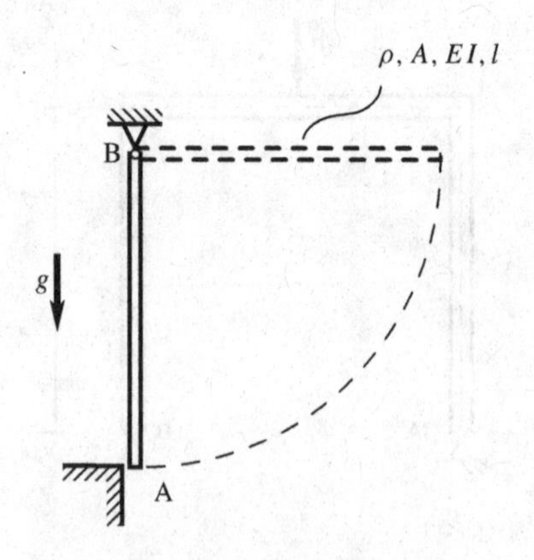

Figure 3.52 Exercise 3.11

3.12 A uniform cantilever beam is forced at the free end by a force $F(t) = \hat{F} + F_0 \cos \Omega t$ and a moment $M(t) = \hat{M} + M_0 \cos \Omega t$. Determine the steady-state response of the beam.

3.13 A beam of rectangular cross-section has a constant width b, while the height varies from $h/2$ to h over its length l. The material has density ρ and Young's modulus E. Determine the approximate modal frequencies and mode-shape-functions when the beam is simply-supported.

3.14 Determine the approximate modal frequencies and mode-shape-functions of a simply-supported beam of length l having a circular cross-section of radius $r(x) = R[1 + (x/l)^2]$.

3.15 A uniform Euler–Bernoulli beam is fixed to a small hub rotating at a constant angular speed Ω in the vertical plane, as shown in Figure 3.53.

(a) Neglecting air resistance, gravity and axial motion of the beam, show that the Lagrangian of the beam is given by

$$\mathcal{L} = \frac{1}{2} \int_0^l \left[\rho A[(\Omega x + w_{,t})^2 + \Omega^2 w^2] - T w_{,x}^2 - EI w_{,xx}^2\right] \mathrm{d}x,$$

where $\mathrm{d}T/\mathrm{d}x = \rho A \Omega^2 x$. Derive the equation of motion for transverse vibrations of the beam.

(b) Using the Ritz and the Galerkin methods, determine the approximate eigenfrequencies of the beam, and plot their variation with Ω.

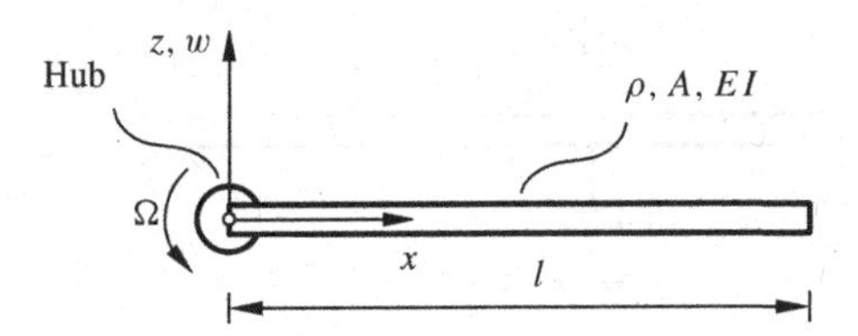

Figure 3.53 Exercise 3.15

3.16 A simply-supported uniform beam of length l is damped by a damper at $a = l/4$, and excited at the mid-span by a harmonic force $F(t) = F_0 \cos \Omega t$, as shown in Figure 3.54. What should the damping constant d be so that for $\Omega = \omega_1$ (where ω_1 is the fundamental frequency of an undamped simply-supported beam), the strain fluctuation is as small as possible? Determine the values of d and a to minimize the strain fluctuation when Ω can be any frequency in the interval $[0, \omega_4]$, and a is restricted to the interval $[0, l/3]$.

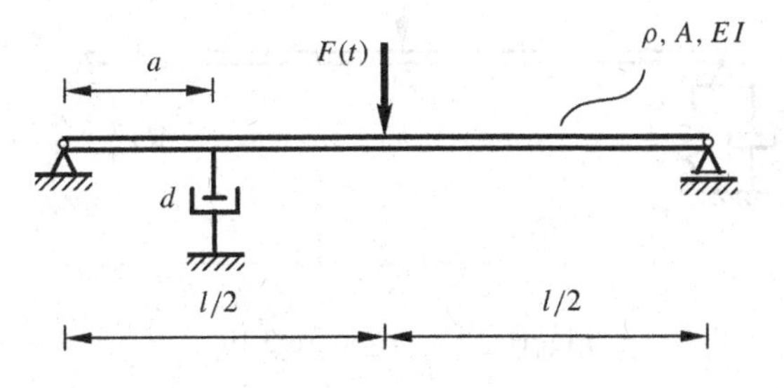

Figure 3.54 Exercise 3.16

3.17 A uniform Euler–Bernoulli beam of length l is supported, as shown in Figure 3.55. The left end of the beam is excited by a harmonic moment $M(t) = M_0 \sin \Omega t$.

(a) Determine the value of the rotational damping coefficient d so that only a positive-traveling harmonic wave is set up in the beam.

(b) If the actuator at the left end can deliver an average power $\overline{P}$, determine the amplitude of the traveling waves.

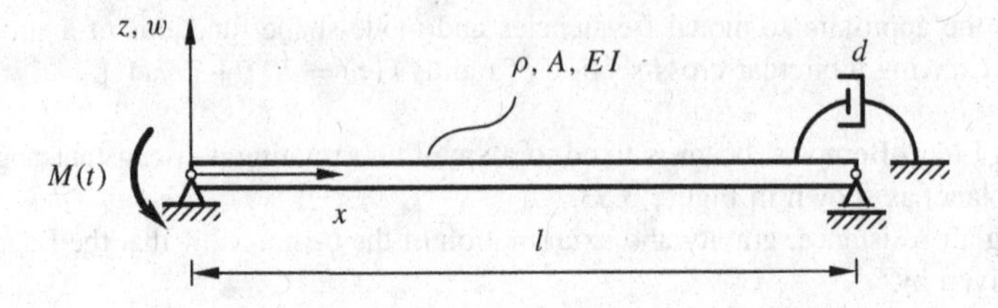

Figure 3.55 Exercise 3.17

3.18 A Timoshenko beam of infinite length is equipped with a linear and a rotational damping element, as shown in Figure 3.56. What should be the values of d_L and d_T so that maximum possible energy is dissipated by the dampers?

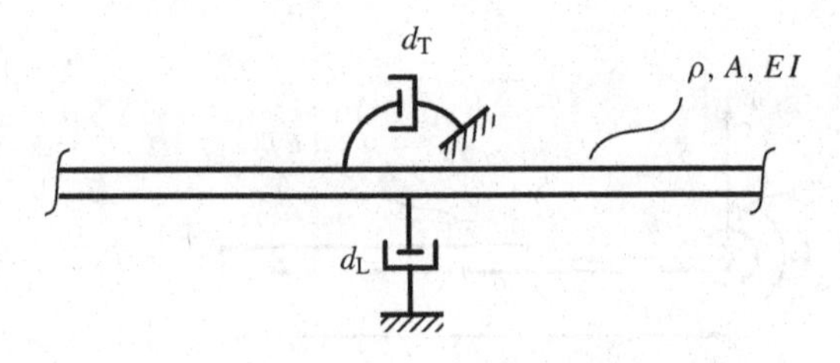

Figure 3.56 Exercise 3.18

3.19 A flexibly supported uniform beam is excited by a harmonic force $F(t) = F_0 \cos \Omega t$ at the mid-span, as shown in Figure 3.57. The excitation frequency $\Omega = \omega_1$, where ω_1 is the (non-trivial) fundamental frequency of the free–free beam. Determine the constants k and c so that the strain fluctuation is as small as possible.

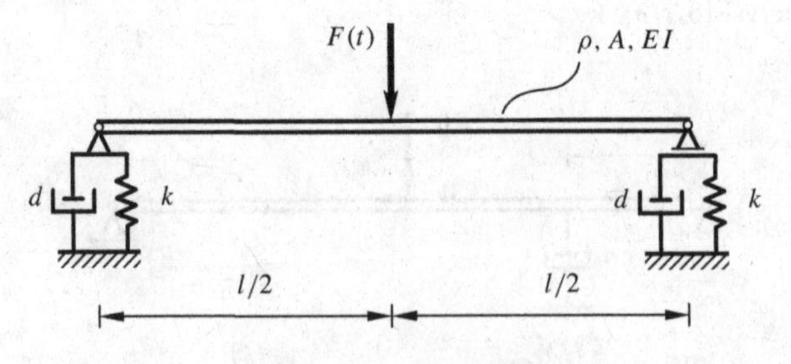

Figure 3.57 Exercise 3.19

3.20 A uniform cylindrical tube of height L has helical stairs of helix diameter $2R$ and helix angle $\alpha = \tan^{-1} l/2\pi R$ welded around it, as shown in Figure 3.58. Model the tube as a fixed–free uniform hollow beam, the stairs as eccentric mass distribution of linear density μ at a radius R from the axis of the tube, and neglect gravity.

(a) Derive the equations of motion of the beam for the two transverse and one torsional modes of vibration.

(b) Determine the approximate natural frequencies of the system.

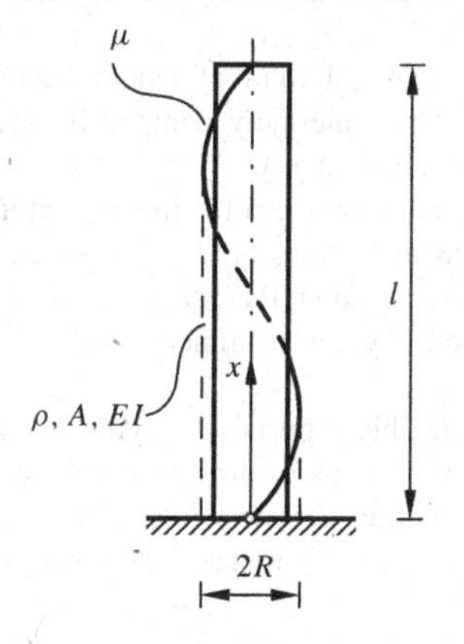

Figure 3.58 Exercise 3.20

3.21 A uniform belt under constant tension T passes at a constant speed v between two roller-stands separated by a distance l, as shown in Figure 3.59. Modeling the belt as an Euler–Bernoulli beam, derive the equation of motion. Assuming clamped boundary condition at the rollers, determine approximately the first critical speed of the belt.

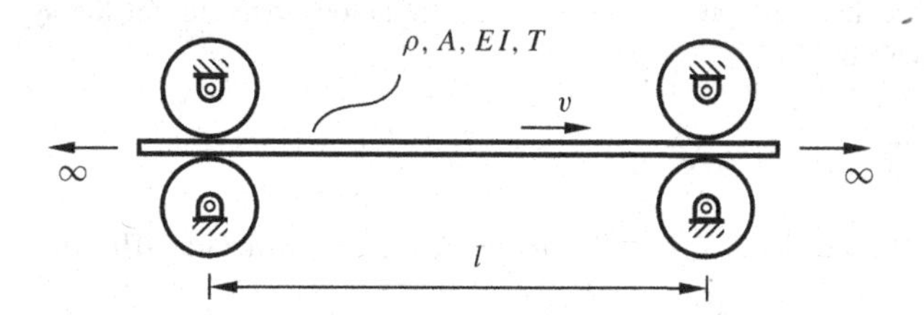

Figure 3.59 Exercise 3.21

3.22 For an axially translating Euler–Bernoulli beam of length l, subject to a constant tension T, and translating at a constant speed v, show that the rate of change of total mechanical energy of the beam is given by

$$\frac{\mathrm{d}\mathcal{E}}{\mathrm{d}t} = \left[EIw_{,xx}(w_{,xt} + vw_{,xx}) - (EIw_{,xxx} - Tw_{,x})(w_{,t} + vw_{,x})\right]\Big|_0^l .$$

3.23 A cantilever uniform cylindrical rotor of length l and flexural stiffness EI rotates at a constant angular speed Ω. Write down the equations of motion and the boundary conditions. Discretize the system using the eigenfunctions of a non-rotating cantilever beam. Determine the critical speeds of the rotor and comment on its stability.

3.24 In Problem 3.23, consider in addition a constant follower force F at the end of the cantilever rotor. Derive the equations of motion and determine the boundary conditions. On the F–Ω parameter plane determine the regions of stability/instability and identify their types.

3.25 A simply-supported uniform cylindrical rotor of length l and flexural stiffness EI rotates at a constant angular speed Ω. The rotor is internally damped having a loss factor η, and externally isotropically damped with damping coefficient ρAr.

(a) Write down the equations of motion of the rotor in the rotating coordinates (see Figure 3.38). (*Hint*: The external damping has to be transformed appropriately.)
(b) Transform the coordinates and write down the equations of motion in the fixed frame.
(c) Determine the conditions for stability, and comment on the types of instability that occur.

3.26 A clamped–clamped uniform flexible pipe is conveying an incompressible inviscid fluid at a constant speed v. Expand the solution in terms of the first two eigenfunctions of a clamped–clamped beam, and discretize the equation of motion. Define $\mu^2 = \rho_f A_f / \rho A$, and $V = \mu v$ (see Section 3.8.6), and study the stability of the system in the μ–V parameter plane. Determine the regions of stability/instability and identify their types.

3.27 A uniform flexible pipe conveying an incompressible inviscid fluid at a constant speed v is clamped at one end and free at the other, as shown in Figure 3.42. Discretize the equation of motion using the eigenfunctions of a cantilever beam. Using the definitions of μ and V as in Exercise 3.26, investigate the stability of the system in the μ–V parameter plane. Determine the regions of stability/instability and identify their types.

3.28 A uniform flexible cantilever pipe is conveying an incompressible fluid with slight viscosity at a constant speed v. Assuming that the viscous drag produces a uniform axial force density fv (force per unit length) throughout the pipe, write down the equation of motion of the system. Repeat the analysis as in Exercise 3.27.

REFERENCES

[1] Timoshenko, S.P., and Goodier, J.N., *Theory of Elasticity*, 3e, McGraw-Hill Book Co., Singapore, 1970.

[2] Anderson, K., and Hagedorn, P., On the Energy Dissipation in Space Dampers in Bundeled Conductors of Overhead Transmission Lines, *J. of Sound and Vibration*, 180(4), 1995, pp. 539-556.

[3] Schmidt, J., *Entwurf von Reglern zur aktiven Schwingungdämpfung an flexible mechanischen Strukturen*, Dissertation, TH Darmstadt, 1987.

[4] Hagedorn, P., *Technische Schwingungslehre: Lineare Schwingungen kontinuierlicher mechanischer Systeme*, Springer-Verlag, Berlin, 1989.

[5] Païdoussis, M.P., *Fluid–Structure Interactions: Slender Structures in Axial Flow, Vol. 1*, Academic Press, London, 1998.

4

Vibrations of membranes

A membrane is a planar two-dimensional pre-tensioned elastic continuum that does not transmit or resist bending moment. In practice, any two-dimensional elastic continuum resists bending moment. However, if the tension is large, and the curvatures are small, the effect of bending moment can be neglected. Thus, the membrane can be imagined as an extension of the string to two dimensions, however, with certain differences which will be pointed out in the course of our discussions. Membranes are found in condenser microphones, ear drums, drum heads, and in various other applications. The objective of this chapter is to study the transverse vibrations and propagation of waves in membranes.

4.1 DYNAMICS OF A MEMBRANE

Since the membrane is two-dimensional, we require two space coordinates to represent the points of the membrane. With two space dimensions, comes the question of choice of a coordinate system in which to represent the space variables. The most common choices are the Cartesian and the polar coordinate systems. Theoretically, the choice can be arbitrary. However, in most practical cases, the boundary of the membrane guides the choice, since it may be easier to represent the boundary conditions and solve the dynamics of a certain membrane shape in a particular coordinate system. For example, the dynamics of a circular membrane can be most conveniently worked out in the polar coordinate system.

Let us begin with the Cartesian coordinate system. Consider the coordinate system shown in Figure 4.1, where the x-y-plane represents the undeformed configuration of the membrane. The configuration of the membrane at any time t will be represented by the field variable $w(x, y, t)$. We will assume in all our analyses that the displacements of all points of the membrane are small and always perpendicular to the x-y-plane, and the thickness h of the membrane remains constant. The principal stresses $\sigma_x(x, y, t)$ and $\sigma_y(x, y, t)$ will be assumed to be equal and constant throughout the membrane.

4.1.1 Newtonian formulation

Consider an infinitesimal element, as shown in Figure 4.1. Define $T = \sigma_x h = \sigma_y h$ as the tension per unit length of the membrane, and $\mu = \rho h$ as the mass per unit area, where ρ

Vibrations and Waves in Continuous Mechanical Systems P. Hagedorn and A. DasGupta

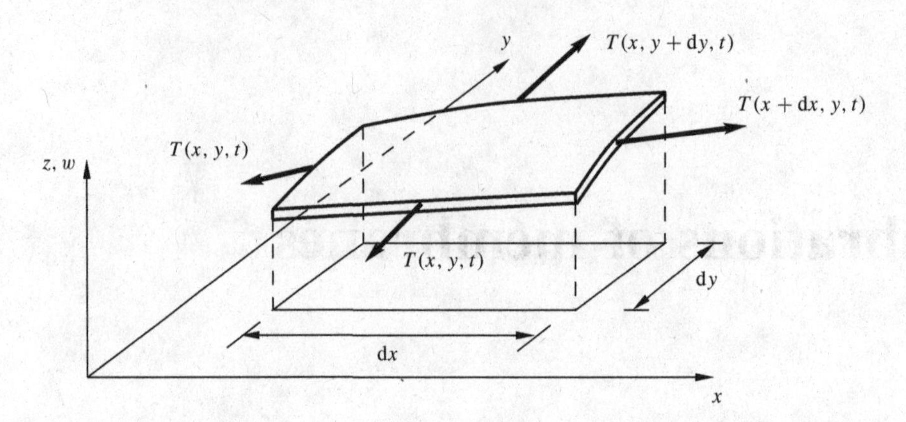

Figure 4.1 An infinitesimal element of a membrane in a Cartesian coordinate system

is the density of the material. Applying Newton's second law to the infinitesimal element yields

$$(\mu \Delta x \Delta y) w_{,tt} = T\Delta y \left(\left.\frac{\partial w}{\partial x}\right|_{x+\Delta x, y} - \left.\frac{\partial w}{\partial x}\right|_{x,y} \right) + T\Delta x \left(\left.\frac{\partial w}{\partial y}\right|_{x, y+\Delta y} - \left.\frac{\partial w}{\partial y}\right|_{x,y} \right). \tag{4.1}$$

Dividing (4.1) by $\Delta x \Delta y$ and taking the limit $\Delta x \to 0$ and $\Delta y \to 0$, we have on rearrangement

$$\mu w_{,tt} - T(w_{,xx} + w_{,yy}) = 0 \tag{4.2}$$

or

$$w_{,tt} - c^2 \nabla^2 w = 0, \tag{4.3}$$

where

$$\nabla^2 = \frac{\partial^2}{\partial x^2} + \frac{\partial^2}{\partial y^2} \tag{4.4}$$

is the Laplacian operator, and

$$c = \sqrt{T/\mu} \tag{4.5}$$

is the speed of transverse waves in the membrane. It is evident that (4.3) is a hyperbolic partial differential equation, and represents the wave equation in two-dimensional Cartesian space.

For a unique solution of (4.2), it is evident that we require certain boundary conditions for space, and initial conditions for time. For simplicity, consider a rectangular membrane stretched between $x \in [0, a]$ and $y \in [0, b]$. If the membrane is fixed on all four edges, the appropriate boundary conditions are

$$w(0, y, t) \equiv 0, \quad w(a, y, t) \equiv 0, \quad w(x, 0, t) \equiv 0, \quad \text{and} \quad w(x, b, t) \equiv 0.$$

All these four boundary conditions are geometric conditions. When an edge, say at $x = a$, is sliding, the boundary condition is obtained from the dynamic condition of zero transverse force at the edge, and can be mathematically written as

$$T w_{,x}(a, y, t) \equiv 0. \tag{4.6}$$

To obtain the equation of motion of a membrane in polar coordinates (r, ϕ), we use the coordinate transformation

$$r = \sqrt{x^2 + y^2} \quad \text{and} \quad \phi = \tan^{-1}\frac{y}{x}.$$

Then, the operator transformations from the Cartesian to polar coordinate system are given by

$$\frac{\partial}{\partial x} = \cos\phi \frac{\partial}{\partial r} - \sin\phi \frac{\partial}{r\partial\phi}, \tag{4.7}$$

$$\frac{\partial}{\partial y} = \sin\phi \frac{\partial}{\partial r} + \cos\phi \frac{\partial}{r\partial\phi}, \tag{4.8}$$

and the equation of motion of the membrane in polar coordinates is obtained as

$$\mu w_{,tt} - T\left(w_{,rr} + \frac{1}{r}w_{,r} + \frac{1}{r^2}w_{,\phi\phi}\right) = 0. \tag{4.9}$$

This also has a compact representation of the form (4.3), where the Laplacian in the polar coordinate system can be written from (4.9) as

$$\nabla^2 = \frac{\partial}{\partial r^2} + \frac{1}{r}\frac{\partial}{\partial r} + \frac{1}{r^2}\frac{\partial^2}{\partial \phi^2}.$$

The boundary condition for a fixed circular membrane can be written as $w(a, \phi, t) \equiv 0$, where a is the radius of the membrane. At the center of the circular membrane, the condition of finiteness of displacement has to be imposed to solve the boundary value problem. This will be discussed further in detail later.

Even though the membrane seems to be similar to a string in two dimensions, there are certain differences that set them apart. The first distinction is observed when one attempts to solve the static deflection of a string and a membrane under a point force. For simplicity,

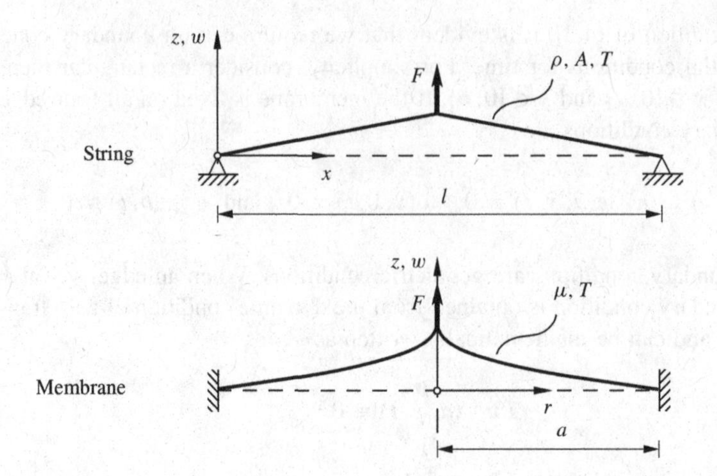

Figure 4.2 Comparison of equilibrium solutions of a string and a circular membrane under a point load

consider a string loaded by a point force at the middle, and a circular membrane loaded at its center. The equilibrium solution of the string is obtained as

$$w(x) = \frac{F}{T}\left[\left(\frac{l}{2} - x\right)\mathcal{H}\left(x - \frac{l}{2}\right) + \frac{1}{2}x\right],$$

where F is the magnitude of the force, T is the string tension, and $\mathcal{H}(\cdot)$ is the Heaviside step function. The solution of the string is a bounded function, as shown in Figure 4.2. On the other hand, the solution for the membrane is obtained as $w(r) = (F/2\pi T)\ln(a/r)$, where T is the tensile force per unit length of the membrane, and a is the radius of the membrane. The solution indicates a logarithmic singularity at $r = 0$, where the solution is unbounded. Thus, mathematically, a membrane cannot support a point force without getting punctured! This behavior can also be understood physically as follows. Consider the equilibrium of a small circular element around the point force. As the radius of the circular element tends to zero, the force available to equilibrate the externally applied point force also tends to zero, since the tensile force per unit length of the membrane is assumed constant. This causes the displacement of the membrane to be unbounded at the point of application of the force. The available distributed force also tends to align itself opposite to the direction of the applied force, making the slope of the membrane $w_{,r}(r) \to \infty$ as $r \to 0$.

4.1.2 Variational formulation

The kinetic energy of a vibrating membrane can be expressed as

$$\mathcal{T} = \frac{1}{2}\int_{A} \mu w_{,t}^2 \, \mathrm{d}A, \tag{4.10}$$

where dA represents the area element in the domain of integration A. From the theory of elasticity (see, for example, [1]), the potential energy can be written, for example, in Cartesian coordinates as

$$\mathcal{V} = \int_{\mathrm{A}} \int_{-h/2}^{h/2} (\sigma_x \epsilon_x + \sigma_y \epsilon_y) \, \mathrm{d}\xi \, \mathrm{d}A, \tag{4.11}$$

where ϵ_x and ϵ_y are the strains in the x and y directions, respectively, $\sigma_x = \sigma_y = T/h$ are the stresses, and ξ is measured along the thickness. Note that the potential energy expression (4.11) is obtained with the assumption that stresses remain unchanged (i.e., strains are small). Similarly to the case of a string, one can write $\epsilon_x \approx w_{,x}^2/2$ and $\epsilon_y \approx w_{,y}^2/2$. Using these expressions in (4.11), and assuming that all quantities are uniform over the thickness, yields

$$\mathcal{V} = \frac{1}{2} \int_{\mathrm{A}} T(w_{,x}^2 + w_{,y}^2) \, \mathrm{d}A = \frac{1}{2} \int_{\mathrm{A}} T \nabla w \cdot \nabla w \, \mathrm{d}A, \tag{4.12}$$

where

$$\nabla w := w_{,x} \hat{\mathbf{i}} + w_{,y} \hat{\mathbf{j}},$$

is the gradient of w, and $\hat{\mathbf{i}}$ and $\hat{\mathbf{j}}$ are the unit vectors along the x- and y-axes, respectively. It may be mentioned that the form of the potential energy in (4.12) using the gradient operator $\nabla(\cdot)$ is valid in any coordinate system, provided the gradient operator is expressed in that coordinate system. Now, the Lagrangian can be written as

$$\mathcal{L} = \mathcal{T} - \mathcal{V} = \frac{1}{2} \int_{\mathrm{A}} \left[\mu w_{,t}^2 - T \nabla w \cdot \nabla w \right] \mathrm{d}A. \tag{4.13}$$

Using Hamilton's principle yields

$$\begin{aligned} &\delta \int_{t_1}^{t_2} \mathcal{L} \mathrm{d}t = 0 \\ \Rightarrow &\int_{\mathrm{A}} \mu w_{,t} \delta w \big|_{t_1}^{t_2} \mathrm{d}A \\ &\quad + \int_{t_1}^{t_2} \int_{A} \left[-\mu w_{,tt} \delta w - T \nabla w \cdot \nabla \delta w \right] \mathrm{d}A \, \mathrm{d}t = 0. \end{aligned} \tag{4.14}$$

Using the condition that $\delta w = 0$ at $t = t_0$ and $t = t_1$, the first integral in (4.14) vanishes. The identity

$$\nabla w \cdot \nabla \delta w = \nabla \cdot (\delta w \nabla w) - (\nabla^2 w) \, \delta w, \tag{4.15}$$

can be used to rewrite (4.14) as

$$\int_{t_1}^{t_2}\int_{A}\left[-\mu w_{,tt}\delta w - T\nabla\cdot(\delta w\nabla w) + T(\nabla^2 w)\delta w\right]\mathrm{d}A\,\mathrm{d}t = 0. \tag{4.16}$$

Now, we use the Gauss divergence theorem

$$\int_{A}\nabla\cdot\mathbf{v}\,\mathrm{d}A = \oint_{B}\mathbf{v}\cdot\hat{\mathbf{n}}\,\mathrm{d}s,$$

where B is the boundary enclosing the domain A, $\hat{\mathbf{n}}$ is the unit outward normal to the boundary, and ds is a boundary element of infinitesimal length, as shown in Figure 4.3. Using this theorem for the second integrand in (4.16), we obtain

$$-\int_{t_1}^{t_2}\oint_{B} Tw_{,n}\,\delta w\,\mathrm{d}s\,\mathrm{d}t + \int_{t_1}^{t_2}\int_{A}\left[-\mu w_{,tt} + T\nabla^2 w\right]\delta w\,\mathrm{d}A\,\mathrm{d}t = 0, \tag{4.17}$$

where we have used the definition $w_{,n} := \nabla w\cdot\hat{\mathbf{n}}$, which is the directional derivative of w along $\hat{\mathbf{n}}$. The vanishing of the second integral in (4.17) yields the equation of motion of the membrane, which is identical to (4.2). The first integral gives the boundary conditions

$$Tw_{,n}\big|_{B} = 0 \quad \text{or} \quad w\big|_{B} = 0.$$

For example, for a rectangular membrane shown in Figure 4.4, the boundary conditions may be written in four parts as

$$Tw_{,x}\big|_{x=a} \equiv 0 \quad \text{or} \quad w\big|_{x=a} \equiv 0,$$

$$Tw_{,y}\big|_{y=b} \equiv 0 \quad \text{or} \quad w\big|_{y=b} \equiv 0,$$

$$-Tw_{,x}\big|_{x=0} \equiv 0 \quad \text{or} \quad w\big|_{x=0} \equiv 0,$$

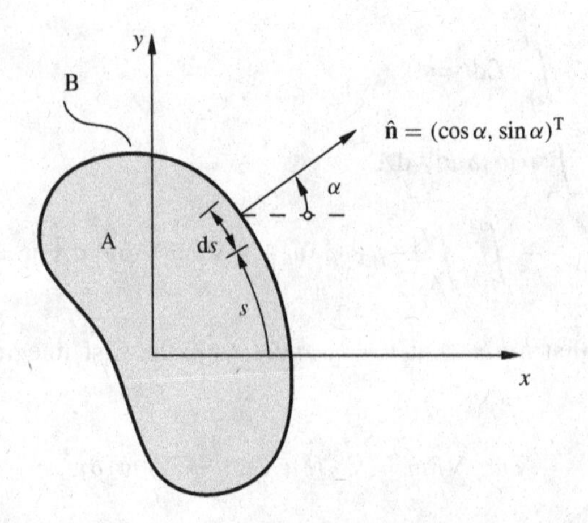

Figure 4.3 A membrane of arbitrary shape in Cartesian coordinates

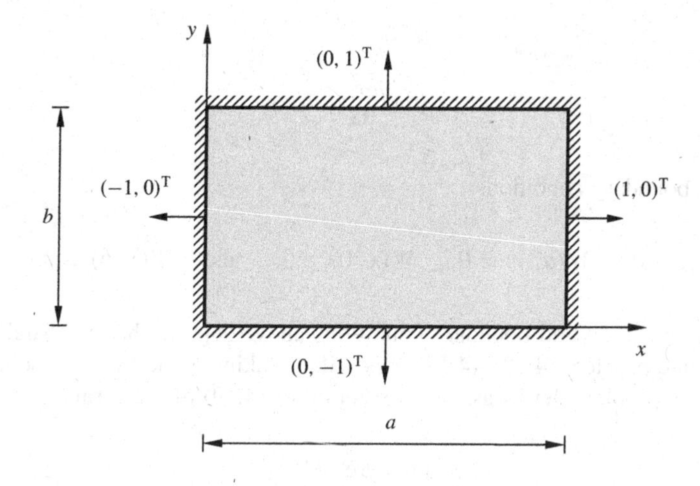

Figure 4.4 A rectangular membrane with boundary normals in Cartesian coordinates

and

$$-Tw_{,y}\Big|_{y=0} \equiv 0 \qquad \text{or} \qquad w\Big|_{y=0} \equiv 0.$$

The variational formulation for a circular membrane can be easily obtained using the operator transformations (4.7)–(4.8) in (4.13).

4.2 MODAL ANALYSIS

We consider here the modal analysis of rectangular and circular membranes, and obtain the corresponding eigenfrequencies and eigenfunctions. The eigenfunctions are then used to construct the solution of the initial value problem using the expansion theorem.

4.2.1 The rectangular membrane

Consider a rectangular membrane with all edges fixed, as shown in Figure 4.4. We first find a solution of the unforced dynamics (4.2) of the membrane of the form

$$w(x, y, t) = W(x, y)\mathrm{e}^{i\omega t}, \tag{4.18}$$

where $W(x, y)$ is the eigenfunction and ω is the corresponding circular eigenfrequency. It is to be remembered that the actual solution is the real part of the solution form (4.18). Substituting (4.18) in (4.2) yields the *Helmholtz equation*

$$W_{,xx} + W_{,yy} + \frac{\omega^2}{c^2}W = 0$$

or

$$\nabla^2 W + \frac{\omega^2}{c^2} W = 0 \tag{4.19}$$

along with the boundary conditions

$$W(0, y) \equiv 0, \quad W(a, y) \equiv 0, \quad W(x, 0) \equiv 0, \quad \text{and} \quad W(x, b) \equiv 0. \tag{4.20}$$

The differential equation and boundary conditions above suggest that a separable solution for the eigenvalue problem (4.19)–(4.20) is possible. Taking a cue from the solution of the string eigenvalue problem, let us assume a solution of (4.19) of the form

$$W(x, y) = B\mathrm{e}^{i(k_x x + k_y y)}, \tag{4.21}$$

where B is a complex constant. Substituting (4.21) in (4.19) yields the *dispersion relation* of the membrane as

$$-k_x^2 - k_y^2 + \frac{\omega^2}{c^2} = 0. \tag{4.22}$$

It is clear from (4.22) that the solutions will be of the form $k_x = \pm\alpha$ and $k_y = \pm\beta$ such that $\alpha^2 + \beta^2 = \omega^2/c^2$. Therefore, the general solution of (4.19) can be written as

$$W(x, y) = (B_1\mathrm{e}^{i\alpha x} + B_2\mathrm{e}^{-i\alpha x})(B_3\mathrm{e}^{i\beta y} + B_4\mathrm{e}^{-i\beta y}), \tag{4.23}$$

or in the form

$$\begin{aligned} W(x, y) = {} & A_1 \cos\alpha x \cos\beta y + A_2 \cos\alpha x \sin\beta y \\ & + A_3 \sin\alpha x \cos\beta y + A_4 \sin\alpha x \sin\beta y. \end{aligned} \tag{4.24}$$

One can also represent the solution as

$$W(x, y) = B_1\mathrm{e}^{i(\alpha x + \beta y)} + B_2\mathrm{e}^{i(-\alpha x + \beta y)} + B_3\mathrm{e}^{i(\alpha x - \beta y)} + B_4\mathrm{e}^{i(-\alpha x - \beta y)}. \tag{4.25}$$

The solution form (4.25) will be interpreted and used later. It is to be noted here that, in the solution form (4.23) and (4.25), the coefficients B_i are in general complex, while the coefficients A_i in (4.24) are real for the solution to be real. At present, we will use the form (4.24) so that the boundary conditions can be put in a convenient form. Using the first and third boundary conditions in (4.20) with the solution form (4.24), one obtains

$$A_1 \cos\beta y + A_2 \sin\beta y = 0, \tag{4.26}$$

$$A_1 \cos\alpha x + A_3 \sin\alpha x = 0. \tag{4.27}$$

It is easy to see that, for satisfaction of (4.26) for all $y \in [0, b]$, and for (4.27) for all $x \in [0, a]$, we must have $A_1 = 0$, $A_2 = 0$, and $A_3 = 0$. Therefore, the solution (4.24) simplifies to

$$W(x, y) = A_4 \sin \alpha x \sin \beta y. \tag{4.28}$$

Using this solution in the second and fourth boundary conditions in (4.20) leads to

$$\sin \alpha a \sin \beta y = 0 \quad \text{and} \quad \sin \alpha x \sin \beta b = 0. \tag{4.29}$$

Satisfaction of (4.29) for all $x \in [0, a]$ and $y \in [0, b]$ requires

$$\alpha = \frac{m\pi}{a} \quad \text{and} \quad \beta = \frac{n\pi}{b}, \qquad m, n = 1, 2, \ldots, \infty. \tag{4.30}$$

The eigenfunctions are then obtained from (4.28) as

$$W^{(m,n)} = \sin \frac{m\pi x}{a} \sin \frac{n\pi x}{b}, \qquad m, n = 1, 2, \ldots, \infty, \tag{4.31}$$

where $W^{(m,n)}$ represents the eigenfunction of the (m, n) mode. It can be easily checked from (4.31) that these eigenfunctions satisfy the orthogonality condition

$$\begin{aligned} \langle W^{(m,n)}(x, y), W^{(r,s)}(x, y) \rangle &= \int_0^a \int_0^b W^{(m,n)}(x, y) W^{(r,s)}(x, y) \, \mathrm{d}x \, \mathrm{d}y \\ &= \frac{ab}{4} \delta_{mr} \delta_{ns}. \end{aligned} \tag{4.32}$$

The first few mode-shapes of the membrane are depicted in Figure 4.5. Using (4.30) in the condition $\alpha^2 + \beta^2 = \omega^2/c^2$ yields the frequency equation

$$\omega_{(m,n)} = \pi c \sqrt{\frac{m^2}{a^2} + \frac{n^2}{b^2}}, \tag{4.33}$$

where $\omega_{(m,n)}$ represents the circular eigenfrequency of the (m, n) mode. It may be noted that, unlike for the string, the eigenfrequencies of the membrane are not, in general, integral multiples of the fundamental frequency. Further, if $l_\mathrm{r} := a/b$ is rational, then there can exist orthogonal modes (m, n) and (r, s) such that $m^2 + l_\mathrm{r}^2 n^2 = r^2 + l_\mathrm{r}^2 s^2$, and hence $\omega_{(m,n)} = \omega_{(r,s)}$. For example, with $l_\mathrm{r} = 4/3$, we have $\omega_{(3,5)} = \omega_{(5,4)}$, $\omega_{(8,3)} = \omega_{(4,6)}$, and so on, and the eigenfunctions are all orthogonal, as can be easily verified. Such independent eigenfunctions corresponding to a single eigenfrequency are known as *degenerate modes*. On the other hand, it can also be concluded from (4.19) that any arbitrary linear combination of these degenerate eigenfunctions, say $W^{(m,n)}$ and $W^{(r,s)}$, is also an eigenfunction corresponding to the eigenvalue $\omega_{(m,n)}$. In other words, the function

$$W(x, y) = W^{(m,n)} \cos \delta + W^{(r,s)} \sin \delta, \tag{4.34}$$

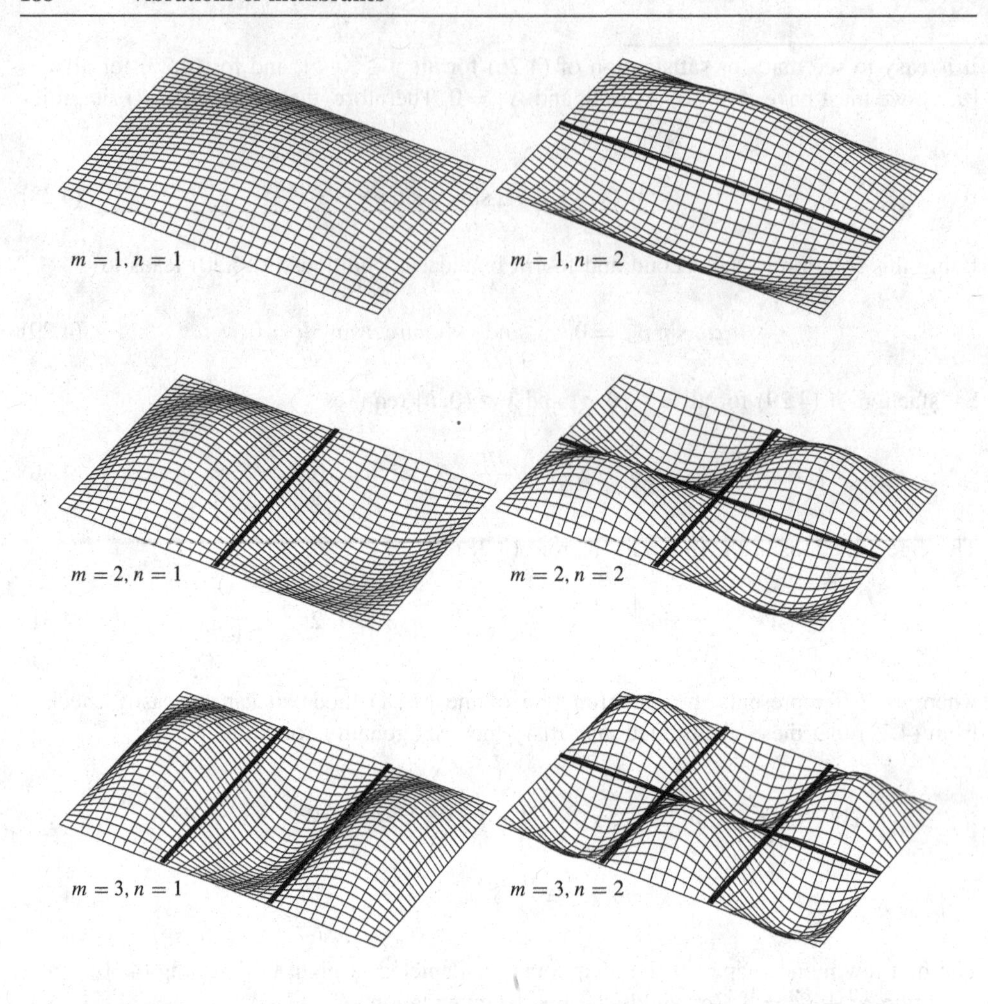

Figure 4.5 First few mode-shapes of a rectangular membrane with fixed boundaries

where δ is arbitrary, is a valid eigenfunction corresponding to $\omega_{(m,n)}$. Thus, there is an *isotropy* in the dynamics in the sub-space of the degenerate modes. In the special case when the membrane is square, i.e., $a/b = 1$, we have $\omega_{(m,n)} = \omega_{(n,m)}$ for all (m, n), $m \neq n$, and hence, all such modes are degenerate. In Figure 4.6, the nodal lines of the eigenfunction (4.34) of the degenerate modes (3,1) and (1,3) of a square membrane are shown for specific values of δ. It is evident that such modal degeneracy does not exist in one-dimensional continua such as strings. This is another distinguishing feature of a membrane when compared to a string. Some interesting phenomena occurring in systems with degenerate modes will be discussed later in this chapter. This isotropy has to do with some kind of physical symmetry of the system. For example, a rectangular membrane is symmetric under a rotation by π.

The general solution of the free vibration problem (or the initial value problem) can now be written using (4.18), (4.31) and (4.33) as

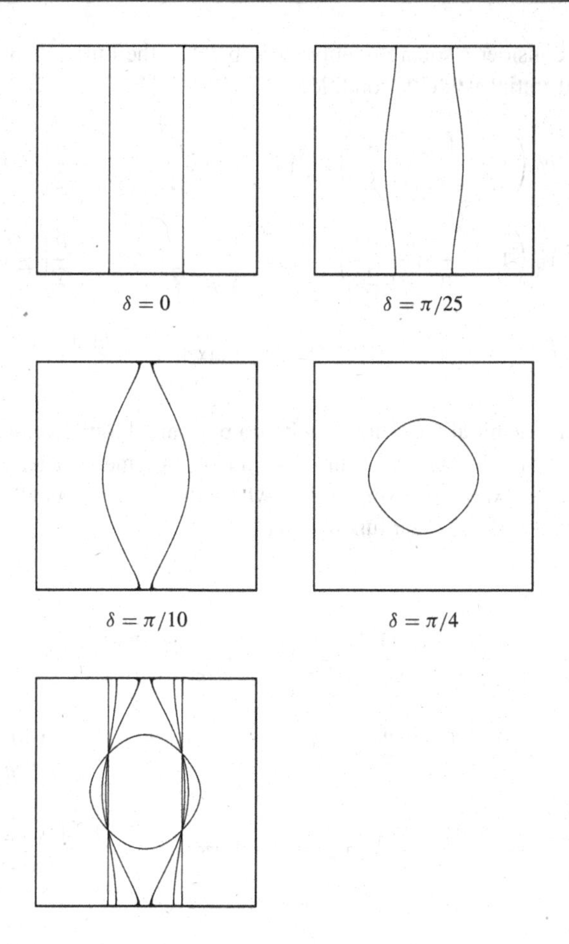

Figure 4.6 Nodal lines of linear combinations of the degenerate modes (3,1) and (1,3) of a square membrane

$$w(x, y, t) = \mathscr{R}\left[\sum_{m=1}^{\infty}\sum_{n=1}^{\infty} A_{(m,n)} \sin\frac{m\pi x}{a} \sin\frac{n\pi y}{b} e^{i\omega_{(m,n)}t}\right]$$

or

$$w(x, y, t) = \sum_{m=1}^{\infty}\sum_{n=1}^{\infty} \sin\frac{m\pi x}{a} \sin\frac{n\pi y}{b} \big[C_{(m,n)} \cos\omega_{(m,n)}t + S_{(m,n)} \sin\omega_{(m,n)}t\big], \qquad (4.35)$$

where $\mathscr{R}[\cdot]$ represents the real part of the argument, $A_{(m,n)} = C_{(m,n)} - i\,S_{(m,n)}$ is a complex constant representing the constants of integration for the initial value problem, and $\omega_{(m,n)}$ is given by (4.33).

As an example, consider a square membrane subject to the initial displacement condition $w(x, y, 0) \equiv 0$, and initial velocity condition

$$w_{,t}(x, y, 0) = \begin{cases} v_0\left(1 - \dfrac{1}{\epsilon a}\left|x - \dfrac{a}{2}\right|\right), & \text{for } \left|y - \dfrac{a}{2}\right| \le \left|x - \dfrac{a}{2}\right| \le \epsilon a, \\ v_0\left(1 - \dfrac{1}{\epsilon a}\left|y - \dfrac{a}{2}\right|\right), & \text{for } \left|x - \dfrac{a}{2}\right| \le \left|y - \dfrac{a}{2}\right| \le \epsilon a, \\ 0, & \text{for } \max\left[\left|x - \dfrac{a}{2}\right|, \left|y - \dfrac{a}{2}\right|\right] > \epsilon a, \end{cases} \tag{4.36}$$

where $\epsilon < 1$. The above initial condition specifies a pyramidal initial velocity over a centrally placed square region of side $2\epsilon a$ on an initially undeformed membrane. The solution of this initial value problem is expressed using (4.35), where the unknown coefficients are obtained using the orthogonality of the eigenfunctions as

$$C_{(m,n)} = 0,$$

$$S_{(m,n)} = \begin{cases} \dfrac{16v_0}{\omega_{(m,n)}\epsilon\pi^3} \dfrac{(-1)^{(m+n)/2}}{m^2 - n^2}\left[\dfrac{\cos n\pi\epsilon \sin m\pi\epsilon}{m} - \dfrac{\cos m\pi\epsilon \sin m\pi\epsilon}{n}\right], & \text{for } m, n \text{ odd, and } m \ne n \\ -\dfrac{4v_0}{\omega_{(m,n)}\epsilon\pi^3 m^3}\left[\sin 2m\pi\epsilon - 2m\pi\epsilon\right], & \text{for } m, n \text{ odd, and } m = n \\ 0, & \text{for } m, n \text{ even.} \end{cases}$$

Using $\epsilon = 0.15$, the successive configurations of the membrane at specific times are shown in Figure 4.7.

4.2.2 The circular membrane

Consider a uniformly stretched circular membrane having a radius $r = a$, as shown in Figure 4.8. Using the field variable $w(r, \phi, t)$ to represent the configuration of the system, the equation of motion and the boundary conditions can be written as

$$\mu w_{,tt} - T\left(w_{,rr} + \frac{1}{r}w_{,r} + \frac{1}{r^2}w_{,\phi\phi}\right) = 0, \tag{4.37}$$

$$w(a, \phi, t) \equiv 0. \tag{4.38}$$

It may be noticed that, while we require two boundary conditions for a unique solution, only the outer boundary condition (4.38) can be specified. This point will be resolved using

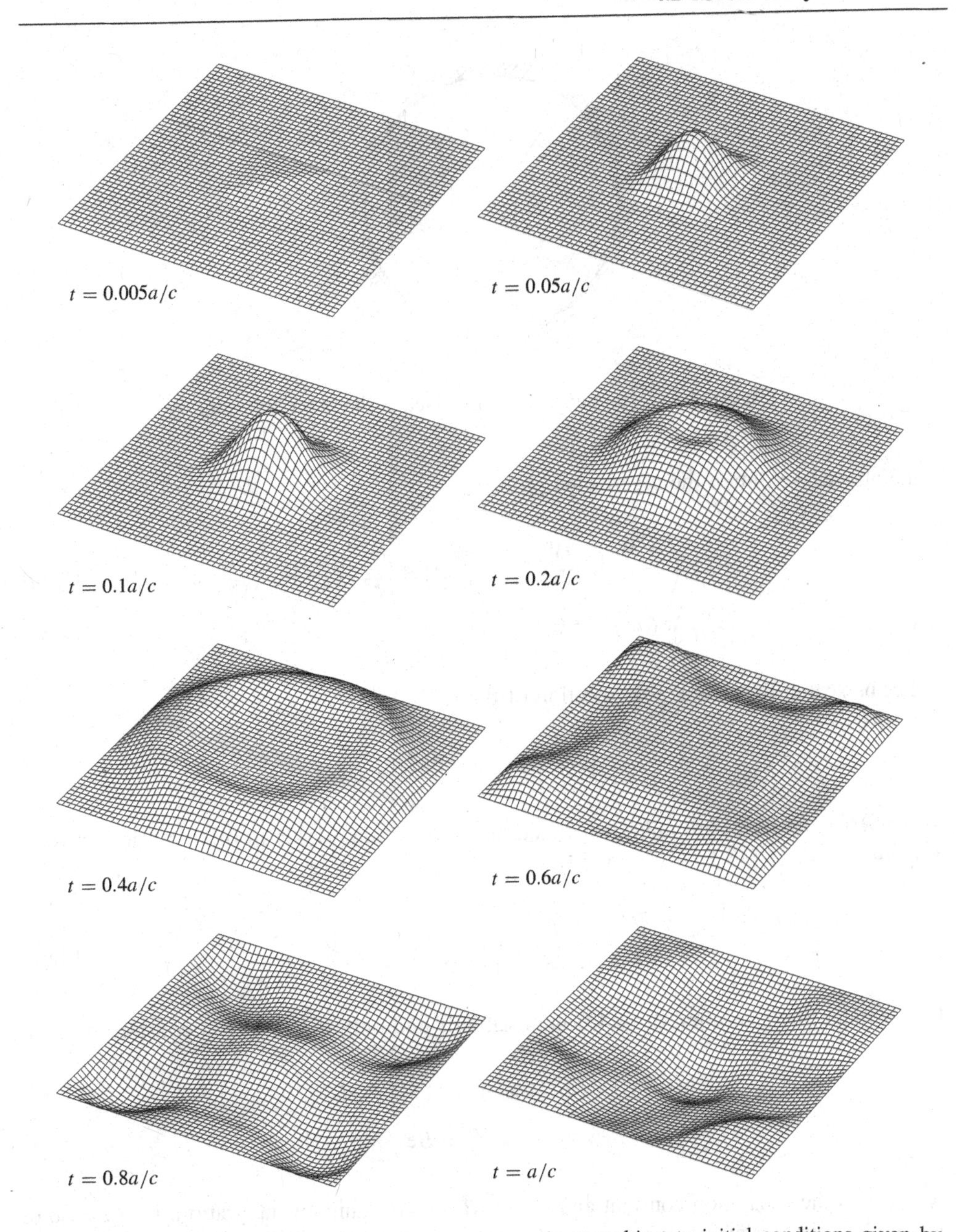

Figure 4.7 Successive configurations of a square membrane subject to initial conditions given by (4.36)

a finiteness condition on the displacement field during the solution of the boundary value problem. Substituting the modal solution

$$w = W(r, \phi)e^{i\omega t} \tag{4.39}$$

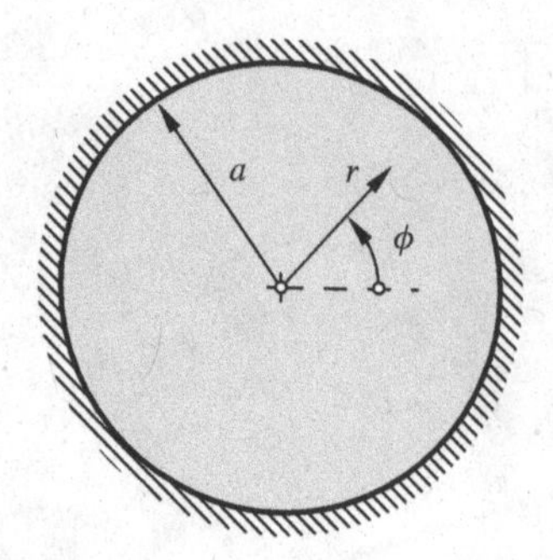

Figure 4.8 A circular membrane fixed at the outer boundary

in the above, we obtain the eigenvalue problem

$$\left(\frac{\partial^2 W}{\partial r^2}+\frac{\partial W}{r\partial r}+\frac{\partial^2 W}{r^2\partial\phi^2}\right)+\frac{\omega^2}{c^2}W=0, \tag{4.40}$$

$$W(a,\phi)\equiv 0. \tag{4.41}$$

Let us search for a separable solution of the form

$$W(r,\phi)=R(r)\Phi(\phi), \tag{4.42}$$

where $R(r)$ and $\Phi(\phi)$ are unknown functions. Substituting (4.42) in (4.40) yields on rearrangement

$$\frac{R''}{R}+\frac{1}{r}\frac{R'}{R}+\frac{1}{r^2}\frac{\Phi''}{\Phi}+\frac{\omega^2}{c^2}=0. \tag{4.43}$$

It is evident that a solution of (4.43) is possible if and only if Φ''/Φ is a constant, i.e.,

$$\Phi''+\nu^2\Phi=0,$$
$$\Rightarrow \Phi=A\mathrm{e}^{i\nu\phi}+B\mathrm{e}^{-i\nu\phi},$$

where ν is the separation constant and A and B are constants of integration. In the case of a complete circular membrane ν must take integer values since the solution must have a 2π periodicity in ϕ, i.e., $\Phi(\phi+2\pi)=\Phi(\phi)$. Therefore, we will take the solution of (4.40) in the form

$$W(r,\phi)=R(r)\mathrm{e}^{im\phi}, \tag{4.44}$$

where $m=0,\pm1,\pm2,\ldots,\pm\infty$. Once again, the actual solution is obtained by taking either the real part or the imaginary part of $W(r,\phi)$ in (4.44). Substituting this solution form in

(4.40)–(4.41), we have on simplification

$$R'' + \frac{1}{r}R' + \left(\gamma^2 - \frac{m^2}{r^2}\right) R = 0, \tag{4.45}$$

$$R(a) = 0, \tag{4.46}$$

where

$$\gamma = \omega / c. \tag{4.47}$$

The differential equation (4.45) is the Bessel differential equation, and its solution can be written directly as

$$R(r) = DJ_m(\gamma r) + EY_m(\gamma r), \tag{4.48}$$

where $J_m(\cdot)$ and $Y_m(\cdot)$ are, respectively, the Bessel functions of the first and second kinds of order m, and D and E are the constants of integration, possibly complex. It is to be noted that $J_m(\cdot) = J_{-m}(\cdot)$ and $Y_m(\cdot) = Y_{-m}(\cdot)$. Hence, we need to take only the positive integer values for m, including zero. The Bessel function $J_m(z)$ is finite for all values of z, whereas $Y_m(z)$ has a logarithmic singularity at $z = 0$. Since the displacement of the membrane at $z = 0$ is required to be finite, we must have $E = 0$ in (4.48),i.e., $R(r) = DJ_m(\gamma r)$. Using the boundary condition (4.46) yields $DJ_m(\gamma a) = 0$, which yields the characteristic equation

$$J_m(\gamma a) = 0. \tag{4.49}$$

Denoting the roots of (4.49) as $a\gamma_{(m,n)}$, the first few values are obtained as

$$\begin{aligned} \gamma_{(0,1)}a &= 2.405, \quad \gamma_{(1,1)}a = 3.832, \\ \gamma_{(0,2)}a &= 5.520, \quad \gamma_{(1,2)}a = 7.016, \\ \gamma_{(0,3)}a &= 8.654, \quad \gamma_{(1,3)}a = 10.173, \end{aligned}$$

where the index m indicates the order of the Bessel function and n denotes the nth root. The circular eigenfrequencies $\omega_{(m,n)}$, are then obtained using (4.47).

From the theory of Bessel functions (see [2]), it is known that

$$J_m(x) \approx \sqrt{\frac{2}{\pi x}} \cos\left(x - (2m+1)\frac{\pi}{4}\right) \qquad \text{for } x \gg 1. \tag{4.50}$$

A comparison of the exact and approximate representations are shown in Figure 4.13. Thus, for large n (i.e., large number of nodal circles), the roots of (4.49) can be obtained using the approximation (4.50) as

$$\begin{aligned} &\cos\left(a\gamma_{(m,n)} - (2m+1)\frac{\pi}{4}\right) = 0 \\ &\Rightarrow \omega_{(m,n)} = (2m + 4n - 1)\frac{\pi c}{4a} \qquad \text{(using (4.47)).} \end{aligned}$$

The above formula works reasonably well even for lower values of n, as can be checked. It may also be concluded that, for a fixed value of m, the difference between any two consecutive exact frequencies is about $\pi c/a$.

Now, the radial part of the shape-function can be expressed as

$$R_{(m,n)}(r) = D_{(m,n)} J_m(\omega_{(m,n)} r/c),$$

where $D_{(m,n)}$ is an unknown constant, possibly complex. Substituting this expression in (4.44) yields the eigenfunction corresponding to the (m, n) mode as

$$W^{(m,n)}(r, \phi) = D_{(m,n)} J_m(\omega_{(m,n)} r/c) e^{im\phi}. \tag{4.51}$$

Writing $D_{(m,n)} = G_{(m,n)} - iH_{(m,n)}$, and taking the real part of (4.51) as the solution of the eigenvalue problem, we have the real eigenfunction of the (m, n) mode as

$$W^{(m,n)}(r, \phi) = J_m(\omega_{(m,n)} r/c)[G_{(m,n)} \cos m\phi + H_{(m,n)} \sin m\phi]. \tag{4.52}$$

It is to be noticed in (4.52) that when $m \neq 0$, $W^{(m,n)}$ is a linear combination of the form

$$W^{(m,n)}(r, \phi) = G_{(m,n)} W_{\mathrm{C}}^{(m,n)} + H_{(m,n)} W_{\mathrm{S}}^{(m,n)}, \tag{4.53}$$

where

$$W_{\mathrm{C}}^{(m,n)} := J_m(\omega_{(m,n)} r/c) \cos m\phi \quad \text{and} \quad W_{\mathrm{S}}^{(m,n)} := J_m(\omega_{(m,n)} r/c) \sin m\phi \tag{4.54}$$

are two independent eigenfunctions, and referred as the cosine and the sine modes, respectively. Using the properties of Bessel functions (see, for example, [3]), it is not difficult to show the orthogonality property

$$\int_0^a \int_0^{2\pi} W_I^{(m,n)} W_J^{(p,q)} r \,\mathrm{d}\phi \,\mathrm{d}r = \pi \frac{a^2}{2} J_{m+1}^2(\omega_{(m,n)} a/c) \delta_{IJ} \delta_{mp} \delta_{nq} \tag{4.55}$$

for any two eigenfunctions $W_I^{(m,n)}$ and $W_J^{(p,q)}$, where $I, J =$C or S, $\delta_{\mathrm{CC}} = \delta_{\mathrm{SS}} = 1$, and $\delta_{\mathrm{CS}} = \delta_{\mathrm{SC}} = 0$. Since $W_{\mathrm{C}}^{(m,n)}$ and $W_{\mathrm{S}}^{(m,n)}$ are independent (in fact, orthogonal), and correspond to the same circular eigenfrequency $\omega_{(m,n)}$, they form a pair of degenerate modes. It is evident that every mode with $m \neq 0$ of the circular membrane is degenerate. Therefore, any function of the form (4.53) is an eigenfunction corresponding to a circular eigenfrequency $\omega_{(m,n)}$. The first few mode-shapes of a circular membrane are shown in Figure 4.9. It may be noted from the figure that a mode-shape (m, n) has m nodal diameters and $n - 1$ nodal circles. The modes with $m = 0$ have axial symmetry, and are known as the *symmetric modes*, while those with $m \neq 0$ are known as the *asymmetric modes*.

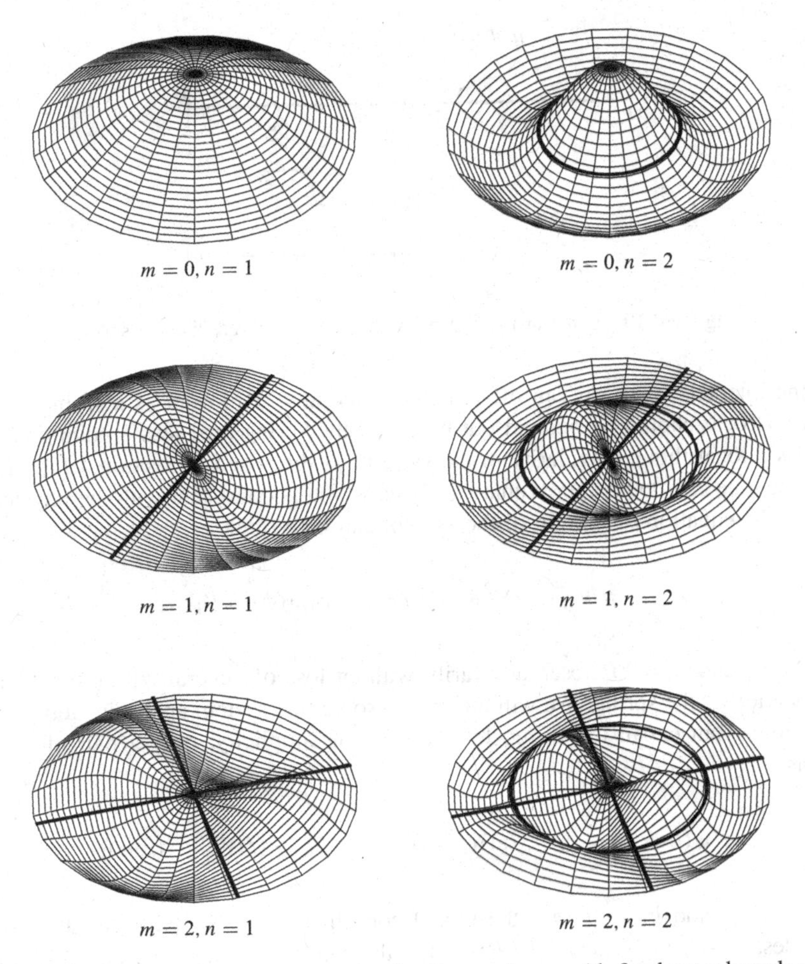

Figure 4.9 First few mode-shapes of a circular membrane with fixed outer boundary

The general solution of the free vibration problem can thus be written using (4.54) as

$$w(r,\phi,t) = \mathscr{R}\left[\sum_{m=0}^{\infty}\sum_{n=1}^{\infty}\left(D_{(m,n)}W_{\mathrm{C}}^{(m,n)} + E_{(m,n)}W_{\mathrm{S}}^{(m,n)}\right)\mathrm{e}^{i\omega_{(m,n)}t}\right]$$

$$\Rightarrow w(r,\phi,t) = \sum_{m=0}^{\infty}\sum_{n=1}^{\infty}\left[\left(A_{(m,n)}\cos\omega_{(m,n)}t + B_{(m,n)}\sin\omega_{(m,n)}t\right)W_{\mathrm{C}}^{(m,n)}\right.$$
$$\left. + \left(G_{(m,n)}\cos\omega_{(m,n)}t + H_{(m,n)}\sin\omega_{(m,n)}t\right)W_{\mathrm{S}}^{(m,n)}\right], \tag{4.56}$$

where we have taken $D_{(m,n)} = A_{(m,n)} - i\,B_{(m,n)}$, and $E_{(m,n)} = G_{(m,n)} - i\,H_{(m,n)}$.

An interesting consequence of modal degeneracy is the phenomenon of *frequency splitting* (sometimes also referred to as *mode splitting*), which can occur when there is a localized interaction of the membrane with another system. It is particularly prominent when the

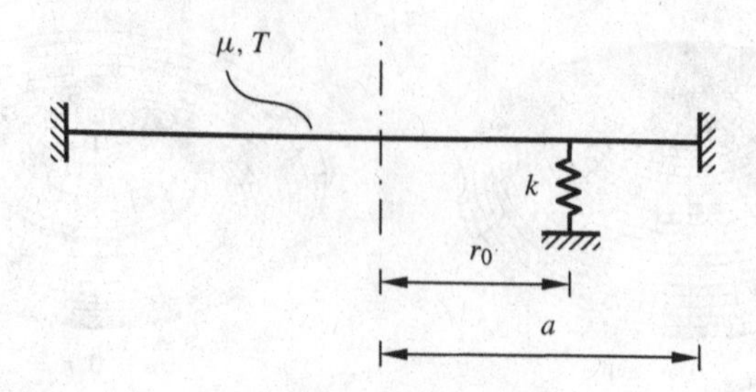

Figure 4.10 Circular membrane with an off-centered discrete spring

membrane interacts with, say, an off-centered discrete system such as a spring, as shown in Figure 4.10. It should be pointed out here that the spring produces a point force which introduces a logarithmic singularity in the system, as discussed in Section 4.1.1. However, in this case, one can do an approximate analysis which is consistent with the physics of the problem. The equation of motion of the system can be written as

$$\mu w_{,tt} - S\nabla^2 w + kw\delta(r - r_0)\delta(\phi) = 0, \tag{4.57}$$

where $r = r_0$ and $\phi = 0$ (chosen arbitrarily without loss of generality) are the coordinates of the location of the spring. We will attempt to solve this system approximately by writing the solution as an expansion using, for the sake of simplicity, only a pair of degenerate modes as

$$w(r, \phi, t) = g_{(m,n)}(t) W_{\mathrm{C}}^{(m,n)}(r, \phi) + h_{(m,n)}(t) W_{\mathrm{S}}^{(m,n)}(r, \phi), \tag{4.58}$$

where $g_{(m,n)}(t)$ and $h_{(m,n)}(t)$ are the modal coordinates corresponding to the cosine and sine modes, respectively, and $m \neq 0$ so that the modes are degenerate. Substituting this solution form in the equation of motion (4.57), taking the inner products with the individual eigenfunctions, and using (4.55), yields on simplification

$$\ddot{g}_{(m,n)} + \left[\frac{T\gamma_{(m,n)}^2}{\pi\mu} + \frac{2kJ_m^2(\gamma_{(m,n)}r_0)}{\pi\mu b^2 J_{m+1}^2(b\gamma_{(m,n)})}\right] g_{(m,n)} = 0, \tag{4.59}$$

$$\ddot{h}_{(m,n)} + \left[\frac{T\gamma_{(m,n)}}{\pi\mu}\right] h_{(m,n)} = 0. \tag{4.60}$$

From the above two differential equations for the modal coordinates, it is clear that the eigenfrequency corresponding to the degenerate cosine and sine modes is no longer the same. It is interesting to note that, in the case of a single spring, the frequency of all degenerate modes $\{(m, n)|m \neq 0,\ n = 1, 2, \ldots, \infty\}$ are now split, and as a consequence, there are no longer degenerate modes.

In a continuous system with degenerate modes, the phenomenon of frequency splitting can occur, in general, due to any interaction destroying the isotropy of the degenerate modes. Such a loss of isotropy can occur, for example, due to an attached discrete system as demonstrated above. In special situations, certain degenerate modes split while the others do not. It is obvious that in one-dimensional continua such as strings, the phenomenon of frequency splitting cannot occur since there is no modal degeneracy.

4.3 FORCED VIBRATION ANALYSIS

In the presence of external transverse forcing on a membrane, the equation of motion of the membrane takes the form

$$\mu w_{,tt} - T\nabla^2 w = q(x, y, t), \tag{4.61}$$

where $q(x, y, t)$ is a given distributed force. The eigenfunctions of the unforced membrane form a basis of the space of functions satisfying the geometric boundary conditions, and the solution of (4.61) for a rectangular membrane of dimension $a \times b$ can be conveniently expressed using the expansion theorem as

$$w(x, y, t) = \sum_m \sum_n p_{(m,n)}(t) \sin\frac{m\pi x}{a} \sin\frac{n\pi y}{b}, \tag{4.62}$$

where $p_{(m,n)}(t)$ is the modal coordinate corresponding to mode (m, n). Substituting (4.62) in (4.61), and using the orthogonality relations (4.32), one can discretize (4.61) to obtain the differential equations for the individual modal coordinates. In the case of a circular membrane, the solution expansion is of the form

$$w(r, \phi, t) = \sum_m \sum_n \left[p_{(m,n)}(t) W_{\mathrm{C}}^{(m,n)}(r, \phi) + q_{(m,n)}(t) W_{\mathrm{S}}^{(m,n)}(r, \phi) \right],$$

where $p_{(m,n)}(t)$ and $q_{(m,n)}(t)$ are, respectively, the modal coordinates corresponding to the cosine and sine components of the mode (m, n).

4.4 APPLICATIONS: KETTLEDRUM AND CONDENSER MICROPHONE

4.4.1 Modal analysis

As an application of the above analysis, consider the modal analysis of a kettledrum. The kettledrum, as shown in Figure 4.11, consists of a hemispherical cavity of radius a that is

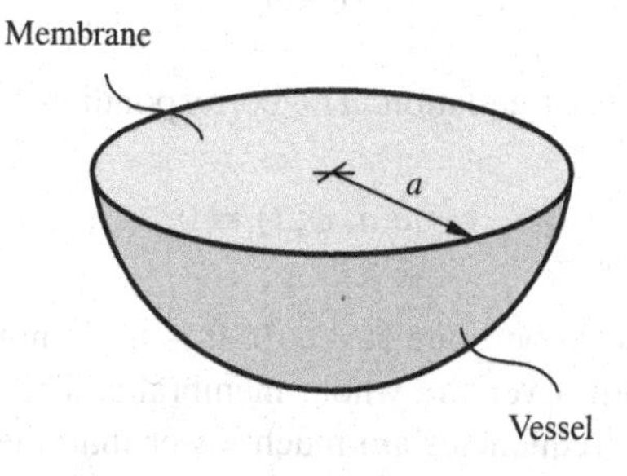

Figure 4.11 A kettledrum

closed by a stretched circular membrane such that the air inside is confined. Certain types of condenser microphones resemble the kettledrum in construction, and hence can be analyzed similarly, as discussed below.

Since the air inside the vessel is confined, the transverse displacement of the membrane will in general change the volume of the air, and hence change the pressure of the confined air. The difference in pressure on either side of the membrane then produces a force on the membrane. In order to estimate this force, consider the change in volume ΔV given by

$$\Delta V(t) = \int_0^a \int_0^{2\pi} w(r, \phi, t) r \, \mathrm{d}\phi \, \mathrm{d}r, \tag{4.63}$$

where $w(r, \phi, t)$ is the transverse deflection of the membrane. In order to relate this volume change to the change in pressure, we use the adiabatic condition

$$pV^{\gamma} = p_0 V_0^{\gamma}, \tag{4.64}$$

where p is the pressure at any time instant, p_0 is the initial pressure, V_0 is the initial volume, and $\gamma = C_{\mathrm{p}}/C_{\mathrm{v}}$ is the ratio of molar specific heats of air at constant pressure and constant volume. Assuming the process to be adiabatic is reasonable since the rate at which the volume changes is much faster than the thermal diffusion rate in air. From (4.64), one can write

$$\begin{aligned}
&(p_0 + \Delta p)(V_0 + \Delta V)^{\gamma} = p_0 V_0^{\gamma} \\
&\Rightarrow \frac{\Delta p}{p_0} + \gamma \frac{\Delta V}{V_0} \approx 0 \quad \text{(expanding and retaining terms up to first order)} \\
&\Rightarrow \Delta p \approx -\frac{\gamma p_0}{V_0} \int_0^a \int_0^{2\pi} w(r, \phi, t) r \, \mathrm{d}\phi \, \mathrm{d}r \qquad \text{(using (4.63))}.
\end{aligned} \tag{4.65}$$

This differential pressure acts as an external force on the membrane. It may be observed that this force depends on the displacement field of the membrane. The dynamics of the membrane can now be written as

$$\mu w_{,tt} - T\nabla^2 w = -\frac{\gamma p_0}{V_0} \int_0^a \int_0^{2\pi} w(r, \phi, t) r \, \mathrm{d}\phi \, \mathrm{d}r, \tag{4.66}$$

which is an integro-differential equation. The corresponding boundary condition is

$$w(a, \phi, t) \equiv 0, \tag{4.67}$$

and the solution is required to be finite at $r = 0$. It is to be noted here that the pressure has been assumed to be uniform over the whole membrane. This assumption is reasonable, if the kettledrum's vibration frequencies are much lower than the first resonance frequency of the air in the cavity. The coupled vibration problem consisting of the air in the cavity, the membrane, and the air outside of the kettle, is more complex.

Consider a modal solution of (4.66) in the form

$$w(r, \phi, t) = W(r, \phi)e^{i\omega t}. \tag{4.68}$$

Substitution of (4.68) in (4.66) yields on rearrangement,

$$\nabla^2 W + \beta^2 W = \frac{\gamma p_0}{T V_0} \int_0^a \int_0^{2\pi} W r \, d\phi \, dr, \tag{4.69}$$

where $\beta = \omega/c$ and $c^2 = T/\mu$. Taking our cue from the solution of the circular membrane, consider the solution of (4.69) as an expansion of the form

$$W(r, \phi) = \sum_{m=0}^{\infty} B_m W_C^m(r, \phi) + C_m W_S^m(r, \phi), \tag{4.70}$$

where

$$W_C^m(r, \phi) = R_m(r) \cos m\phi, \qquad W_S^m(r, \phi) = R_m(r) \sin m\phi, \tag{4.71}$$

and $R_m(r)$ are arbitrary functions of r. It can be easily verified that, for all terms with $m \neq 0$ in the assumed solution form (4.70), the right-hand side of (4.69) is identically zero. Thus, all asymmetric modes leave the volume unchanged.

Let us first consider the terms with $m \neq 0$. For all such terms, (4.69) yields

$$R_m'' + \frac{1}{r} R_m' + \left(\beta^2 - \frac{m^2}{r^2}\right) R_m = 0, \qquad m = 1, 2, \ldots, \infty, \tag{4.72}$$

which is the familiar Bessel differential equation. The solution of (4.72) satisfying the finiteness condition at $r = 0$ can be written as

$$R_m(r) = A_m J_m(\beta r), \qquad m = 1, 2, \ldots, \infty, \tag{4.73}$$

where A_m is an arbitrary constant. Satisfaction of the outer boundary condition requires that $J_m(\beta a) = 0$ for $m = 1, 2, \ldots, \infty$, which yields the eigenfrequencies of the asymmetric modes. It is evident that the eigenfrequencies of the asymmetric modes of the kettledrum are the same as those of a normal circular membrane given in Section 4.2.2.

Next, for the term with $m = 0$ in (4.70) (i.e., $B_0 W_C^0(r, \phi) = B_0 R_0(r)$), (4.69) yields

$$R_0'' + \frac{1}{r} R_0' + \beta^2 R_0 = \frac{2\pi \gamma p_0}{T V_0} \int_0^a R_0 r \, dr. \tag{4.74}$$

In view of the boundary condition (4.67), let us consider a solution of (4.74) as

$$R_0(r) = J_0(\beta r) - J_0(\beta a), \tag{4.75}$$

which clearly satisfies (4.67). Substituting this solution form in (4.74) yields on simplification

$$J_0''(\beta r) + \frac{1}{r}J_0'(\beta r) + \beta^2[J_0(\beta r) - J_0(\beta a)] = \frac{\chi}{a^2}\left[\frac{2}{\beta a}J_1(\beta a) - J_0(\beta a)\right], \tag{4.76}$$

where $\chi := \gamma\pi a^4 p_0/TV_0$, and we have used the property (see, for example, [3])

$$\int_0^z \zeta J_0(\zeta)\,\mathrm{d}\zeta = zJ_1(z).$$

Since

$$J_0''(\beta r) + \frac{1}{r}J_0'(\beta r) + \beta^2 J_0(\beta r) = 0, \tag{4.77}$$

we have from (4.76) the characteristic equation

$$J_0(\beta a) = \frac{\chi}{a^2}\left[\frac{2}{\beta a}J_1(\beta a) - J_0(\beta a)\right]$$

or

$$J_0(\beta a) + \frac{\chi}{\beta^2 a^2}J_2(\beta a) = 0, \tag{4.78}$$

where we have used the property $J_{m-1}(z) + J_{m+1}(z) = (2m/z)J_m(z)$. For different values of χ, the first few solutions of the above characteristic equation are obtained as

$$\chi = 0:\ \beta_{(0,1)}a = 2.4048,\quad \beta_{(0,2)}a = 5.5200,\quad \beta_{(0,3)}a = 8.6537,$$
$$\chi = 1:\ \beta_{(0,1)}a = 2.5437,\quad \beta_{(0,2)}a = 5.5323,\quad \beta_{(0,3)}a = 8.6568,$$
$$\chi = 10:\ \beta_{(0,1)}a = 3.4874,\quad \beta_{(0,2)}a = 5.6753,\quad \beta_{(0,3)}a = 8.6888.$$

The circular eigenfrequencies are obtained from $\omega_{(m,n)} = c\beta_{(m,n)}$. It may be noted from (4.78) that $\chi = 0$ yields the characteristic equation for symmetric modes of a circular membrane. It is observed that the presence of the vessel increases the eigenfrequencies of the symmetric modes. However, the increment in the frequencies decreases as we go to the higher modes. This is due to the fact that the change in volume of the air inside the kettledrum is more for the lower modes than the higher ones for the same peak amplitude (at the center). Finally, the eigenfunctions are obtained from (4.71) and (4.75) as

$$W^{(0,n)} = J_0(\beta_n r) - J_0(\beta_n a), \quad n = 1, 2, \ldots, \infty, \tag{4.79}$$

$$W_{\mathrm{C}}^{(m,n)} = J_m(\beta_n r)\cos m\phi, \quad m, n = 1, 2, \ldots, \infty, \tag{4.80}$$

$$W_{\mathrm{S}}^{(m,n)} = J_m(\beta_n r)\sin m\phi, \quad m, n = 1, 2, \ldots, \infty. \tag{4.81}$$

These eigenfunctions are all orthogonal, as can be easily checked.

4.4.2 Forced vibration analysis

Now we consider the forced vibration response of a condenser microphone, which we assume has a similar construction to a kettledrum. The equation of motion is obtained from (4.66) as

$$\mu w_{,tt} - T\nabla^2 w + \frac{\gamma p_0}{V_0}\int_0^a \int_0^{2\pi} w(r,\phi,t) r \, \mathrm{d}\phi \, \mathrm{d}r = P \mathrm{e}^{i\Omega t}, \tag{4.82}$$

where P is the pressure amplitude (assumed uniform over the membrane) and Ω is the forcing frequency. Such a form of excitation is reasonable since the size of the membrane in a microphone is usually small compared to the wavelengths of the normal human voice.

The homogeneous solution of (4.82) has already been discussed in the previous section. The particular solution of (4.82) can be represented in terms of the eigenfunctions (4.79)–(4.81) as

$$w_{\mathrm{P}}(r,\phi,t) = \left(\sum_{n=1}^{\infty} \alpha_n W^{(0,n)} + \sum_{m=1}^{\infty}\sum_{n=1}^{\infty} [\delta^{\mathrm{C}}_{(m,n)} W_{\mathrm{C}}^{(m,n)} + \delta^{\mathrm{S}}_{(m,n)} W_{\mathrm{S}}^{(m,n)}]\right) \mathrm{e}^{i\Omega t}, \tag{4.83}$$

where α_n, $\delta^{\mathrm{C}}_{(m,n)}$ and $\delta^{\mathrm{S}}_{(m,n)}$ are unknown constants. Substituting this solution form in (4.82), and using (4.69) and the orthogonality of the eigenfunctions, yields

$$\alpha_n = \frac{\langle P, W^{(0,n)}\rangle}{(c^2\beta^2_{(0,n)} - \Omega^2)\langle \mu W^{(0,n)}, W^{(0,n)}\rangle}, \qquad \text{for } n = 1, 2, \ldots, \infty,$$

$$\delta^{\mathrm{C}}_{(m,n)} = 0, \qquad \text{and} \qquad \delta^{\mathrm{S}}_{(m,n)} = 0, \qquad \text{for all } (m,n),\ m \neq 0,$$

where

$$\langle W_1, W_2\rangle = \int_0^a \int_0^{2\pi} W_1 W_2 r \, \mathrm{d}\phi \, \mathrm{d}r,$$

denotes the inner product of two functions $W_1(r,\phi)$ and $W_2(r,\phi)$. Using the properties of the Bessel functions (see, for example, [3])

$$\int x J_0(x) \, \mathrm{d}x = x J_1(x) \qquad \text{and} \qquad \int x J_0^2(x) \, \mathrm{d}x = \frac{x^2}{2}[J_0^2(x) + J_1^2(x)],$$

one can easily write

$$\begin{aligned}\langle P, W^{(0,n)}\rangle &= 2\pi \int_0^a Pr[J_0(\beta_{(0,n)} r) - J_0(\beta_{(0,n)} a)] \, \mathrm{d}r \\ &= 2\pi P\left[\frac{a}{\beta_{(0,n)}} J_1(\beta_{(0,n)} a) - \frac{a^2}{2} J_0(\beta_{(0,n)} a)\right]\end{aligned}$$

and

$$
\begin{aligned}
\langle W^{(0,n)}, W^{(0,n)}\rangle &= 2\pi \int_0^a r[J_0(\beta_{(0,n)}r) - J_0(\beta_{(0,n)}a)]^2 \mathrm{d}r \\
&= 2\pi \left[a^2 J_0^2(\beta_{(0,n)}a) - \frac{2a}{\beta_{(0,n)}} J_0(\beta_{(0,n)}a) J_1(\beta_{(0,n)}a) \right. \\
&\quad \left. + \frac{a^2}{2} J_1^2(\beta_{(0,n)}a)\right].
\end{aligned}
$$

This completes the solution of (4.83).

4.5 WAVES IN MEMBRANES

4.5.1 Waves in Cartesian coordinates

Consider the dynamics of an infinite membrane described by the two-dimensional wave equation

$$
w_{,tt} - c^2(w_{,xx} + w_{,yy}) = 0. \tag{4.84}
$$

For simplicity, consider first a harmonic traveling plane wave solution for the wave equation (4.84). A harmonic plane wave, traveling in the direction $\hat{\mathbf{n}} = \cos\theta\hat{\mathbf{i}} + \sin\theta\hat{\mathbf{j}}$, as shown in Figure 4.12, can be represented as

$$
\begin{aligned}
w(x, y, t) &= De^{i(k\xi - \omega t)} \\
&= De^{i(k\hat{\mathbf{n}}\cdot\mathbf{r} - \omega t)} = De^{i(\mathbf{k}\cdot\mathbf{r} - \omega t)} \qquad (4.85) \\
&= De^{i(kx\cos\theta + ky\sin\theta - \omega t)}, \qquad (4.86)
\end{aligned}
$$

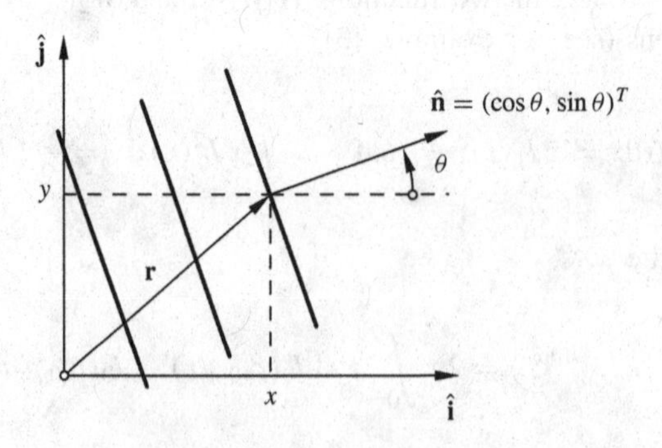

Figure 4.12 Traveling waves in a two-dimensional plane

where D is a complex constant, k is the wave number, $\mathbf{k} = k\hat{\mathbf{n}}$ is defined as the *wave vector*, and $\mathbf{r} = x\hat{\mathbf{i}} + y\hat{\mathbf{j}}$. Substituting (4.86) in (4.84) yields the dispersion relation

$$\omega^2 - c^2k^2 = 0 \qquad \Rightarrow \qquad \omega = \pm kc, \tag{4.87}$$

where the two signs correspond to the two possible directions of travel, namely $\hat{\mathbf{n}}$ and $-\hat{\mathbf{n}}$. Therefore, a general harmonic plane wave solution along $\hat{\mathbf{n}}$ may be represented as

$$w_\theta(x, y, t) = A(k, \theta)\mathrm{e}^{ik(x\cos\theta + y\sin\theta - ct)},$$

where $A(k, \theta)$ is a complex function of k and θ. Using the theory of Fourier transforms, it is not difficult to conclude that a general wave pulse along $\hat{\mathbf{n}}$ may then be represented by

$$w_\theta(x, y, t) = f(\theta, x\cos\theta + y\sin\theta - ct),$$

and a general wave solution can be obtained by superposing $f_\theta(\cdot)$ over all values of θ as

$$w(x, y, t) = \int_0^{2\pi} f(\theta, x\cos\theta + y\sin\theta - ct)\,\mathrm{d}\theta.$$

Some other solution forms that also satisfy the wave equation are

$$w(x, y, t) = (ay + b)f(x - ct), \quad \text{and} \quad w(x, y, t) = (ax + b)f(y - ct),$$

as can be easily checked.

Next, we define the impedance of a membrane to harmonic plane waves. The impedance per unit length of a membrane (with respect to harmonic waves) can be represented as

$$\mathcal{Z} = \frac{\mathscr{A}[F]}{\mathscr{A}[v]}, \tag{4.88}$$

where $\mathscr{A}[\cdot]$ represents the complex amplitude, F is the applied transverse (complex) harmonic force per unit length, and v is the resulting transverse (complex) harmonic velocity of the membrane at the location of the force. Consider a harmonic plane wave represented by (4.86). The transverse force per unit length is given by

$$\begin{aligned} F &= -Tw_{,n} = -T(w_{,x}\cos\theta + w_{,y}\sin\theta) \\ &= -TikA\mathrm{e}^{ik(x\cos\theta + y\sin\theta - ct)}. \end{aligned} \tag{4.89}$$

The transverse velocity of the membrane is given by

$$v = w_{,t} = -i\omega A\mathrm{e}^{ik(x\cos\theta + y\sin\theta - ct)}. \tag{4.90}$$

Using (4.89) and (4.90), the impedance per unit length is obtained from (4.88) as

$$\mathcal{Z} = \frac{Tk}{\omega} = \mu c \qquad \text{(using (4.5) and (4.87))}. \tag{4.91}$$

4.5.2 Waves in polar coordinates

Consider next the equation of motion of an infinite membrane in polar coordinates as

$$w_{,tt} - c^2(w_{,rr} + \frac{1}{r}w_{,r} + \frac{1}{r^2}w_{,\phi\phi}) = 0. \tag{4.92}$$

Substituting the solution form

$$w(r, \phi, t) = R(r)\mathrm{e}^{im\phi}\mathrm{e}^{i\omega t} \tag{4.93}$$

yields on rearrangement

$$R'' + \frac{1}{r}R' + \left(\gamma^2 - \frac{m^2}{r^2}\right) R = 0, \tag{4.94}$$

where $\gamma = \omega/c$. The solution of the Bessel differential equation (4.94) can be written as (see, for example, [3])

$$R(r) = AJ_m(\gamma r) + BY_m(\gamma r), \tag{4.95}$$

where A and B are arbitrary real constants. However, for reasons clarified below, we will represent the solution as

$$R(r) = DH_m^{(1)}(\gamma r) + EH_m^{(2)}(\gamma r), \tag{4.96}$$

where D and E are arbitrary complex constants, and

$$H_m^{(1)}(\gamma r) = J_m(\gamma r) + iY_m(\gamma r),$$
$$H_m^{(2)}(\gamma r) = J_m(\gamma r) - iY_m(\gamma r),$$

are known as the Hankel functions (see, for example, [3]) of first and second kind, respectively. Therefore, the solution (4.93) can be written as

$$w(r, \phi, t) = (DH_m^{(1)}(\gamma r) + EH_m^{(2)}(\gamma r))\mathrm{e}^{im\phi}\mathrm{e}^{i\omega t}. \tag{4.97}$$

It is known from the theory of Bessel functions (see [2]) that, when $x \gg 1$,

$$J_m(x) \approx \sqrt{\frac{2}{\pi x}} \cos\left(x - (2m+1)\frac{\pi}{4}\right), \tag{4.98}$$

$$Y_m(x) \approx \sqrt{\frac{2}{\pi x}} \sin\left(x - (2m+1)\frac{\pi}{4}\right). \tag{4.99}$$

A comparison of the exact and approximate representations of $J_m(x)$ and $Y_m(x)$ is shown in Figure 4.13. Using these approximate representations, it can be easily checked that, for $x \gg 1$,

$$H_m^{(1)}(x) \approx \sqrt{\frac{2}{\pi x}} e^{i(x-(2m+1)\pi/4)},$$

$$H_m^{(2)}(x) \approx \sqrt{\frac{2}{\pi x}} e^{-i(x-(2m+1)\pi/4)},$$

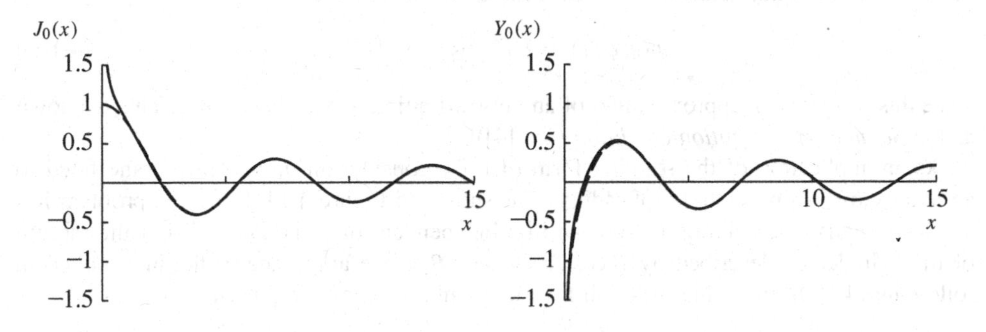

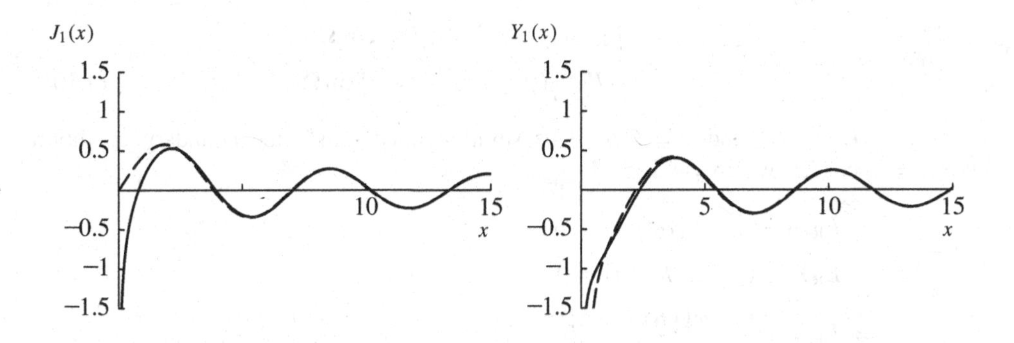

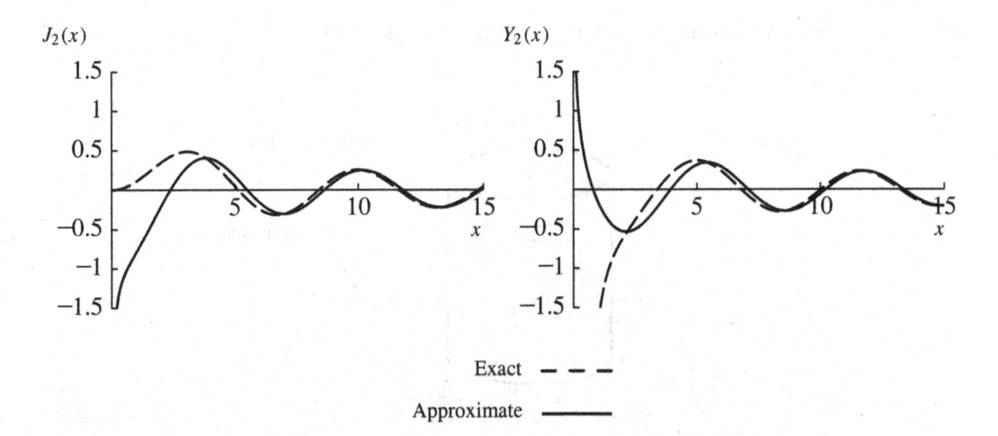

Figure 4.13 Comparison of exact and approximate representations of Bessel functions of first and second kinds

and (4.97) may be approximated for large values of r as

$$w(r, \phi, t) = \sqrt{\frac{2}{\pi r}} \left[D e^{i(r+\omega t-(2m+1)\pi/4)} + E e^{i(-r+\omega t+(2m+1)\pi/4)}\right] e^{im\phi}.$$

It is evident that the coefficient of D represents an incoming harmonic wave from infinity, while the coefficient of E corresponds to a harmonic wave traveling outwards to infinity. Now, if we consider that all the waves are propagating outwards to infinity (i.e., there is no other source of disturbance, or wave reflection), we must have

$$w(r, \phi, t) = E H_m^{(2)}(\gamma r) e^{im\phi} e^{i\omega t}, \tag{4.100}$$

since this form has an approximation of an outward-going wave when $r \gg 1$. This is known as the *Sommerfeld radiation condition* (see [4]).

As an application of the solution form (4.100), consider an infinite membrane fixed to an axially oscillating cylinder of radius a, as shown in Figure 4.14. Since the problem has axial symmetry, the solution (4.100) will be independent of ϕ, i.e., $m = 0$. Let the motion of the cylinder be described by $B \cos \Omega t$, where B is the amplitude of the motion. Then, following (4.100), the actual solution for the membrane can be represented as

$$\begin{aligned} w(r, t) &= \mathscr{R}[E H_0^{(2)}(\gamma r) e^{i\Omega t}] \\ &= [E_{\mathrm{R}} J_0(\gamma r) + E_{\mathrm{I}} Y_0(\gamma r)] \cos \Omega t \\ &\quad + [E_{\mathrm{R}} Y_0(\gamma r) - E_{\mathrm{I}} J_0(\gamma r)] \sin \Omega t, \end{aligned} \tag{4.101}$$

where $E = E_{\mathrm{R}} + i E_{\mathrm{I}}$, and $\gamma = \Omega/c$. This solution must satisfy the boundary condition $w(a, t) = B \cos \Omega t$, which yields

$$\begin{aligned} & E_{\mathrm{R}} J_0(\gamma a) + E_{\mathrm{I}} Y_0(\gamma a) = B, \\ & E_{\mathrm{R}} Y_0(\gamma a) - E_{\mathrm{I}} J_0(\gamma a) = 0 \\ \Rightarrow\ & E_{\mathrm{R}} = \frac{J_0(\gamma a)}{J_0^2(\gamma a) + Y_0^2(\gamma a)} B \quad \text{and} \quad E_{\mathrm{I}} = \frac{Y_0(\gamma a)}{J_0^2(\gamma a) + Y_0^2(\gamma a)} B. \end{aligned}$$

Substituting these expressions in (4.101) completes the solution.

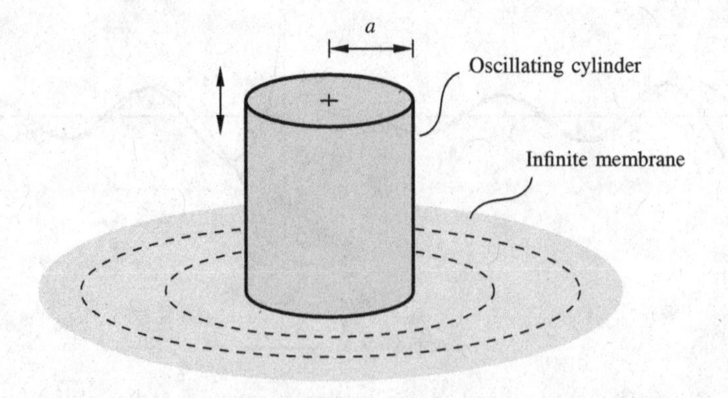

Figure 4.14 An infinite membrane excited by an axially oscillating cylinder

4.5.3 Energetics of membrane waves

The mechanical energy density (energy per unit area) of a membrane can be written using (4.10) and (4.12) as

$$\begin{aligned}\hat{\mathcal{E}} &= \hat{\mathcal{T}} + \hat{\mathcal{V}} \\ &= \frac{1}{2}\mu w_{,t}^2 + \frac{1}{2}T(w_{,x}^2 + w_{,y}^2).\end{aligned} \tag{4.102}$$

The time rate of change of $\hat{\mathcal{E}}$ is obtained as

$$\begin{aligned}\frac{\partial \hat{\mathcal{E}}}{\partial t} &= \mu w_{,t} w_{,tt} + T(w_{,x} w_{,xt} + w_{,y} w_{,yt}) \\ &= \mu w_{,t} w_{,tt} + (T w_{,x} w_{,t})_{,x} - T w_{,xx} w_{,t} + (T w_{,y} w_{,t})_{,y} - T w_{,yy} w_{,t} \\ &= [\mu w_{,tt} - T(w_{,xx} + w_{,yy})] w_{,t} + (T w_{,x} w_{,t})_{,x} + (T w_{,y} w_{,t})_{,y} \\ &= (T w_{,x} w_{,t})_{,x} + (T w_{,y} w_{,t})_{,y} \qquad \text{(using (4.2))}.\end{aligned} \tag{4.103}$$

Let us define a vector

$$\mathcal{I} := -(T w_{,x} w_{,t})\hat{\mathbf{i}} - (T w_{,y} w_{,t})\hat{\mathbf{j}}, \tag{4.104}$$

representing the rate of energy per unit length crossing a line in the membrane. Thus, $\mathcal{I} \cdot \hat{\mathbf{n}} dl$ represents the power flow through a line element of length dl having a unit normal $\hat{\mathbf{n}}$. One may also refer to $\mathcal{I}$ as the *intensity vector* for membrane waves. Using the definition (4.104), one can rewrite (4.103) as

$$\frac{\partial \hat{\mathcal{E}}}{\partial t} + \nabla \cdot \mathcal{I} = 0,$$

where

$$\nabla = \hat{\mathbf{i}}\frac{\partial}{\partial x} + \hat{\mathbf{j}}\frac{\partial}{\partial y}, \tag{4.105}$$

and $\nabla \cdot \mathcal{I}$ is the divergence of $\mathcal{I}$.

Consider a harmonic plane wave on the membrane represented by

$$w(x, y, t) = A\cos(k_x x + k_y y - \omega t), \tag{4.106}$$

where $k_x^2 + k_y^2 = k^2$. The energy density is obtained as

$$\begin{aligned}\hat{\mathcal{E}} &= \left[\frac{1}{2}\mu\omega^2 + \frac{1}{2}Tk^2\right] A^2 \sin^2(k_x x + k_y y - \omega t) \\ &= \mu\omega^2 A^2 \sin^2(k_x x + k_y y - \omega t) \qquad \text{(using (4.5) and (4.87))},\end{aligned}$$

and the average energy density is

$$\langle \hat{\mathcal{E}} \rangle = \frac{1}{2}\mu\omega^2 A^2.$$

The intensity vector can be computed from (4.104) as

$$\begin{aligned}\mathcal{I} &= T\omega A^2 \sin^2(k_x x + k_y y - \omega t)[k_x \hat{\mathbf{i}} + k_y \hat{\mathbf{j}}], \\ &= c^2 \mu\omega A^2 \sin^2(k_x x + k_y y - \omega t)[k_x \hat{\mathbf{i}} + k_y \hat{\mathbf{j}}],\end{aligned}$$

and the net power flow per unit length is obtained as

$$I = |\mathcal{I}| = c^2 \mu\omega k A^2 \sin^2(k_x x + k_y y - \omega t) = \mu c \omega^2 A^2 \sin^2(k_x x + k_y y - \omega t).$$

The average power flow per unit length is given by

$$\langle I \rangle = \frac{\omega}{2\pi} \int_0^{2\pi/\omega} P \, \mathrm{d}t = \frac{1}{2}\mu c \omega^2 A^2 = c \langle \hat{\mathcal{E}} \rangle.$$

Thus, energy in a membrane propagates at the transverse wave speed c. The average power flow per unit length of a planar harmonic wave can also be directly obtained from (see Section 2.3)

$$\langle I \rangle = \frac{1}{2} \mathcal{Z} |w_{,t}|^2 = \frac{1}{2}\mu c \omega^2 A^2.$$

4.5.4 Initial value problem for infinite membranes

The discrete spectrum known from finite membranes is replaced by a continuous spectrum in the case of an infinite membrane, and the expansion theorem (with discrete sums) can no longer be used for representing the solution. In such cases, we have to use the Fourier or Hankel transforms to obtain the motion generated from initial conditions. The field equation of the membrane is the wave equation itself without any boundary conditions. Let us assume that the initial conditions are specified as

$$w(x, y, 0) = w_0(x, y) \quad \text{and} \quad w_{,t}(x, y, 0) = v_0(x, y), \tag{4.107}$$

where $w_0(x, y)$ and $v_0(x, y)$ are the initial configuration and initial velocity profile of the membrane, respectively. Then, the solution of the wave equation is given by (see [5])

$$\begin{aligned} w(x, y, t) = \frac{1}{2\pi c} \Bigg[&\frac{\partial}{\partial t} \int_{r=0}^{ct} \int_{\phi=0}^{2\pi} \frac{w_0(x - r\cos\phi, y - r\sin\phi) r \mathrm{d}\phi \mathrm{d}r}{\sqrt{c^2 t^2 - r^2}} \\ &+ \int_{r=0}^{ct} \int_{\phi=0}^{2\pi} \frac{v_0(x - r\cos\phi, y - r\sin\phi) r \mathrm{d}\phi \mathrm{d}r}{\sqrt{c^2 t^2 - r^2}} \Bigg], \end{aligned}$$

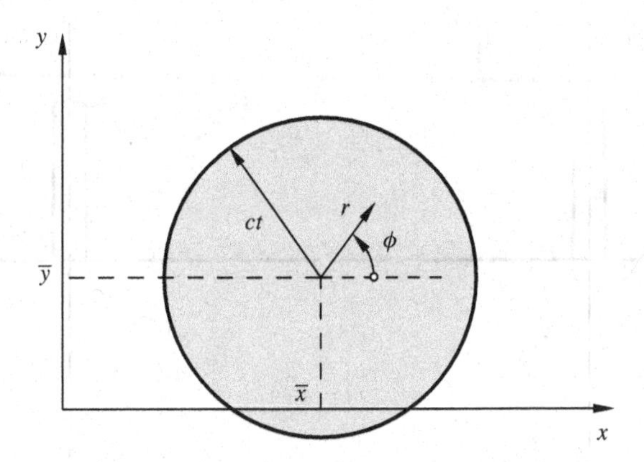

Figure 4.15 Area of dependence for the solution $w(\overline{x}, \overline{y}, t)$

where r and ϕ are the polar coordinates of any point, as shown in Figure 4.15. This form of solution is known as *Poisson's* or *Parseval's formula.* The solution indicates that the displacement $w(x, y, t)$ at any point $(\overline{x}, \overline{y})$ at time t depends on the functions $w_0(x, y)$ and $v_0(x, y)$ within the circle of radius ct centered at (x, y), as shown in Figure 4.15.

Now we compare the wave propagation in a membrane with that in a string, considering two special cases. In the first comparison, we consider the propagation of a cylindrical initial shape of the membrane and the propagation of a rectangular initial shape of the string, as shown in Figure 4.16. In Figure 4.17, we compare the displacement profiles of the membrane and the string resulting from an initial constant velocity distribution over a circle on the undisturbed membrane, and a line segment of the undisturbed string. As observed from these figures, the wave propagation is qualitatively different. The string wave front, unlike the membrane wave front, maintains its shape and size as it travels. This is to be expected, since in the membrane, the wave fronts are circular, and hence the wave energy gets distributed over circles of ever-increasing radius as the wave front propagates.

4.5.5 Reflection of plane waves

Consider a semi-infinite membrane with a fixed boundary on the line $x = 0$, as shown in Figure 4.18. Let a harmonic traveling wave with wave number k be incident on the boundary, and a reflected wave with wave number k' be generated. Then the complete wave field in the membrane can be represented as

$$w(x, y, t) = Ae^{ik(x \cos\alpha + y \sin\alpha - ct)} + Be^{ik'(-x \cos\alpha' - y \sin\alpha' - ct)}, \tag{4.108}$$

where c is the propagation speed of waves in the membrane. Applying the boundary condition $w(0, y, t) \equiv 0$, we obtain

$$Ae^{ik(y \sin\alpha - ct)} + Be^{ik'(y \sin\alpha' - ct)} \equiv 0. \tag{4.109}$$

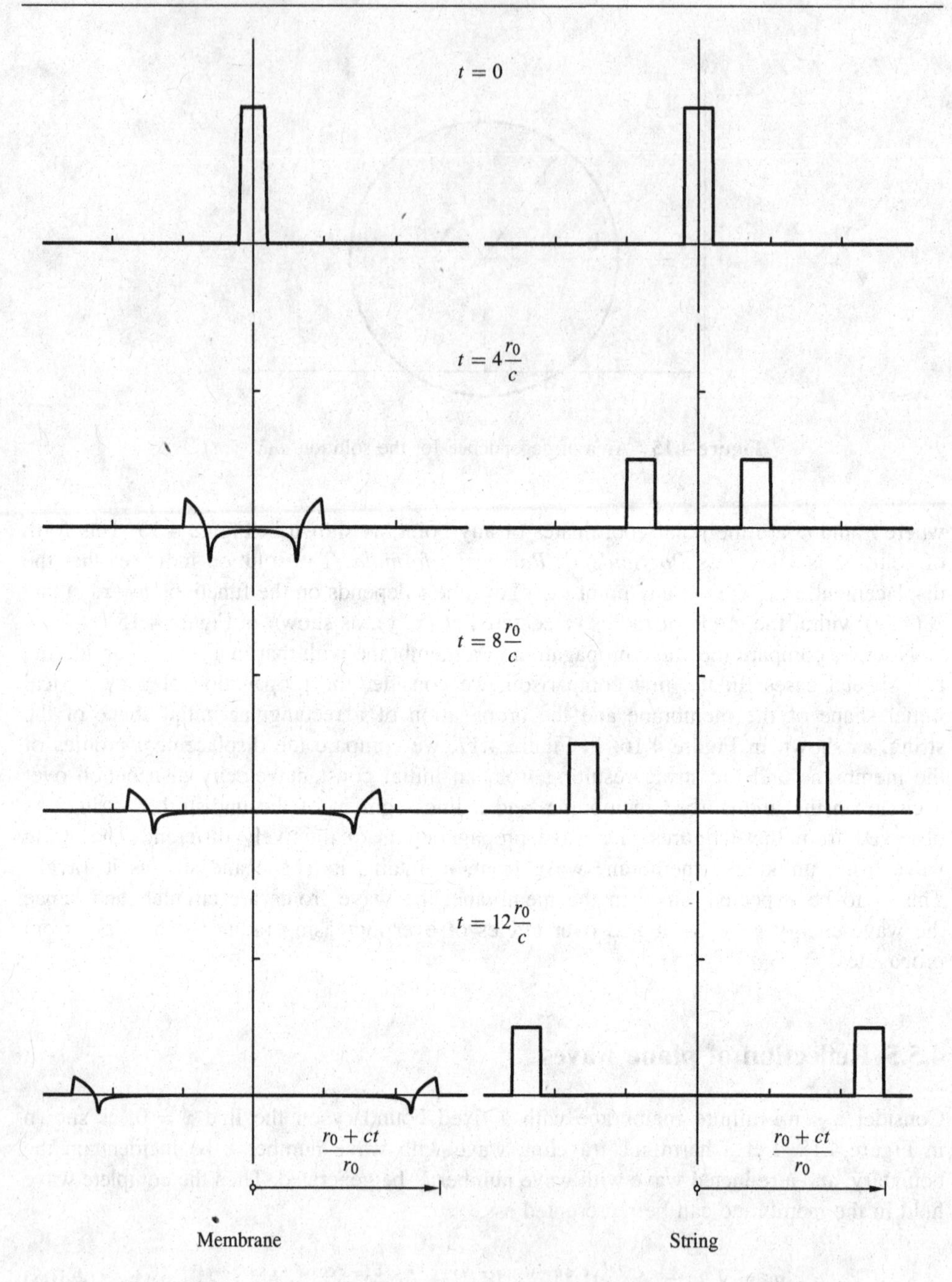

Figure 4.16 Infinite membrane and string with initial displacement condition

Satisfaction of (4.109) identically requires

$$k \sin\alpha = k' \sin\alpha' \qquad \text{and} \qquad ck = ck'$$

$$\Rightarrow \alpha = \alpha' \qquad \text{and} \qquad k = k'.$$

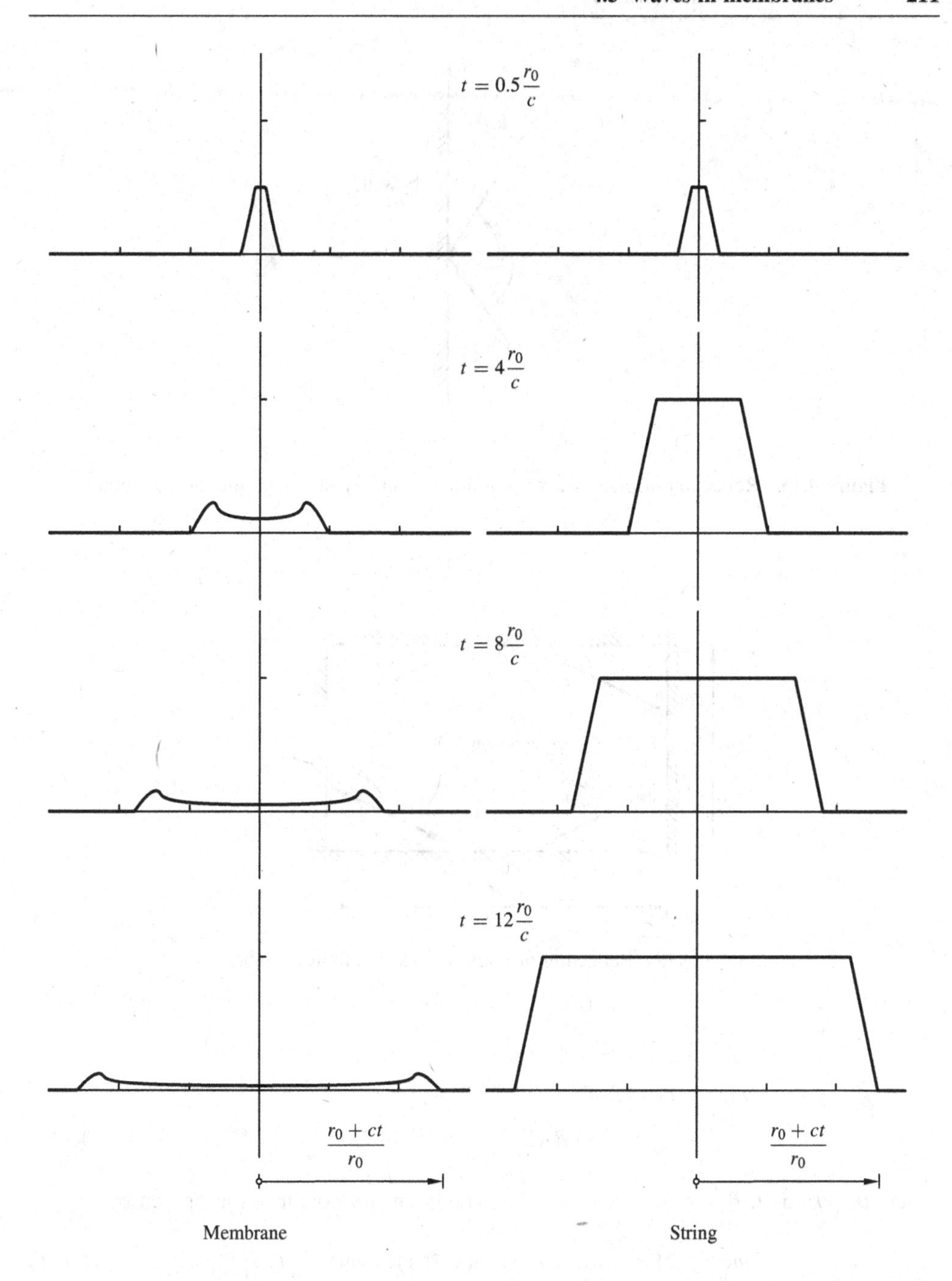

Figure 4.17 Infinite membrane and string with initial velocity condition

It is to be noted that $\alpha = \alpha'$ is the only meaningful implication of $\sin\alpha = \sin\alpha'$. Thus, we obtain the law of reflection for plane waves.

Let us now consider the reflection process in a finite rectangular membrane shown in Figure 4.19. Since there is reflection from all the four edges, one can write the complete

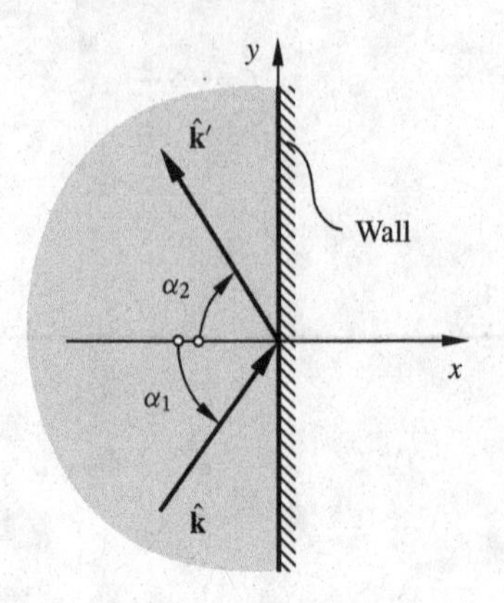

Figure 4.18 Reflection of plane waves at a fixed boundary of a semi-infinite membrane

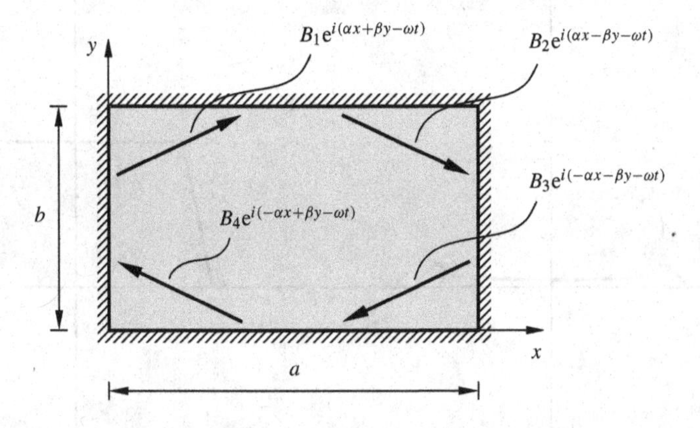

Figure 4.19 Reflection of plane waves in a finite membrane

wave field as

$$
\begin{aligned}
w(x, y, t) = {} & B_1 e^{i(\alpha x+\beta y-\omega t)} + B_2 e^{i(\alpha x-\beta y-\omega t)} \\
& + B_3 e^{i(-\alpha x+\beta y-\omega t)} + B_4 e^{i(-\alpha x-\beta y-\omega t)},
\end{aligned}
\tag{4.110}
$$

where $\alpha = k\cos\theta$, $\beta = k\sin\theta$, and $\omega = kc$. The boundary conditions are given by

$$
w(0, y, t) = w(a, y, t) = w(x, 0, t) = w(x, b, t) \equiv 0. \tag{4.111}
$$

Using the first and the third boundary conditions from (4.111) yields $B_1 = -B_2 = -B_3 = B_4 = B$ (say), and the solution then reduces to

$$
\begin{aligned}
w(x, y, t) &= B\left[e^{i(\alpha x+\beta y)} - e^{i(\alpha x-\beta y)} - e^{i(-\alpha x+\beta y)} + e^{i(-\alpha x-\beta y)}\right] e^{-i\omega t} \\
&= (\tilde{B}\sin\alpha x\,\sin\beta y)e^{i\omega t}.
\end{aligned}
\tag{4.112}
$$

Applying the second and the fourth boundary conditions from (4.111) provides the two conditions $\sin\alpha a = 0$ and $\sin\beta b = 0$, respectively. These conditions were also obtained earlier from (4.29), and hence, the frequencies are obtained from (4.33).

EXERCISES

4.1 Determine the eigenfrequencies and eigenfunctions of a rectangular membrane that is fixed at the boundaries $x = 0$ and $x = a$, and sliding at the boundaries $y = 0$ and $y = b$.

4.2 Determine the eigenfrequencies and eigenfunctions of an annular membrane that is fixed at the boundaries $r = a$ and $r = a/2$.

4.3 A rectangular membrane of width a and length $2a$ is made of two materials of mass densities μ_1 and μ_2 which are joined together, as shown in Figure 4.20. Determine the characteristic equation for the membrane. Also determine the approximate eigenfrequencies and eigenfunctions.

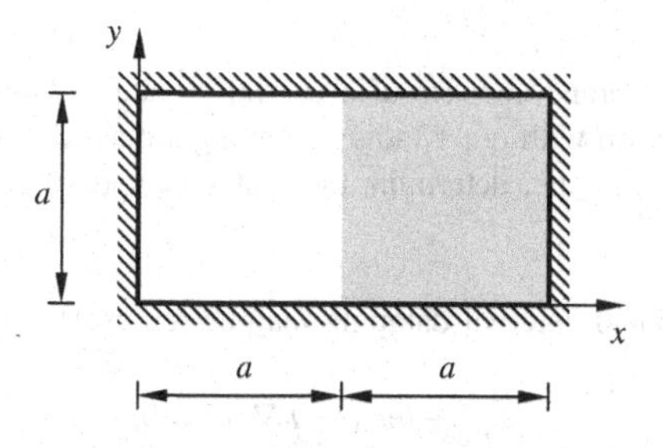

Figure 4.20 Exercise 4.3

4.4 A circular membrane of radius a with fixed boundary has a small particle of mass m attached at the center. Determine the approximate eigenfrequencies of the membrane.

4.5 Determine the approximate fundamental eigenfrequency of the membrane shown in Figure 4.21. Assume that the membrane is fixed at all the boundaries.

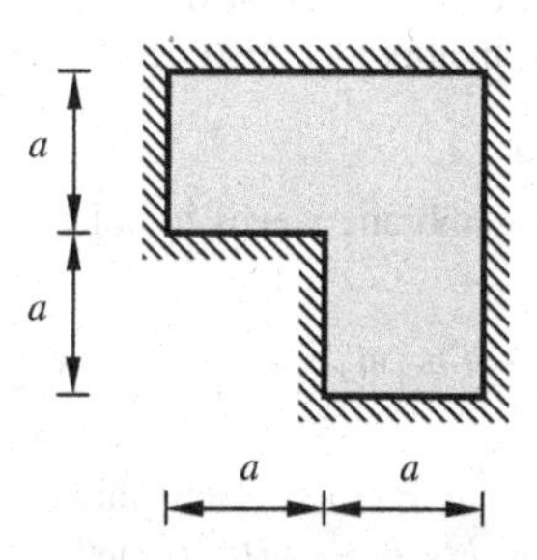

Figure 4.21 Exercise 4.5

4.6 A circular membrane of radius a with fixed boundary has three identical springs of stiffness k attached equidistant from the center at $b = a/2$, as shown in Figure 4.22. Using the eigenfunctions of a normal circular membrane, determine the first six approximate natural frequencies for different

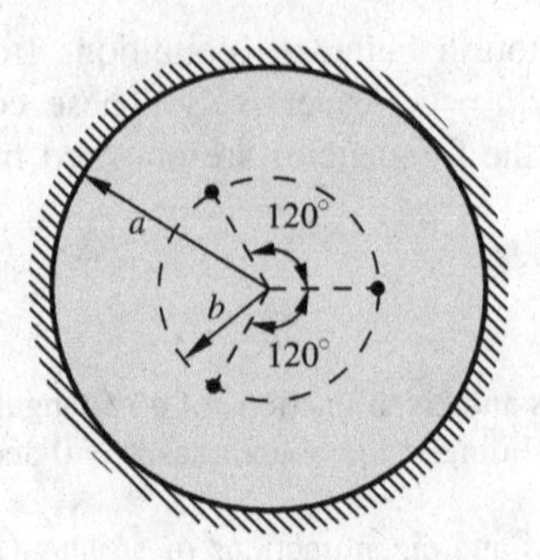

Figure 4.22 Exercise 4.6

values of k starting from $k = 0$, and plot the result. Note the frequencies that do not split. Can you identify the frequencies that do not split if there are N identical springs on the vertices of a regular N-gon?

4.7 A composite circular membrane consists of a central circular region of radius a having a mass density μ_1 and an annular region with outer radius b having a mass density μ_2. Using the eigenfunctions of a normal circular membrane, determine the first six approximate natural frequencies of the composite membrane.

4.8 A membrane with distributed external damping may be represented by

$$\mu w_{,tt} + b w_{,t} - T\nabla^2 w = 0,$$

where b is the damping coefficient. Discretize the equation of motion for a rectangular membrane using the eigenfunctions of the undamped membrane.

4.9 Determine the response of a kettledrum when excited by a unit impulsive force at $r = \bar{r}$ and $\phi = \bar{\phi}$ given by $\delta(t - \tau)\delta(r - \bar{r})\delta(\phi - \bar{\phi})$.

4.10 A circular membrane of radius a carries a rotating transverse force $q(r, \phi, t) = Q\delta(r - r_0)\delta(\phi - \Omega t)$, where Q is a constant, r_0 is the radial location of the force, and Ω is the circular frequency of rotation. Determine the response of the membrane. Also determine the critical rotation speeds of the force at which the membrane resonates.

4.11 Show that the intensity vector of membrane waves in polar coordinates takes the form

$$\mathcal{I} = -T w_{,t} w_{,r} \hat{\mathbf{e}}_r - \frac{T}{r} w_{,t} w_{,\phi} \hat{\mathbf{e}}_\phi.$$

where $\hat{\mathbf{e}}_r$ and $\hat{\mathbf{e}}_\phi$ are, respectively, the unit vectors. Using this expression, show that for symmetric outward radiating waves represented by $w(r, t) = A H_0^{(2)}(kr)e^{i\Omega t}$, the intensity falls as $1/r$.

REFERENCES

[1] Timoshenko, S.P., and Goodier, J.N., *Theory of Elasticity*, 3e, McGraw-Hill Book Co., Singapore, 1970.

[2] Watson, G.N., *A Treatise on the Theory of Bessel Functions*, Cambridge University Press, Cambridge, 1952.
[3] Kreyszig, E., *Advanced Engineering Mathematics*, 2e, Wiley Eastern Pvt. Ltd., New Delhi, 1969.
[4] Sommerfeld, A., *Partial Differential Equations in Physics*, Academic Press, New York, 1964.
[5] Meirovitch, L., *Analytical Methods in Vibrations*, The Macmillan Co., New York, 1967.

5

Vibrations of plates

Plates are two-dimensional elastic continua with finite bending stiffness. Thus, in a way, they are akin to beams in two dimensions. Any planar elastic continuum of a certain thickness can be treated as a plate if the thickness is much smaller than its width. Plates are commonly found in mechanical and civil engineering structures. In this chapter, small-amplitude transverse vibrations of unstretched plates in pure bending are discussed. The effect of transverse shear is neglected, or in other words, the plate is considered to be infinitely stiff in transverse shear. Some attention is also devoted to wave propagation in plates.

5.1 DYNAMICS OF PLATES

Similarly to what we did for beams, we assume that the plates have a neutral plane, which lies in the middle of the thickness of the plate, and remains unstrained. Since no shear distortion is considered, we also assume that material points of the plate situated on a normal to the neutral plane in the undeformed state, also remains on a line normal to the neutral surface in the deformed state. This is known as the *Kirchhoff hypothesis for plates*, and is analogous to the Euler–Bernoulli hypothesis for beams discussed in Chapter 3. The deformation of the neutral plane will essentially represent the transverse deformation of the plate. It will be assumed that the spatial derivatives of the deformation field are much smaller than unity.

5.1.1 Newtonian formulation

Consider a plate of constant thickness h with the undeformed neutral plane coinciding with the x-y-plane. An infinitesimal element of such a plate is shown in Figure 5.1. Since the normal stress σ_{zz} on the top and bottom surfaces is zero (except for the external loading, which is considered to be small), we will assume that it is also negligible inside the plate. Then, we are left with only the normal stress components σ_{xx} and σ_{yy}, and the shear stress components σ_{xy}, σ_{xz} and σ_{yz}.

Vibrations and Waves in Continuous Mechanical Systems P. Hagedorn and A. DasGupta

In order to work in two dimensions, it is convenient to integrate the stresses over the thickness of the plate to yield what are known as the *stress resultants*. These are defined as

$$N_x = \int_{-h/2}^{h/2} \sigma_{xx} \mathrm{d}z, \quad N_y = \int_{-h/2}^{h/2} \sigma_{yy} \mathrm{d}z, \quad N_{xy} = \int_{-h/2}^{h/2} \sigma_{xy} \mathrm{d}z, \tag{5.1}$$

$$Q_x = \int_{-h/2}^{h/2} \sigma_{xz} \mathrm{d}z, \quad Q_y = \int_{-h/2}^{h/2} \sigma_{yz} \mathrm{d}z, \tag{5.2}$$

$$M_x = \int_{-h/2}^{h/2} z\sigma_{xx} \mathrm{d}z, \quad M_y = \int_{-h/2}^{h/2} z\sigma_{yy} \mathrm{d}z, \quad M_{xy} = \int_{-h/2}^{h/2} z\sigma_{xy} \mathrm{d}z, \tag{5.3}$$

where (5.1) gives the normal stress resultants, (5.2) represents the shear stress resultants, and the moment resultants are given in (5.3). These force and moment resultants are shown in Figures 5.1 and 5.2. In these figures, the stress resultants on the invisible surfaces have been omitted for clarity. Thus, the force and moment resultants have, respectively, units of force per unit length and moment per unit length of the plate. When there are no in-plane forces on the plate, we have $N_x = N_y = N_{xy} = 0$. This does not, however, mean that corresponding stresses are zero. The stresses (bending stresses) can be obtained in terms of the strains using Hooke's law as

$$\sigma_{xx} = \frac{E}{1-\nu^2}[\epsilon_{xx} + \nu\epsilon_{yy}], \quad \sigma_{yy} = \frac{E}{1-\nu^2}[\nu\epsilon_{xx} + \epsilon_{yy}], \quad \text{and} \quad \sigma_{xy} = \frac{E}{1+\nu}\epsilon_{xy}. \tag{5.4}$$

Here, an important point needs to be mentioned. Since we have assumed that the plate is infinitely stiff in transverse shear, we must have $\epsilon_{xz} = \epsilon_{yz} = 0$. This implies that Q_x and Q_y cannot be determined from the theory of elasticity. However, they can be determined

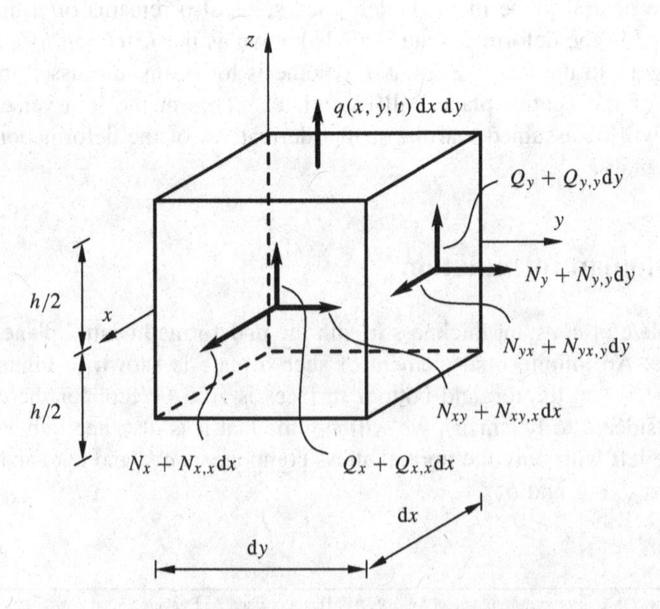

Figure 5.1 Free-body diagram of an infinitesimal plate element

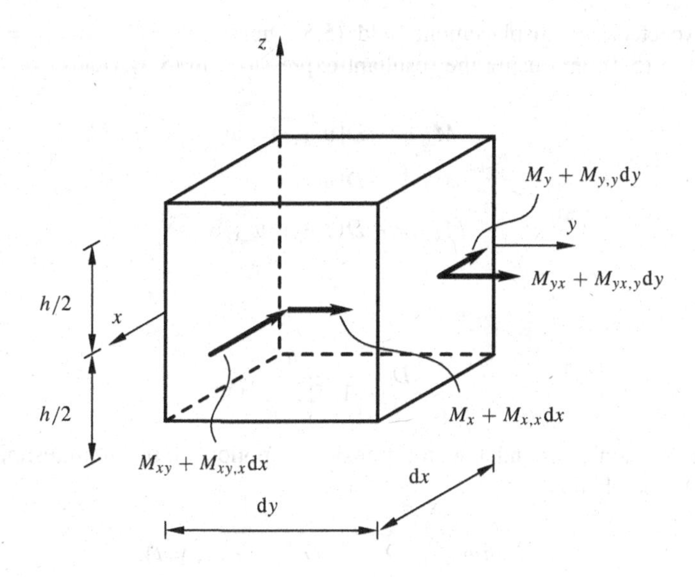

Figure 5.2 Moment resultants on an infinitesimal plate element

from the moment balance equations discussed later, in a way analogous to that occurring in a beam.

Next, we relate the strains to the displacement field of the plate. Let the transverse displacement field of the neutral plane of the plate be denoted by $w(x, y, t)$. Due to this transverse displacement and according to the previously mentioned kinematic relations, the material displacements along x-axis and y-axis directions can be obtained, as shown in Figure 5.3 as, respectively,

$$u(x, y, z, t) = -zw_{,x}(x, y, t) \quad \text{and} \quad v(x, y, z, t) = -zw_{,y}(x, y, t). \tag{5.5}$$

The strain field is then obtained as

$$\epsilon_{xx} = u_{,x} = -zw_{,xx}(x, y, t), \tag{5.6}$$

$$\epsilon_{yy} = v_{,y} = -zw_{,yy}(x, y, t), \tag{5.7}$$

$$\epsilon_{xy} = \frac{1}{2}(u_{,x} + v_{,y}) = -zw_{,xy}(x, y, t). \tag{5.8}$$

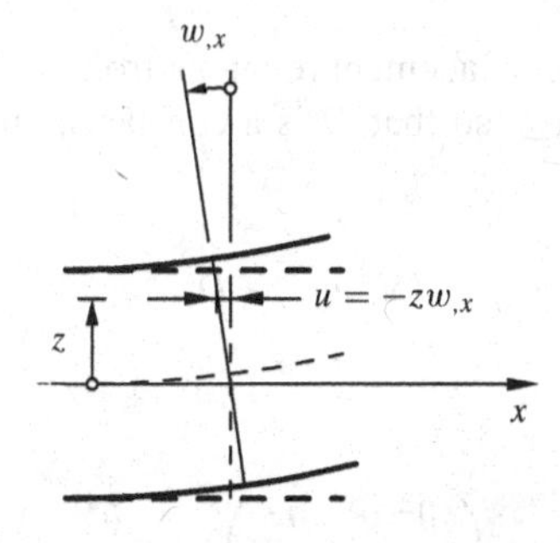

Figure 5.3 Visualization of the in-plane displacement along the x-axis

As was expected, the displacement field (5.5) yields $\epsilon_{xz} = 0$ and $\epsilon_{yz} = 0$. Substituting (5.6)–(5.8) in (5.4), and using the resultant expressions in (5.3) yields

$$M_x = -D[w_{,xx} + \nu w_{,yy}], \quad (5.9)$$

$$M_y = -D[w_{,yy} + \nu w_{,xx}], \quad (5.10)$$

$$M_{xy} = -D(1-\nu)w_{,xy}, \quad (5.11)$$

where

$$D = \frac{Eh^3}{12(1-\nu^2)}.$$

Writing Newton's second law for transverse motion of the infinitesimal element (see Figure 5.1), one obtains

$$\rho h w_{,tt} = Q_{x,x} + Q_{y,y} + q(x, y, t), \quad (5.12)$$

where $q(x, y, t)$ is the external force distribution. The moment equations for the infinitesimal element about the x-axis and y-axis directions yield

$$I w_{,xtt} = -M_{x,x} - M_{xy,y} + Q_x, \quad (5.13)$$

$$I w_{,ytt} = -M_{y,y} - M_{xy,x} + Q_y, \quad (5.14)$$

where $w_{,xtt}$ and $w_{,ytt}$ represent, respectively, the angular accelerations about the y- and x-axis, and

$$I = \int_{-h/2}^{h/2} \rho z^2 \, \mathrm{d}z = \frac{\rho h^3}{12}$$

is the moment of inertia per unit area of the plate. Solving for Q_x and Q_y from the respective equations (5.13) and (5.14), and substituting in (5.12), leads to

$$\rho h w_{,tt} - I(w_{,xxtt} + w_{,yytt}) - (M_{x,xx} + 2M_{xy,xy} + M_{y,yy}) = q(x, y, t). \quad (5.15)$$

Finally, using the moment–displacement relations from (5.9)–(5.11) in (5.15), and assuming the thickness h to be constant (so that D is a constant), one obtains the equation of motion of the plate in the form

$$\rho h w_{,tt} - I\nabla^2 w_{,tt} + D\nabla^4 w = q(x, y, t), \quad (5.16)$$

where

$$\nabla^4 = \nabla^2\nabla^2 = \left(\frac{\partial^2}{\partial x^2} + \frac{\partial^2}{\partial y^2}\right)^2 = \frac{\partial^4}{\partial x^4} + 2\frac{\partial^4}{\partial x^2 \partial y^2} + \frac{\partial^4}{\partial y^4}.$$

Dynamics of plates with non-constant thickness is discussed briefly in Section 5.5. The partial differential equation (5.16) is known as the *Kirchhoff–Rayleigh plate* equation. If the rotary inertia term $I\nabla^2 w_{,tt}$ is neglected, we obtain the normal *Kirchhoff plate* equation

$$\rho h w_{,tt} + D\nabla^4 w = q(x, y, t). \tag{5.17}$$

The next task is to formulate the boundary conditions for plates. Let us first consider a rectangular plate with boundaries at $x = 0, a$ and $y = 0, b$. If a boundary, say at $x = a$, is clamped, then the appropriate boundary conditions are

$$w\big|_{x=a} \equiv 0 \quad \text{and} \quad w_{,y}\big|_{x=a} \equiv 0,$$

both of which are geometric conditions. On the other hand, if the edge $x = a$ is simply-supported, we have one geometric and one natural boundary condition given by, respectively,

$$w\big|_{x=a} \equiv 0 \quad \text{and} \quad M_x\big|_{x=a} = -D[w_{,xx} + \nu w_{,yy}]_{x=a} \equiv 0,$$

where the moment M_x is given by (5.9). Since for the simply-supported edge $w_{,yy}|_{x=a} \equiv 0$, the moment condition above reduces to $w_{,xx}|_{x=a} \equiv 0$.

Next, consider the straight edge $x = a$ as a free boundary. In this case, we intuitively expect the boundary conditions to be $Q_x|_{x=a} \equiv 0$, $M_x|_{x=a} \equiv 0$, and $M_{xy}|_{x=a} \equiv 0$. However, as shown by Kirchhoff, the fourth-order differential equation of the plate eigenvalue problem will not have, in general, a solution satisfying these three conditions at the free edge $x = a$. The correct boundary conditions for a free edge were given by Kirchhoff as

$$M_x\big|_{x=a} \equiv 0 \quad \text{and} \quad V_x := (Q_x + M_{xy,y})\big|_{x=a} \equiv 0,$$

where V_x is known as the *edge force* (see [1]). The edge force can be understood physically through Figure 5.4. Apart from the shear force Q_x (not shown in the figure) at the edge, the edge force contribution due to the moment M_{xy} can be calculated at the point A (see figure) as $[(M_{xy} + \epsilon M_{xy,y})/\epsilon - M_{xy}/\epsilon]_{x=a} = M_{xy,y}|_{x=a}$. These conditions can be more conveniently arrived at using the variational formulation, as discussed in Appendix C in detail. Expressed in terms of derivatives of the transverse displacement, the moment and edge force boundary conditions for the free edge $x = a$ are obtained as (see Appendix C), respectively,

$$\begin{aligned} M\big|_{x=a} &= D[w_{,xx} + \nu w_{,yy}]_{x=a} = 0 \\ V\big|_{x=a} &= D[(\nabla^2 w)_{,x} + (1-\nu) w_{,xyy}]_{x=a} = 0. \end{aligned}$$

In the case of circular plates, polar coordinates are more suitable than Cartesian coordinates. The equation of motion in (r, ϕ) coordinates is given by (5.16) or (5.17) where

$$\nabla^2 = \frac{\partial^2}{\partial r^2} + \frac{1}{r}\frac{\partial}{\partial r} + \frac{\partial^2}{\partial \phi^2}.$$

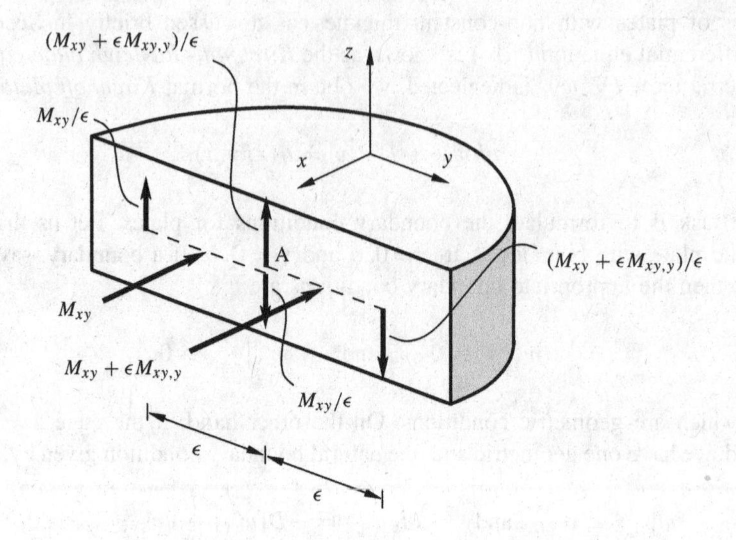

Figure 5.4 Calculation of edge force on a plate due to moment M_{xy}

The boundary conditions for circular plates are derived again in Appendix C. For example, for a free boundary at $r = R$, the boundary conditions are obtained as, respectively,

$$M_r\Big|_{r=R} = \left[\nabla^2 w - (1-\nu)\frac{1}{r}\left(w_{,r} + \frac{1}{r}w_{,\phi\phi}\right)\right]_{r=R} = 0, \tag{5.18}$$

$$V_r\Big|_{r=R} = \left[(\nabla^2 w)_{,r} + (1-\nu)\frac{1}{r}\left(\frac{1}{r}w_{,\phi\phi}\right)_{,r}\right]_{r=R} = 0. \tag{5.19}$$

5.2 VIBRATIONS OF RECTANGULAR PLATES

5.2.1 Free vibrations

Consider an unforced rectangular Kirchhoff plate governed by the equation of motion

$$\rho h w_{,tt} + D\nabla^4 w = 0. \tag{5.20}$$

We will search for a modal solution of the form

$$w(x, y, t) = W(x, y)e^{i\omega t}, \tag{5.21}$$

where $W(x, y)$ is an unknown function and ω is the circular frequency. Substituting (5.21) in the equation of motion (5.20), we obtain

$$(\nabla^4 - \gamma^4)W = 0, \tag{5.22}$$

where

$$\gamma^4 := \omega^2 \rho h / D. \tag{5.23}$$

One can factorize the operator in (5.22) as

$$(\nabla^2 + \gamma^2)(\nabla^2 - \gamma^2)W = 0. \tag{5.24}$$

Define two functions $W_1(x, y)$ and $W_2(x, y)$ such that

$$(\nabla^2 + \gamma^2)W_1 = 0, \tag{5.25}$$

$$(\nabla^2 - \gamma^2)W_2 = 0. \tag{5.26}$$

Since the operators $(\nabla^2 + \gamma^2)$ and $(\nabla^2 - \gamma^2)$ commute, it is not difficult to show that

$$W(x, y) = W_1(x, y) + W_2(x, y) \tag{5.27}$$

is always a solution of (5.22). However, the converse is not true. There are solutions of (5.22) that cannot be written in the form (5.27).

First, consider the partial differential equation (5.25). This can be easily recognized to be the Helmholtz equation, which also appeared in the study of the dynamics of membranes. The solution of (5.25) can be written as (see (4.24))

$$\begin{aligned} W_1(x, y) = {} & A_1 \sin \alpha x \, \sin \beta y + A_2 \sin \alpha x \, \cos \beta y \\ & + A_3 \cos \alpha x \, \sin \beta y + A_4 \cos \alpha x \, \cos \beta y, \end{aligned} \tag{5.28}$$

where A_i are the arbitrary constants of integration, and α and β are such that

$$\alpha^2 + \beta^2 = \gamma^2. \tag{5.29}$$

Assume a separable solution of (5.26) in the form

$$W_2(x, y) = X(x)Y(y). \tag{5.30}$$

Substituting this expression in (5.26) yields on rearrangement

$$\frac{1}{X}\frac{\mathrm{d}^2 X}{\mathrm{d}x^2} + \frac{1}{Y}\frac{\mathrm{d}^2 Y}{\mathrm{d}y^2} - \gamma^2 = 0. \tag{5.31}$$

It is evident that a non-trivial solution of (5.31) exists if and only if

$$\frac{1}{X}\frac{\mathrm{d}^2X}{\mathrm{d}x^2} = \overline{\alpha}^2 \quad \Rightarrow \quad \frac{\mathrm{d}^2X}{\mathrm{d}x^2} - \overline{\alpha}^2 X = 0, \tag{5.32}$$

$$\frac{1}{Y}\frac{\mathrm{d}^2Y}{\mathrm{d}y^2} = \overline{\beta}^2 \quad \Rightarrow \quad \frac{\mathrm{d}^2Y}{\mathrm{d}y^2} - \overline{\beta}^2 R = 0, \tag{5.33}$$

where $\overline{\alpha}$ and $\overline{\beta}$ are constants such that

$$\overline{\alpha}^2 + \overline{\beta}^2 = \gamma^2. \tag{5.34}$$

The general solutions of (5.32) and (5.33) are obtained as, respectively,

$$X(x) = C_1 \sinh \overline{\alpha}x + C_2 \cosh \overline{\alpha}x,$$
$$Y(y) = C_3 \sinh \overline{\beta}y + C_4 \cosh \overline{\beta}y,$$

where C_i are the arbitrary constants of integration. Therefore, a solution $W(x, y)$ of the type (5.27) can be written as

$$\begin{aligned} W(x, y) = {} & A_1 \sin\alpha x \ \sin\beta y + A_2 \sin\alpha x \ \cos\beta y \\ & + A_3 \cos\alpha x \ \sin\beta y + A_4 \cos\alpha x \ \cos\beta y \\ & + A_5 \sinh\overline{\alpha}x \ \sinh\overline{\beta}y + A_6 \sinh\overline{\alpha}x \ \cosh\overline{\beta}y \\ & + A_7 \cosh\overline{\alpha}x \ \sinh\overline{\beta}y + A_8 \cosh\overline{\alpha}x \ \cosh\overline{\beta}y, \end{aligned} \tag{5.35}$$

where A_5–A_8 are defined from the product of $X(x)$ and $Y(y)$ in (5.30). This, however, is not the most general solution of (5.22). The actual eigenfunction does of course also depend on the boundary conditions.

5.2.1.1 Simply-supported plate

To begin with, let us consider the elementary case of a plate with simply-supported edges. The boundary conditions in this case are

$$w\big|_{x=0,a} \equiv 0, \quad w_{,xx}\big|_{x=0,a} \equiv 0, \quad w\big|_{y=0,b} \equiv 0, \quad \text{and} \quad w_{,yy}\big|_{y=0,b} \equiv 0.$$

The boundary conditions for the function $W(x, y)$ are then obtained as

$$W\big|_{x=0,a} \equiv 0, \quad W_{,xx}\big|_{x=0,a} \equiv 0, \quad W\big|_{y=0,b} \equiv 0, \quad \text{and} \quad W_{,yy}\big|_{y=0,b} \equiv 0. \tag{5.36}$$

It can be easily verified that these boundary conditions can be identically satisfied by (5.35) if and only if A_i, $i = 2, \ldots, 8$, vanish. Therefore, we have

$$W(x, y) = A_1 \sin \alpha x \sin \beta y, \tag{5.37}$$

where, according to the boundary conditions (5.36), α and β must be such that $\sin \alpha a = 0$ and $\sin \beta b = 0$. This yields

$$\alpha_m a = m\pi \quad \text{and} \quad \beta_n b = n\pi.$$

The circular eigenfrequencies are then obtained using (5.29) in (5.23) as

$$\omega_{(m,n)} = \pi^2 \left(\frac{m^2}{a^2} + \frac{n^2}{b^2} \right) \sqrt{\frac{D}{\rho h}}, \qquad m = 1, 2, \ldots, \infty. \tag{5.38}$$

It may be recalled that the eigenfunctions of a fixed rectangular membrane are also of the form (5.37), though the eigenfrequencies are of course different. Finally, the general solution of the free vibration problem of a simply-supported plate is obtained as

$$w(x, y, t) = \sum_{m,n=1}^{\infty} A_{(m,n)} \sin \frac{m\pi x}{a} \sin \frac{n\pi y}{b} \sin \left[t\pi^2 \left(\frac{m^2}{a^2} + \frac{n^2}{b^2} \right) \sqrt{\frac{D}{\rho h}} + \psi_{(m,n)} \right],$$

where $A_{(m,n)}$ and $\psi_{(m,n)}$ are arbitrary constants which are determined from the initial conditions. It is easy to check that the eigenfunctions satisfy the orthogonality relation

$$\langle W_{(m,n)} W_{(r,s)} \rangle = \int_0^a \int_0^b W_{(m,n)} W_{(r,s)} \, \mathrm{d}x \, \mathrm{d}y = \frac{ab}{4} \delta_{mr} \delta_{ns}.$$

The orthogonality relations for plates of arbitrary shapes are discussed later.

Let us now consider a rectangular plate simply-supported at the edges $x = 0$ and $x = a$, and clamped at the edges $y = 0$ and $y = b$. The boundary conditions in this case are given by

$$w\big|_{x=0,a} \equiv 0, \quad w_{,xx}\big|_{x=0,a} \equiv 0, \quad w\big|_{y=0,b} \equiv 0, \quad \text{and} \quad w_{,yy}\big|_{y=0,b} \equiv 0.$$

The corresponding boundary conditions for the function $W(x, y)$ are, therefore, obtained as

$$W\big|_{x=0,a} = 0, \qquad W_{,xx}\big|_{x=0,a} = 0, \tag{5.39}$$

$$W\big|_{y=0,b} = 0, \qquad W_{,y}\big|_{y=0,b} = 0. \tag{5.40}$$

It may be checked that only the trivial solution is of the form (5.35) for the above boundary conditions. Taking our cue from the simply-supported plate case discussed above, let us search for a separable solution of (5.22) with the given boundary conditions as

$$W(x, y) = \sin\frac{m\pi x}{a}Y(y), \tag{5.41}$$

where $Y(y)$ is an unknown function of y. Substituting (5.41) in (5.22) yields

$$Y'''' - 2\frac{m^2\pi^2}{a^2}Y'' + \left(\frac{m^4\pi^4}{a^4} - \gamma^4\right)Y = 0$$

or

$$\left(\frac{d^2}{dy^2} - \frac{m^2\pi^2}{a^2} - \gamma^2\right)\left(\frac{d^2}{dy^2} - \frac{m^2\pi^2}{a^2} + \gamma^2\right)Y = 0.$$

This has a solution of the form

$$Y(y) = C_1\cosh\alpha y + C_2\sinh\alpha y + C_3\cos\beta y + C_4\sin\beta y, \tag{5.42}$$

where

$$\alpha = \sqrt{\gamma^2 + \frac{m^2\pi^2}{a^2}} \quad \text{and} \quad \beta = \sqrt{\gamma^2 - \frac{m^2\pi^2}{a^2}}. \tag{5.43}$$

Hence, the solution (5.41) can be written as

$$W(x, y) = (C_1\cosh\alpha y + C_2\sinh\alpha y + C_3\cos\beta y + C_4\sin\beta y)\sin\frac{m\pi x}{a}. \tag{5.44}$$

It is evident that the solution form (5.44) already satisfies the boundary conditions (5.39). Using (5.44) in the boundary conditions in (5.40) yields the set of homogeneous linear equations

$$\begin{bmatrix} 1 & 0 & 1 & 0 \\ 0 & \alpha & 0 & \beta \\ \cosh\alpha b & \sinh\alpha b & \cos\beta b & \sin\beta b \\ \alpha\sinh\alpha b & \alpha\cosh\alpha b & -\beta\sin\beta b & \beta\cos\beta b \end{bmatrix}\begin{Bmatrix} C_1 \\ C_2 \\ C_3 \\ C_4 \end{Bmatrix} = 0. \tag{5.45}$$

Non-trivial solutions of $(C_1, C_2, C_3, C_4)^{\mathrm{T}}$ are obtained if and only if the determinant of the matrix in (5.45) vanishes, i.e.,

$$2\alpha\beta(\cos\beta b\cosh\alpha b - 1) + (\beta^2 - \alpha^2)\sin\beta b\sinh\alpha b = 0. \tag{5.46}$$

This is the characteristic equation for the rectangular plate with two simply-supported and two clamped edges. Using the definitions (5.43) in (5.46), one obtains the roots of (5.46) of the form $\gamma_{(m,n)}$ for the mode (m, n). The eigenfrequencies are then obtained using (5.23). For, example, for a square plate of side of length a, we obtain the first three roots as

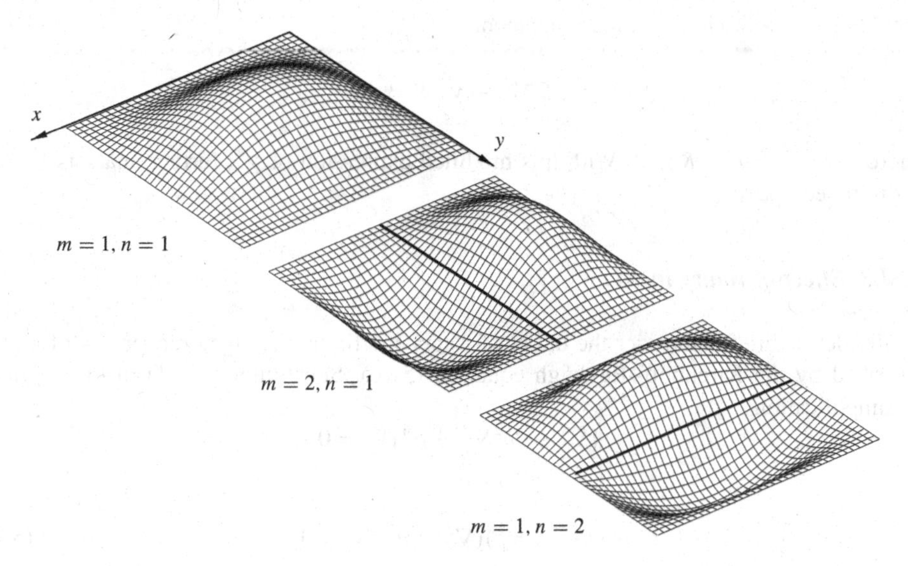

Figure 5.5 Mode-shapes of a square plate simply-supported at $x = 0, a$, and clamped at $y = 0, a$

$\gamma^2_{(1,1)} = 28.946/a^2$, $\gamma_{(2,1)} = 54.743/a^2$, and $\gamma_{(1,2)} = 69.327/a^2$. Substituting these roots in (5.45), one can solve for $(C_1, C_2, C_3, C_4)^T$. Finally, the eigenfunctions are obtained from (5.44). The first three mode-shapes of the square plate are shown in Figure 5.5.

Alternatively, one can also solve the free vibration problem (5.20) approximately using either the Ritz or the Galerkin method discussed in previous chapters. In these methods, we expand the solution as

$$w(x, y, t) = \sum_{m,n=1}^{\infty} p_{(m,n)}(t)\psi_{(m,n)}(x, y),$$

where $\psi(x, y)$ are shape-functions that have to satisfy at least the geometric boundary conditions of the problem in the case of the Ritz method, or all the boundary conditions in the case of the Galerkin method. A convenient approach to generate shape-functions is to use the product of beam eigenfunctions, i.e.,

$$\psi_{(m,n)}(x, y) = W_m(x)W_n(y),$$

where $W_m(x)$ and $W_n(y)$ are the eigenfunctions of beams with the same boundary conditions as of the plate in the x- and y-axis directions, respectively. More discussions on the generation of shape-functions for rectangular plates can be found in [2].

5.2.1.2 Plate on a flexible foundation

Next, consider an elastic plate on a flexible foundation having a stiffness per unit area of K. The equation of motion of the plate can be easily obtained as

$$\rho h w_{,tt} + Kw + D\nabla^4 w = 0. \tag{5.47}$$

Substituting (5.21) yields on rearrangement

$$\nabla^4 W - \gamma^4 W = 0,$$

where $\gamma^4 = (\omega^2 \rho h - K)/D$. With this modified expression of γ^4, further analysis follows as discussed above.

5.2.1.3 Effect of rotary inertia

Finally, let us briefly consider the effect of rotary inertia on the eigenfrequencies of a plate described by the Kirchhoff–Rayleigh equation (5.16). Substituting (5.21) in (5.16) yields on simplification

$$[\nabla^4 + 2\delta^2\nabla^2 - \mu^4]W = 0$$

or

$$(\nabla^2 + \gamma_1^2)(\nabla^2 - \gamma_2^2)W = 0. \tag{5.48}$$

where $2\delta^2 := \omega^2 I/D$, $\mu^4 := \omega^2 \rho h/D$, $\gamma_1^2 = \sqrt{\delta^4 + \mu^4} + \delta^2$, and $\gamma_2^2 = \sqrt{\delta^4 + \mu^4} - \delta^2$. The similarity between (5.48) and (5.24) is evident. For a simply-supported rectangular plate, one can easily obtain the eigenfunctions as

$$W_{(m,n)} = \sin \alpha_m x \sin \beta_n y,$$

where $\alpha_m = m\pi/a$, $\beta_n = n\pi/b$. The condition (5.29) now reads

$$\alpha_m^2 + \beta_n^2 = \gamma_1^2$$
$$\Rightarrow \frac{I}{2D}\omega^2 + \left(\sqrt{\frac{I}{2D} + \frac{\rho h}{D}}\right)\omega - \left(\frac{m^2\pi^2}{a^2} + \frac{n^2\pi^2}{b^2}\right) = 0,$$

where ω is the circular eigenfrequency of the Kirchhoff–Rayleigh plate.

5.2.2 Orthogonality of plate eigenfunctions

Consider the differential equation (5.22) for a plate for two eigenfunctions as

$$D\nabla^4 W_{(j,k)} - \omega_{(j,k)}^2 \rho h W_{(j,k)} = 0, \tag{5.49}$$

$$D\nabla^4 W_{(r,s)} - \omega_{(r,s)}^2 \rho h W_{(r,s)} = 0. \tag{5.50}$$

Multiplying (5.49) by $W_{(r,s)}$ and (5.50) by $W_{(j,k)}$, and integrating the difference of the two equations over the domain of the plate yields

$$\int_A [DW_k \nabla^4 W_{(j,k)} - DW_{(j,k)}\nabla^4 W_{(r,s)} - (\omega_{(j,k)}^2 - \omega_{(r,s)}^2)\rho h W_{(j,k)} W_{(r,s)}]\, \mathrm{d}A = 0. \tag{5.51}$$

Writing $\nabla^4 W = \nabla \cdot [\nabla(\nabla \cdot \nabla W)]$, and using the Gauss divergence theorem, one can rewrite (5.51) as

$$D \oint_{\mathrm{B}} \left[W_{(j,k)} \hat{\mathbf{n}} \cdot \nabla(\nabla^2 W_{(r,s)}) - W_{(r,s)} \hat{\mathbf{n}} \cdot \nabla(\nabla^2 W_{(j,k)}) \right.$$
$$\left. + \nabla^2 W_{(j,k)} \hat{\mathbf{n}} \cdot \nabla W_{(r,s)} - \nabla^2 W_{(r,s)} \hat{\mathbf{n}} \cdot \nabla W_{(j,k)} \right] \mathrm{d}s$$
$$- (\omega_{(j,k)}^2 - \omega_{(r,s)}^2) \int_{\mathrm{A}} \rho h W_{(j,k)} W_{(r,s)} \, \mathrm{d}A = 0, \tag{5.52}$$

where B is the boundary of the area A, and $\hat{\mathbf{n}}$ is the unit normal vector at the boundary. Therefore, for $\omega_{(j,k)} \neq \omega_{(r,s)}$ the condition for orthogonality of the eigenfunctions of a plate is

$$\oint \left[W_{(j,k)} \hat{\mathbf{n}} \cdot \nabla(\nabla^2 W_{(r,s)}) - W_{(r,s)} \hat{\mathbf{n}} \cdot \nabla(\nabla^2 W_{(j,k)}) \right.$$
$$\left. - \nabla^2 W_{(r,s)} \hat{\mathbf{n}} \cdot \nabla W_{(j,k)} + \nabla^2 W_{(j,k)} \hat{\mathbf{n}} \cdot \nabla W_{(r,s)} \right] \mathrm{d}s = 0. \tag{5.53}$$

Finally, the orthogonality relation for plates or arbitrary shapes is given by

$$\int_{\mathrm{A}} \rho h W_{(j,k)} W_{(r,s)} \, \mathrm{d}A = 0.$$

One can check explicitly that for standard boundary conditions such as for simply-supported, clamped, and free boundaries, the condition (5.53) is satisfied. It may noted that (5.53) represents the condition of self-adjointness of the boundary value problem.

5.2.3 Forced vibrations

The undamped forced dynamics of a Kirchhoff plate can be represented by

$$\rho h w_{,tt} + D \nabla^4 w = q(x, y, t), \tag{5.54}$$

where $q(x, y, t)$ represents the external force distribution on the plate. Using the expansion theorem, the solution of (5.54) can be expressed in terms of the eigenfunctions determined from the modal analysis as

$$w(x, y, t) = \sum_{m,n=1}^{\infty} p_{(m,n)}(t) W_{(m,n)}(x, y), \tag{5.55}$$

where $p_{(m,n)}(t)$ are the unknown time-varying modal coordinates. In practice, one can truncate the series (5.55) to finite terms for an approximate solution. Substituting (5.55) in (5.54), and taking the inner product on both sides with $W_{(r,s)}$ yields, on account of orthogonality of the eigenfunctions, the modal dynamics of the mode (r, s) as

$$\ddot{p}_{(r,s)} + \omega^2_{(r,s)} p_{(r,s)} = Q_{(r,s)}(t), \tag{5.56}$$

where

$$Q_{(r,s)} = \langle q(x, y, t), W_{(r,s)} \rangle = \int_0^a \int_0^b q(x, y, t) W_{(r,s)}(x, y)\, \mathrm{d}y\, \mathrm{d}x.$$

One can easily obtain the solution for the modal coordinates from (5.56) using, for example, the Laplace transform or Green's function. Finally, the complete solution is obtained from (5.55).

In many practical situations, the forcing term in (5.54) is harmonic and separable in the form $Q(x, y, t) = f(x, y) \cos \Omega t = \mathscr{R}[f(x, y)\mathrm{e}^{i\Omega t}]$. To determine the corresponding stationary solution, first consider the forced dynamics

$$D\nabla^4 w + \rho h w_{,tt} = \delta(x - \overline{x}, y - \overline{y})\mathrm{e}^{i\Omega t}, \tag{5.57}$$

where $\delta(x - \overline{x}, y - \overline{y}) := \delta(x - \overline{x})\delta(y - \overline{y})$, and $(\overline{x}, \overline{y})$ are the coordinates of any arbitrary point on the plate. Searching for a solution in the form $w(x, y, t) = W(x, y)\mathrm{e}^{i\Omega t}$, (5.57) yields

$$D\nabla^4 W - \rho h \Omega^2 W = \delta(x - \overline{x}, y - \overline{y}). \tag{5.58}$$

It may be easily recognized that the solution of (5.58) with the corresponding boundary conditions is Green's function for the vibration problem of a plate excited by a concentrated time-harmonic unit force, which may be determined using the eigenfunction expansion

$$W(x, y, \overline{x}, \overline{y}) = \sum_{m,n=1}^{\infty} a_{(m,n)}(\overline{x}, \overline{y}) W_{(m,n)}(x, y), \tag{5.59}$$

where $a_{(m,n)}(\overline{x}, \overline{y})$ are unknown functions. It is not difficult to show that

$$W(x, y, \overline{x}, \overline{y}) = \sum_{m,n=1}^{\infty} \frac{W_{(m,n)}(\overline{x}, \overline{y})}{(\omega^2_{(m,n)} - \Omega^2)\langle \rho h W_{(m,n)}, W_{(m,n)} \rangle} W_{(m,n)}(x, y).$$

The stationary solution for $w(x, y, t)$ can then be written as

$$w(x, y, t) = \mathscr{R}\left[\left(\int_0^a \int_0^b f(\overline{x}, \overline{y}) W(x, y, \overline{x}, \overline{y}) \mathrm{d}\overline{x}\mathrm{d}\overline{y}\right) \mathrm{e}^{i\Omega t}\right].$$

5.3 VIBRATIONS OF CIRCULAR PLATES

5.3.1 Free vibrations

In the case of circular Kirchhoff plates, the equation of motion in the polar coordinate system can be written as

$$\rho h w_{,tt} + D\nabla^4 w = 0, \tag{5.60}$$

where

$$\nabla^2 = \frac{\partial^2}{\partial r^2} + \frac{1}{r}\frac{\partial}{\partial r} + \frac{1}{r^2}\frac{\partial^2}{\partial \phi^2}.$$

Substituting a time-harmonic solution of the form

$$w(r, \phi, t) = W(r, \phi)\mathrm{e}^{i\omega t} \tag{5.61}$$

in (5.60) yields

$$\nabla^4 W - \gamma^4 W = 0$$

or

$$(\nabla^2 + \gamma^2)(\nabla^2 - \gamma^2)W = 0, \tag{5.62}$$

where

$$\gamma^4 = \frac{\rho h \omega^2}{D}. \tag{5.63}$$

Assuming that the plate boundary conditions allow a separable solution for $W(r, \phi)$, one can write

$$W(r, \phi) = R(r)\mathrm{e}^{im\phi}, \tag{5.64}$$

where we have used the condition of periodicity of the solution in ϕ. Using this in (5.62) yields

$$\left[\frac{\mathrm{d}^2}{\mathrm{d}r^2} + \frac{1}{r}\frac{\mathrm{d}}{\mathrm{d}r} + \left(\gamma^2 - \frac{m^2}{r^2}\right)\right]\left[\frac{\mathrm{d}^2}{\mathrm{d}r^2} + \frac{1}{r}\frac{\mathrm{d}}{\mathrm{d}r} - \left(\gamma^2 + \frac{m^2}{r^2}\right)\right]R_m = 0. \tag{5.65}$$

One can find solutions of $R_m(r)$ as

$$R_m(r) = A_m(r) + B_m(r), \tag{5.66}$$

where

$$\frac{\mathrm{d}^2 A_m}{\mathrm{d}r^2} + \frac{1}{r}\frac{\mathrm{d}A_m}{\mathrm{d}r} + \left(\gamma^2 - \frac{m^2}{r^2}\right) A_m = 0, \tag{5.67}$$

$$\frac{\mathrm{d}^2 B_m}{\mathrm{d}r^2} + \frac{1}{r}\frac{\mathrm{d}B_m}{\mathrm{d}r} - \left(\gamma^2 + \frac{m^2}{r^2}\right) B_m = 0. \tag{5.68}$$

The general solution of the Bessel differential equation (5.67) can be written as

$$A_m(r) = C_1 J_m(\gamma r) + C_2 Y_m(\gamma r), \tag{5.69}$$

where C_1 and C_2 are arbitrary constants, and $J_m(\cdot)$ and $Y_m(\cdot)$ are, respectively, the Bessel functions of first and second kinds of order m. The differential equation (5.68) is known as the *modified Bessel differential equation*, and the solution can be written as

$$B_m(r) = C_3 I_m(\gamma r) + C_4 K_m(\gamma r), \tag{5.70}$$

where C_3 and C_4 are arbitrary constants, and $I_m(\cdot)$ and $K_m(\cdot)$ are known as, respectively, the *modified Bessel functions of first and second kinds* of order m. Therefore, the solution $R_m(r)$ according to (5.66) is

$$R_m(r) = C_1 J_m(\gamma r) + C_2 Y_m(\gamma r) + C_3 I_m(\gamma r) + C_4 K_m(\gamma r). \tag{5.71}$$

The value of γ, and the arbitrary constants in (5.71) are to be determined from the boundary conditions. It may be mentioned that, depending on the boundary conditions, it may not be always possible to construct solutions in the manner discussed above for the boundary value problem described by (5.65) and the corresponding boundary conditions.

Let us consider a circular plate of radius a with a clamped boundary. The boundary conditions can be expressed as

$$w(a, \phi, t) \equiv 0 \quad \Rightarrow \quad R_m(a) = 0, \tag{5.72}$$

$$w_{,r}(a, \phi, t) \equiv 0 \quad \Rightarrow \quad R'_m(a) = 0. \tag{5.73}$$

It is known that $Y_m(x)$ and $K_m(x)$ are unbounded at $x = 0$. Therefore, the requirement of finiteness of the solution at $r = 0$ imposes the conditions $C_2 = C_4 = 0$. Substituting these conditions in (5.71), we finally obtain from the boundary conditions (5.72)–(5.73) the set of homogeneous equations

$$\begin{bmatrix} J_m(\gamma a) & I_m(\gamma a) \\ J'_m(\gamma a) & I'_m(\gamma a) \end{bmatrix} \begin{Bmatrix} C_1 \\ C_3 \end{Bmatrix} = 0. \tag{5.74}$$

For non-trivial solutions of C_1 and C_3, we must have

$$J_m(\gamma a)I'_m(\gamma a) - J'_m(\gamma a)I_m(\gamma a) = 0. \tag{5.75}$$

This is the characteristic equation which yields countably infinitely many solutions of γa. Denoting the modes by (m, n), the first few values of $\gamma_{(m,n)}a$ are obtained as

$$\gamma_{(0,1)}a = 3.196, \qquad \gamma_{(0,2)}a = 6.306,$$

$$\gamma_{(1,1)}a = 4.611, \qquad \gamma_{(1,2)}a = 7.799,$$

$$\gamma_{(2,1)}a = 5.905, \qquad \gamma_{(1,2)}a = 9.197.$$

The circular eigenfrequencies are then obtained from (5.63) as

$$\omega_{(m,n)} = \gamma^2_{(m,n)}\sqrt{\frac{D}{\rho h}}.$$

Finally, the radial factors of the eigenfunctions are obtained by solving for $(C_1, C_3)^{\mathrm{T}}$ from (5.74) and substituting in (5.71) as

$$R_{(m,n)}(r) = C[I_m(\gamma_{(m,n)}a)J_m(\gamma_{(m,n)}r) - J_m(\gamma_{(m,n)}a)I_m(\gamma_{(m,n)}r)], \tag{5.76}$$

and the eigenfunctions of the circular plate are obtained from the real and imaginary parts of (5.64). The first few mode-shapes of the plate are shown in Figure 5.6. The general solution of the initial value problem of a circular plate can now be written as

$$w(r, \phi, t) = \sum_{m,n=0}^{\infty} [D_{(m,n)} \cos m\phi + E_{(m,n)} \sin m\phi] R_{(m,n)}(r) \mathrm{e}^{i\omega_{(m,n)}t}.$$

It may be observed that, corresponding to each circular eigenfrequency $\omega_{(m,n)}$ with $m \neq 0$, there are two eigenmodes, namely the cosine and the sine modes given by, respectively,

$$W^{(m,n)}_{\mathrm{C}}(r, \phi) = R_{(m,n)}(r) \cos m\phi \qquad \text{and} \qquad W^{(m,n)}_{\mathrm{S}}(r, \phi) = R_{(m,n)}(r) \sin m\phi.$$

Thus, for $m \neq 0$, the plate has modal degeneracy, the consequences of which have been discussed previously in Chapter 4.

For $\gamma a \gg 1$, one can approximate the characteristic equation (5.75) as

$$J_m(\gamma a) - J'_m(\gamma a) = 0. \tag{5.77}$$

Using (4.98), one can further approximate (5.77) as $\tan[\gamma a - (2m + 1)\pi/4] = -1$. This leads to the approximate solution

$$\gamma_{(m,n)}a \approx (m + 2n)\frac{\pi}{2}, \qquad \text{for } n \gg 1,$$

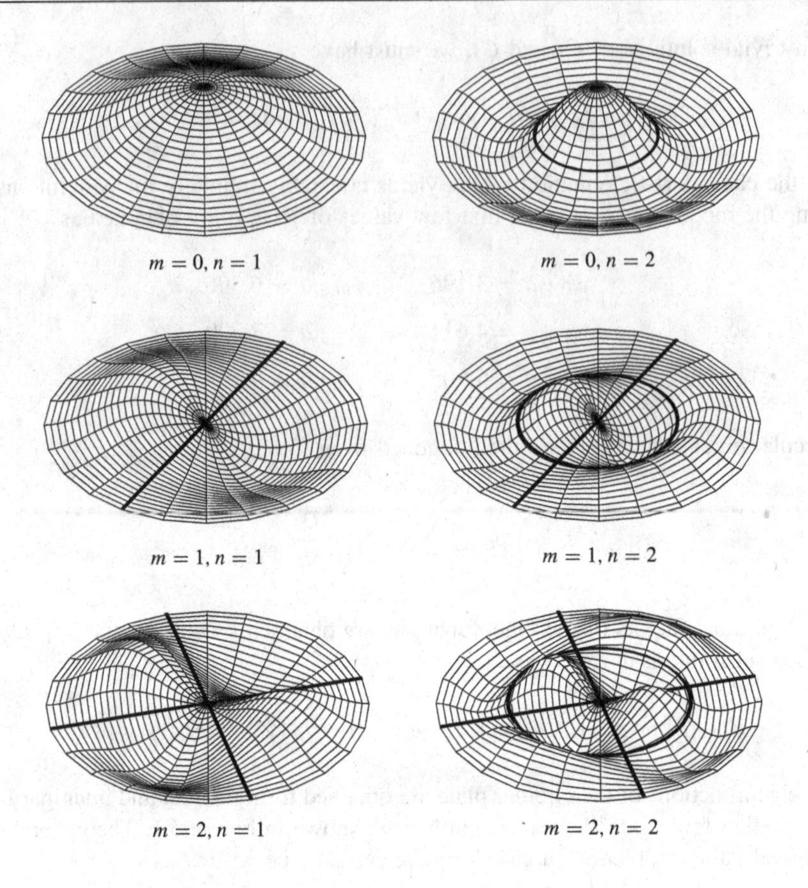

Figure 5.6 Mode-shapes of a circular plate clamped at $r = a$

which yields

$$\omega_{(m,n)} \approx (m + 2n)^2 \frac{\pi^2}{4a^2} \sqrt{\frac{D}{\rho h}}.$$

We have considered here only a simple boundary condition for circular plates. There are many other kinds of boundary condition of practical interest, and the corresponding results can be found in, for example, [1].

5.3.2 Forced vibrations

The method of eigenfunction expansion discussed in Section 5.2.3 can be extended to solve the forced vibration problem for circular plates. The general solution is of the form

$$w(r, \phi, t) = \sum_{m,n=0}^{\infty} [p_{(m,n)}(t) \cos m\phi + q_{(m,n)}(t) \sin m\phi] R_{(m,n)}(r),$$

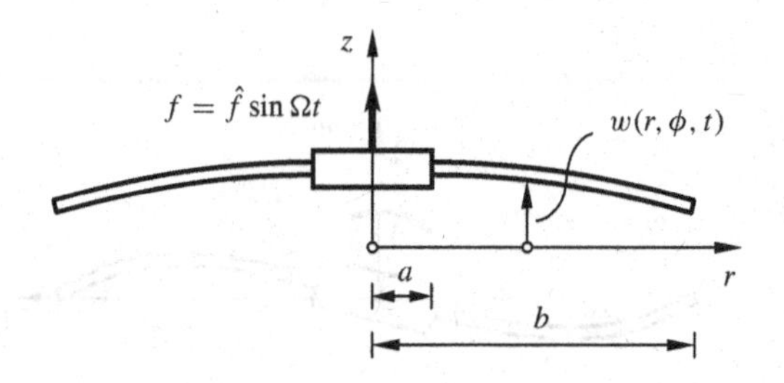

Figure 5.7 Annular plate with a harmonic force on the collar

where $p_{(m,n)}(t)$ and $q_{(m,n)}(t)$ are the modal coordinates. Substituting this expression in the equation of motion and using the orthogonality of the eigenfunctions, one can easily obtain the differential equations for the modal coordinates. This will not be elaborated further. Here, we will only discuss certain special cases of boundary excitation of annular plates.

Consider an annular plate of inner radius a and outer radius b, and mounted on a rigid massless collar, as shown in Figure 5.7. The outer edge of the plate is assumed to be free. The collar is excited by a force $f = \hat{f} \sin \Omega t$. Since this is a case of boundary excitation, the problem is defined by the governing equation of motion

$$\rho h w_{,tt} + D\nabla^4 w = 0, \tag{5.78}$$

along with the boundary conditions

$$w_{,r}(r, \phi, t)\big|_{r=a} \equiv 0, \tag{5.79}$$

$$V_r(a, \phi, t) = -\frac{\hat{f}}{2\pi a} \sin \Omega t, \tag{5.80}$$

$$M_r(b, \phi, t) \equiv 0, \tag{5.81}$$

$$V_r(b, \phi, t) \equiv 0, \tag{5.82}$$

where the expressions of moment M_r and edge force V_r are obtained in terms of $w(r, \phi, t)$ from (5.18) and (5.19), respectively.

The solution of the forced problem can be written in the form

$$w(r, \phi, t) = \sum_{m=0}^{\infty} [A_m J_m(\beta r) + B_m Y_m(\beta r) + C_m I_m(\beta r) + D_m K_m(\beta r)] \cos(m\phi + \gamma_m) \sin \Omega t.$$

However, from the symmetry of the forcing, it is not difficult to conclude that the solution will be independent of ϕ, i.e., $m = 0$. Therefore, the solution takes the form

$$w(r, \phi, t) = [A_0 J_0(\beta r) + B_0 Y_0(\beta r) + C_0 I_0(\beta r) + D_0 K_0(\beta r)] \sin \Omega t. \tag{5.83}$$

Substituting (5.83) in (5.79)–(5.82), we obtain four linear equations which can be solved easily for the four unknowns A_0, B_0, C_0, and D_0.

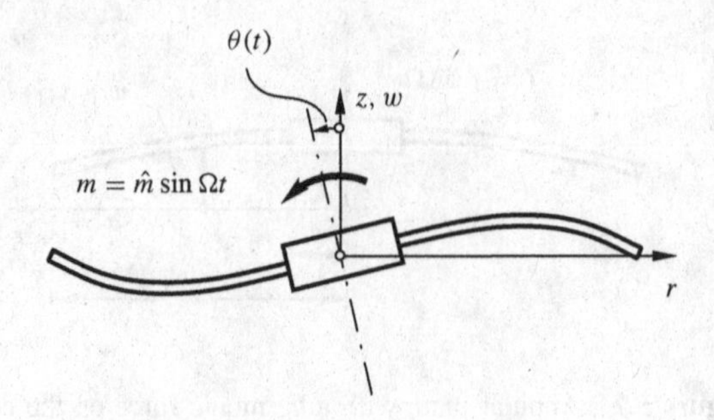

Figure 5.8 Annular plate with a harmonic moment on the collar

Another case of boundary forcing is shown in Figure 5.8 where the collar is given a small harmonic moment $m = \hat{m} \sin \Omega t$ perpendicular to the plane of the paper. Let us represent the corresponding small angular motion of the collar by $\theta = \hat{\theta} \sin \Omega t$, where $\hat{\theta}$ is an unknown. It is to be noted that since there is no damping, the angular motion of the collar will be in phase (or in phase opposition) with the force. The equation of motion in this case is given by (5.78), while the boundary conditions at the inner boundary can be written as

$$w_{,r}(r, \phi, t)\big|_{r=a} = \hat{\theta} \cos \phi, \tag{5.84}$$

$$[w - a w_{,r}]_{r=a} = 0, \tag{5.85}$$

where we have assumed that $\sin \theta \approx \theta$ and $\cos \theta = 1$. The boundary conditions at $r = b$ remain the same as given by (5.81) and (5.82).

Observing the boundary condition (5.84) in this case, it can be easily concluded that the solution must be of the form

$$w(r, \phi, t) = [A_1 J_1(\beta r) + B_1 Y_1(\beta r) + C_1 I_1(\beta r) + D_1 K_1(\beta r)] \cos \phi \sin \Omega t. \tag{5.86}$$

Substituting (5.86) in the boundary conditions again yields four linear equations in the four unknowns A_1, B_1, C_1, and D_1. Solving these equations, we obtain the solution in terms of the unknown $\hat{\theta}$. Finally, $\hat{\theta}$ can be determined from the moment balance equation

$$\hat{m} \sin \Omega t = \int_0^{2\pi} [M_r(a, \phi, t) - a V_r(a, \phi, t)] \cos \phi \, d\phi,$$

where the expressions of M_r and V_r are obtained from (5.18) and (5.19), respectively.

5.4 WAVES IN PLATES

Consider the equation of motion of an infinite Kirchhoff plate described by

$$\rho h w_{,tt} + D(w_{,xxxx} + 2w_{,xxyy} + w_{,yyyy}) = 0. \tag{5.87}$$

Assume a harmonic traveling plane wave solution of the form

$$w(x, y, t) = Ae^{i(k_x x + k_y y - \omega t)}. \tag{5.88}$$

Substitution of this solution into (5.87) yields the dispersion relation

$$\rho h \omega^2 - Dk^4 = 0,$$

where $k^2 := k_x^2 + k_y^2$. The phase velocity of harmonic waves in a Kirchhoff plate is then obtained as

$$c_{\mathrm{P}}^{\mathrm{K}} = \frac{\omega}{k} = \sqrt{\frac{Dk^2}{\rho h}}.$$

It is clear from the phase velocity expression that the plate is a dispersive medium. Moreover, the phase velocity increases without bound with the wave number. This is obviously unrealistic. Similar to what we saw in beams, the consideration of rotational inertia in the equation of motion limits the phase speed as follows.

Consider the Kirchhoff–Rayleigh plate described by

$$\rho h w_{,tt} - I(w_{,xxt} + w_{,yyt}) + D(w_{,xxxx} + 2w_{,xxyy} + w_{,yyyy}) = 0, \tag{5.89}$$

where $I = \rho h^3/12$. Substituting the harmonic wave solution in (5.89) yields the dispersion relation

$$\rho h \omega^2 + \rho \frac{h^3}{12} k^2 \omega^2 - Dk^4 = 0, \tag{5.90}$$

and the phase velocity of harmonic waves in a Kirchhoff–Rayleigh plate is obtained as

$$c_{\mathrm{P}}^{\mathrm{KR}} = \sqrt{\frac{Dk^2}{\rho h + \rho h^3 k^2/12}} = \sqrt{\frac{E}{\rho(1-\nu^2)} \frac{h^2 k^2}{(12 + h^2 k^2)}} \tag{5.91}$$

In the small wave number limit, when $k \ll 1$, one can neglect $\rho h^3 k^2/12$ in comparison to ρh in (5.91), and observe that $c_{\mathrm{P}}^{\mathrm{KR}} \rightarrow c_{\mathrm{P}}^{\mathrm{K}}$. On the other hand, for large wave numbers (i.e., $k \rightarrow \infty$), we obtain

$$c_{\mathrm{P}}^{\mathrm{KR}} \rightarrow \bar{c} := \sqrt{\frac{E}{\rho(1-\nu^2)}}.$$

As in the case of beams, one can also consider the effect of shear deformation on the dynamics of plates as done independently by Mindlin (see [3]) and Reissner (see [4]). This leads to higher-order plate equations, which will not be pursued here. However, consider the dispersion relation for a Mindlin plate, which is given by

$$\left(1 - \frac{c_{\mathrm{P}}^2}{\kappa^2 c_{\mathrm{S}}^2}\right)\left(\frac{\bar{c}^2}{c_{\mathrm{P}}^2} - 1\right) = \frac{12}{h^2 k^2}, \tag{5.92}$$

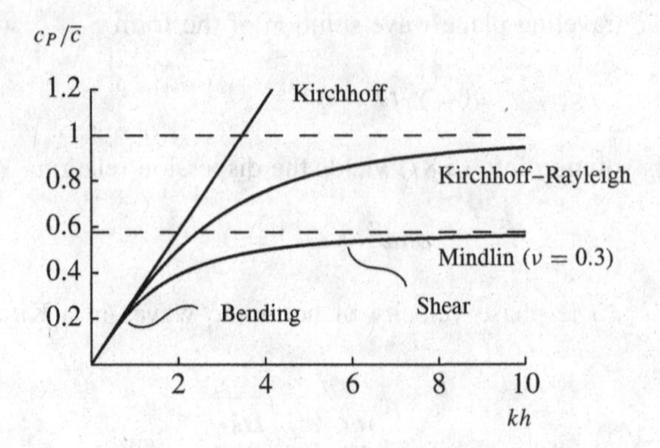

Figure 5.9 Phase velocities for different plate models as a function of the wave number

where

$$c_S = \sqrt{\frac{E}{2\rho(1+\nu)}},$$

is known as the *shear wave speed* (see Chapter 8), and κ^2 is the shear correction factor, which is of the order of $\kappa^2 \sim 5/6$. It may be noted that (5.92) is quadratic in c_P^2. Therefore, we obtain two phase speeds c_{P1} and c_{P2} ($c_{P1} < c_{P2}$). These two phase speeds have similarity to the two phase speeds obtained for a Timoshenko beam in Chapter 3. If we intend to neglect the shear deformation, we make the plate infinitely stiff in shear, i.e., $c_S \to \infty$. Taking this limit in (5.92) yields the dispersion relation (5.90), as can be easily checked. Thus, we can conclude that the branch of the dispersion relation (or phase speed) of the Mindlin plate matching with the Kirchhoff–Rayleigh plate must correspond to the pure bending mode. In the limit $k \to \infty$, the phase speeds of shear and bending waves are obtained from (5.92) as, respectively,

$$c_{P1} \to \kappa^2 c_S \quad \text{and} \quad c_{P2} \to \bar{c}.$$

Thus, in the large wave number limit, the lower branch c_{P1} corresponds to the shear mode. The variation of phase velocities with the wave number for the different plate models discussed above are shown in Figure 5.9. In the figure, only the lower branch c_{P1} of the two branches contained in (5.92) is shown.

5.5 PLATES WITH VARYING THICKNESS

So far we have only considered plates of constant thickness h. However, when the thickness of the plate varies as $h(x, y)$, the section modulus also depends on the space coordinates x and y, i.e., $D = D(x, y)$. In this case, the equation of motion of a Kirchhoff plate becomes (see [5])

$$\rho h w_{,tt} + \nabla^2[D\nabla^2 w] - (1-\nu)[D_{,xx}w_{,yy} - 2D_{,xy}w_{,xy} + D_{,yy}w_{,xx}] = q(x, y, t).$$

In the case of circular plates, if the thickness varies radially as $h = h(r)$, the equation of motion takes the form

$$\rho h w_{,tt} + \nabla^2[D\nabla^2 w] - (1-\nu)\left[D_{,rr}\left(\frac{1}{r}w_{,r} + \frac{1}{r^2}w_{,\phi\phi}\right) + \frac{1}{r}D_{,r}w_{,rr}\right] = q(r,\phi,t).$$

These equations can be easily obtained following the derivation procedure discussed in Section 5.1. However, for obtaining the correct boundary conditions, the variational formulation discussed in Appendix C is preferable.

For free vibration analysis of a circular plate of radially varying thickness, consider the modal solution

$$w(r,\phi,t) = R(r)\mathrm{e}^{im\phi}\mathrm{e}^{i\omega t}.$$

Substituting in the equation of motion yields

$$R''''(r) + f_3(r)R'''(r) + f_2(r)R''(r) + f_1(r)R'(r) + [f_0(r) - \omega^2\rho h]R(r) = 0,$$

where the functions $f_i(r)$ can be easily obtained from the given thickness variation $h(r)$. This differential equation with appropriate boundary conditions may be solved, for example, using Galerkin's method to obtain the eigenfrequencies and the eigenfunctions.

EXERCISES

5.1 A square plate of side length a is simply-supported at the four edges, and carries a particle of mass m at the center. The particle is connected to a spring of stiffness k. Determine the eigenfrequencies and eigenfunctions of the plate.

5.2 A square plate of side length a is supported by an elastic foundation of stiffness density b (stiffness per unit area). The plate is simply-supported at the four edges. Determine the eigenfrequencies and eigenfunctions of the plate.

5.3 A circular plate of radius a is simply-supported at the boundary. Determine the reaction forces at the boundary for different modes of vibration of the plate.

5.4 A square plate of side a is supported at the boundary of a flexible foundation of stiffness density b (stiffness per unit length). Determine the approximate eigenfrequencies of the plate for finite values of b.

5.5 Determine the eigenfrequencies and eigenfunctions of an annular plate of inner radius a and outer radius b. Take both inner and outer boundaries to be clamped.

5.6 A rectangular plate of length $2a$ and width a is simply-supported at the four edges, and on the line AB (i.e., transverse displacement on AB is zero), as shown in Figure 5.10. Determine the eigenfrequencies and eigenfunctions of the plate.

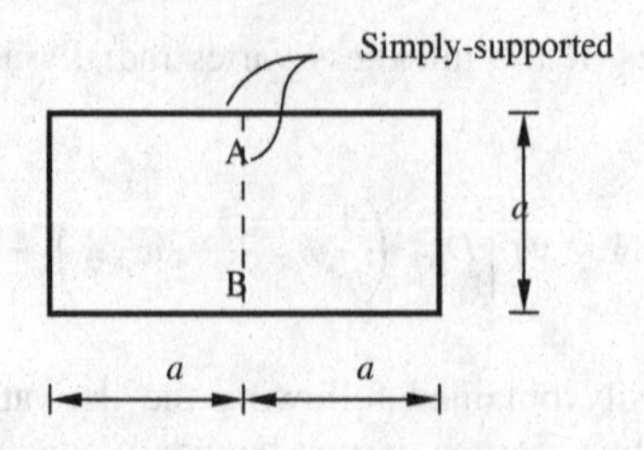

Figure 5.10 Exercise 5.6

5.7 A circular plate of radius a is simply-supported on a circle of radius b, as shown in Figure 5.11. Determine the optimum ratio b/a for which the plate is most firmly supported in the mode (0, 1) (i.e., the corresponding frequency is maximized).

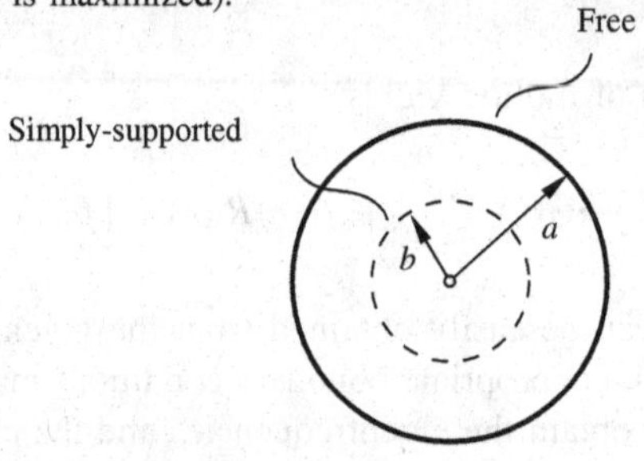

Figure 5.11 Exercise 5.7

5.8 A drumhead of a kettledrum is made of an unstretched circular plate of radius a clamped at the boundaries. Determine the eigenfrequencies and eigenfunctions of the $(0, n)$ modes of the kettledrum.

5.9 A circular plate of radius a is clamped at the boundary $r = a$ and is connected to a spring of spring constant k at $r = 0$. Determine the two values of k that will match the first two eigenfrequencies of $(0, n)$ modes of the kettledrum obtained in Exercise 5.8.

5.10 A circular plate of radius a is clamped at the boundary $r = a$. A particle of mass m is dropped from a height h exactly on the center of the plate. Determine the contact force between the particle and the plate as a function of time. When will the particle lose contact with the plate?

5.11 An annular plate of inner radius a and outer radius b is clamped at the boundary $r = b$, and clamped to a massless collar (at $r = a$) sliding without friction on a guide, as shown in Figure 5.12.

(a) Determine the eigenfrequencies and eigenfunctions of the system.

(b) If the collar is excited by a harmonic force $Q(t) = A \cos \Omega t$, determine the response of the plate.

5.12 Two square plates of side a are simply-supported in a horizontal position, with one vertically above the other. The centers of the two plates are connected by a spring of stiffness k. Determine the first two eigenfrequencies and eigenfunctions of the system.

5.13 Repeat Exercise 4.6 for a circular plate of radius a which is free at the boundary.

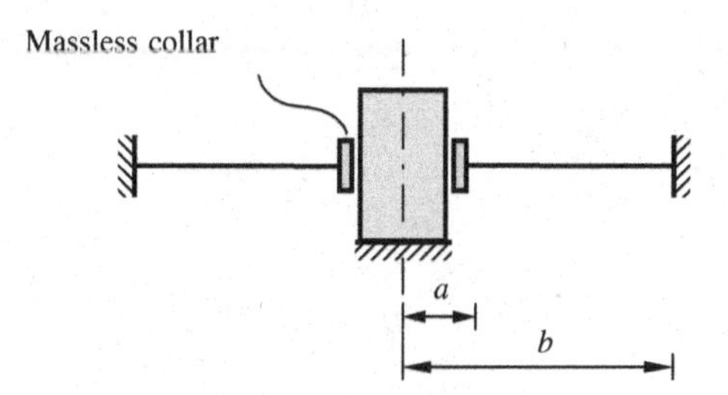

Figure 5.12 Exercise 5.11

5.14 A rectangular plate of length a and width b is simply-supported at the four edges. The plate is excited by a constant traveling point force $q(x, y, t) = Q_0\delta(x - vt)\delta(y - b/2)$, where Q_0 is the constant magnitude of the force and v is the speed of travel.

(a) Determine the response of the plate.

(b) At which values of v will the plate resonate?

5.15 A circular plate of radius a is clamped at the boundary. A constant point force is circulating on the plate, i.e., $q(r, \phi, t) = Q_0\delta(r - r_0)\delta(\phi - \Omega t)$, where Q_0 is the constant magnitude, r_0 is the radius of the circular path of the force, and Ω is the angular speed of circulation.

(a) Determine the response of the plate.

(b) At what values of Ω will the plate resonate?

5.16 A square plate of side a is simply-supported at the edges on a rigid frame. The frame is made to oscillate harmonically with a circular frequency Ω in the transverse direction along the normal to the plate. Determine the response of the plate.

5.17 For the system described in Exercise 5.16, determine the response if the frame is given harmonic angular oscillations about a center line parallel to an edge.

5.18 A circular plate of radius a is clamped at the boundary on a rigid circular frame. The frame is made to oscillate harmonically with a circular frequency Ω along the normal to the plate. Determine the response of the plate.

REFERENCES

[1] Leissa, A., *Vibration of Plates*, Acoustical Society of America, New York, 1993.

[2] Bhat, R.B., Natural Frequencies of Rectangular Plates Using Characteristic Orthogonal Polynomials in Rayleigh–Ritz Method, *J. of Sound and Vibration*, 102(4), 1985, pp. 493–499.

[3] Mindlin, R. D., Influence of Rotary Inertia and Shear on Flexural Motions of Isotropic Elastic Plates, *J. of Applied Mechanics, Trans. of ASME*, 73, 1951, pp. 31–38.

[4] Reissner, E.,The Effect of Transverse Shear Deformation on the Bending of Elastic Plates, *J. of Applied Mechanics, Trans. of ASME*, 67, 1945, A-69 to A-77.

[5] Timoshenko, S.P., and Woinowsky-Krieger, S., *Theory of Plates and Shells*, McGraw-Hill Book Co., New York, 1959.

6

Boundary value and eigenvalue problems in vibrations

This chapter summarizes the ideas related to the boundary value and eigenvalue problems presented in the previous chapters, and puts them in a general abstract framework. Such a framework is useful in unifying the concepts associated with the matrix eigenvalue problem for discrete systems and the operator eigenvalue problem for continuous systems. This naturally leads to the expansion theorem and its associated convergence properties. These ideas are then used to formulate some of the important approximate methods for discretizing continuous systems.

6.1 SELF-ADJOINT OPERATORS AND EIGENVALUE PROBLEMS FOR UNDAMPED FREE VIBRATIONS

6.1.1 General properties and expansion theorem

It is known from the elementary theory of vibrations that undamped free vibrations of a linear discrete system with n degrees of freedom are, in general, described by a system of n ordinary differential equations of the type

$$\mathbf{M}\ddot{\mathbf{q}} + \mathbf{K}\mathbf{q} = \mathbf{0}, \tag{6.1}$$

where $\mathbf{q}(t) = (q_1, q_2, \ldots, q_n)^{\mathrm{T}}$ is the vector of the generalized coordinates, and $\mathbf{M}$ and $\mathbf{K}$ are, respectively, the mass and stiffness matrices, both of which are *positive definite*. On the other hand, the dependence of the configuration of continuous systems on spatial coordinates and time brings in the concept of field variables. For example, in three-dimensional Cartesian coordinates, one usually represents the field variables as $u(x, y, z, t)$, $v(x, y, z, t)$, and $w(x, y, z, t)$ for displacements along the directions x, y, and z, respectively. It is then natural that the dynamics of continuous systems is governed by a set of partial differential equations.

Vibrations and Waves in Continuous Mechanical Systems P. Hagedorn and A. DasGupta

In the following, for simplicity of exposition, we first consider a continuum in which the displacements are only in one direction, say in the z-axis direction, and represented by $w(x, y, z, t)$. The equation of motion of such a continuum can be written in the form

$$\mathcal{M}[w_{,tt}(x, y, z, t)] + \mathcal{K}[w(x, y, z, t)] = 0, \qquad (x, y, z) \in \mathcal{G}, \tag{6.2}$$

where the domain $\mathcal{G}$ represents the (finite) spatial extent of the system, and $\mathcal{M}[\cdot]$ and $\mathcal{K}[\cdot]$ are now linear partial differential operators involving derivatives up to order $2q$ and $2p$, respectively, with respect to x, y, and z only (and not time). It is usually the case that $p > q$. In addition to the equation of motion (6.2) defined in the domain $\mathcal{G}$, we consider boundary conditions given by linear homogeneous partial differential equations

$$\mathcal{L}_i[w(x, y, z, t)] = 0, \qquad (x, y, z) \in \partial\mathcal{G}, \qquad i = 1, 2, \ldots, 2m, \tag{6.3}$$

where $\partial\mathcal{G}$ represents the boundary of the domain $\mathcal{G}$ and $\mathcal{L}_i[\cdot]$ are partial differential operators involving spatial derivatives of order up to $2p - 1$. The boundary conditions involving spatial derivatives up to order $p - 1$ are referred to as *geometric* or *essential boundary conditions*, while the others are known as *dynamic* or *natural boundary conditions*.

The boundary value problem defined by (6.2) and (6.3) in the case of continuous systems, replaces (6.1) for discrete systems. Almost all of the boundary value problems considered in the preceding chapters are of the type (6.2)–(6.3). In the present chapter, we summarize the essential features of this type of boundary value problems, and discuss some of the important numerical methods for solving them.

Consider a solution of (6.2) of the form

$$w(x, y, z, t) = W(x, y, z)e^{i\omega t}, \tag{6.4}$$

where ω is a constant. Substituting (6.4) in (6.2) and (6.3) yields

$$\mathcal{K}[W(x, y, z)] = \omega^2\mathcal{M}[W(x, y, z)], \qquad (x, y, z) \in \mathcal{G}, \tag{6.5}$$

$$\mathcal{L}_i[W(x, y, z)] = 0, \qquad (x, y, z) \in \partial\mathcal{G}, \qquad i = 1, 2, \ldots, 2m. \tag{6.6}$$

The set of equations (6.5)–(6.6) defines the *eigenvalue problem* for the system described by (6.2)–(6.3). For special values of ω, known as the *eigenvalues*, there exist non-trivial $W(x, y, z)$, known as *eigenfunctions*, that satisfy (6.5)–(6.6).

Functions of (x, y, z) defined in $\mathcal{G}$, which are sufficiently smooth (differentiable at least $2p$ times) so that they belong to the domain of definition of the operators $\mathcal{M}[\cdot]$ and $\mathcal{K}[\cdot]$, and which in addition satisfy all the boundary conditions (6.6), are called *comparison functions*. If, for any two comparison functions $W_1(x, y, z)$ and $W_2(x, y, z)$,

$$\int_{\mathcal{G}} \mathcal{M}[W_1]W_2\, d\mathcal{G} = \int_{\mathcal{G}} \mathcal{M}[W_2]W_1\, d\mathcal{G}, \tag{6.7}$$

then the operator $\mathcal{M}[\cdot]$ is termed *symmetric* or *self-adjoint*. If both the operators $\mathcal{M}[\cdot]$ and $\mathcal{K}[\cdot]$ are self-adjoint, then (6.5)–(6.6) is called a *self-adjoint eigenvalue problem*.

The equations of motion considered in the preceding chapters for free undamped vibrations almost always lead to self-adjoint eigenvalue problems (exceptions being the traveling string and beam, rotating shaft, and cantilever pipe conveying a fluid). For example, in the case of the transverse vibrations of the string

$$\mathcal{M}[\cdot] := \rho A(x)[\cdot] \qquad \text{and} \qquad \mathcal{K}[\cdot] := -\frac{\mathrm{d}}{\mathrm{d}x}\left[T(x)\frac{\mathrm{d}}{\mathrm{d}x}\right][\cdot].$$

For two comparison functions $W_1(x)$ and $W_2(x)$ defined over $\mathcal{G} = [0, l]$, one has

$$\begin{aligned}\int_0^l \mathcal{K}[W_1]W_2\,\mathrm{d}x &= -\int_0^l W_2\frac{\mathrm{d}}{\mathrm{d}x}\left[T(x)\frac{\mathrm{d}W_1}{\mathrm{d}x}\right]\mathrm{d}x \\ &= -\left[W_2T(x)\frac{\mathrm{d}W_1}{\mathrm{d}x}\right]_0^l + \int_0^l T(x)\frac{\mathrm{d}W_1}{\mathrm{d}x}\frac{\mathrm{d}W_2}{\mathrm{d}x}\,\mathrm{d}x.\end{aligned} \tag{6.8}$$

For a string with one fixed and one sliding end, the boundary conditions are $W(0) = 0$ and $T(l)W'(l) = 0$. This reduces to zero the boundary terms on the right-hand side of (6.8), so that only the integral, which is symmetric in W_1 and W_2, remains. Thus, the operator $\mathcal{K}[\cdot]$ is symmetric or self-adjoint. In this case, the operator $\mathcal{M}[\cdot]$ is also symmetric, independently of the boundary conditions, since

$$\int_0^l \mathcal{M}[W_1]W_2\,\mathrm{d}x = \int_0^l \rho A(x)W_1W_2\mathrm{d}x = \int_0^l \mathcal{M}[W_2]W_1\,\mathrm{d}x.$$

Hence, the problem of the free vibrations of the string with the boundary conditions under consideration leads to a self-adjoint eigenvalue problem. In a similar way, self-adjointness can be defined and verified for the problems of beam and plate vibrations. Intuitively, self-adjointness of the operators $\mathcal{M}[\cdot]$ and $\mathcal{K}[\cdot]$ is similar to symmetry of the matrices $\mathbf{M}$ and $\mathbf{K}$ in (6.1). This, therefore, may be understood as related to the conservation of energy and the absence of gyroscopic terms.

In the case of one-dimensional continua, the differential operators $\mathcal{M}[\cdot]$ and $\mathcal{K}[\cdot]$ are of the type

$$\mathcal{M}[W(x)] = \sum_{j=0}^{2q} a_j(x)W^{[j]}(x),$$

$$\mathcal{K}[W(x)] = \sum_{j=0}^{2p} c_j(x)W^{[j]}(x), \qquad p > q,$$

where the superscript $[j]$ indicates spatial differentiation of order j. A necessary and sufficient condition for self-adjointness of these operators is that they should be of the form

$$\mathcal{M}[W(x)] = \sum_{j=0}^{q}(-1)^j\left[f_j(x)W^{[j]}(x)\right]^{[j]}, \tag{6.9}$$

$$\mathcal{K}[W(x)] = \sum_{j=0}^{p} (-1)^j \left[g_j(x) W^{[j]}(x)\right]^{[j]}, \tag{6.10}$$

with appropriate boundary conditions.

A self-adjoint differential operator $\mathcal{M}[\cdot]$ is called *positive definite* if, for all comparison functions $W(x, y, z)$, there exists a constant $\gamma > 0$ such that the inequality

$$\int_{\mathcal{G}} \mathcal{M}[W]W \, \mathrm{d}\mathcal{G} \geq \gamma \|W\|^2$$

holds, where $\|W\|$ is a suitable norm. For example, a commonly used norm is defined by

$$\|W\|^2 := \int_{\mathcal{G}} W^2 \, \mathrm{d}\mathcal{G}. \tag{6.11}$$

The norm (6.11) is induced by the *scalar product* (or *inner product*) of comparison functions defined by

$$\langle W_1, W_2 \rangle := \int_{\mathcal{G}} W_1 W_2 \, \mathrm{d}\mathcal{G},$$

where W_1 and W_2 are any two comparison functions. With this scalar product, the infinite-dimensional linear vector space of the comparison functions is a *Hilbert space*. If both $\mathcal{M}[\cdot]$ and $\mathcal{K}[\cdot]$ are strictly positive definite, then the eigenvalue problem is also termed positive definite. Using positive definite operators $\mathcal{M}[\cdot]$ or $\mathcal{K}[\cdot]$, one may also define scalar products of the form (6.7), giving rise to different Hilbert spaces. For example, using the scalar products

$$\langle W_1, W_2 \rangle_{\mathcal{M}} := \int_{\mathcal{G}} \mathcal{M}[W_1] W_2 \, \mathrm{d}\mathcal{G} \quad \text{and} \quad \langle W_1, W_2 \rangle_{\mathcal{K}} := \int_{\mathcal{G}} \mathcal{K}[W_1] W_2 \, \mathrm{d}\mathcal{G}, \tag{6.12}$$

one can define the respective *energy norms* as

$$\|W\|_{\mathcal{M}}^2 = \int_{\mathcal{G}} \mathcal{M}[W]W \, \mathrm{d}\mathcal{G} \quad \text{and} \quad \|W\|_{\mathcal{K}}^2 = \int_{\mathcal{G}} \mathcal{K}[W]W \, \mathrm{d}\mathcal{G}. \tag{6.13}$$

All positive definite eigenvalue problems of the type (6.5)–(6.6) share the following important properties, stated in the following theorem:

(a) All eigenvalues of the problem (6.5)–(6.6) are positive, and they form an infinite sequence $0 < \omega_1^2 \leq \omega_2^2 \leq \omega_3^3 \leq \ldots$.

(b) Any two eigenfunctions $W_i(x, y, z)$ and $W_j(x, y, z)$, associated with two different eigenvalues, are orthogonal with respect to $\mathcal{M}[\cdot]$, and also with respect to $\mathcal{K}[\cdot]$, i.e., for $\omega_i^2 \neq \omega_j^2$,

$$\int_{\mathcal{G}} \mathcal{M}[W_i] W_j \, \mathrm{d}\mathcal{G} = 0 \quad \text{and} \quad \int_{\mathcal{G}} \mathcal{K}[W_i] W_j \, \mathrm{d}\mathcal{G} = 0. \tag{6.14}$$

(c) The eigenfunctions $W_1(x, y, z)$, $W_2(x, y, z)$, $W_3(x, y, z)$, ... form a basis of the function space of the comparison functions (a Hilbert space), i.e., any comparison function $W(x, y, z)$ has a unique representation of the type

$$W(x, y, z) = \sum_{j=1}^{\infty} \alpha_i W_i(x, y, z), \tag{6.15}$$

where α_i are appropriate constants. This statement, which appears to generalize Fourier series, is known as the *expansion theorem*.

The properties stated above are completely analogous to those known for matrix eigenvalue problems with symmetric positive definite matrices (such as for (6.1)). The expansion theorem, therefore, assures that the free vibrations for any set of initial conditions can always be represented as a superposition of the eigensolutions. The convergence of expansion (6.15) always holds with respect to the energy norms in (6.13), which correspond, respectively, to the scalar products in (6.12). However, in many cases there may also be uniform pointwise convergence. The types of convergence holding in any specific case will depend on the particular properties of the operators $\mathcal{M}[\cdot]$ and $\mathcal{K}[\cdot]$, and also on the dimension of the domain $\mathcal{G}$. Relatively strong convergence properties can, for example, be stated for one-dimensional continua. Details on the types of convergence and proof of the expansion theorem are given in the mathematical literature (see, for example, [1], [2], and [3]).

Up to now we have considered displacement fields comprising displacements in a single direction only, where the displacements were represented by a single displacement component $w(x, y, z, t)$. In the general case, the material points of a continuous system can experience displacements in an arbitrary direction, which may be described by the three field variables $u(x, y, z, t)$, $v(x, y, z, t)$, $w(x, y, z, t)$, where $(x, y, z) \in \mathcal{G}$. In this case, vector-valued expressions (u, v, w) and $(u_{,tt}, v_{,tt}, w_{,tt})$, replace the scalar differential expressions $\mathcal{M}[w_{,tt}]$, $\mathcal{K}[w]$, and $\mathcal{L}[w]$ in (6.5)–(6.6). Sometimes, the displacement components u, v and w are also expressed through other variables. For example, in the Timoshenko beam, by w and ψ. In all of these cases, properties analogous to those given for the expansion theorem hold. The orthogonality of the eigenfunctions has to be defined accordingly in each case. There are problems in which, because of the presence of time derivatives in the boundary conditions, eigenvalues also appear in the boundary conditions. For such problems, the orthogonality conditions can be formulated if the scalar products are now defined with boundary terms, in addition to the integral expressions (see, for example, Section 1.4.4).

We re-examine these concepts using the example of coupled axial-transverse vibrations of a beam. In studying beam vibrations, we considered only transverse displacements $w(x, t)$ of the points on the beam's axis, while for bar vibrations, only axial displacements $u(x, t)$ were considered. The differential equations for the linear axial and transverse vibrations, so far, were always uncoupled. Using the example shown in Figure 6.1, we will show that, strictly speaking, the uncoupling of the axial from the transverse vibrations also depends on the boundary conditions (a similar case also occurs in the vibrations of frames). Here, obviously bending and axial vibrations are coupled for $\alpha \neq k\pi/2$, $k = 0, 1, \ldots$. In what follows, we will apply the previously introduced concept of self-adjointness to this

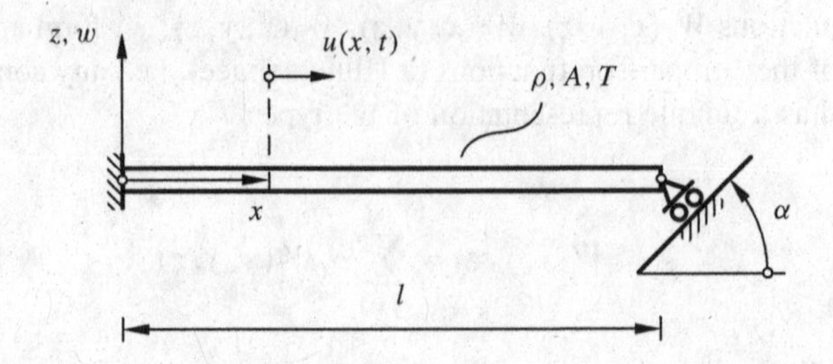

Figure 6.1 Coupling between axial and transverse vibrations of a beam through the boundary conditions

example, slightly generalizing the definition in the process. The equations of motion are

$$\rho A w_{,tt} + EI w_{,xxxx} = 0, \tag{6.16}$$

$$\rho A u_{,tt} - EA u_{,xx} = 0, \tag{6.17}$$

with the (geometric) boundary conditions at the left end as

$$w(0,t) \equiv 0, \qquad w_{,x}(0,t) \equiv 0, \qquad \text{and} \qquad u(0,t) \equiv 0. \tag{6.18}$$

At the right-hand end, one obviously has

$$w_{,xx}(l,t) \equiv 0. \tag{6.19}$$

The other boundary conditions at the right-hand end will soon be discussed. Separation of variables according to

$$w(x,t) = W(x)\mathrm{e}^{i\omega t} \qquad \text{and} \qquad u(x,t) = U(x)\mathrm{e}^{i\omega t},$$

leads to the eigenvalue problem

$$EIW'''' - \omega^2 \rho A W = 0, \tag{6.20}$$

$$-EAU'' - \omega^2 \rho A U = 0, \tag{6.21}$$

along with the so far defined boundary conditions

$$W(0) = 0, \qquad W'(0) = 0, \qquad W''(l) = 0, \qquad \text{and} \qquad U(0) = 0. \tag{6.22}$$

With the definition

$$\mathbf{e}(x) := \begin{Bmatrix} W(x) \\ U(x) \end{Bmatrix}, \tag{6.23}$$

one can write (6.20)–(6.21) as

$$\mathcal{K}[\mathbf{e}] - \omega^2 \mathcal{M}[\mathbf{e}] = 0, \tag{6.24}$$

where the operators $\mathcal{K}[\cdot]$ and $\mathcal{M}[\cdot]$ are defined as

$$\mathcal{K} := \begin{bmatrix} EI\dfrac{\mathrm{d}^4}{\mathrm{d}x^4} & 0 \\ 0 & -EA\dfrac{\mathrm{d}^2}{\mathrm{d}x^2} \end{bmatrix} \quad \text{and} \quad \mathcal{M} := \begin{bmatrix} \rho A & 0 \\ 0 & \rho A \end{bmatrix}. \tag{6.25}$$

The operator $\mathcal{K}[\cdot]$ is self-adjoint if and only if

$$\int_{\mathcal{G}} \mathbf{e}_1^{\mathrm{T}} \mathcal{K}[\mathbf{e}_2]\, \mathrm{d}\mathcal{G} = \int_{\mathcal{G}} \mathbf{e}_2^{\mathrm{T}} \mathcal{K}[\mathbf{e}_1]\, \mathrm{d}\mathcal{G},$$

for arbitrary $\mathbf{e}_1(x)$ and $\mathbf{e}_2(x)$ satisfying all the boundary conditions. We have

$$\begin{aligned} \int_{\mathcal{G}} \mathbf{e}_2^{\mathrm{T}} \mathcal{K}[\mathbf{e}_1]\, \mathrm{d}\mathcal{G} &= \int_0^l EI\, W_1'''' \, W_2 \, \mathrm{d}x - \int_0^l EA\, U_1'' \, U_2 \, \mathrm{d}x \\ &= EI\, W_1''' \, W_2 \big|_0^l - EI\, W_1'' \, W_2' \big|_0^l - EA\, U_1' U_2 \big|_0^l \\ &\quad + \int_0^l EI\, W_1'' \, W_2'' \, \mathrm{d}x + \int_0^l EA\, U_1' U_2' \mathrm{d}x. \end{aligned} \tag{6.26}$$

The integrals in the last line of (6.26) are obviously symmetric in $\mathbf{e}_1$ and $\mathbf{e}_2$. Therefore, the self-adjointness of $\mathcal{K}[\cdot]$ is assured, provided the sum of all the boundary terms in (6.26) vanishes. Using the boundary conditions (6.22) in the boundary terms in (6.26) leads to the condition

$$\begin{aligned} & EI W_1'''(l) W_2(l) - EA U_1'(l) U_2(l) = 0 \\ \Rightarrow\ & \frac{EA U_1'(l)}{EI W_1'''(l)} = \frac{W_2(l)}{U_2(l)}. \end{aligned} \tag{6.27}$$

Thus, (6.27) must hold for $\mathcal{K}[\cdot]$ to be self-adjoint. Since the left-hand side of (6.27) depends only on $\mathbf{e}_1$, and the right-hand side depends only on $\mathbf{e}_2$, equality of the two sides for all comparison functions of the problem implies that both quotients must be constant. At the right end of the beam in Figure 6.1, one can easily relate the displacements in the transverse and axial directions as

$$\begin{aligned} \tan\alpha &= \frac{W_2(l)}{U_2(l)} \\ \Rightarrow W(l) &- \tan\alpha U_2(l) = 0. \end{aligned} \tag{6.28}$$

Using (6.28) in (6.27) and rearranging, one obtains

$$EAU_1'(l) - EIW_1'''(l) = 0. \tag{6.29}$$

It it thus evident that the boundary conditions (6.28)–(6.29) must also hold for self-adjointness of the operator $\mathcal{K}[\cdot]$. The boundary condition (6.28) obviously is a geometric one. On the other hand, the boundary condition (6.29) is a natural one, and is equivalent to saying that the reaction of the support at the right end is orthogonal to the inclined plane supporting the right end. It is easily observed that the components of the reaction from the right support correspond to the shear force $Q = EIW'''(l)$ in the transverse direction and the normal force $N = EAU'(l)$ in the axial direction of the beam.

Let us briefly also consider the case of the beam in Figure 6.1 being pinned instead of clamped at the left end. Then, the immediately recognizable boundary conditions are

$$W(0) = 0, \qquad W''(0) = 0, \qquad U(0) = 0, \qquad \text{and} \qquad W''(l) = 0. \tag{6.30}$$

Also, with these boundary conditions (6.30), it follows from (6.26) and (6.27) that the two boundary conditions (6.28)–(6.29) must also hold.

We have so far discussed the self-adjointness of the operator $\mathcal{K}[\cdot]$ exclusively. The self-adjointness of $\mathcal{M}[\cdot]$ being obvious, the eigenvalue problem is self-adjoint. In the following, we study the eigenvalue problem for the second of the cases discussed above (i.e., with pinned support at the left boundary).

The general solution of (6.20)–(6.21) is given by

$$W(x) = C_1 \sin\beta x + C_2 \cos\beta x + C_3 \sinh\beta x + C_4 \cosh\beta x, \tag{6.31}$$

$$U(x) = D_1 \sin\gamma x + D_2 \cos\gamma x, \tag{6.32}$$

where $\beta^4 := \omega^2\rho A/EI$ and $\gamma^2 := \omega^2\rho/E$. The boundary conditions $W(0) = 0$, $W''(0) = 0$, and $W''(l) = 0$ yield $C_2 = 0$, $C_4 = 0$, and $C_3 = \sin\beta l/\sinh\beta l$ (for $\sinh\beta l \neq 0$), so that

$$W(x) = C_1\left[\sin\beta x + \frac{\sin\beta l}{\sinh\beta l}\sinh\beta x\right]. \tag{6.33}$$

Further, $U(0) = 0$ gives $D_2 = 0$, i.e.,

$$U(x) = D_1 \sin\gamma x. \tag{6.34}$$

Using the remaining boundary conditions (6.28)–(6.29) leads to

$$(2\sin\beta l)C_1 + (\tan\alpha \sin\gamma l)D_1 = 0,$$

$$-EI\beta^3 \tan\alpha\left[-\cos\beta l + \frac{\sin\beta l}{\sinh\beta l}\cosh\beta l\right]C_1 - (\gamma EA\cos\gamma l)D_1 = 0.$$

Finally, the condition for non-triviality of solutions of $(C_1, D_1)^{\mathrm{T}}$ gives the characteristic equation

$$-2\gamma EA\cos\gamma l + EI\beta^3\tan^2\alpha\sin\gamma l(-\cot\beta l + \coth\beta l) = 0. \tag{6.35}$$

With the definition $r_{\mathrm{g}}^2 := I/A$, one can write $\gamma = r_{\mathrm{g}}\beta^2$, so that (6.35) for $\beta \neq 0$, can be written as

$$2 + \frac{\beta l}{s_{\mathrm{r}}}\tan^2\alpha\tan\frac{\beta^2 l^2}{s_{\mathrm{r}}}[\cot\beta l - \coth\beta l] = 0, \tag{6.36}$$

where $s_{\mathrm{r}} := l/r_{\mathrm{g}}$ is the slenderness ratio.

The bending vibrations and the axial vibrations are naturally completely decoupled for $\alpha = 0$ and $\alpha = \pi/2$, and the eigenvalues and eigenfunctions can easily be derived for these two particular cases. For values of α in between these two bounds, the normal modes will be such that the beam will oscillate simultaneously in transverse and axial directions, and the eigenvalues have to be obtained numerically from (6.36). The first few eigenvalues (in terms of βl) as functions of α are shown in Figure 6.2 for a beam with $s_{\mathrm{r}} = 30$. It can be observed that the eigenfrequency corresponding to the first bending mode B1 at $\alpha = 0$ (i.e., uncoupled from the axial vibrations) goes over to zero when $\alpha = \pi/2$. This zero eigenfrequency at $\alpha = \pi/2$ corresponds to the rigid-body rotational motion of the beam. The eigenfrequency corresponding to the first longitudinal mode of vibration L1 at $\alpha = 0$, on the other hand, goes over to the eigenfrequency corresponding to the third bending mode B3 (if the rigid body mode is counted) at $\alpha = \pi/2$, and so on. The eigenfunctions are, of course, easy to determine once the eigenvalues have been found from (6.36). The eigenfunctions satisfy orthogonality conditions of the form

$$\int_0^l EIW_j''W_k''\,\mathrm{d}x + \int_0^l EAU_j'U_k'\,\mathrm{d}x = 0, \qquad \text{for } \omega_j^2 \neq \omega_k^2.$$

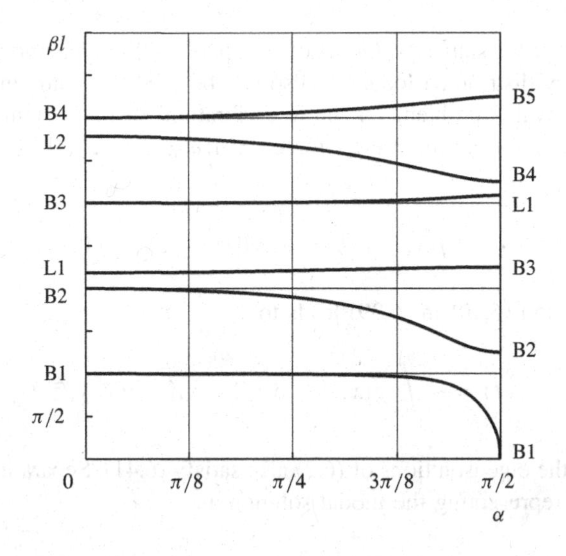

Figure 6.2 Eigenvalues for the system of Figure 6.1 with simply-supported left boundary

6.1.2 Green's functions and integral formulation of eigenvalue problems

Under external excitation, the equation of motion (6.2), together with an inhomogeneity term $q(x, y, z, t)$, takes the form

$$\mathcal{M}[w_{,tt}(x, y, z, t)] + \mathcal{K}[w(x, y, z, t)] = q(x, y, z, t), \qquad (x, y, z) \in \mathcal{G}, \tag{6.37}$$

along with the corresponding boundary conditions, where $q(x, y, z, t)$ represents the distributed excitation force. In the particular case of statics, the functions $w(\cdot)$ and $q(\cdot)$ do not depend on time, so that (6.37) is reduced to

$$\mathcal{K}[w(x, y, z)] = q(x, y, z), \qquad (x, y, z) \in \mathcal{G}. \tag{6.38}$$

The *influence function*, or Green's function, $g(x, y, z; \overline{x}, \overline{y}, \overline{z})$ for the static problem is defined as the solution of the boundary value problem (6.38) with

$$q(x, y, z) := \delta(x - \overline{x}, y - \overline{y}, z - \overline{z}).$$

Green's function represents the deflection of the continuum at a location (x, y, z) due to a concentrated unit force at a location $(\overline{x}, \overline{y}, \overline{z})$, and has the well-known symmetry properties.

The statics problem (6.38) for an arbitrary $q(x, y, z, t)$ can now be solved using Green's function $g(\cdot; \cdot)$ for the statics problem. Due to linearity of the problem, the displacement field $w(x, y, z)$, caused by an arbitrary distributed load $q(x, y, z)$, is given by an integral (superposition of infinitely many solutions)

$$w(x, y, z) = \int_{\mathcal{G}} g(x, y, z; \overline{x}, \overline{y}, \overline{z}) q(\overline{x}, \overline{y}, \overline{z})\, \mathrm{d}\overline{x}\, \mathrm{d}\overline{y}\, \mathrm{d}\overline{z}. \tag{6.39}$$

Green's function for the statics problem not only permits the calculation of the displacements due to an arbitrary distributed load, but also can be used to obtain an interesting reformulation of the eigenvalue problem for the free vibrations of the system. Using d'Alembert's formulation, one may treat the inertia force $-\mathcal{M}[w_{,tt}(x, y, z, t)]$ as a distributed external force, i.e.,

$$q(x, y, z, t) = -\mathcal{M}[w_{,tt}(x, y, z, t)]. \tag{6.40}$$

Using the expression (6.40) in (6.39) leads to

$$w(x, y, z, t) = -\int_{\mathcal{G}} g(x, y, z; \overline{x}, \overline{y}, \overline{z}) \mathcal{M}[w_{,tt}(\overline{x}, \overline{y}, \overline{z}, t)]\, \mathrm{d}\overline{x}\, \mathrm{d}\overline{y}\, \mathrm{d}\overline{z}. \tag{6.41}$$

In particular, all the eigensolutions of (6.2) also satisfy (6.41). Separating the time and space coordinates, and representing the modal solution as

$$w(x, y, z, t) = W(x, y, z) \mathrm{e}^{i\omega t},$$

one obtains from (6.41) the integral equation for the eigenfunctions $W(x, y, z)$ as

$$W(x, y, z) = \omega^2 \int_{\mathcal{G}} g(x, y, z; \overline{x}, \overline{y}, \overline{z}) \mathcal{M}[W(\overline{x}, \overline{y}, \overline{z})] \, d\overline{x} \, d\overline{y} \, d\overline{z}, \tag{6.42}$$

which contains ω^2 as a parameter.

Frequently, the operator $\mathcal{M}[\cdot]$ simply consists of the identity operator with a coefficient $\mu(\cdot)$. In such cases, (6.42) simplifies to

$$W(x, y, z) = \omega^2 \int_{\mathcal{G}} \mu(\overline{x}, \overline{y}, \overline{z}) g(x, y, z; \overline{x}, \overline{y}, \overline{z}) W(\overline{x}, \overline{y}, \overline{z}) \, d\overline{x} \, d\overline{y} \, d\overline{z}. \tag{6.43}$$

The formulation of the eigenvalue problem of the free undamped vibrations based on the integral equation (6.43) is completely equivalent to the previous formulation through the partial differential equations (6.5), and the corresponding boundary conditions (6.6). It is easier to prove the expansion theorem and the related properties given in Section 6.1.1 using the integral equation formulation rather than the partial differential equation formulation. An exposition on eigenvalue theory for symmetric integral equations can be found, for example, in [4].

The integral equation (6.43) is, however, not only of interest in the theory of boundary value/eigenvalue problems, but also is of immediate use in obtaining bounds for the eigenvalues. Similarly to the matrix iteration method for matrix eigenvalue problems (see, for example, [5]), here also an iterative scheme using the definition

$$W^{(n+1)}(P) := \int_{\mathcal{G}} \mu(\overline{P}) g(P; \overline{P}) W^{(n)}(\overline{P}) \, d\overline{P} \tag{6.44}$$

can be formulated to approximate the eigenfunctions and eigenvalues. In (6.44) one has $P = (x, y, z)$ and $\overline{P} = (\overline{x}, \overline{y}, \overline{z})$. As in the discrete case, here also it is convenient to normalize the functions, for example, by replacing (6.44) by

$$W^{(n+1)}(P) := \frac{\int_{\mathcal{G}} \mu(\overline{P}) g(P; \overline{P}) W^{(n)}(\overline{P}) \, d\overline{P}}{\| \int_{\mathcal{G}} \mu(\overline{P}) g(P; \overline{P}) W^{(n)}(\overline{P}) \, d\overline{P} \|} \tag{6.45}$$

It is, however, not necessary to carry out this normalization in each iteration step. Often it suffices to rescale the functions after, say, every third or fourth iteration step.

As $n \to \infty$ in (6.44), the function $W^{(n)}$ converges to the first eigenfunction W_1 under very mild conditions, and ω_1^2 is then given by

$$\omega_1^2 = \lim_{n \to \infty} \frac{\| W^{(n)} \|}{\| W^{(n+1)} \|} = \lim_{n \to \infty} \frac{W^{(n)}}{W^{(n+1)}}, \tag{6.46}$$

where the quotient $W^{(n)} / W^{(n+1)}$ no longer depends on (x, y, z) as $n \to \infty$. Not only the first eigenpair, but also higher-order eigenpairs can be found using this type of iteration, if use is made of the orthogonality conditions, in complete analogy to the discrete eigenvalue problems (see, for example, [5]).

As an example, consider the eigenvalue problem of a fixed–fixed string, which was solved in Chapter 1 using the partial differential equation formulation. Green's function for the static problem in this case is the solution of

$$TW''(x) = -\delta(x - \overline{x}), \qquad W(0) = 0, \qquad \text{and} \qquad W(l) = 0. \tag{6.47}$$

Green's function therefore is

$$g(x, \overline{x}) = \begin{cases} \dfrac{x(l - \overline{x})}{Tl}, & 0 \le x < \overline{x} \le l, \\[2ex] \dfrac{\overline{x}(l - x)}{Tl}, & 0 \le \overline{x} < x \le l, \end{cases} \tag{6.48}$$

as can easily be checked. Choosing the starting function for the iteration (6.44) as

$$W^{(1)}(x) = \frac{x}{l},$$

one obtains from (6.44)

$$W^{(2)}(x) = \frac{\mu}{Tl^2}\left\{\int_0^x \xi[l - x]\xi \, d\xi + \int_x^l x[l - \xi]\xi \, d\xi\right\} = \frac{\mu l^2}{6T}\left[\frac{x}{l} - \frac{x^3}{l^3}\right], \tag{6.49}$$

$$W^{(3)}(x) = \mu \int_0^l g(x; \xi)\, W^{(2)}(\xi)\, d\xi = \frac{\mu l^2}{6T}\frac{7\mu l^2}{60T}\left[\frac{x}{l} - \frac{10}{7}\frac{x^3}{l^3} + \frac{3}{7}\frac{x^5}{l^5}\right], \tag{6.50}$$

and so on. It is interesting to note that while $W^{(1)}(x)$ satisfies only one of the boundary conditions, $W^{(2)}$ satisfies both boundary conditions. Not only does the integration (6.44) smooth the function, but also, through the kernel $g(x; \overline{x})$, assures that the functions $W^{(n)}$ for $n > 1$ satisfy all the boundary conditions of the problem. We have omitted here the normalization indicated in (6.45).

The quotient $\|W^{(2)}\|/\|W^{(3)}\|$ formed using (6.49) and (6.50) can easily be computed in the present case using the norm definition (6.11) as

$$\|W^{(j)}\| = \sqrt{\int_0^l \left[W^{(j)}(x)\right]^2 dx}, \qquad j = 2, 3.$$

Thus, one obtains the approximation

$$\omega_1^2 \approx \frac{\|W^{(2)}\|}{\|W^{(3)}\|} = (3.1543)^2 \frac{T}{\mu l^2},$$

which agrees very well with the exact value $\omega_1^2 = \pi^2 T/(\mu l^2)$.

While it is possible to give a closed-form expression of Green's function for the wave equation with different boundary conditions, or for a beam with constant cross-section, this is usually no longer the case for more complex problems. The integrals then have to be computed numerically, or a representation of Green's function has to be sought in the form of function series. In that case, the utility of the above-discussed iterative method for the determination of eigenfunctions is greatly reduced.

6.1.3 Bounds for eigenvalues: Rayleigh's quotient and other methods

The most important methods for bracketing the eigenvalues of self-adjoint eigenvalue problems are related to Rayleigh's quotient and to the Ritz method, both of which have already been dealt with in the previous chapters. The essential relations will be summarized here once more.

The equations of motion (6.2) with the self-adjoint operators $\mathcal{M}[\cdot]$ and $\mathcal{K}[\cdot]$, and the corresponding boundary conditions can always be derived from Hamilton's principle

$$\delta \int_{t_1}^{t_2} \left(\mathcal{T}[w_{,t}] - \mathcal{V}[w]\right) \mathrm{d}t = 0, \tag{6.51}$$

where $\mathcal{T}$ and $\mathcal{V}$ are, respectively, the kinetic and potential energy expressions for the system. In the integrand in (6.51), the kinetic energy $\mathcal{T}[w_{,t}]$ is a quadratic functional of the function $w_{,t}(x, y, z, t)$ and the potential energy $\mathcal{V}[w]$ is a quadratic functional of $w(x, y, z, t)$. Both these energy functionals contain derivatives with respect to the space coordinates only, and not with respect to time. For comparison functions (i.e., functions which are differentiable a sufficient number of times, and satisfy all the boundary conditions of the problem) the relations

$$\mathcal{T}[w_{,t}] = \frac{1}{2}\int_{\mathcal{G}} \mathcal{M}[w_{,t}]w_{,t}\,\mathrm{d}\mathcal{G} \qquad \text{and} \qquad \mathcal{V}[w] = \frac{1}{2}\int_{\mathcal{G}} \mathcal{K}[w]w\,\mathrm{d}\mathcal{G} \tag{6.52}$$

hold between the energy functionals and the differential operators $\mathcal{M}[\cdot]$ and $\mathcal{V}[\cdot]$. While the operators $\mathcal{M}[\cdot]$ and $\mathcal{K}[\cdot]$, respectively, contain spatial derivatives of the order $2q$ and $2p$, the energy functionals $\mathcal{T}[\cdot]$ and $\mathcal{V}[\cdot]$ are defined for functions which are, respectively, at least q- and p-times differentiable (with respect to the space coordinates). For example, in the longitudinal vibrations of a bar, the energy expressions are given by

$$\mathcal{T}[u_{,t}] = \frac{1}{2}\int_0^l \rho A u_{,t}^2\,\mathrm{d}x \qquad \text{and} \qquad \mathcal{V}[u] = \frac{1}{2}\int_0^l EAu_{,x}^2\,\mathrm{d}x, \tag{6.53}$$

while the differential operators in the equation of motion are given by

$$\mathcal{M}[u_{,tt}] = \rho A u_{,tt} \qquad \text{and} \qquad \mathcal{K}[u] = [EAu_{,x}]_{,x}.$$

The energy expressions can, therefore, be computed even using functions for which the operators $\mathcal{M}[\cdot]$ and $\mathcal{K}[\cdot]$ are not defined. In other words, there exists a larger class of functions for which the energy functionals $\mathcal{T}[\cdot]$ and $\mathcal{V}[\cdot]$ can be computed.

Consider the eigenvalue problem

$$-\omega^2\mathcal{M}[W] + \mathcal{K}[W] = 0, \qquad (x, y, z) \in \mathcal{G}, \tag{6.54}$$

$$\mathcal{L}_j[W] = 0, \qquad j = 1, 2, \ldots, 2m, \qquad (x, y, z) \in \partial\mathcal{G}. \tag{6.55}$$

Rayleigh's quotient for this problem can now be defined using the energy expressions as

$$\mathcal{R}[W] := \frac{\mathcal{V}[W]}{\mathcal{T}[W]}, \tag{6.56}$$

where $W(x)$ is any admissible function, i.e., a function which is at least p-times differentiable and satisfies all the geometric boundary conditions of the problem. The energy expressions in (6.56) can also be calculated through the scalar products (6.52), however, using only comparison functions.

Analogous to the discrete case, Rayleigh's principle for continuous positive definite eigenvalue problems is expressed as

$$\omega_1^2 = \min_{W,\, \|W\|=1} \mathcal{R}[W]. \tag{6.57}$$

Due to the orthogonality relations, the kth eigenvalue can also be characterized recursively as

$$\omega_k^2 = \min_{\substack{W \\ \langle\mathcal{M}[W],\, W_1\rangle = 0, \\ \langle\mathcal{M}[W],\, W_2\rangle = 0, \\ \vdots \\ \langle\mathcal{M}[W],\, W_{k-1}\rangle = 0,}} \mathcal{R}[W], \tag{6.58}$$

where W_j, $j = 1, \ldots, (k-1)$ are the first $(k-1)$ eigenfunctions. The minimization in (6.58) is carried out over all admissible functions. Similarly as in the discrete case, the maximum–minimum characterization also holds for the eigenvalues. For example, the second eigenvalue is of the type

$$\omega_2^2 = \max_F \left\{ \min_{\substack{W \\ \langle\mathcal{M}[W],\, F\rangle = 0}} \mathcal{R}[W] \right\}, \tag{6.59}$$

where $F(\cdot)$ and $W(\cdot)$ are admissible functions. For the kth eigenvalue, one has correspondingly

$$\omega_k^2 = \max_{F_1, F_2, \ldots, F_{k-1}} \left\{ \min_{\substack{W \\ \langle\mathcal{M}[W],\, F_1\rangle = 0, \\ \langle\mathcal{M}[W],\, F_2\rangle = 0, \\ \vdots \\ \langle\mathcal{M}[W],\, F_{k-1}\rangle = 0.}} \mathcal{R}[W]. \right\} \tag{6.60}$$

The proofs are analogous to those for the case of matrix eigenvalue problems (see, for example, [5]).

Finding an upper bound for the first eigenvalue from (6.57) is particularly simple. It suffices to compute Rayleigh's quotient for any admissible function, and this always gives an upper bound. A rather crude guess for the first eigenfunction in this manner often leads to a relatively close bound for ω_1^2. Examples illustrating this point were given in previous chapters (see, for example, Section 1.7.1).

In general, it is far more difficult to find *lower* bounds on the first eigenvalue, and Rayleigh's quotient can also be useful in solving this problem. This can be done in the following way. Let an eigenvalue problem be defined by the energy functionals $\mathcal{T}[W]$ and $\mathcal{V}[W]$, and the geometric boundary conditions $\mathcal{L}_j[W] = 0$, $j = 1, 2, \ldots, s$. Let a different (but related) eigenvalue problem be given by $\overline{\mathcal{T}}[\overline{W}]$, $\overline{\mathcal{V}}[\overline{W}]$, and $\overline{\mathcal{L}}_j[\overline{W}] = 0$, $j = 1, 2, \ldots, \overline{s}$, with $\overline{s} \leq s$. This second problem should be defined such that the class $\mathscr{Z}$ of the admissible functions for the original problem is contained in $\overline{\mathscr{Z}}$, the class of admissible functions for the second problem (the orders of spatial derivatives in $\mathcal{T}$ and $\overline{\mathcal{T}}$ and $\mathcal{V}$ and $\overline{\mathcal{V}}$ should be equal). If in addition

$$\overline{\mathcal{T}}[\overline{W}] \geq \mathcal{T}[\overline{W}] \qquad \text{and} \qquad \overline{\mathcal{V}}[\overline{W}] \leq \mathcal{V}[\overline{W}]$$

holds for all $\overline{W} \in \overline{\mathscr{Z}}$, then one has

$$\overline{\omega}_1^2 = \min_{\overline{W} \in \overline{\mathscr{Z}}} \overline{\mathcal{R}}[\overline{W}] \leq \min_{W \in \mathscr{Z}} \overline{\mathcal{R}}[W] \leq \min_{W \in \mathscr{Z}} \mathcal{R}[W] = \omega_1^2. \tag{6.61}$$

If for a given eigenvalue problem one succeeds in finding a related problem with the properties defined above, and for which the exact solution is known, then, according to (6.61), a known $\overline{\omega}_1^2$ gives a lower bound for ω_1^2. In addition, in a similar way, it is also possible to find bounds according to (6.61) for the higher-order eigenvalues. This procedure is not explained in detail here, and the reader is referred to the specialized literature (see, for example, [6] and [7]). Finally, we point out that, in this manner, Weinstein solved the eigenvalue problem of a completely clamped vibrating rectangular plate, using not just a single auxiliary problem, but a whole sequence of intermediate problems. By taking the limits for this sequence of intermediate problems he was able to construct a converging infinite sequence of upper and lower bounds for the original problem. This method is known as *Weinstein's method of intermediate problems*.

A different possibility of finding bounds for the eigenvalues is given by the inclusion theorem, which is known from matrix eigenvalue problems in the vibrations of discrete systems (see, for example, [5]). A simple inclusion theorem can also be formulated for the vibrations of continuous systems described by self-adjoint operators of the type (6.9)–(6.10). In (6.9) with $q = 0$, we assume that the operator $\mathcal{M}[\cdot]$ is of the type

$$\mathcal{M}[W(\cdot)] = \mu(\cdot)W(\cdot), \tag{6.62}$$

where $\mu(\cdot) > 0$. This is naturally an important case for most practical applications where $\mu(\cdot)$ has the meaning of a mass density. The condition (6.62) on the operator $\mathcal{M}[\cdot]$ corresponds to a diagonal mass matrix in the matrix eigenvalue problem. The inclusion theorem for this type of problems can be stated in the following form. Let $W(x, y, z)$ be any comparison function with the property that the function

$$f(x, y, z) := \frac{\mathcal{K}[W]}{\mathcal{M}[W]} \tag{6.63}$$

assumes values which are within finite bounds for all functions in its domain of definition $\mathcal{G}$, and that it does not change sign. Then the interval $[\min_{\mathcal{G}} f(x, y, z), \max_{\mathcal{G}} f(x, y, z)]$ contains at least one eigenvalue, i.e.,

$$\min_{\mathcal{G}} f(x, y, z) \leq \omega_s^2 \leq \max_{\mathcal{G}} f(x, y, z). \tag{6.64}$$

Normally the bounds given by (6.64) for ω_s^2 are particularly narrow if the comparison function is chosen such that the function

$$\tilde{W}(\cdot) := \frac{\mathcal{K}[W]}{\mu(\cdot)}$$

satisfies as many of the boundary conditions as possible. This inclusion theorem proved by Collatz (see [1]) can be generalized further in several different ways.

Using the inclusion theorem in conjunction with the iteration procedure (6.44) can be particularly convenient. This is because Green's function in the integral automatically assures that $W^{(n+1)}$ satisfies all the boundary conditions as long as $W^{(n)}$ is sufficiently smooth. One can therefore compute $W^{(n+1)}$ from an arbitrarily chosen $W^{(n)}$ according to (6.44), and then replace the right-hand side of (6.63) simply by the ratio $W^{(n)}/W^{(n+1)}$, i.e.,

$$f(x, y, z) := \frac{W^{(n)}}{W^{(n+1)}} \tag{6.65}$$

Note that in this case the normalization as in (6.45) cannot be carried out, since then the extreme values of $f(\cdot)$ no longer furnish the correct bounds for the eigenvalues. Similarly as is known from matrix iteration, in the present case also, one could normally iterate repeatedly according to (6.44), and only normalize once in a few steps whenever required for numerical accuracy.

The formal relation between (6.63) and (6.65) can be easily recognized. The integration of a comparison function $W(\cdot)$ with Green's function $g(\cdot; \cdot)$ as a kernel can be understood as the inversion of the differential operator $\mathcal{K}[\cdot]$. Thus, the inverse operator $\mathcal{K}^{-1}[\cdot]$ can be defined as

$$\mathcal{K}^{-1}[W(P)] := \int_{\mathcal{G}} g(P; \overline{P}) W(\overline{P})\, \mathrm{d}\overline{P},$$

such that (6.44) can be written as $W^{(n+1)} = \mathcal{K}^{-1}[\mu W^{(n)}]$. With $W = W^{(n+1)}$, the definition (6.63) for the function $f(\cdot)$ leads to

$$f(x, y, z) = \frac{\mathcal{K}[W^{(n+1)}]}{\mu W^{(n+1)}} = \frac{\mathcal{K}[\mathcal{K}^{-1}[\mu W^{(n)}]]}{\mu W^{(n+1)}} = \frac{W^{(n)}}{W^{(n+1)}},$$

which coincides with (6.65).

The methods of Southwell and Dunkerley (see, for example, [5]) can also be generalized in an obvious way for the case of elastic continua.

6.2 FORCED VIBRATIONS

6.2.1 Equations of motion

Forced vibrations of a damped linear discrete system with n degrees of freedom are, in general, described by a system of ordinary differential equations of the type

$$\mathbf{M}\ddot{\mathbf{q}} + \mathbf{D}\dot{\mathbf{q}} + \mathbf{G}\dot{\mathbf{q}} + \mathbf{K}\mathbf{q} + \mathbf{N}\mathbf{q} = \mathbf{f}(t) \tag{6.66}$$

(see, for example, [5]). Here, as in (6.1), $\mathbf{q}(t) = (q_1, q_2, \ldots, q_n)^{\mathrm{T}}$ is the vector of the generalized coordinates. The matrix $\mathbf{M}$, which is associated with the kinetic energy, is symmetric and positive definite, i.e., $\mathbf{M} = \mathbf{M}^{\mathrm{T}} > 0$. Further, the matrix $\mathbf{K}$, which corresponds to the linear restoring forces and is generated by a potential, will be assumed to be positive definite, i.e., $\mathbf{K} = \mathbf{K}^{\mathrm{T}} > 0$. In addition, in (6.66), there are forces linear in $\dot{\mathbf{q}}$, which can be grouped in two types. The matrix $\mathbf{D} = \mathbf{D}^{\mathrm{T}}$ corresponds to damping (if $\mathbf{D} \geq 0$) and $\mathbf{G} = -\mathbf{G}^{\mathrm{T}}$ to gyroscopic forces. In some engineering systems *negative damping* occurs, which is then referred to as *self-excitation*. In that case the matrix $\mathbf{D}$ is no longer positive semi-definite, and the quadratic form $\dot{\mathbf{q}}^{\mathrm{T}}\mathbf{D}\dot{\mathbf{q}}$ then assumes negative values, at least for some $\dot{\mathbf{q}}$. A simple example can be given from rotor dynamics, where the forces acting on the shaft due to hydrodynamic bearings can generate such terms. The gyroscopic terms in (6.66) are also present in rotor systems, in particular when the vibrations are described with respect to non-inertial, rotating coordinate systems (see, for example, Section 3.8.4).

In addition to the restoring forces $\mathbf{K}\mathbf{q}$, which correspond to the potential energy expression $(1/2)\mathbf{q}^{\mathrm{T}}\mathbf{K}\mathbf{q}$, sometimes there are also forces which in the first approximation are linear in $\mathbf{q}$ but non-conservative. This type of force, for example, is observed in cantilever pipes conveying fluids (see Section 3.8.6), and can also be present in hydrodynamic bearings. In this case the forces linear in $\mathbf{q}$ can be split up into two parts. One part corresponds to the symmetric matrix $\mathbf{K}$, and the other part is given by the skew-symmetric matrix $\mathbf{N} = -\mathbf{N}^{\mathrm{T}}$, which also may give rise to self-excitation. The non-conservative forces given by the matrix $\mathbf{N}$ are also termed *circulatory forces*. Finally, the inhomogeneity $\mathbf{f}(t)$ on the right-hand side of (6.66) corresponds to the given excitation forces.

In the case of continuous systems where the displacements are only in one direction, a single field variable $w(\cdot)$ replaces the finite-dimensional vectors $\mathbf{q}$. Further, instead of the matrices multiplying the vectors $\mathbf{q}$, $\dot{\mathbf{q}}$, and $\ddot{\mathbf{q}}$, there now are partial differential operators operating on w, $w_{,t}$, and $w_{,tt}$. The equation of motion now takes the form

$$\mathcal{M}[w_{,tt}] + \mathcal{D}[w_{,t}] + \mathcal{G}[w_{,t}] + \mathcal{K}[w] + \mathcal{N}[w] = f(x, y, z, t), \qquad (x, y, z) \in \mathcal{G}. \tag{6.67}$$

Here, all the differential operators contain only spatial derivatives of $w(x, y, z, t)$, $w_{,t}(x, y, z, t)$, or $w_{,tt}(x, y, z, t)$. The differential operators may, however, contain coefficients depending on the spatial coordinates. In addition, we assume that the boundary conditions associated with (6.67) on the boundary $\partial\mathscr{G}$ are such that the simplified homogeneous problem containing only the operators $\mathcal{M}[\cdot]$ and $\mathcal{K}[\cdot]$ is positive definite. The operators $\mathscr{G}[\cdot]$ and $\mathcal{N}[\cdot]$ have the properties

$$\int_{\mathscr{G}} \mathcal{G}[w_{i,t}]w_{j,t}\,\mathrm{d}\mathscr{G} = -\int_{\mathscr{G}} \mathcal{G}[w_{j,t}]w_{i,t}\,\mathrm{d}\mathscr{G}, \tag{6.68}$$

$$\int_{\mathscr{G}} \mathcal{N}[w_i]w_j\,\mathrm{d}\mathscr{G} = -\int_{\mathscr{G}} \mathcal{N}[w_j]w_i\,\mathrm{d}\mathscr{G}, \tag{6.69}$$

where $w_i(x, y, z, t)$ and $w_j(x, y, z, t)$ are arbitrary, but sufficiently often differentiable functions satisfying all the boundary conditions. It is obvious that when $i = j$, the expressions in (6.68)–(6.69) are identically zero.

Different methods are available for the solution of the partial differential equations (6.67). They are usually based on discretization of the continuous system, and some of the most important methods will be described briefly later in Section 6.3. In the following section, we briefly describe a solution method based on Green's function.

6.2.2 Green's function for inhomogeneous vibration problems

Consider the boundary value problem described by (6.67) with the corresponding boundary conditions. Let the excitation force (in complex notation) be of the form

$$\underline{f}(P, t) = \delta(P - \overline{P})\mathrm{e}^{i\Omega t} \tag{6.70}$$

i.e., the excitation is a concentrated time-harmonic force acting at a point $\overline{P}$. In this section, complex quantities will be distinguished by an underline. Under the action of the harmonic force (6.70), the response (except for the resonant case) is of the form

$$\underline{w}(P, t) = \underline{g}(P; \overline{P}; \Omega)\mathrm{e}^{i\Omega t}, \tag{6.71}$$

where Green's function $\underline{g}(P; \overline{P}; \Omega)$ for the inhomogeneous vibration problem satisfies

$$-\Omega^2\mathcal{M}[\underline{g}] + i\Omega\mathcal{D}[\underline{g}] + i\Omega\mathcal{G}[\underline{g}] + \mathcal{K}[\underline{g}] + \mathcal{N}[\underline{g}] = \delta(P - \overline{P}), \tag{6.72}$$

along with the corresponding boundary conditions. In the differential operators contained in (6.72), $\underline{g}(P; \overline{P}; \Omega)$ is differentiated with respect to the coordinates of the point P. If Green's function for the inhomogeneous problem (6.72) is known, then the solution to the problem of forced vibrations with an arbitrary time-varying concentrated force $f(P, t)$ in (6.67) can be found by means of a simple integration.

Using the Fourier transform $\underline{\mathscr{F}}(P, \Omega)$ of the time function $f(P, t)$ in (6.67), one can obtain the Fourier transform of the solution of (6.67) in the form

$$\underline{\mathscr{W}}(P, \Omega) = \int_{\mathscr{G}} \underline{g}(P; \overline{P}; \Omega)\underline{\mathscr{F}}\,\mathrm{d}\overline{P}. \tag{6.73}$$

Finally, one obtains the solution in the time domain using the inverse Fourier transformation

$$w(P,t) = \frac{1}{2\pi}\int_{-\infty}^{\infty} \mathscr{W}(P,\Omega)e^{i\Omega t}\,d\Omega. \tag{6.74}$$

Only in very few cases is it possible to give a closed-form analytical expression for the function $g(P;\overline{P};\Omega)$. Usually it will have to be formulated as a series expansion or will have to be approximated by means of discretization. On the other hand, the complex function $g(P;\overline{P};\Omega)$ corresponds to the dynamic mobility, and as such is amenable to a direct measurement in real physical structures.

6.3 SOME DISCRETIZATION METHODS FOR FREE AND FORCED VIBRATIONS

6.3.1 Expansion in function series

In what follows, we discuss methods in which the free or forced vibration solution of a continuous system is represented as a function series of the form

$$w(x,y,z,t) = \sum_{j=1}^{\infty} F_j(x,y,z)q_j(t). \tag{6.75}$$

Here, $F_j(x,y,z)$, $j = 1,2,\ldots,\infty$, are functions of the space coordinates (shape-functions), which satisfy at least a part of the boundary conditions. The time functions $q_j(t)$ are unknown, and have to be determined. Replacing the infinite series (6.75) by a finite sum reduces the infinite-dimensional problem (6.67) to a finite-dimensional problem of the type (6.66). The infinite-dimensional system is thus mapped to a finite-dimensional one. Both the Ritz and Galerkin methods, which have already been introduced in the preceding chapters, as well as the finite element method are part of a class of methods that use this representation.

The approximate solution

$$\tilde{w}(x,y,z,t) = \sum_{j=1}^{n} F_j(x,y,z)q_j(t) \tag{6.76}$$

is associated with the instantaneous error distribution defined through

$$\tilde{e}(x,y,z,t) := \mathcal{M}[\tilde{w}_{,tt}] + \mathcal{D}[\tilde{w}_{,t}] + \mathcal{G}[\tilde{w}_{,t}] + \mathcal{K}[\tilde{w}] + \mathcal{N}[\tilde{w}] - f(x,y,z,t), \tag{6.77}$$

which is a function of the spatial coordinates and time. The functions $q_j(t)$ are now to be determined in such a manner that the error $\tilde{e}(x,y,z,t)$ (or an average value thereof) is minimized in some sense. The different methods to do that differ in the sense in which this error is minimized.

Next, consider equations of the type

$$\int_{\mathscr{G}} \tilde{e}(P,t)H_i(P)\,dP = 0, \qquad i = 1,2,\ldots,m, \tag{6.78}$$

where $H_i(P), i = 1, 2, \ldots, m$, are arbitrary functions. One may interpret (6.78) as forcing to zero the *projection* of the error $\tilde{e}(P, t)$ on the functions $H_i(P)$. Imposing conditions (6.78) for different chosen functions $H_i(P)$, $i = 1, 2, \ldots, m$, sets all the m projections equal to zero. If m goes to infinity, and if the sequence of functions $H_i(P)$, $i = 1, 2, \ldots, \infty$, is complete in some sense, then with $q_j(t)$, $j = 1, 2, \ldots, \infty$, determined from the infinitely many projections (6.78), the error $\tilde{e}(P, t)$ will vanish identically. All the methods discussed below are of this type, and they only differ in the choice of the functions $F_j(P)$, $j = 1, 2, \ldots, n$ and $H_i(P)$, $i = 1, 2, \ldots, m$.

According to (6.76)–(6.77) the error $\tilde{e}(P, t)$ depends linearly on $q_j(t)$, $j = 1, 2, \ldots, n$, and their time derivatives. Due to linearity, (6.76) implies

$$\mathcal{M}[\tilde{w}_{,tt}] = \mathcal{M}\left[\sum_{j=1}^{n} F_j(P)\,\ddot{q}_j(t)\right] = \sum_{j=1}^{n} \mathcal{M}[F_j(P)]\ddot{q}_j(t), \tag{6.79}$$

$$\mathcal{D}[\tilde{w}_{,t}] = \sum_{j=1}^{n} \mathcal{D}[F_j(P)]\dot{q}_j(t), \tag{6.80}$$

and so on. Thus, for each function $H_i(P)$, the condition (6.78) gives one linear differential equation in $q_j(t), \dot{q}_j(t)$, and $\ddot{q}_j(t)$, $j = 1, 2, \ldots, n$. With $m = n$, we obtain a discrete system of the type (6.66), and therefore, in general, we have a sufficient (and necessary) number of differential equations for the determination of the functions $q_j(t)$, $j = 1, 2, \ldots, n$. In this case, the elements of the matrices **M** and **D** are given by

$$m_{ij} = \int_{\mathcal{G}} \mathcal{M}[F_j(P)]H_i(P)\,\mathrm{d}P \quad \text{and} \quad d_{ij} = \int_{\mathcal{G}} \mathcal{D}[F_j(P)]H_i(P)\,\mathrm{d}P. \tag{6.81}$$

The functions $H_j(P)$ are, to a large extent, arbitrary. They do not have to satisfy any differentiability or boundary conditions, since only the integrations indicated in (6.78) have to be carried out.

In what follows, we discuss different possibilities for choice of functions $H_k(P)$. All the different resulting approaches have the following in common: introducing an additional shape-function in (6.76) (and, correspondingly, an additional function $H_k(P)$) increases the number of degrees of freedom of the discretized system from n to $n + 1$, resulting only in an additional row and column in the matrices **M**, **D**, **G**, **K**, and **N**. The $n \times n$ matrix elements calculated previously remain unaffected.

6.3.2 The collocation method

The collocation method is the simplest discretization method mapping the continuous problem (6.67) to a discrete one using the expansion (6.76) together with (6.78). Here, the points P_i, $i = 1, 2, \ldots, n$, are chosen in $\mathcal{G}$, and the functions $H_i(P)$ in (6.78) are chosen as

$$H_i(P) = \delta(P - P_i), \qquad k = 1, 2, \ldots, n. \tag{6.82}$$

This choice leads to, for example, matrices $\mathbf{M}$ and $\mathbf{D}$ according to (6.81) as

$$m_{ij} = \mathcal{M}[F_j(P_i)] \qquad \text{and} \qquad d_{ij} = \mathcal{D}[F_j(P_i)]. \tag{6.83}$$

This choice of the functions $H_i(P)$ according to (6.82) causes the approximate solution (6.76) to satisfy the partial differential equation of the problem exactly at the points P_i, $i = 1, 2, \ldots, n$. This of course does not say anything about the deviations of the approximate solution from the exact one (in particular also at the collocation points, where the errors may be quite large). No exact error bounds for the approximate solution can be given in the collocation method, and this is a considerable disadvantage of this approach. The usefulness of this approximation depends strongly on the choice of the shape-functions $F_j(P)$, $j = 1, 2, \ldots, n$. On the other hand, if one has a good intuitive feeling about the eigenfunctions (for example, as in free vibration problems), it is possible to choose shape-functions giving useful approximations at least for the first eigenvalue and eigenfunction. It should, however, be observed that the matrix $\mathbf{M}$ obtained with the collocation method may not be symmetric, and similarly, the symmetry properties of the other matrices may also be lost.

It is also possible to choose m points $m > n$, where n is the number of shape-functions $F_j(\cdot)$ in (6.76). If the error is set equal to zero at the location P_k, then

$$\sum_{j=1}^{n} \left[m_{ij}\ddot{q}_j + d_{ij}\dot{q}_j + g_{ij}\dot{q}_j + c_{ij}q_j + n_{ij}q_j\right] = f_i(t), \qquad i = 1, 2, \ldots, m. \tag{6.84}$$

It is of course not possible to find functions $q_j(t)$ simultaneously satisfying these m equations exactly. In matrix form, (6.84) can be written as

$$\mathbf{M}\ddot{\mathbf{q}} + \mathbf{D}\dot{\mathbf{q}} + \mathbf{G}\dot{\mathbf{q}} + \mathbf{K}\mathbf{q} + \mathbf{N}\mathbf{q} = \mathbf{f}(t), \tag{6.85}$$

where the matrices $\mathbf{M}$, $\mathbf{D}$, $\mathbf{G}$, $\mathbf{K}$, and $\mathbf{N}$, are now of the order $m \times n$, and $\mathbf{q}$ is of the type $n \times 1$, and $\mathbf{f}$ of the type $m \times 1$. The system can, however, be satisfied in an average sense. For example, multiplying (6.85) from the left with $\mathbf{M}^{\mathrm{T}}$ gives a system with the structure (6.66) where all matrices are now square. While the new matrix $\tilde{\mathbf{M}} = \mathbf{M}^{\mathrm{T}}\mathbf{M}$ is symmetric, the matrix $\tilde{\mathbf{D}} = \mathbf{M}^{\mathrm{T}}\mathbf{D}$ may not be symmetric. It can of course always be decomposed into a symmetric and a skew-symmetric part. The reduction of a system of m equations in $n\,(< m)$ unknowns to a system of n equations (in the n unknowns) corresponds to the Gauss least-square minimization.

The disadvantages of the collocation method, mentioned above, need to be weighed against the computational simplicity of the method. The coefficients of the system matrices of the discretized system are found simply by differentiation and substitution, for example, as given by (6.83). No additional integrations or other operations are necessary.

In what follows, we consider a benchmark eigenvalue problem which will be used to compare the different discretization procedures presented in this chapter. To this end, we consider the problem of the axial vibrations of a bar with non-constant cross-section, which is fixed

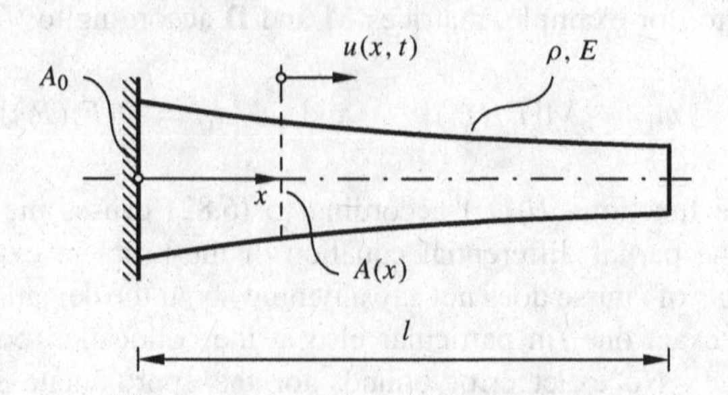

Figure 6.3 A fixed–free circular bar of varying cross-section

at the left end and free at the right end, as shown in Figure 6.3. We assume that the cross-sectional area is of the form

$$A(x) = A_0 \left[1 - \left(\frac{x}{2l}\right)^k\right]^m, \tag{6.86}$$

where k and m are integer constants. We further assume Young's modulus E and the density ρ to be constant, so that the free longitudinal vibrations are described by

$$\rho A(x) u_{,tt}(x,t) - [EA(x)u_{,x}]_{,x} = 0, \tag{6.87}$$

along with the boundary conditions

$$u(0,t) \equiv 0 \qquad \text{and} \qquad u_{,x}(l,t) \equiv 0. \tag{6.88}$$

First, we derive an *exact* solution of the problem, which will be used for comparison. Using the separation of variables and writing the solution as

$$u(x,t) = U(x)p(t), \tag{6.89}$$

one obtains the ordinary differential equations

$$\ddot{p}(t) + \omega^2 p(t) = 0,$$

$$U''(x) - \frac{mk}{2l} \frac{1}{1 - \left[\frac{x}{2l}\right]^k} \left[\frac{x}{2l}\right]^{k-1} U'(x) + \frac{\omega^2}{c^2} U(x) = 0,$$

where $c^2 = E/\rho$.

In what follows, we restrict our attention to the special case $m = k = 1$, for which the substitutions

$$s := (-x + 2l)\frac{\omega}{c} \qquad \text{and} \qquad \overline{U}(s) := U(x(s))$$

lead to the boundary value problem

$$s^2\overline{U}''(s) + s\overline{U}'(s) + s^2\overline{U}(s) = 0, \tag{6.90}$$

$$\overline{U}(s) = 0, \quad \text{and} \quad \overline{U}'(l\omega/c) = 0. \tag{6.91}$$

The general solution of Bessel's differential equation (6.90) is given by

$$\overline{U}(s) = AJ_0(s) + BY_0(s), \tag{6.92}$$

and using the boundary conditions in (6.91) leads to the characteristic equation

$$J_0(2l\omega/c)Y_1(l\omega/c) - J_1(l\omega/c)Y_0(2l\omega/c) = 0.$$

The numerically *exact* solutions for the first three circular eigenfrequencies are obtained as

$$\omega_1 = 1.794011\frac{c}{l}, \qquad \omega_2 = 4.802061\frac{c}{l}, \qquad \text{and} \qquad \omega_3 = 7.908962\frac{c}{l}. \tag{6.93}$$

Using these results, the corresponding eigenfunctions can be easily derived. For example, for the first eigenfunction, the values of the constants in (6.92) are $A_1 = 0.640457$ and $B_1 = 1.638424$.

We now compare the exact values with the approximate results obtained using the collocation method. As shape-functions, we use the polynomials

$$F_j(x) = \left(1 - \frac{x}{l}\right)^{j+1}, \qquad j = 1, 2, \ldots, n, \tag{6.94}$$

which satisfy both the boundary conditions in (6.91). We define the *collocation points* through the coordinates

$$x_j = jl/n, \qquad j = 1, 2, \ldots, n.$$

This leads to matrices $\mathbf{M}$ and $\mathbf{K}$ of the discretized system with coefficients

$$m_{ij} = \rho A(x_i)\, F_j(x_i) = \rho A_0\left(1 - \frac{i}{2n}\right)\left[\left(1 - \frac{i}{n}\right)^{j+1} - 1\right],$$

$$k_{ij} = -\frac{\partial}{\partial x}\left[EA_0\left(1 - \frac{x}{2l}\right)\frac{-j-1}{l}\left(1 - \frac{x}{l}\right)^{j}\right]_{x=x_i}$$

$$= -\frac{EA_0}{2l^2}(j+1)\left[\left(1 - \frac{i}{n}\right)^{j} + 2j\left(1 - \frac{i}{2n}\right)\left(1 - \frac{i}{n}\right)^{j-1}\right].$$

If three shape-functions are used (i.e., $n = 3$), the (asymmetric) matrices **M** and **K** give the circular eigenfrequencies

$$\overline{\omega}_1 = 1.796683\frac{c}{l}, \qquad \overline{\omega}_2 = 4.448599\frac{c}{l}, \qquad \text{and} \qquad \overline{\omega}_3 = 7.553602\frac{c}{l}. \tag{6.95}$$

A comparison with the exact values in (6.93) shows that the error is very small in the first eigenfrequency, and even the third eigenfrequency is found within 5% of the correct value. If the first three eigenfrequencies are to be approximated with a better accuracy, in general one has to choose $n > 3$. As a common practice, n is taken as twice the number of accurately desired eigenfrequencies (i.e., $n = 6$ is expected to yield the first three eigenfrequency with 'reasonable' accuracy). The approximated eigenfunctions can now be easily determined. The first two approximate eigenfunctions thus obtained are compared with the exact eigenfunctions in Figure 6.4. The first approximated eigenfunction cannot be distinguished from the corresponding exact eigenfunction within the graphical accuracy of the plot.

6.3.3 The method of subdomains

It is a disadvantage of the collocation method that the local errors may become very large in some regions when the error $\tilde{e}(P, t)$ is forced to zero at the chosen collocation points. It is often more convenient to minimize an average error, in the sense of the projections given by (6.78), using $H_k(P)$ as 'classical' functions instead of the generalized functions (Dirac delta functions) used in the collocation method. In this alternative approach, the domain $\mathcal{G}$ is divided into subdomains $\mathcal{G}_i$, $i = 1, 2, \ldots, n$, such that

$$\mathcal{G} = \cup_i \mathcal{G}_i. \tag{6.96}$$

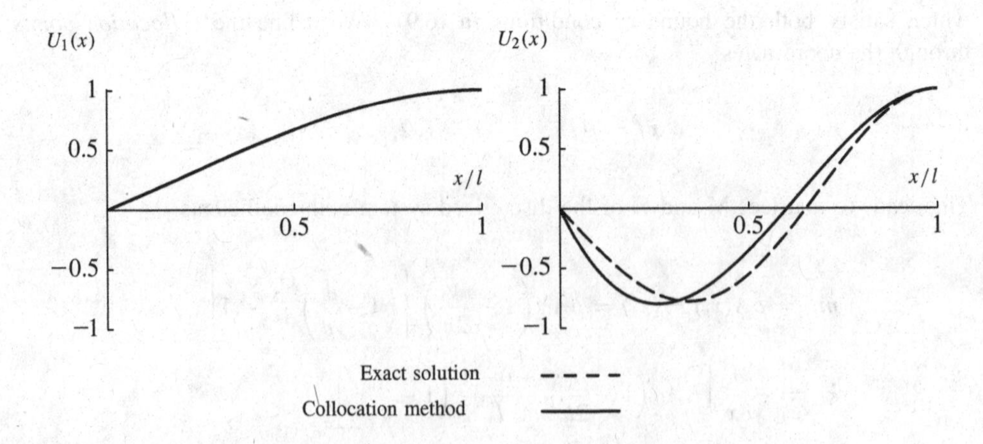

Figure 6.4 Comparison of the first two approximate eigenfunctions obtained from the collocation method with the exact eigenfunctions

The functions $H_j(P)$ are then defined such that they vanish outside of the domain $\mathcal{G}_j$. Within $\mathcal{G}_j$, $H_j(P)$ can, for example, be set identically equal to unity. The shape-functions $F_j(P)$ are not affected by this, and can assume values different from zero in the whole domain $\mathcal{G}$.

This now leads to matrices **M** and **D** with coefficients

$$m_{ij} = \int_{\mathcal{G}_i} \mathcal{M}[F_j(P)]\,\mathrm{d}P \qquad \text{and} \qquad d_{ij} = \int_{\mathcal{G}_i} \mathcal{D}[F_j(P)]\,\mathrm{d}P \tag{6.97}$$

instead of those given by (6.81). The finer the subdivision of the domain (i.e., the smaller the subdomains $\mathcal{G}_i$), the larger will their number be, and the more the discretization formulas given by (6.97) will be equivalent to (6.83), if the points P_j, $j = 1, 2, \ldots, n$, in the collocation method are, respectively, chosen from the domains $\mathcal{G}_j$.

On increasing the number of shape-functions and the number of subdomains in (6.96), in this method, in general, all the functions $H_j(P)$ get changed. This implies that all elements of the matrices **M, D, G, K,** and **N** need to be recomputed. The only exception is the case in which only certain subdomains $\mathcal{G}_j$ are, respectively, divided further into new subdomains, the others being maintained as before.

6.3.4 Galerkin's method

A frequent choice of the functions $H_j(P)$, $j = 1, 2, \ldots, n$, on which the error $\tilde{e}(P, t)$ is to be projected according to (6.78), is the shape-functions themselves, i.e.,

$$H_j(P) = F_j(P), \qquad j = 1, 2, \ldots, n. \tag{6.98}$$

With this choice, the functions $H_j(P)$, $j = 1, 2, \ldots, n$, now automatically satisfy much stronger differentiability conditions than needed in general in (6.78), since all $F_j(P)$ must be sufficiently often differentiable, so that the spatial derivatives contained in $\mathcal{M}[\cdot]$, $\mathcal{D}[\cdot]$, $\mathcal{G}[\cdot]$, $\mathcal{K}[\cdot]$, and $\mathcal{N}[\cdot]$ can be carried out.

The matrices of the discretized system (6.66) are now given by, for example,

$$m_{ij} = \int_{\mathcal{G}} \mathcal{M}[F_j(P)]F_i(P)\,\mathrm{d}P \qquad \text{and} \qquad d_{ij} = \int_{\mathcal{G}} \mathcal{D}[F_j(P)]F_i(P)\,\mathrm{d}P,$$

and so on, and

$$f_i = \int_{\mathcal{G}} f(P, t)F_i(P)\,\mathrm{d}P.$$

In general, the matrices **M** and **K** did not turn out to be symmetric in the methods so far explained in Section 6.3. In the present case, however, they will often be symmetric. This is certainly the case if the homogeneous boundary value problem, defined by $\mathcal{M}[\cdot]$ and $\mathcal{K}[\cdot]$ and the corresponding boundary conditions, is self-adjoint, and the shape-functions $F_j(P)$, $j = 1, 2, \ldots, n$, are chosen from the set of comparison functions (i.e., they are at least

$2p$-times differentiable, and satisfy all the boundary conditions), resulting in $\mathbf{M}^{\mathrm{T}} = \mathbf{M}$ and $\mathbf{K}^{\mathrm{T}} = \mathbf{K}$. Further, if the system is proportionally damped, i.e.,

$$\mathcal{D}[\cdot] = \alpha\mathcal{M}[\cdot] + \beta\mathcal{K}[\cdot]$$

(see Section 1.8.1), then the matrix $\mathbf{D}$ will also be symmetric. On the other hand, non-proportional damping will lead to a non-symmetric matrix $\tilde{\mathbf{D}}$, which can always be split into a symmetric matrix $\mathbf{D}$, and a skew-symmetric matrix $\mathbf{G}$. In many engineering problems, the operator $\mathcal{N}[\cdot]$, and hence the matrix $\mathbf{N}$, does not arise. The convergence of Galerkin's method can be proved for the case when comparison functions are used as shape-functions in self-adjoint boundary value problems (see [2] and [3]). The method in this case converges in the sense of the $\mathscr{L}^2$-norm.

Galerkin's method is frequently very useful for the solution of continuous systems. In many continuous systems with spatially varying parameters, this variation is usually slow, and the eigenfunctions are similar to that of a simplified related problem with equations of motion having constant coefficients. In these cases the eigenfunctions of the simplified problem may be used as shape-functions for the original problem. It may then be possible to obtain very good results with a relatively small number of shape-functions chosen in this manner. Often the results are so simple that the dependence of the first few eigenfrequencies and other system properties on the system parameters can be studied analytically.

We now compare the exact solutions obtained for the example (6.87)–(6.88) with the results from a discretization carried out with Galerkin's method. The functions (6.94) are again used as shape-functions, since they are comparison functions for the problem. This leads to the matrices $\mathbf{M}$ and $\mathbf{K}$ with the coefficients

$$\begin{aligned} m_{ij} &= \int_0^l \rho A_0 \left[1 - \frac{x}{2l}\right] \left\{\left[1 - \frac{x}{l}\right]^{i+1} - 1\right\} \left\{\left[1 - \frac{x}{l}\right]^{j+1} - 1\right\} \mathrm{d}x \\ &= \rho A_0 l \left\{ \frac{1}{i+j+3} - \frac{1}{i+2} - \frac{1}{j+2} + 1 \right. \\ &\quad \left. - \frac{1}{2[i+j+3][i+j+4]} + \frac{1}{2[i+3][i+2]} + \frac{1}{2[j+3][j+2]} - \frac{1}{4} \right\}, \end{aligned}$$

and

$$\begin{aligned} k_{ij} &= -\int_0^l EA_0 \left[1 - \frac{x}{2l}\right] \left\{\left[1 - \frac{x}{l}\right]^{j+1} - 1\right\}'' \left\{\left[1 - \frac{x}{l}\right]^{i+1} - 1\right\} \mathrm{d}x \\ &= \frac{EA_0}{l} \frac{[j+1][i+1][2i+2j+3]}{2[i+j+1][i+j+2]}. \end{aligned}$$

Note that $\mathbf{M}$ and $\mathbf{K}$ now are symmetric. With $n = 3$ shape-functions, the approximate solutions for the first three circular eigenfrequencies turn out to be

$$\overline{\omega}_1 = 1.794013\frac{c}{l}, \qquad \overline{\omega}_1 = 4.813378\frac{c}{l}, \qquad \text{and} \qquad \overline{\omega}_1 = 8.569588\frac{c}{l}. \tag{6.99}$$

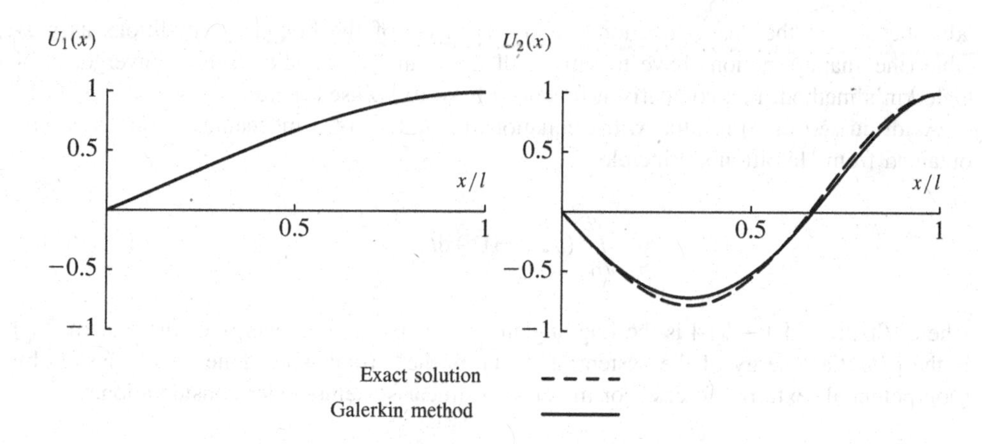

Figure 6.5 Comparison of the first two approximate eigenfunctions obtained from Galerkin's method ($n = 3$) with the exact eigenfunctions

While the first two of these values agree well with the exact results in (6.93), the deviation in $\bar{\omega}_3$ is about 8.35%. Here the error is even larger than obtained using the collocation method (compare with (6.95)), notwithstanding the fact that the collocation method is computationally much cheaper. This is, however, an accident, and depends on the choice of the shape-functions and the collocation points. For all the methods discussed here, as a rule of thumb, it can be said that only the first $n/2$ eigenvalues of the discretized problem give useful approximations (n being the number of shape-functions used in the expansion). The approximate shape-functions can be easily calculated using the approximate eigenvalues. The first two approximate eigenfunctions with the corresponding exact solutions are compared in Figure 6.5. With the present example also, the exact and the approximate solution of the first eigenfunction cannot be graphically distinguished within the given precision of the plot.

6.3.5 The Rayleigh–Ritz method

In the approaches discussed till now under Section 6.3, the approximation

$$\tilde{w}(x, y, z, t) = \sum_{j=1}^{n} F_j(x, y, z)q_j(t) \tag{6.100}$$

was substituted directly in the partial differential equation, and the projections of the error $\tilde{e}(x, y, z, t)$ (given by (6.77)) on the functions $H_k(x, y, z)$, $k = 1, 2, \ldots, n$, were then set equal to zero. The shape-functions $F_j(x, y, z)$, therefore, have to be at least $2p$-times differentiable, where $2p$ is the maximal order of spatial differentiation appearing in $\mathcal{K}[\cdot]$. The order of spatial differentiation in $\mathcal{M}[\cdot]$ has been defined as $2q$ in Section 6.1, and it has been assumed that $q < p$. Nothing has been said so far about the orders of the operators $\mathcal{D}[\cdot]$, $\mathcal{G}[\cdot]$, and $\mathcal{N}[\cdot]$. They are, however, in general, not larger than $2p$. In addition, it is advantageous if the shape-functions satisfy as many of the boundary conditions as possible (the shape-functions have to satisfy all the boundary conditions for convergence of Galerkin's method, i.e., comparison functions have to be used there).

As discussed in Appendix A, the equations of motion of a mechanical systems can be obtained from Hamilton's principle

$$\int_{t_1}^{t_2} (\delta\mathcal{L} + \overline{\delta\mathcal{W}})\, \mathrm{d}t = 0, \tag{6.101}$$

where, $\mathcal{L}[\cdot] := \mathcal{T}[\cdot] - \mathcal{V}[\cdot]$ is the Lagrangian, $\mathcal{T}[\cdot]$ is the kinetic energy of the system, $\mathcal{V}[\cdot]$ is the potential energy of the system, and $\overline{\delta\mathcal{W}}$ is the virtual work done on the system by (non-potential) external forces. For the class of linear systems under consideration,

$$\mathcal{L}[w, w_{,t}] = \mathcal{T}[w_{,t}] - \mathcal{V}[w], \tag{6.102}$$

where $\mathcal{T}[w_{,t}]$ and $\mathcal{V}[w]$ are, respectively, positive definite quadratic forms in $w_{,t}$ and w, and are obtained by integrating over the whole domain $\mathscr{G}$ the kinetic and potential energy densities (see, for example, (6.53)). In general $\overline{\delta\mathcal{W}}$ is a function of w, $w_{,t}$, and t.

In the Rayleigh–Ritz method, the expansion (6.76) is substituted directly into (6.101). The expression of the Lagrangian in (6.102) is then replaced by

$$\mathcal{L}[\tilde{w}, \tilde{w}_{,t}] = \mathcal{T}[\tilde{w}_{,t}] - \mathcal{V}[\tilde{w}], \tag{6.103}$$

where

$$\mathcal{T}[\tilde{w}_{,t}] = \frac{1}{2}\sum_{i,j=1}^{n} m_{ij}\dot{q}_i\dot{q}_j \quad \text{and} \quad \mathcal{V}[\tilde{w}] = \frac{1}{2}\sum_{i,j=1}^{n} k_{ij}q_iq_j.$$

The coefficients m_{ij} and k_{ij} are then functionals of the shape-functions $F_i(x, y, z)$, $i = 1, 2, \ldots, n$. Further, replacing w and $w_{,t}$ in $\overline{\delta\mathcal{W}}$, respectively, by the expressions for $\tilde{w}$ and $\tilde{w}_{,t}$ obtained from (6.76), gives

$$\overline{\delta\mathcal{W}} = \sum_{i=1}^{n} Q_i\delta q_i.$$

Here, obviously the generalized forces Q_i, $i = 1, 2, \ldots, n$, also depend on the shape-functions. The generalized force terms can be grouped in two parts which are, respectively, of order one and zero in the variables $q_i(t)$ and $\dot{q}_i(t)$, $i = 1, 2, \ldots, n$.

A given continuous system is thus discretized without the equations of motion so far having been given explicitly. The discretized equations of motion now follow from Hamilton's principle. For example, using Lagrange's equations

$$\frac{\mathrm{d}}{\mathrm{d}t}\left(\frac{\partial\mathcal{L}}{\partial\dot{q}_i}\right) - \frac{\partial\mathcal{L}}{\partial q_i} = Q_i, \qquad i = 1, 2, \ldots, n, \tag{6.104}$$

one obtains the matrix differential equations. The elements m_{ij} of matrix **M** and the elements k_{ij} of **K** may also be obtained directly by substituting (6.76) into the energy expressions. When $\mathcal{L}$ is of the form (6.103) we always obtain $\mathbf{M}^{\mathrm{T}} = \mathbf{M}$, and $\mathbf{K}^{\mathrm{T}} = \mathbf{K}$.

In equation (6.104), all the forces represented by **D**, **G**, **N**, and **f** are contained in Q_i, $i = 1, 2, \ldots, n$. The gyroscopic terms corresponding to the skew-symmetric matrix **G** are, however, conservative, and can, therefore, always be derived from suitable terms in the Lagrangian. The kinetic energy in that case is a sum of a quadratic form in $w_{,t}$ and a bilinear form in w and $w_{,t}$ (see Section 3.8.4 for an example, or [5]). A Rayleigh dissipation function can be formulated for the linear damping terms, and the damping forces can be directly obtained from its derivatives, as is known from the discrete systems.

Obviously, in the Rayleigh–Ritz method, the differentiability requirements for the shape-functions are much lower than in the case of Galerkin's method. In the latter method, the functions $F_i(P)$, $i = 1, 2, \ldots, n$, have to be at least $2p$-times differentiable, since $\tilde{w}$ is substituted into the operator $\mathcal{K}[\cdot]$. However, in the case of the Rayleigh–Ritz method, it is sufficient that the functions be p-times differentiable. This is due to the fact that the energy functional $\mathcal{V}[\tilde{w}]$, which is a quadratic form, only contains differentiations of order at most p.

With respect to satisfaction of the boundary conditions, the requirements placed on the shape-functions in the Rayleigh–Ritz method are fewer compared to Galerkin's method. As already known from Rayleigh's principle, it suffices that the shape-functions satisfy the essential (i.e., the geometric) boundary conditions. In Galerkin's method, on the other hand, the shape-functions have to be comparison functions. If in the Rayleigh–Ritz method comparison functions are used instead of admissible functions, then both approaches will give exactly the same results for the discretized equations of motion. This implies that, in the Rayleigh–Ritz method, one has more freedom in the choice of the shape-functions than in Galerkin's method, which is particularly important in planar problems (for example, plates), and in three-dimensional systems. This will become more obvious in Section 6.3.6.

We now reconsider the example (6.87)–(6.88) using the Rayleigh–Ritz method. We deliberately use shape-functions that satisfy only the geometric boundary conditions. Choosing the monomials

$$F_j(x) = \left(\frac{x}{l}\right)^j,$$

the kinetic and the potential energy expressions are obtained as

$$\mathcal{T} = \frac{1}{2}\sum_{i,j=1}^{n}\int_0^l \rho A_0\left[1 - \frac{x}{2l}\right]\left[\frac{x}{l}\right]^i\left[\frac{x}{l}\right]^j \dot{p}_i(t)\,\dot{p}_j(t)\,\mathrm{d}x,$$

$$\mathcal{V} = \frac{1}{2}\sum_{i,j=1}^{n}\int_0^l EA_0\left[1 - \frac{x}{2l}\right]\frac{i}{l}\left[\frac{x}{l}\right]^{i-1}\frac{j}{l}\left[\frac{x}{l}\right]^{j-1} p_i(t)\,p_j(t)\,\mathrm{d}x.$$

A short calculation immediately gives the elements of the matrices **M** and **K** as

$$m_{ij} = \rho A_0 l\,\frac{i+j+3}{2[i+j+1][i+j+2]} \qquad \text{and} \qquad k_{ij} = \frac{EA_0}{l}\,\frac{ij(i+j+1)}{2[i+j-1][i+j]}.$$

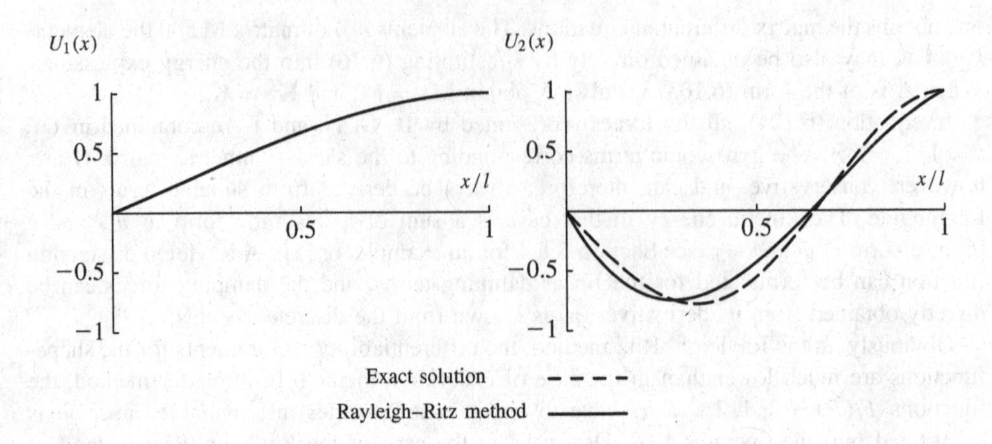

Figure 6.6 Comparison of first two mode-shapes from the Rayleigh–Ritz method ($n = 3$) and exact solution

The circular eigenfrequencies of the discrete system obtained in this manner with $n = 3$ shape-functions are

$$\varpi_1 = 1.79402\frac{c}{l}, \qquad \varpi_2 = 4.981290\frac{c}{l}, \qquad \text{and} \qquad \varpi_3 = 10.041803\frac{c}{l}. \tag{6.105}$$

The first eigenfrequency is practically identical to the exact solutions, the second one is a coarse approximation with an error within 4% of the exact value, while the third one is in error exceeding 20% of the exact value. The exact and the approximate solution for first and second eigenfunctions obtained with $n = 3$ are compared in Figure 6.6. In this case also, the quality of the approximation can be easily improved by increasing the number of shape-functions. Using shape-functions fulfilling additional boundary conditions also can considerably improve the results.

6.3.6 The finite-element method

For systems with complicated geometry, it is often futile to search for admissible functions for the Rayleigh–Ritz method defined over the whole domain $\mathcal{G}$. Even if it is possible to find such functions, they may still be computationally impractical due to their complexity. The finite-element method described below makes it possible to treat problems of very complex geometry, at the same time offering a very simple and systematic approach.

The basic idea of the method is related to the method of subdomains. As in the latter method, the domain $\mathcal{G}$ is divided into subdomains $\mathcal{G}_i$, $i = 1, 2, \ldots, n$, satisfying (6.96). The shape-functions $F_i(P)$, $i = 1, 2, \ldots, n$, in the expansion (6.76) are now chosen in such a way that $F_j(P)$ assumes values different from zero in the subdomain $\mathcal{G}_j$ only, vanishing elsewhere. These functions $F_j(P)$, $j = 1, 2, \ldots, n$, defined over the respective subdomains $\mathcal{G}_j$, $j = 1, 2, \ldots, n$, are called *finite elements*. Even complex geometries and complex boundaries can be dealt with in this manner with very simple shape-functions. Polynomials of low order can be used as finite elements in general, leading to simple calculations for the energy expressions, and often rendering superfluous the numerical integration with respect to P.

As a consequence of the convenient form of the finite-element method, the classical Rayleigh–Ritz method has fallen into oblivion with many engineers. Due to this, the finite-element method is frequently being used today even for problems for which the classical Rayleigh–Ritz method would permit a very easy solution. It should be kept in mind that sometimes for simple geometries excellent results may be obtained with very few shape-functions, which may even show the relevant parametric functional dependencies in closed form. In the same problems, possibly large numbers of finite elements may be needed to obtain a similar accuracy, with the danger that simple qualitative relations may be completely missed. The essential basic ideas of the finite-element method will be explained in what follows. For simplicity of exposition, we focus on one-dimensional continua only.

We first deal with boundary value problems described by second-order partial differential equations, as known for the transverse vibrations of taut strings and longitudinal vibrations of bars. For the string, the kinetic and potential energies are given by

$$\mathcal{T}[u_{,t}] = \frac{1}{2}\int_0^l \rho A(x) u_{,t}^2 \, \mathrm{d}x \quad \text{and} \quad \mathcal{V}[u] = \frac{1}{2}\int_0^l T(x) u_{,x}^2 \, \mathrm{d}x, \tag{6.106}$$

and for the bar by

$$\mathcal{T}[u_{,t}] = \frac{1}{2}\int_0^l \rho A(x) u_{,t}^2 \, \mathrm{d}x \quad \text{and} \quad \mathcal{V}[u] = \frac{1}{2}\int_0^l EA(x) u_{,x}^2 \, \mathrm{d}x. \tag{6.107}$$

In what follows, we first deal with a string of length l with boundary conditions $w(0,t) \equiv 0$ and $w_{,x}(l,t) \equiv 0$.

We divide the interval $[0, l]$ into n elements of equal length $h = l/n$. The boundary points of these elements are termed *nodes*. It may be convenient in some cases to choose elements of unequal length. A finer division may be appropriate at those locations where the system parameters undergo strong variations.

Each shape-function $F_i(x)$, $i = 1, 2, \ldots, (n-1)$, is now defined such that it is equal to zero outside the interval $[(i-1)h, (i+1)h]$. The shape-functions have to be chosen in such way that their first derivatives are piecewise continuous. The simplest shape-functions are, therefore, piecewise linear, and we define them as

$$\begin{aligned} F_i(x) &= 1 + \frac{x - ih}{h}, \qquad (i-1)h \le x \le ih, \\ F_i(x) &= 1 - \frac{x - (i+1)h}{h}, \qquad ih \le x \le (i+1)h, \end{aligned} \tag{6.108}$$

for $i = 1, 2, \ldots, n$. One such shape-function is shown in Figure 6.7. The last function $F_n(x)$ is defined differently, in the form

$$F_n(x) = 1 + \frac{x - nh}{h} \qquad [n-1]h \le x \le 1, \tag{6.109}$$

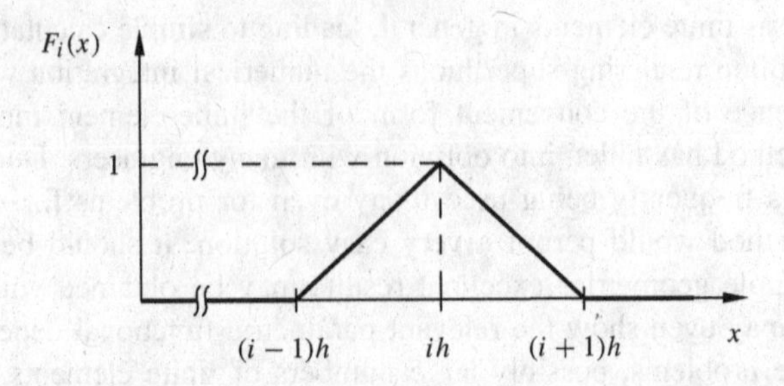

Figure 6.7 Visualization of the finite-element method (linear elements)

in order to permit non-vanishing displacements at the sliding end of the string. It is to be understood from the definitions (6.108)–(6.109) that the functions $F_i(x)$, $i = 1, 2, \ldots, n$, vanish outside the intervals specified in (6.108)–(6.109).

The shape-functions thus defined have a number of interesting properties. At the nodal points $x = jh$, $j = 1, 2, \ldots, (n-1)$, for example, the approximate solution

$$\tilde{w}(x,t) = \sum_{i=1}^{n} F_i(x) q_i(t) \tag{6.110}$$

assumes the values

$$\tilde{w}(jh,t) = q_j(t). \tag{6.111}$$

The *generalized coordinate* $q_j(t)$ is, therefore, exactly equal to the displacement at the jth node. Though the functions $F_i(x)$, $i = 1, 2, \ldots, n$, are not orthogonal, they satisfy the relations

$$\int_0^1 F_i(x) F_j(x)\,\mathrm{d}x = 0, \qquad |i-j| > 1, \tag{6.112}$$

and even

$$\int_0^1 f_0(x) F_i(x) F_j(x)\,\mathrm{d}x = 0, \qquad |i-j| > 1, \tag{6.113}$$

for arbitrary $f_0(x)$. The condition (6.113) obviously is similar to an orthogonality condition. In case (6.113) holds for all $i \neq j$, then the functions $F_i(x)$, $i = 1, 2, \ldots, n$, would be orthogonal to each other with respect to $f_0(x)$. Unfortunately, equation (6.113) does not hold for all shape-functions, but it holds exactly for those with $|i-j| > 1$ independently of the particular function $f_0(x)$. This property is reflected in the structure of the system matrices in that the resulting matrices are almost diagonal, or more precisely have a *banded* structure.

Since the shape-functions (6.108)–(6.109) are piecewise linear, (6.110) and (6.111) imply that $\tilde{w}(x,t)$ at each time instant t is a piecewise linear function of x, with the values at

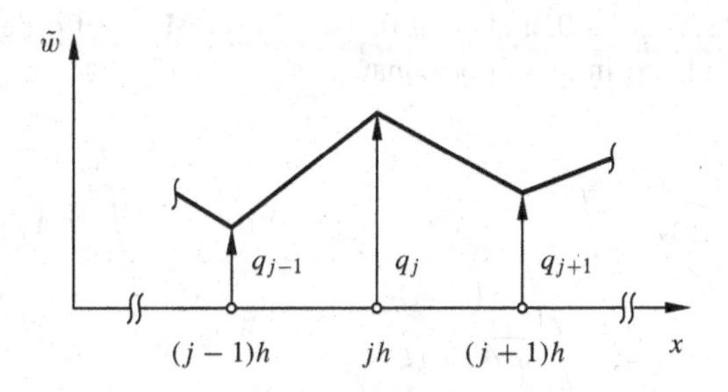

Figure 6.8 Node-point coordinates of finite elements

the points of discontinuity of slope being equal to $q_j(t)$, as shown in Figure 6.8. Within an interval $[(j-1)h, jh]$, the function $\tilde{w}(x,t)$ is thus given by

$$\tilde{w}(x,t) = \frac{jh - x}{h} q_{j-1} + \frac{x - (j-1)h}{h} q_j \tag{6.114}$$

It is convenient to introduce a local coordinate defined by

$$\xi := \frac{jh - x}{h}, \qquad (j-1)h \le x \le jh, \tag{6.115}$$

in each of these intervals. Thus, ξ assumes the value one at the left boundary and zero at the right boundary. Equation (6.114) can then be written as

$$\tilde{w}(x,t) = L_1(\xi)\, q_{j-1}(t) + L_2(\xi)\, q_j(t) \tag{6.116}$$

with the linear *interpolation functions*

$$L_1(\xi) := \xi \qquad \text{and} \qquad L_2(\xi) := 1 - \xi. \tag{6.117}$$

The energy expressions in (6.106) (or (6.107)) can now be obtained from the contributions of the single elements. For example, for the string, one obtains

$$\mathcal{T} = \sum_{j=1}^{n} \frac{1}{2} \int_{(j-1)h}^{jh} \mu(x)\, \tilde{w}_{,t}^2(x,t)\, \mathrm{d}x \qquad \text{and} \qquad \mathcal{V} = \sum_{j=1}^{n} \frac{1}{2} \int_{(j-1)h}^{jh} T(x)\, \tilde{w}_{,x}^2(x,t)\, \mathrm{d}x. \tag{6.118}$$

Obviously, the kinetic and potential energy terms in (6.118) are, respectively, quadratic forms in $\dot{q}_j$ and q_j, so that instead of (6.118), one can also write

$$\mathcal{T} = \sum_{j=1}^{n} \frac{1}{2} (\dot{q}_{j-1}, \dot{q}_j) \mathbf{M}_j (\dot{q}_{j-1}, \dot{q}_j)^{\mathrm{T}} \qquad \text{and} \qquad \mathcal{V} = \sum_{j=1}^{n} \frac{1}{2} (q_{j-1}, q_j) \mathbf{C}_j (q_{j-1}, q_j)^{\mathrm{T}},$$

where one substitutes $q_0 \equiv 0$, and $\dot{q}_0 \equiv 0$. The matrix $\mathbf{M}_j$ is of order 2×2, and its elements can easily be calculated in local coordinates as

$$m_{j(1,1)} = \int_{(j-1)h}^{jh} \rho A(x)\, L_1^2\,(\xi(x))\, \mathrm{d}x = -\int_1^0 \overline{\rho A}_j(\xi)\, \xi^2 h\, \mathrm{d}\xi$$

$$= h \int_0^1 \overline{\rho A}_j(\xi)\, \xi^2\, \mathrm{d}\xi, \tag{6.119}$$

$$m_{j(1,2)} = m_{j(2,1)} = \int_{(j-1)h}^{jh} \rho A(x)\, L_1\,(\xi(x))\; L_2\,(\xi(x))\, \mathrm{d}x$$

$$= h \int_0^1 \overline{\rho A}_j(\xi)\, \xi(1-\xi)\, \mathrm{d}\xi, \tag{6.120}$$

$$m_{j(2,2)} = h \int_0^1 \overline{\rho A}_j(\xi)\, \xi(1-\xi)^2\, \mathrm{d}\xi, \tag{6.121}$$

where

$$\overline{\rho A}_j(\xi) := \rho A\,(x(\xi)), \qquad (j-1)h \le x \le jh.$$

Similarly, the elements of the 2×2 matrix $\mathbf{K}_j$ are given by

$$k_{j(1,1)} = \int_{(j-1)h}^{jh} T(x) \left[\frac{\mathrm{d}L_1\,(\xi(x))}{\mathrm{d}x}\right]^2 \mathrm{d}x = \int_0^1 \overline{T}_j(\xi)\, \frac{1}{h^2} \left[\frac{\mathrm{d}L_1}{\mathrm{d}\xi}\right]^2 h\, \mathrm{d}\xi$$

$$= \frac{1}{h} \int_0^1 \overline{T}_j(\xi)\, \mathrm{d}\xi, \tag{6.122}$$

$$k_{j(1,2)} = k_{j(2,1)} = \int_{(j-1)h}^{jh} T(x) \frac{\mathrm{d}L_1\,(\xi(x))}{\mathrm{d}x} \frac{\mathrm{d}L_2\,(\xi(x))}{\mathrm{d}x}\, \mathrm{d}x$$

$$= -\frac{1}{h} \int_0^1 \overline{T}_j(\xi)\, \mathrm{d}\xi, \tag{6.123}$$

$$k_{j(2,2)} = \int_{(j-1)h}^{jh} T(x) \left[\frac{L_2\,(\xi(x))}{\mathrm{d}x}\right]^2 \mathrm{d}x = \frac{1}{h} \int_0^1 \overline{T}_j(\xi)\, \mathrm{d}\xi, \tag{6.124}$$

where

$$\overline{T}_j(\xi) := T\,(x(\xi)), \qquad (j-1)h \le x \le jh.$$

In (6.118), the terms quadratic in $\dot{q}_j$ (respectively, q_j) occur in two terms, while the mixed products $\dot{q}_{j-1}\dot{q}_j$ (respectively, $q_{j-1}q_j$) appear only in one term.

Let us write the kinetic energy in matrix form as

$$\mathcal{T} = \frac{1}{2}\dot{\mathbf{q}}^{\mathrm{T}}\mathbf{M}\dot{\mathbf{q}},$$

where $\dot{\mathbf{q}} := (\dot{q}_1, \ldots, \dot{q}_n)^{\mathrm{T}}$. This leads to the global mass matrix formed by superposition of the elemental mass matrices $\mathbf{M}_j$ in such a way that two consecutive matrices $\mathbf{M}_{j-1}$ and $\mathbf{M}_j$ are aligned along the diagonal, i.e., $\mathbf{M}$ has a structure given by

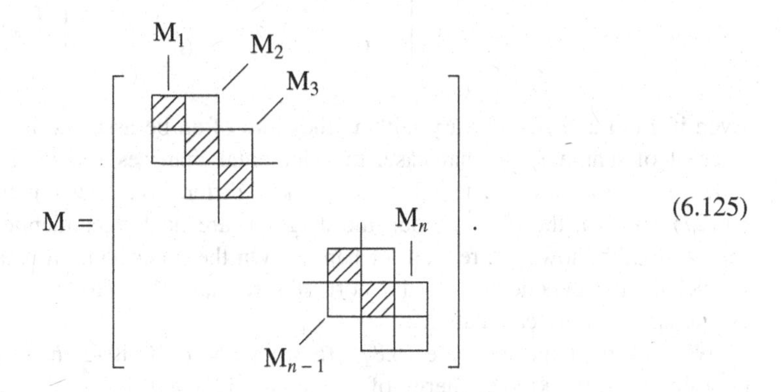

(6.125)

In (6.125), the hatched parts of the boxes stand for the superposition of parts of two elemental matrices. Due to the boundary condition $w(0, t) \equiv 0$ at the left end, the element $m_{1(2,2)}$ of the sub-matrix $\mathbf{M}_1$ contributes to the superposition with $m_{2(1,1)}$ of $\mathbf{M}_2$. On the other hand, no superposition is carried out at the last element of the diagonal due to the boundary condition $w_{,x}(l, t) \equiv 0$ at the right end. The global stiffness matrix $\mathbf{K}$ is formed from the elemental matrices $\mathbf{K}_j$, $j = 1, 2, \ldots, n$, in an analogous way. In the present case, the matrices $\mathbf{K}$ and $\mathbf{M}$ have a banded structure. Only the elements on the diagonal, the first sub-diagonal, and the first super-diagonal have elements different from zero, i.e., the 'bandwidth' is three.

The integrals (6.119)–(6.124) can be easily computed. For example, in the case when both $T(x)$ and $\rho A(x)$ are constants, we obtain

$$\mathbf{K}_j = \frac{T}{h}\begin{bmatrix} 1 & -1 \\ -1 & 1 \end{bmatrix} \quad \text{and} \quad \mathbf{M}_j = \frac{1}{6}h\rho A \begin{bmatrix} 2 & 1 \\ 1 & 2 \end{bmatrix}. \tag{6.126}$$

This leads to the global mass and stiffness matrices as, respectively,

$$\mathbf{M} = \frac{h\rho A}{6}\begin{bmatrix} 4 & 1 & 0 & & \ldots & 0 \\ 1 & 4 & 1 & & & \\ 0 & 1 & 4 & & & \\ & & & \ddots & & \\ \vdots & & & & & 0 \\ & & & & 4 & 1 \\ 0 & & & 0 & 1 & 2 \end{bmatrix} \tag{6.127}$$

and

$$\mathbf{K} = \frac{T}{h}\begin{bmatrix} 2 & -1 & 0 & \dots & & 0 \\ -1 & 2 & -1 & & & \\ 0 & -1 & 2 & & & \\ & & & \ddots & & \\ \vdots & & & & & 0 \\ & & & & 2 & -1 \\ 0 & & & 0 & -1 & 1 \end{bmatrix}. \tag{6.128}$$

Even if $T(x)$ and $\rho A(x)$ vary with x, they can often be assumed to be constant over each interval of length h. In that case, the elemental matrices in (6.126) will depend on the index j, so that the elemental matrices now differ from element to element. Hence, in (6.127)–(6.128), the blocks along the diagonal are no longer identical. The structure and the bandwidth, however, remain unchanged. On the other hand, it is also often not difficult to include the dependence of the normal force and the mass distribution on the spatial coordinate x in the calculations.

We now treat the example (6.87)–(6.88) with the finite-element method. In the local coordinates ξ, the kinetic energy of an element is given by

$$\mathcal{T}_i = \frac{h}{2}\int_0^1 \rho A(\xi_i)\left[(1-\xi_i)\,\dot{q}_{i-1} + \xi_i\,\dot{q}_i\right]^2 \mathrm{d}\xi_i,$$

where

$$\rho A(\xi_i) = \rho A_0\left[1 - h\frac{\xi_i + i - 1}{2l}\right] = \rho A_0\left[1 - \frac{\xi_i + i - 1}{2n}\right].$$

This leads to the elemental mass matrix

$$\mathbf{M}_i = \frac{\rho A_0 l}{24n^2}\begin{bmatrix} 3+8n-4i & 1+4n-2i \\ 1+4n-2i & 1+8n-4i \end{bmatrix}. \tag{6.129}$$

The potential energy of an element is given by

$$\mathcal{V}_i = \frac{1}{2}\int_{h(i-1)}^{hi} EA\,u_{i,x}^2\,\mathrm{d}x = \frac{h}{2}\int_0^1 EA\left[-\frac{1}{h}q_{i-1} + \frac{1}{h}q_i\right]^2 \mathrm{d}\xi,$$

which leads to the elemental stiffness matrix

$$\mathbf{K}_i = \frac{EA_0}{l}\left(n + \frac{1}{4} - \frac{i}{2}\right)\begin{bmatrix} 1 & -1 \\ -1 & 1 \end{bmatrix}.$$

For $n = 6$ elements, for example, the global matrices are obtained as

$$\mathbf{M} = \frac{\rho A_0 l}{864}\begin{bmatrix} 45+43 & 21 & 0 & 0 & 0 & 0 \\ 21 & 41+39 & 19 & 0 & 0 & 0 \\ 0 & 19 & 37+35 & 17 & 0 & 0 \\ 0 & 0 & 17 & 33+31 & 15 & 0 \\ 0 & 0 & 0 & 15 & 29+27 & 13 \\ 0 & 0 & 0 & 0 & 13 & 25 \end{bmatrix}$$

$$= \frac{\rho A_0 l}{864}\begin{bmatrix} 88 & 21 & 0 & 0 & 0 & 0 \\ 21 & 80 & 19 & 0 & 0 & 0 \\ 0 & 19 & 72 & 17 & 0 & 0 \\ 0 & 0 & 17 & 64 & 15 & 0 \\ 0 & 0 & 0 & 15 & 56 & 13 \\ 0 & 0 & 0 & 0 & 13 & 25 \end{bmatrix} \tag{6.130}$$

and

$$\mathbf{K} = \frac{EA_0}{4l}\begin{bmatrix} 23+21 & -21 & 0 & 0 & 0 & 0 \\ -21 & 21+19 & -19 & 0 & 0 & 0 \\ 0 & -19 & 19+17 & -17 & 0 & 0 \\ 0 & 0 & -17 & 17+15 & -15 & 0 \\ 0 & 0 & 0 & -15 & 15+13 & -13 \\ 0 & 0 & 0 & 0 & -13 & 13 \end{bmatrix}$$

$$= \frac{EA_0}{864\,l}\begin{bmatrix} 9504 & -4536 & 0 & 0 & 0 & 0 \\ -4536 & 8640 & -4104 & 0 & 0 & 0 \\ 0 & -4104 & 7776 & -3672 & 0 & 0 \\ 0 & 0 & -3672 & 6912 & -3240 & 0 \\ 0 & 0 & 0 & -3240 & 6048 & -2808 \\ 0 & 0 & 0 & 0 & -2808 & 2808 \end{bmatrix}. \tag{6.131}$$

The solution of the eigenvalue problem defined by the matrices (6.130)–(6.131) gives the following approximations for the first three circular eigenfrequencies:

$$\overline{\omega}_1 = 1.79800\frac{c}{l}, \qquad \overline{\omega}_2 = 4.92199\frac{c}{l}, \qquad \text{and} \qquad \overline{\omega}_3 = 8.46764\frac{c}{l}. \tag{6.132}$$

A comparison with the exact solution given in (6.93) shows that the first eigenvalue is practically exact, the second one is obtained with an error of about 2%, while the error for the third one is about 7%. The first two approximate eigenfunctions obtained in this manner are compared with the exact eigenfunctions in Figure 6.9. Increasing the number of elements from $n = 6$ to $n = 10$ gives $\overline{\omega}_3 = 8.11092c/l$. This estimate (with $n = 10$), which is off from the exact value by about 2.6%, is not much better than the one obtained with the

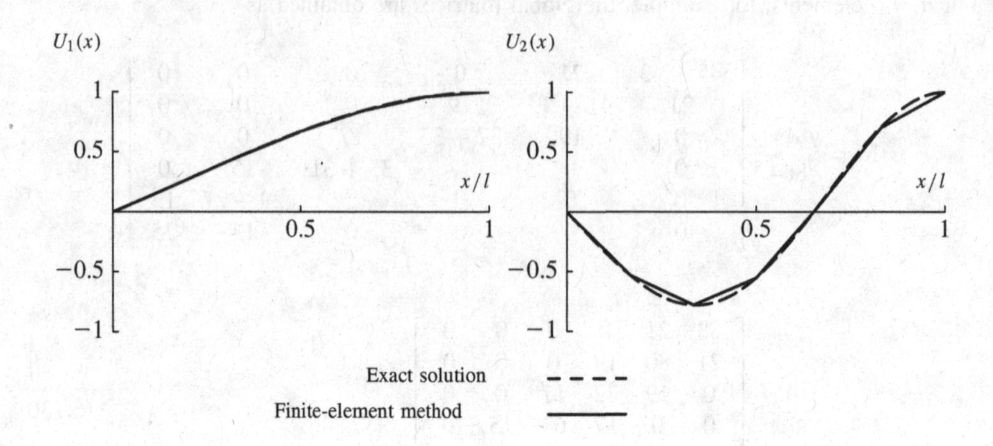

Figure 6.9 Comparison of first two mode-shapes from finite-element method and exact solution

much simpler collocation method with three shape-functions only. On the other hand, while almost nothing is known about convergence in the collocation method, in the finite-element method all the convergence results known for the Ritz method are carried over. Thus, one knows for example that in all generality the approximate eigenfrequencies are always larger than or equal to the exact values.

As already mentioned above, one may take the mass distribution $\rho A(x)$ and the stiffness $EA(x)$ to be constant within an element. It may often be assumed that with a sufficiently large number of elements the error caused by this simplification goes to zero. Strictly speaking, however, the finite-element method with this simplification no longer is a particular case of the Ritz method applied to the original problem. Hence, the convergence properties mentioned before can no longer be assured.

In the numerical example considered above, let us take the cross-section in an element to be constant, and equal to the value at the center of the respective element, i.e.,

$$A_i = A_0 \left[1 - \frac{i - 1/2}{2n} \right].$$

This leads to the elemental mass matrix

$$\mathbf{M}_i = \frac{\rho A_0 l}{24 n^2} \begin{bmatrix} 2 + 8n - 4i & 1 + 4n - 2i \\ 1 + 4n - 2i & 2 + 8n - 4i \end{bmatrix} \tag{6.133}$$

instead of (6.129). The elemental stiffness matrix is not affected by this simplification in the present example. This is because the integral in (6.129) is carried out for functions linear in ξ only (the cross-sectional area A is linear in ξ), and the integral of a linear function is equal to its mean value multiplied by the length of the domain of integration. Thus, for

$n = 6$ elements, the global mass matrix given by

$$\mathbf{M} = \frac{\rho A_0 l}{864} \begin{bmatrix} 46+42 & 21 & 0 & 0 & 0 & 0 \\ 21 & 42+38 & 19 & 0 & 0 & 0 \\ 0 & 19 & 38+34 & 17 & 0 & 0 \\ 0 & 0 & 17 & 34+30 & 15 & 0 \\ 0 & 0 & 0 & 15 & 30+26 & 13 \\ 0 & 0 & 0 & 0 & 13 & 26 \end{bmatrix}$$

$$= \frac{\rho A_0 l}{864} \begin{bmatrix} 88 & 21 & 0 & 0 & 0 & 0 \\ 21 & 80 & 19 & 0 & 0 & 0 \\ 0 & 19 & 72 & 17 & 0 & 0 \\ 0 & 0 & 17 & 64 & 15 & 0 \\ 0 & 0 & 0 & 15 & 56 & 13 \\ 0 & 0 & 0 & 0 & 13 & 26 \end{bmatrix}, \tag{6.134}$$

now replaces (6.130). The approximate values for the first three circular eigenfrequencies are now

$$\overline{\omega}_1 = 1.79455\frac{c}{l}, \qquad \overline{\omega}_2 = 4.90979\frac{c}{l}, \qquad \text{and} \qquad \overline{\omega}_3 = 8.44198\frac{c}{l}. \tag{6.135}$$

These values are almost identical to the ones obtained in (6.132) by the exact integration over the intervals. It is important to note that the circular eigenfrequencies in (6.135) are lower than those obtained in (6.132) (and hence closer to the exact values in (6.93)). However, it should not be concluded that taking the properties constant over the elements leads to a better result. The reason for this decrease in the circular eigenfrequencies is that, by taking constant properties over the elements, we have unknowingly added some extra mass to the system (since the stiffness matrix remains the same).

Up to now, only linear interpolation polynomials (6.117) have been used. They were chosen such that $L_1(0) = 0$, $L_1(1) = 1$, $L_2(0) = 1$, and $L_2(1) = 0$. Frequently it is convenient to use higher-order polynomials instead of the linear ones. For example, one may use quadratic polynomials

$$L(\xi) = c_1 + c_2\xi + c_3\xi^2 \tag{6.136}$$

or cubic polynomials

$$L(\xi) = c_1 + c_2\xi + c_3\xi^2 + c_4\xi^3. \tag{6.137}$$

Since the quadratic polynomials (6.136) contain three additional coefficients, it is natural to introduce an internal node at $\xi = 1/2$ in addition to the two external nodes $\xi = 0$ and $\xi = 1$.

The displacement is then represented by

$$\begin{aligned}\tilde{w}(x,t) &= L_1(\xi)\,\overline{q}_{j-1}(t) + L_2(\xi)\,\overline{q}_{j-1/2} + L_3(\xi)\,\overline{q}_j(t)\\ &= \mathbf{l}^{\mathrm{T}}(\xi)\,\overline{\mathbf{q}}_j(t), \qquad (j-1)h \le x \le jh,\end{aligned} \tag{6.138}$$

where

$$\begin{aligned}\mathbf{l}(\xi) &:= (L_1(\xi), L_2(\xi), L_3(\xi))^{\mathrm{T}},\\ \overline{\mathbf{q}}_j(t) &:= (\overline{q}_{j-1}(t), \overline{q}_{j-1/2}(t), \overline{q}_j(t))^{\mathrm{T}}.\end{aligned}$$

If the coefficients of the quadratic polynomial (6.136) are determined from the conditions

$$\begin{aligned} &L_1(0) = 0, \qquad L_1(1/2) = 0, \qquad L_1(1) = 1,\\ &L_2(0) = 0, \qquad L_2(1/2) = 1, \qquad L_2(1) = 0,\\ &L_3(0) = 1, \qquad L_3(1/2) = 0, \qquad L_3(1) = 0,\end{aligned} \tag{6.139}$$

or more compactly from

$$[\mathbf{l}(1), \mathbf{l}(1/2), \mathbf{l}(0)] = \mathbf{E}_{3\times 3}, \tag{6.140}$$

where $\mathbf{E}_{3\times 3}$ is the 3×3 identity matrix, then the components of $\overline{\mathbf{q}}(t)$ in (6.138) obviously correspond to the displacements at the three nodes. The conditions (6.140) lead to the three polynomials

$$L_1(\xi) = \xi(2\xi - 1), \qquad L_2(\xi) = 4\xi(1-\xi), \qquad L_3(\xi) = 1 - 3\xi + 2\xi^2.$$

One can then evaluate the integrals in the sums given in (6.118) to obtain

$$\begin{aligned}\int_{(j-1)h}^{jh} \rho A(x)\,\tilde{w}_{,t}^2(x,t)\,\mathrm{d}x &= \dot{\overline{\mathbf{q}}}^{\mathrm{T}}\mathbf{M}_j\dot{\overline{\mathbf{q}}},\\ \int_{(j-1)h}^{jh} T(x)\,\tilde{w}_{,x}^2(x,t)\,\mathrm{d}x &= \overline{\mathbf{q}}^{\mathrm{T}}\mathbf{K}_j\overline{\mathbf{q}},\end{aligned}$$

where

$$\mathbf{M}_j = h\int_0^1 \overline{\rho A}_j(\xi)\mathbf{l}(\xi)\mathbf{l}^{\mathrm{T}}(\xi)\,\mathrm{d}\xi, \tag{6.141}$$

$$\mathbf{K}_j = \frac{1}{h}\int_0^1 \overline{T}_j(\xi)\mathbf{l}'(\xi)\mathbf{l}'^{\mathrm{T}}(\xi)\,\mathrm{d}\xi, \tag{6.142}$$

and the prime in $\mathbf{l}'(\xi)$ stands for the derivative with respect to the local variable ξ. The global mass matrix is obtained from (6.141) (in analogy to (6.125)) as

$$\mathbf{M} = \qquad (6.143)$$

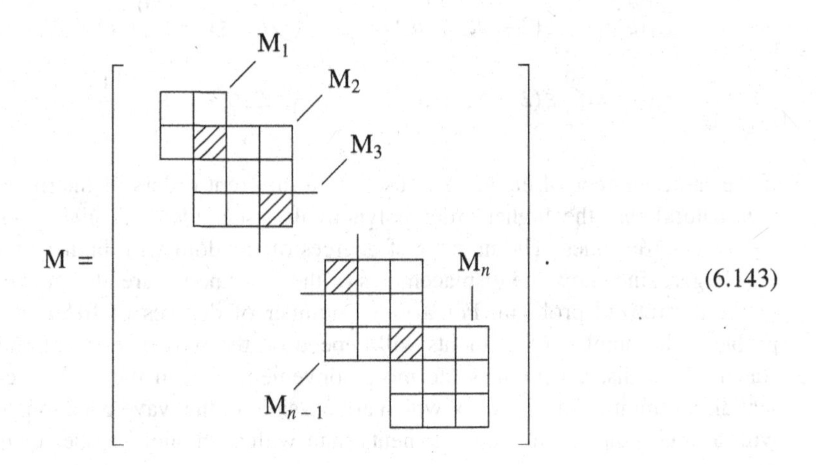

The first row and the first column were again omitted from the matrix resulting from the superposition, due to the boundary condition $w(0, t) \equiv 0$. The superposition and addition of the elemental matrices is now restricted to the external nodes only. The inner nodes are not affected, since they belong to the respective elements only. It can be seen that the mass matrix (6.143) now has a 'bandwidth' of five. The global stiffness matrix has a structure analogous to (6.143). For the case where $T(x)$ and $EA(x)$ are constants, a simple calculation gives

$$\mathbf{M}_j = \frac{h\rho A}{30}\begin{bmatrix} 4 & 2 & -1 \\ 2 & 16 & 2 \\ -1 & 2 & 4 \end{bmatrix} \quad \text{and} \quad \mathbf{K}_j = \frac{T}{3h}\begin{bmatrix} 7 & -8 & 1 \\ -8 & 16 & -8 \\ 1 & -8 & 7 \end{bmatrix}.$$

The procedure described above is completely analogous also for the case of cubic polynomials. In the cubic polynomial case, two internal nodes have to be introduced (instead of one, as in the quadratic case), and the free coefficients in the shape-functions (6.137) can be found by choosing

$$[\mathbf{l}(1), \mathbf{l}(2/3), \mathbf{l}(1/3), \mathbf{l}(0)] = \mathbf{E}_{4\times 4}. \qquad (6.144)$$

Here, naturally one has

$$\mathbf{l}(\xi) := [L_1(\xi), L_2(\xi), L_3(\xi), L_4(\xi)]^{\mathrm{T}},$$

and all $L_k(\xi)$ are of the type (6.137). The vector $\overline{\mathbf{q}}(t)$ is now four-dimensional, and $\tilde{w}(x, t)$ can again be written as

$$\tilde{w}(x, t) = \mathbf{l}^{\mathrm{T}}(\xi)\overline{\mathbf{q}}_j(t).$$

The matrix equation (6.144) now results in four cubic polynomials

$$L_1(\xi) = \frac{1}{2}\xi(2 - 9\xi + 9\xi^2), \qquad L_2(\xi) = -\frac{9}{2}\xi(1 - 4\xi + 3\xi^2),$$

$$L_3(\xi) = \frac{9}{2}\xi(2 - 5\xi + 3\xi^2), \qquad L_4(\xi) = 1 - \frac{11}{2}\xi + 9\xi^2 - \frac{9}{2}\xi^3.$$

If the same number of elements is used with different orders of interpolating polynomials, it is natural that the higher-order polynomials, as a rule, will give higher precision than the lower-order ones. The number of degrees of freedom with higher-order polynomials is also larger, since now the displacements of the inner nodes are also generalized coordinates of the discretized problem. For a given number of degrees of freedom in the discretized problem, the number of elements will depend on the type of element chosen. It cannot be stated which discretization is the more convenient one, in the general case. It is obvious that discontinuities in $w_{,x}(x,t)$, which are possible in the wave equation, are better modeled with a large number of linear elements than with a smaller number of quadratic or cubic elements.

After this short introduction to the finite-element method for the string and bar vibrations, let us briefly consider the case of a vibrating beam. The kinetic energy for the beam is defined in exactly the same way as for the string. Of course, there are differences when it comes to the potential energy. The potential energy for a beam subjected to an axial force distribution is given by

$$\mathcal{V} = \frac{1}{2}\int_0^l \left[EI(x)w^2_{,xx}(x,t) - N(x)w^2_{,x}\right] \mathrm{d}x,$$

i.e., there are now second-order derivatives with respect to x (in a string, only first-order derivatives were present). In the Ritz method, of which the finite-element method is a special case, the assumed shape-functions for a beam should be twice differentiable. This condition, therefore, excludes linear elements, since their first derivatives are discontinuous at the nodes. In order to assure the continuity of w and $w_{,x}$ at the nodes, four adjustable constants are required in the interpolating polynomials. This means that the polynomials to be used as shape-functions should be of at least third order.

For a beam, it is convenient to choose as the generalized coordinates, the displacements w_{j-1} and w_j, and the rotations of the cross-section θ_{j-1} and θ_j, at the two nodes of the jth element. In order to have identical dimensions for all generalized coordinates, we multiply the rotations by h, i.e., the vector of generalized coordinates for the jth element is now given by

$$\bar{\mathbf{q}}_j(t) := \left[w_{j-1}(t), h\,\theta_{j-1}(t), w_j(t), h\,\theta_j(t)\right]^{\mathrm{T}}, \tag{6.145}$$

as shown in Figure 6.10. With

$$\mathbf{l}(\xi) = \left[L_1(\xi), L_2(\xi), L_3(\xi), L_4(\xi)\right]^{\mathrm{T}}, \tag{6.146}$$

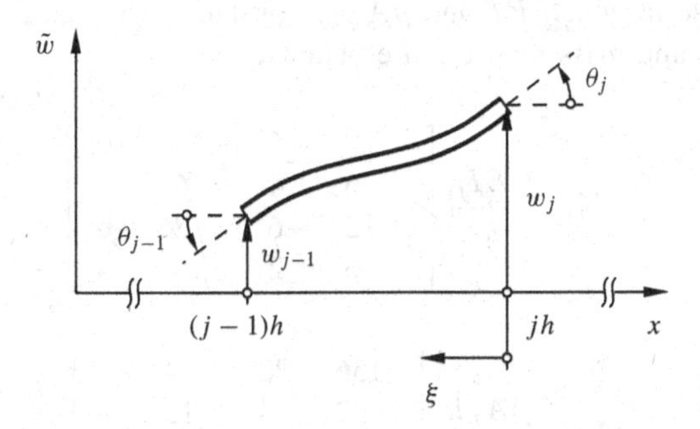

Figure 6.10 Node-point coordinates of finite elements for a beam

one obtains

$$\tilde{w}(x,t) = \mathbf{l}^{\mathrm{T}}(\xi)\,\overline{\mathbf{q}}_j(t), \qquad (j-1)h \le x \le jh. \tag{6.147}$$

The polynomials $L_k(\xi)$ in (6.146) are again of the type (6.137), where, however, the constants are determined from the conditions

$$\left[\mathbf{l}(1), -\mathbf{l}'(1), \mathbf{l}(0), -\mathbf{l}'(0)\right] = \mathbf{E}_{4\times 4}. \tag{6.148}$$

The negative signs in the second and fourth columns on the left-hand side of (6.148) stem from the fact that the local coordinate ξ is measured from right to left, while the global coordinate x goes from left to right (see Figure 6.10). Using (6.146) in the conditions (6.148), one obtains the *Hermite polynomials*

$$L_1(\xi) = 3\xi^2 - 2\xi^3, \qquad L_2(\xi) = \xi^2 - \xi^3,$$
$$L_3(\xi) = 1 - 3\xi^2 + 2\xi^3, \qquad L_4(\xi) = -\xi + 2\xi^2 - \xi^3.$$

For the potential energy of a beam under an axial force distribution, one therefore has

$$\mathcal{V} = \frac{1}{2}\int_0^l \left[EI w_{,xx}^2(x,t) - N w_{,x}^2\right] \mathrm{d}x = \frac{1}{2}\sum_{j=1}^{n} \overline{\mathbf{q}}_j^{\mathrm{T}} \mathbf{K} \overline{\mathbf{q}}_j, \tag{6.149}$$

with the local stiffness matrices

$$\mathbf{K}_j = h\int_0^1 \left[\frac{EI}{h^4}\mathbf{l}''\mathbf{l}''^{\mathrm{T}} - \frac{N}{h^2}\mathbf{l}'\mathbf{l}'^{\mathrm{T}}\right] \mathrm{d}\xi, \qquad j = 1, 2, \ldots, n, \tag{6.150}$$

where the primes in (6.150) denote derivative with respect to ξ. In a similar way, the kinetic energy can be written as

$$\mathcal{T} = \frac{1}{2}\int_0^l \rho A(x)\, w_{,t}^2(x,t)\,\mathrm{d}x = \frac{1}{2}\sum_{j=1}^{n} \dot{\overline{\mathbf{q}}}_j^{\mathrm{T}} \mathbf{M}_j \dot{\overline{\mathbf{q}}}_j.$$

For the special case in which EI and ρA are constant over each element, and $N = 0$, the elemental stiffness and mass matrices are obtained as

$$\mathbf{K}_j = \frac{EI_j}{h^3}\begin{bmatrix} 12 & 6 & -12 & 6 \\ 6 & 4 & -6 & 2 \\ -12 & -6 & 12 & -6 \\ 6 & 2 & -6 & 4 \end{bmatrix}, \tag{6.151}$$

$$\mathbf{M}_j = \frac{\rho A_j h}{420}\begin{bmatrix} 156 & 22 & 54 & -13 \\ 22 & 4 & 13 & -3 \\ 54 & 13 & 156 & -22 \\ -13 & -3 & -22 & 4 \end{bmatrix}. \tag{6.152}$$

Different expressions are, however, obtained if the parameters ρA and EI depend on x.

The global stiffness and mass matrices are again obtained by assembling the elemental matrices according to the scheme (6.125). However, now the marked regions of the boxes stand for sub-matrices of order 2×2. The 'bandwidth' of the matrices is now four. The first two rows and columns are to be eliminated from the global matrices if the boundary conditions at the left end are $w(0, t) \equiv 0$ and $w_{,x}(0, t) \equiv 0$. Nothing needs to be done at the lower end of the global matrices if the boundary at the right end is free (natural boundary conditions).

Additional aspects need to be considered for applying the finite-element method to spatial and higher-dimensional continua. While in a one-dimensional continuum such as a string, the boundary consists of only two points, in a membrane, the boundary is in general a curve, which often cannot be represented correctly by finite elements. In the example shown in Figure 6.11, the domain $\mathcal{G}$ is divided into triangular subdomains $\mathcal{G}_j$. The functions $F_j(x, y)$ are defined in such a way that they assume values different from zero only at points within the triangles.

Local variables ξ_1, ξ_2, and ξ_3, for example, can be chosen as shown in Figure 6.12. The local coordinates of a point $P = (\xi_1, \xi_2, \xi_3)$ within the triangle with nodes marked 1, 2, and 3, are defined in the following way. Let A_j be the area of the marked region in Figure 6.12, which is *cut-off* by the line parallel to the jth side of the triangle passing through the point P. If A is the total area of the triangle, the jth coordinate of the point P is then defined by $\xi_j = A_j/A$. The three sides of the triangle then obviously correspond to the lines defined by $\xi_1 = 0$, $\xi_2 = 0$, and $\xi_3 = 0$. On the other hand, for the corners, one of the coordinates is equal to unity, and the other two are equal to zero.

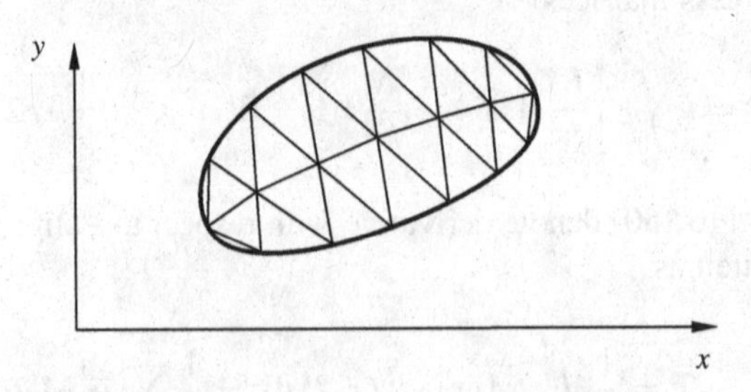

Figure 6.11 Meshing of a plane region with finite elements

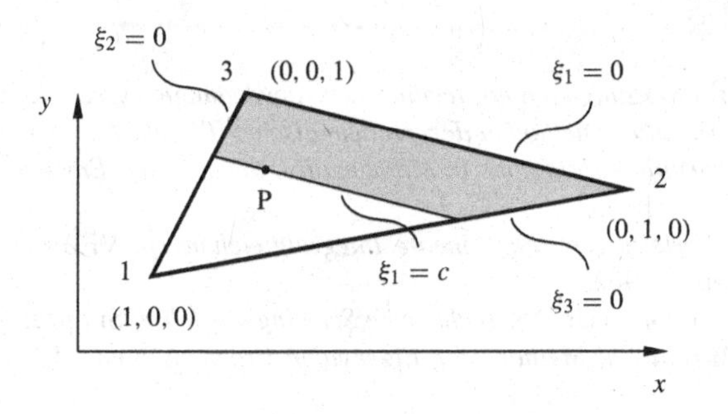

Figure 6.12 A triangular finite element with definition of local coordinates

For a membrane, it is sufficient to guarantee the continuity of the function $w(x, y, t)$. This is easily achieved by choosing linear interpolation polynomials of first order of the type

$$L_i(\xi_1, \xi_2, \xi_3) = \xi_i, \qquad i = 1, 2, 3,$$

and defining

$$w(x, y, t) = \sum_{i=1}^{3} L_i(P)\overline{q}_i(t)$$

over each element. The conversion from local to global coordinates of course also has to be done here (see, for example, [8]), and the generalized coordinates $\overline{\mathbf{q}}_j(t)$ correspond exactly to the nodal displacements, as shown in Figure 6.13. It may be mentioned here that the ordering (numbering) of the elements, which was trivial in the linear continua, will now influence the bandwidth of the global matrices.

In planar continua also higher-order interpolating polynomials can naturally be used with the triangular elements. In that case, internal nodes are introduced. Further, not only triangular elements, but also other shapes may be used. For more details, the reader is referred to the specialized literature on the subject.

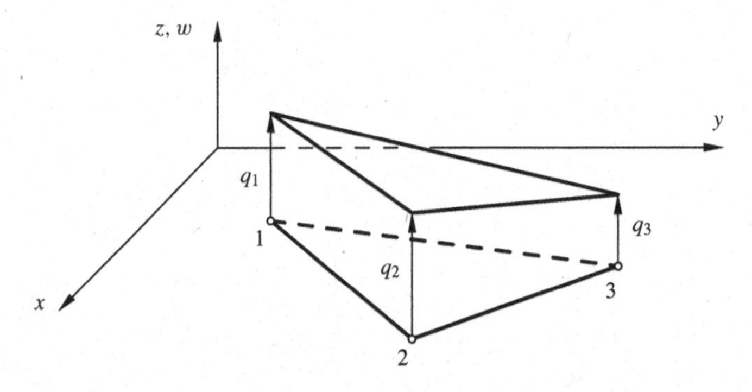

Figure 6.13 Linear triangular finite element (for example, for a membrane)

REFERENCES

[1] Collatz, L., *Eigenwertaufgaben mit technischen Anwendungen*, Geert & Portig, Leipzig, 1963.
[2] Michlin, S.G., *Variationsmethoden der mathematischen Physik*, Akademie-Verlag, Berlin, 1962.
[3] Rektorys, K., *Variational Methods in Mathematics, Science and Engineering*, Reidel, Dordrecht, 1975.
[4] Michlin, S.G., *Vorlesungen über Lineare Integralgleichungen*, VEB Deutscher Verlag der Wissenschaften, Berlin, 1962.
[5] Hagedorn, P., and Otterbein, S., *Technische Schwingungslehre*, Springer-Verlag, Berlin, 1982.
[6] Gould, S.H., *Variational Methods for Eigenvalue Problems*, Oxford University Press, London, 1955.
[7] Weinstein, A. and Stenger, W., *Methods of Intermediate Problems for Eigenvalues, Theory and Ramifications*, Academic Press, New York, 1972.
[8] Meirovitch, L., *Computational Methods in Structural Dynamics*, Sijthoff & Noordhoff, Alphen aan den Rijn, 1980.

7

Waves in fluids

In this chapter, two kinds of waves in fluids are discussed, namely acoustic waves in compressible fluids (liquids and gases) and surface waves in incompressible liquids. In both cases, it is assumed that the fluid is homogeneous, isotropic and inviscid, unless otherwise specified. While the study of acoustic waves finds applications in building acoustics, sonar and underwater communication among others, the study of surface waves is important in understanding, for example, sloshing of liquids in partially filled containers.

7.1 ACOUSTIC WAVES IN FLUIDS

7.1.1 The acoustic wave equation

In this section, we will derive the equations of motion of acoustic waves in fluids from the elementary fluid-mechanical equations and also from Hamilton's principle.

Consider an inviscid fluid under a pressure field $p(x, y, z, t)$. The motion of the fluid is governed by the Euler equation (see, for example, [1])

$$\rho \frac{\mathrm{D}\mathbf{v}}{\mathrm{D}t} = -\nabla p, \tag{7.1}$$

where $\mathbf{v}(x, y, z, t) = [v_x, v_y, v_z]^{\mathrm{T}}$ is the velocity field of the fluid, $\rho(x, y, z, t)$ is the density field, $\nabla p := [p_{,x}, p_{,y}, p_{,z}]^{\mathrm{T}}$ is the gradient of the scalar function $p(x, y, z, t)$, and $\mathrm{D}\mathbf{v}/\mathrm{D}t$ is the material or substantial derivative of $\mathbf{v}$, which is defined as

$$\begin{aligned} \frac{\mathrm{D}\mathbf{v}}{\mathrm{D}t} &= \mathbf{v}_{,t} + (\mathbf{v} \cdot \nabla)\mathbf{v} \\ &= \frac{\partial \mathbf{v}}{\partial t} + v_x \frac{\partial \mathbf{v}}{\partial x} + v_y \frac{\partial \mathbf{v}}{\partial y} + v_z \frac{\partial \mathbf{v}}{\partial z}. \end{aligned} \tag{7.2}$$

Note that we have not considered the effect of gravity or any other given force fields in (7.1) which stratify the fluid (in terms of its density). Usually, the effects of such stratification on

Vibrations and Waves in Continuous Mechanical Systems P. Hagedorn and A. DasGupta

acoustic waves is negligible (unless one studies wave propagation over large length scales, such as in oceans). The second governing equation for the fluid is the mass conservation or continuity equation

$$\rho_{,t} = -\nabla \cdot (\rho \mathbf{v}), \tag{7.3}$$

where $\nabla \cdot (\rho \mathbf{v}) := (\rho v_x)_{,x} + (\rho v_y)_{,y} + (\rho v_z)_{,z}$ is the divergence of the vector $\rho \mathbf{v}$.

Consider small oscillations of the fluid particles in an otherwise stationary fluid. Assuming the velocities to be small, one can linearize (7.2) as

$$\frac{\mathrm{D}\mathbf{v}}{\mathrm{D}t} \approx \mathbf{v}_{,t}. \tag{7.4}$$

Let

$$p(x, y, z, t) = p_0 + \overline{p}(x, y, z, t), \tag{7.5}$$

$$\rho(x, y, z, t) = \rho_0 + \overline{\rho}(x, y, z, t), \tag{7.6}$$

where p_0 and ρ_0 are, respectively, the ambient pressure and density (both assumed constant) of the fluid when it is static, and $\overline{p}(x, y, z, t)$ and $\overline{\rho}(x, y, z, t)$ are small variations on the respective ambient values, due to small oscillations of the fluid. Substituting (7.4)–(7.6) in (7.1) and (7.3), and dropping all higher order terms, we obtain the linearized governing equations for the fluid motion as

$$\rho_0 \mathbf{v}_{,t} = -\nabla \overline{p} \qquad \text{(from (7.1))}, \tag{7.7}$$

$$\overline{\rho}_{,t} = -\rho_0 \nabla \cdot \mathbf{v} \qquad \text{(from (7.3))}. \tag{7.8}$$

Thus, we have one vector equation (7.7), and one scalar equation (7.8) in three unknowns, namely $\mathbf{v}(x, y, z, t)$, $\overline{p}(x, y, z, t)$, and $\overline{\rho}(x, y, z, t)$. Therefore, we require another scalar equation to be able to solve the system uniquely. This scalar equation is provided by a relation between the pressure and density fields. Such a relation follows from the thermodynamic properties of the fluid under consideration.

Since the compressibility of liquids is low and the thermal conductivity is high, one may assume that the compression process is isothermal. On the other hand, in the case of gases, high compressibility and poor thermal conductivity makes the compression process adiabatic. To have a common description for liquids and gases, therefore, we assume a general relation between pressure and density as

$$p = p(\rho), \tag{7.9}$$

such that, around the nominal density of ρ_0, one can write

$$\begin{aligned} p_0 + \overline{p} &= p(\rho_0 + \overline{\rho}) \\ \Rightarrow p_0 + \overline{p} &\approx p(\rho_0) + \left.\frac{\mathrm{d}p}{\mathrm{d}\rho}\right|_{\rho_0} \overline{\rho} \\ \Rightarrow \overline{p} &= c^2 \overline{\rho}, \end{aligned} \tag{7.10}$$

where

$$c^2 := \left.\frac{\mathrm{d}p}{\mathrm{d}\rho}\right|_{\rho=\rho_0} \tag{7.11}$$

is a constant. Substituting the expression of $\overline{p}$ from (7.10) in (7.8), and differentiating the resulting equation with respect to time yields the linearized equation

$$\overline{p}_{,tt} = -c^2 \rho_0 \nabla \cdot \mathbf{v}_{,t}$$
$$\Rightarrow \overline{p}_{,tt} = c^2 \nabla \cdot (\nabla \overline{p}) \quad \text{(using (7.7))}$$

or

$$\overline{p}_{,tt} - c^2 \nabla^2 \overline{p} = 0. \tag{7.12}$$

This is the acoustic wave equation, where c, as defined by (7.11), is the acoustic wave speed in the medium under consideration.

Assuming that the motion of the fluid is irrotational, i.e., $\nabla \times \mathbf{v} \equiv \mathbf{0}$, one can express $\mathbf{v}$ as

$$\mathbf{v} = \nabla \psi, \tag{7.13}$$

where $\psi(x, y, z, t)$ is a scalar field and known as the *velocity potential*. In terms of the velocity potential, one may rewrite (7.7) as

$$\rho_0 (\nabla \psi)_{,t} = -\nabla \overline{p}$$
$$\Rightarrow \rho_0 \psi_{,t} = -\overline{p} \tag{7.14}$$

or

$$\rho_0 \psi_{,t} = -c^2 \overline{\rho} \quad \text{(using (7.10))}. \tag{7.15}$$

Note that in (7.14), any integration constant (which will be purely a function of t) can be absorbed in $\psi_{,t}$. Differentiating (7.15) with respect to time and using (7.8) yields

$$\psi_{,tt} = c^2 \nabla \cdot \mathbf{v} = c^2 \nabla \cdot (\nabla \psi) \quad \text{(using (7.13))},$$

or

$$\psi_{,tt} - c^2 \nabla^2 \psi = 0, \tag{7.16}$$

which is the wave equation for the velocity potential. Taking the gradient of (7.16), one can easily show that each of the velocity vector components also satisfies the wave equation. However, it should be remembered that this observation is true only when the motion of the fluid is irrotational.

Now, we consider (7.10), and discuss the specific relations for liquids and gases. In the case of liquids, the change in pressure and the corresponding volume strain are related through the *bulk modulus* B of the liquid as (see, for example, [1])

$$\overline{p} = -B\frac{\overline{V}}{V_0}, \tag{7.17}$$

where $\overline{V}$ is the small change in the volume and V_0 is the original volume. The relation (7.17) is based on the assumption of isothermal compression of a liquid. For a fixed mass of liquid occupying a volume V_0, one can write

$$\begin{aligned}\rho_0 V_0 &= (\rho_0 + \overline{\rho})(V_0 + \overline{V}) \\ \Rightarrow \frac{\overline{V}}{V_0} &\approx -\frac{\overline{\rho}}{\rho_0},\end{aligned} \tag{7.18}$$

where we have retained only terms up to first order. Therefore, from (7.17) and (7.18), we have

$$\overline{p} = \frac{B}{\rho_0}\overline{\rho}. \tag{7.19}$$

Comparing (7.19) with (7.10), we obtain the velocity of sound in liquids as

$$c = \sqrt{\frac{B}{\rho_0}}.$$

For example, for water with $B = 2.2 \times 10^9$ N/m^2, and $\rho_0 = 1000$ kg/m^3, the speed of sound, is obtained as $c = 1483.2$ m/s.

In the case of gases, the process of compression and rarefaction during the propagation of sound is so fast compared to thermal diffusion, that the process can be considered adiabatic. For an adiabatic process, it is known that

$$\frac{p}{\rho^\gamma} = \frac{p_0}{\rho_0^\gamma}, \tag{7.20}$$

where $\gamma := C_p/C_v$ and C_p and C_v are, respectively, the molar specific heats of the gas at constant pressure and constant volume. Now, using (7.20) in the definition (7.11), we obtain the expression of speed of sound in gases as

$$c = \sqrt{\frac{\gamma p_0}{\rho_0}}. \tag{7.21}$$

One can also express the sound speed in terms of the gas temperature using the ideal gas *equation of state*

$$p = \rho \frac{R}{M} T, \tag{7.22}$$

where R=8.3143 kJ/(K·mole) is the universal gas constant, M is the molar mass of the gas, and T is the absolute temperature of the gas. Then, using the relation $p_0/\rho_0 = RT_0/M$ from (7.22), one can rewrite (7.21) as

$$c = \sqrt{\frac{\gamma R T_0}{M}}, \tag{7.23}$$

where T_0 is the ambient temperature of the undisturbed medium. As an example, consider air, which has $\gamma = 1.4$, and is composed of approximately 4/5 parts of nitrogen (molecular mass 28×10^{-3} kg/mole) and approximately 1/5 part of oxygen (molecular mass 32×10^{-3} kg/mole). Then the average molecular mass of air is given as

$$M_{\text{air}} = \frac{4}{5}\, 28 \times 10^{-3} + \frac{1}{5}\, 32 \times 10^{-3} = 28.8 \times 10^{-3} \text{ kg/mole}.$$

Then, at 293 K (i.e., 20°C), the speed of sound in air is obtained from (7.23) as $c = 343.8$ m/s.

7.1.1.1 The variational formulation

The kinetic and potential energy densities of a fluid are, respectively, (see [2])

$$\hat{\mathcal{T}} = \frac{1}{2}\rho_0 \nabla\psi \cdot \nabla\psi \tag{7.24}$$

$$\hat{\mathcal{V}} = \frac{1}{2}\frac{\rho_0}{c^2}(\psi_{,t})^2. \tag{7.25}$$

The Lagrangian density is then given by

$$\hat{\mathcal{L}} = \hat{\mathcal{T}} - \hat{\mathcal{V}} = \frac{1}{2}\rho_0 \nabla\psi \cdot \nabla\psi - \frac{1}{2}\frac{\rho_0}{c^2}(\psi_{,t})^2. \tag{7.26}$$

From Hamilton's principle, we have

$$\delta \int_{t_1}^{t_2} \int_V \hat{\mathcal{L}}\, \mathrm{d}V\, \mathrm{d}t = 0$$

or

$$
\begin{aligned}
&\int_{t_1}^{t_2}\int_{\mathrm{V}}\left[\rho_0\nabla\psi\cdot\nabla(\delta\psi)-\frac{\rho_0}{c^2}\psi_{,t}(\delta\psi)_{,t}\right]\mathrm{d}V\,\mathrm{d}t=0\\
\Rightarrow &\int_{t_1}^{t_2}\int_{\mathrm{V}}\left[\nabla\cdot(\delta\psi\nabla\psi)-(\nabla^2\psi)\delta\psi-\frac{1}{c^2}\psi_{,t}(\delta\psi)_{,t}\right]\mathrm{d}V\,\mathrm{d}t=0.
\end{aligned}
\tag{7.27}
$$

Now, the Gauss divergence theorem states that (see, for example, [3])

$$
\int_{\mathrm{V}}\nabla\cdot\mathbf{v}\,\mathrm{d}V=\int_{\mathrm{S}}\mathbf{v}\cdot\hat{\mathbf{n}}\,\mathrm{d}A,
\tag{7.28}
$$

where S is the surface bounding the volume V, and $\hat{\mathbf{n}}$ is the unit surface normal. Using (7.28), one can rewrite (7.27) as

$$
\begin{aligned}
&\int_{t_1}^{t_2}\int_{\mathrm{S}}\delta\psi\nabla\psi\cdot\hat{\mathbf{n}}\,\mathrm{d}A\,\mathrm{d}t-\int_{\mathrm{V}}\left(\frac{1}{c^2}\psi_{,t}\delta\psi\right)\Big|_{t_1}^{t_2}\mathrm{d}V\\
&\quad+\int_{t_1}^{t_2}\int_{\mathrm{V}}\left(-\nabla^2\psi+\frac{1}{c^2}\psi_{,tt}\right)\delta\psi\,\mathrm{d}V\,\mathrm{d}t=0\\
\Rightarrow &\int_{t_1}^{t_2}\int_{\mathrm{S}}\psi_{,n}\,\delta\psi\,\mathrm{d}A\,\mathrm{d}t+\int_{t_1}^{t_2}\int_{\mathrm{V}}\left(-\nabla^2\psi+\frac{1}{c^2}\psi_{,tt}\right)\delta\psi\,\mathrm{d}V\,\mathrm{d}t=0,
\end{aligned}
\tag{7.29}
$$

where $\psi_{,n} := (\nabla\psi)\cdot\hat{\mathbf{n}}$ represents the normal velocity at the surface S. The boundary conditions are obtained from the first integral in (7.29), while the second integral yields the acoustic wave equation (7.16). When a fluid is in contact with a rigid surface S, the boundary condition is given by $\psi_{,n}|_{\mathrm{S}} = 0$. We will come across other types of boundary conditions later in this chapter.

7.1.2 Planar acoustic waves

Following the discussions in Section 4.5.1, it can be easily checked that traveling planar waves in three-dimensional Cartesian space can be represented using the velocity potential

$$
\psi(x,y,z,t)=f(\hat{\mathbf{n}}\cdot\mathbf{r}-ct),
$$

where $f(z)$ is a scalar function, $\hat{\mathbf{n}}$ is a unit vector representing the direction of wave propagation, $\mathbf{r}=(x,y,z)^{\mathrm{T}}$ is the position vector, and c is the wave speed. For harmonic plane waves, the velocity potential is

$$
\psi(x,y,z,t)=A\mathrm{e}^{i(\mathbf{k}\cdot\mathbf{r}-\omega t)}=A\mathrm{e}^{ik(\hat{\mathbf{n}}\cdot\mathbf{r}-ct)},
\tag{7.30}
$$

where A is the complex amplitude, $\mathbf{k} = k\hat{\mathbf{n}}$ is the wave vector, and k is the wave number. The velocity and pressure fields are obtained as, respectively,

$$\mathbf{v} = \nabla\psi = Aik\hat{\mathbf{n}}e^{ik(\hat{\mathbf{n}}\cdot\mathbf{r}-ct)}, \tag{7.31}$$

$$\overline{p} = -\rho_0\psi_{,t} = \rho_0 Aikce^{ik(\hat{\mathbf{n}}\cdot\mathbf{r}-ct)}. \tag{7.32}$$

The real parts of the complex quantities in (7.30), (7.31) and (7.32) represents the actual quantities.

Substituting the velocity potential (7.30) in (7.16) yields the dispersion relation

$$\omega^2 - c^2k^2 = 0. \tag{7.33}$$

The phase and group velocities are obtained as $c_P = c_G = c$. Thus, a fluid medium is non-dispersive to acoustic waves.

The specific mechanical impedance of a fluid medium to acoustic harmonic waves is defined as

$$\mathcal{Z} = \frac{\mathscr{A}[\overline{p}]}{\mathscr{A}[v_n]}, \tag{7.34}$$

where $\mathscr{A}[\cdot]$ represents the complex amplitude, $\overline{p}$ is obtained from (7.14), and v_n is defined as

$$v_n = \psi_{,n} = \hat{\mathbf{n}}\cdot\nabla\psi = ikAe^{ik(\hat{\mathbf{n}}\cdot\mathbf{r}-ct)}.$$

Using the expressions of $\overline{p}$ and v in (7.34), the specific impedance is obtained as

$$\mathcal{Z} = \rho_0 c. \tag{7.35}$$

7.1.3 Energetics of planar acoustic waves

The total mechanical energy density (energy per unit volume) of a fluid medium subject to a small-amplitude disturbance velocity potential ψ can be written as

$$\hat{\mathcal{E}} = \hat{\mathcal{T}} + \hat{\mathcal{V}} = \frac{1}{2}\rho_0\left[\nabla\psi\cdot\nabla\psi + \frac{1}{c^2}(\psi_{,t})^2\right]. \tag{7.36}$$

Differentiating (7.36) with respect to time yields

$$\begin{aligned}
\frac{\partial\hat{\mathcal{E}}}{\partial t} &= \rho_0\left[\nabla\psi\cdot\nabla\psi_{,t} + \frac{1}{c^2}\psi_{,t}\psi_{,tt}\right] \\
&= \rho_0\left[\nabla\cdot(\psi_{,t}\nabla\psi) + \left(-\nabla^2\psi + \frac{1}{c^2}\psi_{,tt}\right)\psi_{,t}\right] \\
&= \nabla\cdot(\rho_0\psi_{,t}\nabla\psi) = \nabla\cdot(-\overline{p}\mathbf{v}).
\end{aligned} \tag{7.37}$$

Note here that the time derivative $D(\cdot)/Dt = \partial(\cdot)/\partial t$, since there is no ambient velocity of the fluid. Defining the acoustic intensity vector for acoustic waves as

$$\mathcal{I} := \overline{p}\mathbf{v}, \tag{7.38}$$

one can rewrite (7.37) as

$$\frac{\partial \hat{\mathcal{E}}}{\partial t} + \nabla \cdot \mathcal{I} = 0. \tag{7.39}$$

The acoustic intensity vector represents the acoustic power per unit area flowing in the direction represented by the vector.

Consider a harmonic planar wave represented by the velocity potential

$$\psi = A\cos[k(\hat{\mathbf{n}} \cdot \mathbf{r} - ct)].$$

The corresponding mechanical energy density can be written as

$$\hat{\mathcal{E}} = \rho_0 k^2 A^2 \sin^2[k(\hat{\mathbf{n}} \cdot \mathbf{r} - ct)],$$

and the average energy density is then obtained as

$$\langle \hat{\mathcal{E}} \rangle = \frac{1}{2}\rho_0 k^2 A^2. \tag{7.40}$$

The acoustic intensity vector for the plane harmonic wave is given by

$$\mathcal{I} = -\rho_0 \psi_{,t} \nabla\psi = \left(\rho_0 c k^2 A^2 \sin^2[k(\hat{\mathbf{n}} \cdot \mathbf{r} - ct)]\right)\hat{\mathbf{n}}. \tag{7.41}$$

The magnitude of the intensity vector (termed the *intensity*) represents the incident power per unit area, and is given by

$$\mathcal{I} = |\mathcal{I}| = \rho_0 c k^2 A^2 \sin^2[k(\hat{\mathbf{n}} \cdot \mathbf{r} - ct)].$$

The average intensity is then obtained as

$$\langle \mathcal{I} \rangle = \frac{\omega}{2\pi}\int_0^{2\pi/\omega} \mathcal{I}\mathrm{d}t = \frac{1}{2}\rho_0 c k^2 A^2 = c\langle \hat{\mathcal{E}} \rangle. \tag{7.42}$$

Thus, the acoustic energy propagates at the acoustic wave speed c in a fluid.

If we represent p and $\mathbf{v}$ in complex notation as in (7.32) and (7.31), then $\langle \mathcal{I} \rangle$ can also be directly obtained using

$$\langle \mathcal{I} \rangle = \frac{1}{2}\mathscr{R}[\overline{p}^* \mathbf{v}], \tag{7.43}$$

where $\overline{p}^*$ represents the complex conjugate of $\overline{p}$. Using the concept of specific impedance from (7.34), one can also write the average intensity as

$$\langle \mathcal{I} \rangle = \frac{1}{2}\mathcal{R}[\mathcal{Z}^* \mathbf{v}^* \cdot \mathbf{v}] = \frac{1}{2}\rho_0 c k^2 A^2. \tag{7.44}$$

Apart from the intensity, one can also quantitatively characterize the strength of a sound field by the root mean square value (or effective value) of the pressure field, which is defined as

$$p_{\mathrm{rms}} = \left(\frac{1}{2}\mathcal{R}[p^* p]\right)^{1/2}. \tag{7.45}$$

In the case of plane acoustic waves, $p_{\mathrm{rms}} = \rho_0 A k$. It can be easily shown that

$$\langle \mathcal{I} \rangle = p_{\mathrm{rms}}^2 / \rho_0 c. \tag{7.46}$$

In many practical situations concerned with sound measurement and detection it is convenient to non-dimensionalize the pressure level of sound with a standard pressure level. This is expressed as the *sound pressure level* (SPL), which is defined as

$$L_p := 20 \log_{10} \frac{p_{\mathrm{rms}}}{p_{\mathrm{rms}}^{\mathrm{ref}}} \quad \mathrm{dB}, \tag{7.47}$$

where $p_{\mathrm{rms}}^{\mathrm{ref}} = 20\ \mu\mathrm{Pa}$ is known as the *reference effective pressure*, and dB stands for decibel, which is a dimensionless scale. This reference effective pressure is the minimum effective pressure at 2 kHz detectable by a normal human ear. Using (7.46), the reference intensity level is obtained as $\langle \mathcal{I} \rangle^{\mathrm{ref}} \approx 10^{-12}\ \mathrm{W/m^2}$.

7.1.4 Reflection and refraction of planar acoustic waves

We now study the process of reflection and refraction of planar acoustic waves. Consider for simplicity, a plane wave in the x-y-plane incident on the y-z-plane, as shown in Figure 7.1. In the figure, $\hat{\mathbf{k}} = (\cos\alpha, \sin\alpha)^{\mathrm{T}}$ and $\hat{\mathbf{k}}' = (-\cos\alpha', \sin\alpha')^{\mathrm{T}}$, where α and α' are, respectively, the angles of incidence and reflection. The velocity potential of the incident wave can be written as

$$\psi_1(x, y, t) = A e^{ik(\cos\alpha x + \sin\alpha y - ct)}, \tag{7.48}$$

where k is the wave number and c is the acoustic wave speed in the medium. Let the velocity potential corresponding to the reflected wave be represented by

$$\psi_2(x, y, t) = B e^{ik'(-\cos\alpha' x + \sin\alpha' y - ct)}, \tag{7.49}$$

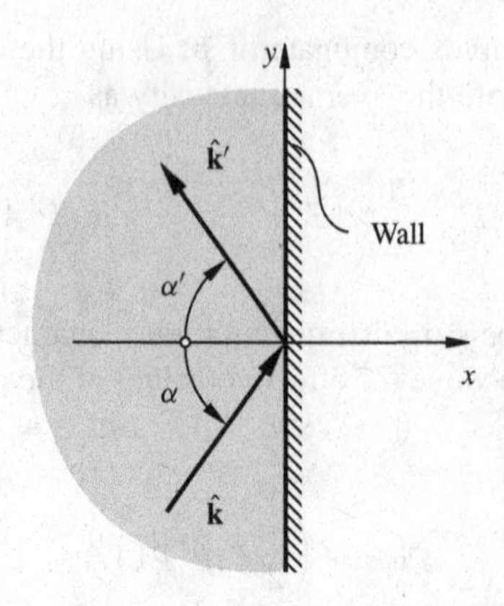

Figure 7.1 Reflection of a plane wave from a rigid wall

where k' is the wave number of the reflected wave. Thus, the total wave field is governed by the velocity potential

$$\psi(x, y, t) = \psi_1(x, y, t) + \psi_2(x, y, t). \tag{7.50}$$

At the y-z-plane, the velocity of the fluid along the normal to the plane must be zero, i.e.,

$$v_x\Big|_{x=0} = \psi_{,x}\Big|_{x=0} \equiv 0. \tag{7.51}$$

Using (7.48)–(7.50) in (7.51) yields

$$k\cos\alpha A e^{ik(\sin\alpha y - ct)} - k'\cos\alpha' B e^{ik'(\sin\alpha' y - ct)} \equiv 0. \tag{7.52}$$

This condition can be identically satisfied if and only if

$$\begin{aligned} k\sin\alpha &= k'\sin\alpha' \qquad \text{and} \qquad kc = k'c \\ \Rightarrow k &= k' \qquad \text{and} \qquad \alpha' = \alpha, \end{aligned} \tag{7.53}$$

which gives the law of reflection. It may be noted that $\sin\alpha = \sin\alpha'$ has another solution, which is $\alpha' = \pi - \alpha$. However, this has to be discarded since it is physically impossible. Using the conditions from (7.53) in (7.52), we obtain $A = B$. Thus, the amplitude of the wave after reflection at the rigid wall is the same as that of the incident wave.

Next, we consider the process of refraction occurring when a wave travels from one medium to another, as shown in Figure 7.2. The properties of the second medium are considered to be different from those of the first medium. In such a situation, we also have a partial reflection back into the first medium. Let the velocity potential for the wave field in the first medium be represented by

$$\psi_1(x, y, t) = A e^{ik_1(\cos\alpha_1 x + \sin\alpha_1 y - c_1 t)} + B e^{ik_1'(-\cos\alpha_1' x + \sin\alpha_1' y - c_1 t)}, \tag{7.54}$$

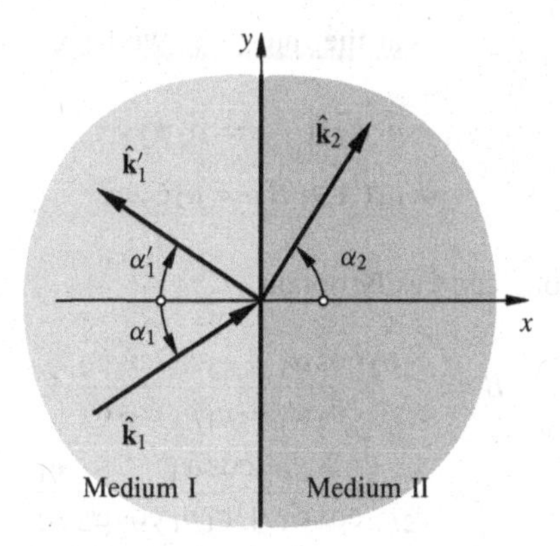

Figure 7.2 Reflection and refraction of a plane wave at an interface of two media

where the first term is for the incident wave, the second term represents the partially reflected wave, and c_1 is the velocity of sound in the first medium. The refracted wave in the second medium can be represented by the velocity potential

$$\psi_2(x, y, t) = Ce^{ik_2(\cos\alpha_2 x + \sin\alpha_2 y - c_2 t)}, \tag{7.55}$$

where c_2 is the speed of sound in the second medium. Now, the conditions at the boundary of the two media are given by

$$v_x^{\mathrm{I}}(0, y, t) = v_x^{\mathrm{II}}(0, y, t) \quad \Rightarrow \quad \psi_{1,x}\big|_{x=0} = \psi_{2,x}\big|_{x=0}, \tag{7.56}$$

$$p^{\mathrm{I}}(0, y, t) = p^{\mathrm{II}}(0, y, t) \quad \Rightarrow \quad \psi_{1,t}\big|_{x=0} = \psi_{2,t}\big|_{x=0}, \tag{7.57}$$

where the superscripts I and II indicate the first and the second medium, respectively, and the pressure has been expressed in terms of the velocity potential using (7.14). Using (7.48)–(7.50) and (7.55) in (7.56) yields

$$k_1 \cos\alpha_1\, Ae^{ik_1(\sin\alpha_1 y - c_1 t)} + k'_1 \cos\alpha_2\, Be^{ik'_1(\sin\alpha'_1 y - c_1 t)}$$
$$= k_2 \cos\alpha_2\, Ce^{ik_2(\sin\alpha_2 y - c_2 t)}. \tag{7.58}$$

This condition can be identically satisfied if and only if

$$k_1 \sin\alpha_1 = k'_1 \sin\alpha'_1 = k_2 \sin\alpha_2 \quad \text{and} \quad k_1 c_1 = k'_1 c_1 = k_2 c_2,$$
$$\Rightarrow k_1 = k'_1, \quad \alpha_1 = \alpha_2 \quad \text{and} \quad \frac{\sin\alpha_2}{\sin\alpha_1} = \frac{k_1}{k_2} = \frac{c_2}{c_1}. \tag{7.59}$$

Thus, we obtain the familiar Snell's law of refraction in (7.59). Using the conditions from (7.59) in (7.58), we obtain

$$k_1 \cos\alpha_1 (A - B) = k_2 C \cos\alpha_2. \tag{7.60}$$

From the pressure condition (7.57) at the interface, we have

$$\rho_1 \psi_{1,t}\big|_{x=0} = \rho_2 \psi_{2,t}\big|_{x=0}$$
$$\Rightarrow \rho_1(A+B) = \rho_2 C. \tag{7.61}$$

Using (7.60) and (7.61) one can easily obtain

$$B = \frac{c_2\rho_2\cos\alpha_1 - c_1\rho_1\cos\alpha_2}{c_2\rho_2\cos\alpha_1 + c_1\rho_1\cos\alpha_2}A, \tag{7.62}$$

$$C = \frac{2c_2\rho_2\cos\alpha_1}{c_2\rho_2\cos\alpha_1 + c_1\rho_1\cos\alpha_2}\frac{\rho_1}{\rho_2}A. \tag{7.63}$$

The reflected and the refracted waves are now completely determined.

Let us now consider the case when $\sin\alpha_2 = (c_2/c_1)\sin\alpha_1 > 1$. As is obvious, this can happen only when $c_2 > c_1$. Then, we have

$$\cos\alpha_2 = \sqrt{1-\sin^2\alpha_2} = i\sqrt{\sin^2\alpha_2 - 1} := i\beta. \tag{7.64}$$

Using this expression and (7.59) in the velocity potential of the refracted wave, we have

$$\psi_2(x, y, t) = C\mathrm{e}^{-k_2\beta x}\mathrm{e}^{ik_1(y\sin\alpha_1 - c_1 t)}. \tag{7.65}$$

It may be noted here that the wave in the second medium is evanescent along the x-axis direction, and travels along the y-axis direction with the speed c_1. Such a wave is known as an *inhomogeneous wave*. Since evanescent waves do not carry energy, there is no energy transport in the x-axis direction. Hence, in this case, the incident energy is completely reflected back into the first medium (*total reflection*).

7.1.5 Spherical waves

Any disturbance in a three-dimensional medium causes, in general, a spatial wave. The simplest of such spatial waves is the spherical wave. Consider the three-dimensional wave equation in spherical coordinates (r, θ, ϕ) given by

$$\psi_{,tt} - c^2\left(\psi_{,rr} + \frac{2}{r}\psi_{,r} + \frac{1}{r^2\sin^2\theta}\psi_{,\phi\phi} + \frac{1}{r^2}\psi_{,\theta\theta} + \frac{1}{r^2}\cot\theta\,\psi_{,\theta}\right) = 0, \tag{7.66}$$

where $\psi(r, \theta, \phi, t)$ is the velocity potential. For symmetrical spherical waves, the velocity potential should depend on r only, i.e., $\psi = \psi(r, t)$. In that case, (7.66) simplifies to

$$\psi_{,tt} - c^2\left(\psi_{,rr} + \frac{2}{r}\psi_{r}\right) = 0$$
$$\Rightarrow (r\psi)_{,tt} - c^2(r\psi)_{,rr} = 0. \tag{7.67}$$

This is clearly the one-dimensional wave equation for $r\psi$. Therefore, a general solution of (7.67) can be written as

$$r\psi(r,t) = f(r-ct) + g(r+ct)$$
$$\Rightarrow \psi(r,t) = \frac{1}{r}[f(r-ct) + g(r+ct)],$$

where $f(z)$ and $g(z)$ are arbitrary functions.

As mentioned previously, any wave pulse can be represented by superposition of harmonic waves. Consider an outgoing harmonic spherical wave solution of (7.67) in the form

$$\psi(r,t) = \frac{A}{r}\,\mathrm{e}^{i(kr-\omega t)} \tag{7.68}$$

where A is the amplitude of the harmonic wave. The velocity and pressure fields can be calculated as, respectively,

$$\mathbf{v}(r,t) = \nabla\psi = A\left[-\frac{1}{r^2} + \frac{ik}{r}\right]\mathrm{e}^{i(kr-\omega t)}\hat{\mathbf{e}}_\mathrm{r}$$

$$\overline{p}(r,t) = -\rho_0\psi_{,t} = A\rho_0\frac{i\omega}{r}\mathrm{e}^{i(kr-\omega t)},$$

where $\hat{\mathbf{e}}_\mathrm{r}$ is the unit vector in the outward radial direction.

The specific mechanical impedance of a medium for spherical harmonic waves can be obtained as

$$\mathcal{Z} = \frac{\mathscr{A}[\overline{p}]}{\mathscr{A}[v]} = \rho_0 c\left(\frac{k^2r^2}{1+k^2r^2} - i\frac{kr}{1+k^2r^2}\right). \tag{7.69}$$

It is interesting to note that, as $kr \to \infty$, one obtains $\mathcal{Z} \to \rho_0 c$, which is the specific impedance of the medium for a plane wave. The average intensity of the spherical wave at any radius r can be determined using (7.69) in the definition (7.44) as

$$\langle\mathcal{I}\rangle = \frac{1}{2}\mathscr{R}[\mathcal{Z}^*\mathbf{v}^*\cdot\mathbf{v}] = \frac{1}{2}\rho_0 c\frac{k^2}{r^2}A^2.$$

Thus, the average intensity of spherical harmonic waves falls as $1/r^2$. The average power flow can be obtained as

$$\langle\mathcal{P}\rangle = 4\pi r^2\langle\mathcal{I}\rangle = 2\pi\rho_0 ck^2A^2.$$

It is evident from the above that the real part of $\mathcal{Z}$ is associated with the power flow. A physical interpretation of the real and imaginary parts of $\mathcal{Z}$ is discussed with the following example.

Let us consider the spherical harmonic waves generated from a harmonically *breathing* sphere of radius R, as shown in Figure 7.3, and determine the associated acoustic field. Assume that the surface of the sphere oscillates radially with a velocity

$$\mathbf{v}_\mathrm{S} = \hat{v}\mathrm{e}^{i\omega t}\hat{\mathbf{e}}_\mathrm{r}, \tag{7.70}$$

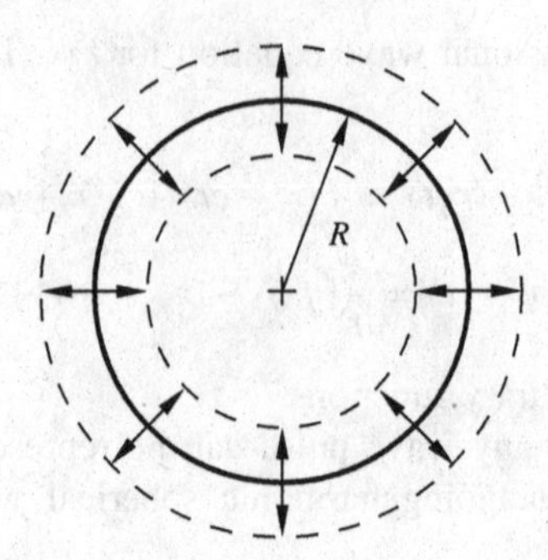

Figure 7.3 A uniformly radially oscillating (*breathing*) sphere

where $\hat{v}$ is the amplitude of vibration. Let the corresponding velocity potential for the fluid be given by

$$\psi(r,t) = \frac{\hat{\psi}}{r}\mathrm{e}^{i(-kr+\omega t)}, \tag{7.71}$$

where $\hat{\psi}$ is an unknown amplitude. The velocity potential (7.71) has been chosen with the consideration that the waves are all outgoing (see Sommerfeld radiation condition in Section 4.5.2), and having the same frequency as that of the surface of the sphere. The velocity field in the fluid is then obtained as

$$\mathbf{v}_{\mathrm{F}} = \left(\hat{\psi}\left[\frac{1}{r^2} + \frac{ik}{r}\right]\mathrm{e}^{i(-kr+\omega t)}\right)\hat{\mathbf{e}}_r.$$

Matching the velocity of the fluid on the surface of the sphere $\mathbf{v}_{\mathrm{F}}|_{r=R}$, and the velocity of the surface $\mathbf{v}_{\mathrm{S}}$ yields

$$\begin{aligned}\hat{v}\mathrm{e}^{i\omega t} &= \hat{\psi}\left[\frac{1}{R^2} + \frac{ik}{R}\right]\mathrm{e}^{-ikR}\\ \Rightarrow \hat{\psi} &= \hat{v}\frac{R^2}{1+ikR}\mathrm{e}^{ikR}.\end{aligned} \tag{7.72}$$

Using this, the velocity and pressure fields in the fluid can be expressed as, respectively,

$$\mathbf{v}(r,t) = \frac{R^2}{r^2}\frac{1+ikr}{1+ikR}\hat{v}\mathrm{e}^{ikR}\mathrm{e}^{i(-kr+\omega t)}\hat{\mathbf{e}}_r, \tag{7.73}$$

$$\overline{p}(r,t) = \frac{\rho_0 i\omega}{r}\frac{R^2}{1+ikR}\hat{v}\mathrm{e}^{ikR}\mathrm{e}^{i(-kr+\omega t)}. \tag{7.74}$$

The actual solution is obtained by taking the real part of the above expressions.

The specific impedance of the medium, as seen by the sphere, can be easily obtained from (7.73)–(7.74) as

$$\mathcal{Z} = \frac{\mathscr{A}[\overline{p}]\big|_{r=R}}{\mathscr{A}[v]\big|_{r=R}} = \left(\frac{\rho_0 c k^2 R^2}{1+k^2R^2}\right) + i\left(\frac{\rho_0 c k R}{1+k^2R^2}\right) = \mathcal{Z}_\mathrm{R} + i\mathcal{Z}_\mathrm{I}. \tag{7.75}$$

In general, the real part $\mathcal{Z}_\mathrm{R}$ is known as *radiation resistance*, while the imaginary part $\mathcal{Z}_\mathrm{I}$ is known as *radiation reactance*. As can be easily checked, the pressure on the sphere surface due to a harmonic radial velocity of the form (7.70) can be written as

$$p(t) = \mathcal{Z}_\mathrm{R} v + \frac{\mathcal{Z}_\mathrm{I}}{\omega}\frac{\mathrm{d}v}{\mathrm{d}t} = d_\mathrm{r} v + m_\mathrm{r}\frac{\mathrm{d}v}{\mathrm{d}t}, \tag{7.76}$$

where we have assumed $\mathcal{Z}_\mathrm{I} > 0$. This equation provides a physical interpretation of the real and imaginary parts of the specific impedance $\mathcal{Z}$. It is evident that $d_\mathrm{r} := \mathcal{Z}_\mathrm{R}$ can be interpreted as the damping coefficient, and is known as the *radiation damping* coefficient. It was observed before that $\mathcal{Z}_\mathrm{R}$ is associated with the power flow through the medium. In this case, the power flow through the medium will be equal to the power lost by the sphere through radiation damping. On the other hand $m_\mathrm{r} := \mathcal{Z}_\mathrm{I}/\omega = \mathcal{Z}_\mathrm{I}/ck$ behaves like a mass, and hence represents an *added mass* (or *virtual mass*) per unit area of the sphere surface. For the breathing sphere, the variations of the real and imaginary parts of the specific impedance with the non-dimensional wave number kR are shown in Figure 7.4. The variations of the radiation damping and added mass coefficients (both non-dimensionalized) with the non-dimensional wave number kR for a sphere of radius R are shown in Figure 7.5. It is observed that for small wave numbers (i.e., large wavelengths), the fluid behaves more like an added mass. On the other hand, for large wave numbers, the behavior of the fluid is more like that of a damper. Alternatively, it may be said that the sphere radiates more effectively at large wave numbers. For $\mathcal{Z}_\mathrm{I} < 0$, the expression of $p(t)$ in (7.76) should be modified to

$$p(t) = \mathcal{Z}_\mathrm{R} v - \omega\mathcal{Z}_\mathrm{I}\int_0^t v(\tau)\,\mathrm{d}\tau. \tag{7.77}$$

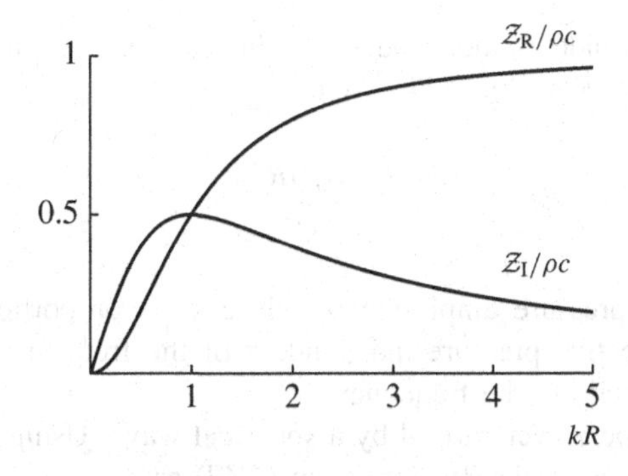

Figure 7.4 Non-dimensional specific radiation resistance $\mathcal{Z}_\mathrm{R}/\rho c$ and reactance $\mathcal{Z}_\mathrm{I}/\rho c$ for a breathing sphere

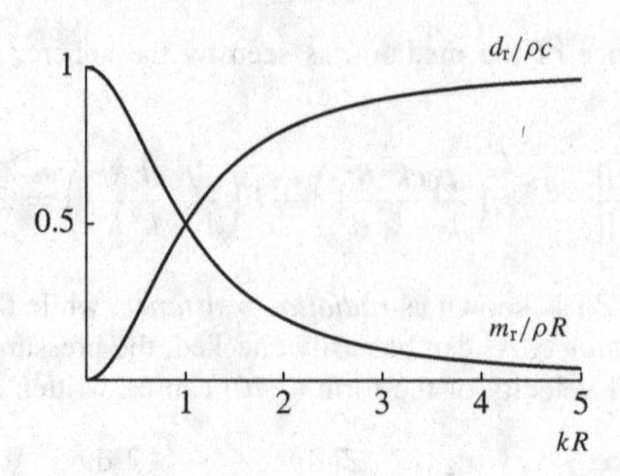

Figure 7.5 Non-dimensional added mass coefficient $m_r/\rho R$, and non-dimensional radiation damping coefficient $d_r/\rho c$ for a breathing sphere

Then $-\omega\mathcal{Z}_I$ has the interpretation of a stiffness. It may be reiterated that the above interpretations of mass, stiffness and damping are valid only for harmonic breathing motion of the sphere. One more point of importance is the distinction between ordinary damping and radiation damping. While in ordinary damping mechanical energy is irreversibly converted to thermal energy, in radiation damping mechanical energy just flows out into an infinite medium, and is therefore lost.

Let us consider the pressure field around a breathing sphere. The root mean square value (or effective value) of the pressure at any radius r is obtained using the definition (7.45) as

$$\overline{p}_{\text{rms}} = \frac{\rho_0\omega}{\sqrt{2}}\frac{R^2}{\sqrt{1+k^2R^2}}\frac{\hat{v}}{r}. \tag{7.78}$$

It is interesting to consider the influence of the wave number k (specifically kR) on the pressure amplitude. For $kR \gg 1$ (i.e., $R \gg \lambda$), we have

$$\overline{p}_{\text{rms}} \approx \frac{\rho_0\omega}{k\sqrt{2}}R\frac{\hat{v}}{r} = \frac{\rho_0 c R}{\sqrt{2}}\frac{\hat{v}}{r}. \tag{7.79}$$

Thus, the pressure amplitude is independent of the frequency in this case. On the other hand, when $kR \ll 1$ (i.e., $R \ll \lambda$), (7.78) yields

$$\overline{p}_{\text{rms}} \approx \frac{\rho_0\omega R^2}{\sqrt{2}}\frac{\hat{v}}{r}, \tag{7.80}$$

which implies that the pressure amplitude in this case is proportional to the frequency. Hence, to make the effective pressure independent of the frequency, one has to vary the velocity amplitude inversely to the frequency.

Next, we determine the power carried by a spherical wave. Using (7.73) and (7.74), one can compute the average acoustic intensity from (7.43) as

$$\langle\mathcal{I}\rangle = \langle|\boldsymbol{\mathcal{I}}|\rangle = \frac{1}{2}\rho_0\omega k\frac{R^4}{1+k^2R^2}\frac{\hat{v}^2}{r^2}. \tag{7.81}$$

Since the intensity of a wave is the power per unit area, one can compute the average power carried by the spherical wave as

$$\langle \mathcal{P} \rangle = 4\pi r^2 \langle \mathcal{I} \rangle = 2\pi \rho_0 \omega k \hat{v}^2 \frac{R^4}{1 + k^2 R^2}. \tag{7.82}$$

For $kR \gg 1$, we have from (7.82)

$$\langle \mathcal{P} \rangle \approx 2\pi \rho_0 c \hat{v}^2 R^4. \tag{7.83}$$

On the other hand, when $kR \ll 1$, (7.82) yields

$$\langle \mathcal{P} \rangle \approx 2\pi \rho_0 \omega k \hat{v}^2 R^4 = 2\pi \frac{\rho_0}{c} \omega^2 \hat{v}^2 R^4. \tag{7.84}$$

For $kR \gg 1$, the acoustic power is independent of the frequency. Hence, to deliver uniform acoustic power over a frequency band, we need to ensure that $R \gg c/\omega_{\min}$, where $\omega_{\min}$ is the minimum circular frequency of the band.

A breathing sphere constitutes what is defined as a simple source of sound, or an *acoustic monopole*. An acoustic monopole may be considered to be a spherical source with $R \to 0$, but with finite volume flow rate amplitude of the fluid. The volume flow rate amplitude, which is also referred to as the *source strength* of the monopole, can be computed as

$$\hat{Q} := \left| \int_S \mathbf{v} \cdot \hat{\mathbf{n}} \mathrm{dS} \right| = 4\pi R^2 \hat{v}.$$

Using this definition and the condition $R \to 0$, the velocity potential obtained from (7.71) and (7.72) takes the form

$$\psi(r, t) = \frac{\hat{Q}}{4\pi r} \mathrm{e}^{i(-kr + \omega t)}.$$

A complex sound source may be modeled, in the first approximation, as a distribution of such acoustic monopoles.

7.1.6 Cylindrical waves

There are many problems of interest where the sound waves can be approximated by spherical waves. There are, however, also situations where the waves are generated from an extended surface. To the first approximation, they can be considered to be generated from a cylindrical surface with the radius of the cylinder oscillating uniformly throughout its length.

Consider the wave equation in cylindrical polar coordinates (r, ϕ, z) as

$$\psi_{,tt} - c^2 \left[\frac{1}{r} (r \psi_{,r})_{,r} + \frac{1}{r^2} \psi_{,\phi\phi} + \psi_{,zz} \right] = 0. \tag{7.85}$$

Assuming uniformity in the ϕ and z coordinate directions, we have $\psi = \psi(r,t)$. The wave equation (7.85) then simplifies to

$$\psi_{,tt} - c^2 \left[\frac{1}{r}(r\psi_{,r})_{,r} \right] = 0. \tag{7.86}$$

Substituting

$$\psi(r,t) = F(r)\mathrm{e}^{i\omega t} \tag{7.87}$$

in (7.86) yields on simplification

$$F''(r) + \frac{1}{r}F'(r) + \frac{\omega^2}{c^2}F(r) = 0, \tag{7.88}$$

which is Bessel's differential equation. Defining $k = \omega/c$, the general solution of (7.88) can be written as (see, for example, [3])

$$F(r) = BJ_0(kr) + CY_0(kr), \tag{7.89}$$

where $J_0(kr)$ and $Y_0(kr)$ are, respectively, the zeroth-order Bessel functions of first and second kinds.

Let us assume that there are no other sources of disturbance in the medium. Then, as discussed in Section 4.5.2, the solution of (7.86) must satisfy the Sommerfeld radiation condition. This is ensured by first expressing the solution (7.89) in terms of the Hankel functions (see Section 4.5.2) as $F(r) = DH_0^{(1)}(kr) + EH_0^{(2)}(kr)$, and then taking $D = 0$. Here, $H_0^{(1)}(\cdot)$ and $H_0^{(2)}(\cdot)$ are the Hankel functions of first and second kinds, respectively. Thus, the solution (7.87) is of the form

$$\psi(r,t) = EH_0^{(2)}(kr)\mathrm{e}^{i\omega t}, \tag{7.90}$$

where

$$H_0^{(2)}(kr) = J_0(kr) - iY_0(kr).$$

For $kr \gg 1$, we can approximate $H_0^{(2)}(kr)$ as

$$H_0^{(2)}(kr) \approx \sqrt{\frac{2}{\pi kr}}\mathrm{e}^{i(-kr+\pi/4)}, \tag{7.91}$$

so that (7.90) can be written as

$$\psi(r,t) \approx E\sqrt{\frac{2}{\pi kr}}\mathrm{e}^{i(-kr+\omega t+\pi/4)} \qquad \text{for } kr \gg 1, \tag{7.92}$$

which is an outgoing harmonic wave radiated from a cylindrical surface. The calculations of wave impedance and energetics of cylindrical waves can be performed similarly as discussed for the case of spherical waves in Section 7.1.5.

7.1.7 Acoustic radiation from membranes and plates

The problem of sound emission from machines or other vibrating flexible surfaces is an important and complex engineering problem. This problem requires a detailed study of the interplay between the dynamics of the surface and the dynamics of the surrounding fluid. The dynamics of vibrating flexible surfaces in the simplest cases can be modeled as either a membrane or a plate. In most of these cases, no analytical solutions are possible, and only numerical solutions are sought. Here, we consider only a simplified problem consisting of radiation of sound in air from a vibrating flexible surface. To keep the analysis simple, we disregard the effect of the forces generated by the air on the surface. This implies that the surface provides only a kinematic boundary condition for the motion of the air above it. The effect of fluid loading on the vibrating surface will be discussed briefly at the end of this section.

Consider a membrane or a plate of infinite extent in the horizontal x-y-plane of the Cartesian coordinate system, as shown in Figure 7.6. The motion of the flexible surface, assumed to be independent of the dynamics of the air, is taken as

$$w(x, y, t) = A \sin k_x x \, \sin k_y y \, e^{i\Omega t}, \tag{7.93}$$

where the real part of $w(x, y, t)$ represents the actual motion. It may be mentioned that such a motion of an infinite surface can be considered to be the superposition of two counter-propagating harmonic waves. The normal velocity of the surface is given by

$$w_{,t} = \hat{v} \sin k_x x \, \sin k_y y \, e^{i\Omega t}, \tag{7.94}$$

where $\hat{v} = i\Omega A$.

Let us study the sound generated above the surface, i.e., in the region $z > 0$. The acoustic wave equation governing the motion of the air above the surface may be written as

$$\psi_{,tt} - c^2(\psi_{,xx} + \psi_{,yy} + \psi_{,zz}) = 0, \tag{7.95}$$

where $\psi(x, y, z, t)$ represents the velocity potential and c is the speed of sound in air. Since the velocity of the air normal to the surface must satisfy the boundary condition (7.94), let us consider a velocity potential of the form

$$\psi = Z(z) \sin k_x x \, \sin k_y y \, e^{i\Omega t}, \tag{7.96}$$

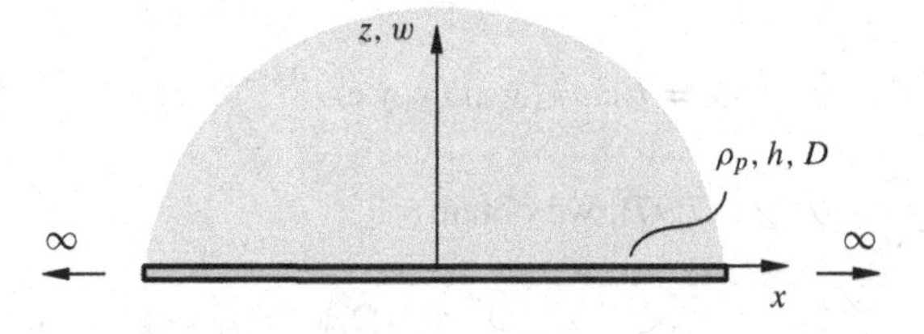

Figure 7.6 Coordinate system for an infinite surface radiating sound

where $Z(z)$ is an unknown function of z. The velocity boundary condition on the surface can now be represented as

$$v_z\big|_{z=0} = \psi_{,z}\big|_{z=0} = w_{,t},$$
$$\Rightarrow Z'(0) = \hat{v} \quad \text{(using (7.94) and (7.96))}. \tag{7.97}$$

Substituting (7.96) in (7.95) yields

$$Z'' + \left(\frac{\Omega^2}{c^2} - k_x^2 - k_y^2\right) Z = 0$$

or

$$Z'' + (k^2 - k_\mathrm{T}^2)Z = 0, \tag{7.98}$$

where $k^2 := \Omega^2/c^2$ and $k_T^2 := k_x^2 + k_y^2$. Now, there can be two possibilities:

$$k^2 - k_\mathrm{T}^2 > 0 \qquad \text{and} \qquad k^2 - k_\mathrm{T}^2 < 0.$$

Corresponding to these two possibilities, let us represent (7.98) as, respectively,

$$Z'' + k_z^2 Z = 0, \qquad \text{where } k_z^2 = k^2 - k_\mathrm{T}^2, \tag{7.99}$$

or

$$Z'' - k_z^2 Z = 0, \qquad \text{where } k_z^2 = k_\mathrm{T}^2 - k^2. \tag{7.100}$$

The solution of (7.99) can be represented as

$$Z(z) = a\mathrm{e}^{ik_z z} + b\mathrm{e}^{-ik_z z}, \tag{7.101}$$

where a and b are the arbitrary constants of integration. Now, if we consider an infinite air medium with no sound source other than the surface, then all acoustic waves must propagate along the positive z-axis direction above the surface. Hence, we must have $a = 0$. Therefore, the velocity potential (7.96) can be written as

$$\psi = b \sin k_x x \, \sin k_y y \, \mathrm{e}^{i(-k_z z + \Omega t)}. \tag{7.102}$$

From the boundary condition (7.97), we obtain

$$b = i\frac{\hat{v}}{k_z} = -\frac{\Omega A}{k_z}.$$

Therefore, the actual velocity and pressure fields are given by, respectively,

$$\mathbf{v}(x, y, z, t) = \begin{Bmatrix} -\Omega A \dfrac{k_x}{k_z} \cos k_x x \, \sin k_y y \, \cos(-k_z z + \Omega t) \\ -\Omega A \dfrac{k_y}{k_z} \sin k_x x \, \cos k_y y \, \cos(-k_z z + \Omega t) \\ \Omega A \sin k_x x \, \sin k_y y \, \sin(-k_z z + \Omega t) \end{Bmatrix},$$

$$p(x, y, z, t) = \rho_0 \frac{\Omega A}{k_z} \sin k_x x \, \sin k_y y \, \sin(-k_z z + \Omega t).$$

Consider next the case represented by (7.100) for which the solution is obtained as

$$Z(z) = a e^{k_z z} + a e^{-k_z z}.$$

In the region $z > 0$, if the solution is to be finite, we must have $a = 0$. Using this condition, the velocity potential satisfying the boundary condition (7.97) is easily obtained as

$$\psi = -\frac{i\omega A}{k_z} \sin k_x x \, \sin k_y y \, \mathrm{e}^{-k_z z} \mathrm{e}^{i\Omega t}. \tag{7.103}$$

In this case, the actual velocity and pressure fields are given by

$$\mathbf{v}(x, y, z, t) = \begin{Bmatrix} \Omega A \dfrac{k_x}{k_z} \cos k_x x \, \sin k_y y \, \mathrm{e}^{-k_z z} \sin \Omega t \\ \Omega A \dfrac{k_y}{k_z} \sin k_x x \, \cos k_y y \, \mathrm{e}^{-k_z z} \sin \Omega t \\ -\Omega A \sin k_x x \, \sin k_y y \, \mathrm{e}^{-k_z z} \sin \Omega t \end{Bmatrix},$$

$$p(x, y, z, t) = -\rho_0 \frac{\Omega A}{k_z} \sin k_x x \, \sin k_y y \, \mathrm{e}^{-k_z z} \cos \Omega t.$$

Let us now compute the average intensity of the acoustic waves on a plane parallel to the x-y-plane for the two cases discussed above. This calculation can be easily performed as

$$\langle \mathcal{I}(x, y) \rangle = \frac{1}{2} \mathscr{R}[p^* v_z] = -\frac{1}{2} \mathscr{R}[\rho_0 \psi_{,t}^* \psi_{,z}].$$

For the second case (i.e., $k^2 - k_{\mathrm{T}}^2 < 0$) $\langle \mathcal{I}(x, y) \rangle = 0$, while for the first case

$$\langle \mathcal{I}(x, y) \rangle = \frac{\rho_0 \Omega}{2 k_z} \hat{v}^2 \sin^2 k_x x \, \sin^2 k_y y. \tag{7.104}$$

One can also write (7.104) as

$$\langle \mathcal{I}(x, y)\rangle = \rho_0 c \frac{k}{k_z}\frac{\hat{v}}{2} \sin^2 k_x x \, \sin^2 k_y y = \rho_0 c \frac{k}{k_z} v_{\text{rms}}^2,$$

where $v_{\text{rms}}^2 = \mathcal{R}[\mathbf{v}^* \cdot \mathbf{v}]/2$. The average intensity of a plane wave generated by an oscillating *rigid* surface is given by

$$\langle \mathcal{I}_{\text{w}}\rangle = p_{\text{rms}} v_{\text{rms}} = \rho_0 c v_{\text{rms}}^2.$$

Comparing $\langle \mathcal{I}\rangle$ with $\langle \mathcal{I}_{\text{w}}\rangle$, one can define the *radiation efficiency* σ of a membrane or a plate as

$$\sigma := \frac{\langle \mathcal{I}\rangle}{\langle \mathcal{I}_{\text{w}}\rangle} = \frac{k}{k_z}.$$

When $k < k_{\text{T}}$, $\sigma = 0$, while in the case $k > k_{\text{T}}$, we have

$$\sigma = \frac{k}{\sqrt{k^2 - k_{\text{T}}^2}}.$$

The dependence of the radiation efficiency on the non-dimensional factor k/k_{T} is shown in Figure 7.7. For a *finite* membrane or plate, the figure is somewhat modified since there is some radiation even for $k < k_{\text{T}}$.

Let us now examine the situation $k < k_{\text{T}}$ in which case the acoustic wave becomes evanescent. Consider the $x - z$-plane $k_y y = \pi/2$ on which the velocity field is given by

$$\mathbf{v}(x, y, z, t) = \begin{Bmatrix} \Omega A \dfrac{k_x}{k_z} \cos k_x x \, \mathrm{e}^{-k_z z} \, \sin \Omega t \\ 0 \\ -\Omega A \sin k_x x \, \mathrm{e}^{-k_z z} \, \sin \Omega t \end{Bmatrix}. \tag{7.105}$$

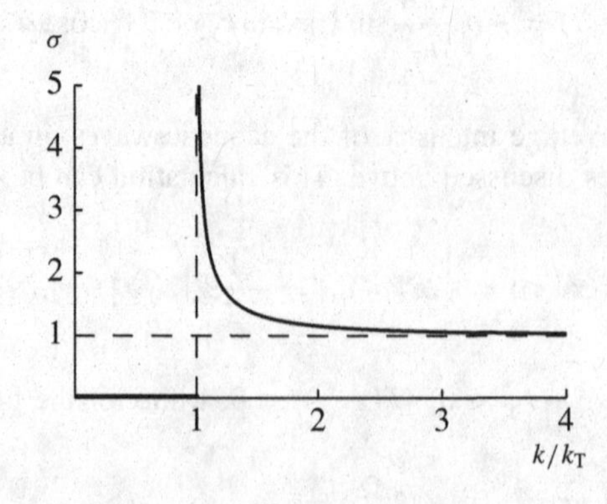

Figure 7.7 Radiation efficiency of a vibrating flexible surface as a function of the wave number

In order to determine the path of the air particles during motion, we compute the streamlines, which are the curves tangential to the velocity vector at every point.

In the two-dimensional x-z-plane $k_y y = \pi/2$, the tangent to the streamlines at any point can be determined as

$$\frac{\mathrm{d}z}{\mathrm{d}x} = \frac{v_z}{v_x} = -\frac{k_z}{k_x} \tan k_x x. \tag{7.106}$$

The solution of (7.106) yields the streamlines as

$$z = \frac{k_z}{k_x} \ln[\cos k_x x] + a,$$

where a is the constant of integration. The streamlines over a vibrating flexible surface are shown in Figure 7.8. The surface velocity profile on the corresponding $x - z$-plane is indicated by the dashed line. It is clear from (7.105) that the magnitude of the velocity vector falls off exponentially with increasing z. Therefore, the motion of the air is limited to the neighborhood of the oscillating surface. This is sometimes referred to as *acoustic* or *hydrodynamic short-circuit*. Since the velocity field is harmonic in time, the direction of the velocity vector reverses harmonically along the streamlines. It is to be noted that the individual air particles do not travel along the complete streamline, but only oscillate along a very small part of it.

In the above analysis, we assumed a motion of the surface of the form (7.93) in which k_x, k_y and Ω seem unrelated. However, they are related by the dynamics of the surface, which may be modeled, for example, as a membrane or a plate. The membrane equation of motion is given by

$$\mu w_{,tt} - T\nabla^2 w = 0,$$

where μ is the mass of the membrane per unit area and T is the tension per unit length. It must be pointed out here that we have made an approximation in this model by excluding the fluid reaction pressure on the membrane. Substituting (7.93) in the membrane equation of motion yields

$$k_\mathrm{T}^2 = \frac{\mu}{T}\,\Omega^2. \tag{7.107}$$

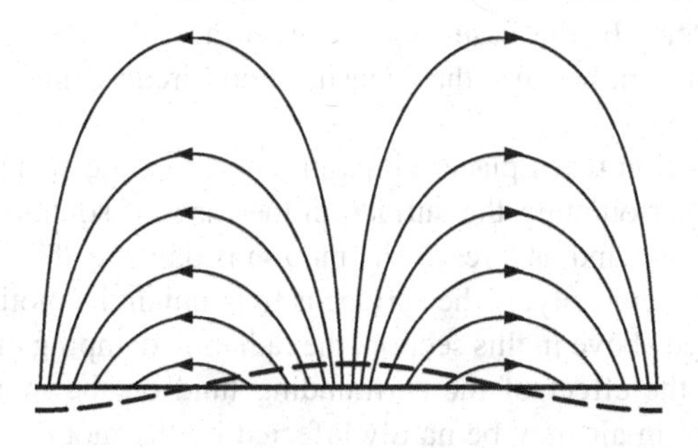

Figure 7.8 Fluid streamlines over a vibrating flexible surface

Now, for the case of acoustic radiation, we have

$$k_{\mathrm{T}}^2 < \Omega^2/c^2. \tag{7.108}$$

Therefore, using (7.107) in (7.108) yields

$$\frac{\mu}{T}\,\Omega^2 < \frac{\Omega^2}{c^2} \quad \Rightarrow \quad T > T^{\mathrm{c}} := \mu c^2, \tag{7.109}$$

where T^{c} is the critical value of tension per unit length for acoustic radiation from a membrane. It can be easily concluded from (7.109) that the wave speed in the membrane $c_{\mathrm{M}} = \sqrt{T/\mu}$ must be greater than the sound speed c in the fluid for the membrane to radiate sound.

In the case of a uniform plate, the equation of motion is given by

$$\rho_{\mathrm{P}} h w_{,tt} + D\nabla^4 w = 0,$$

where ρ_{P} is the density of the plate material, h is the thickness, and $D = Eh^3/[12(1-\nu^2)]$. Substituting (7.93) in the plate equation of motion, we obtain

$$k_{\mathrm{T}}^4 = \frac{\rho_{\mathrm{P}} h}{D}\Omega^2. \tag{7.110}$$

Using (7.110) in (7.108) yields the radiation condition for a plate as

$$\Omega > \Omega^{\mathrm{c}} := c^2\sqrt{\frac{\rho_{\mathrm{P}} h}{D}}, \tag{7.111}$$

where Ω^{c} is known as the *coincidence frequency*. The expression of phase velocity of bending waves in a plate can be obtained from (7.110) as $c_{\mathrm{P}} = \Omega/k_{\mathrm{T}} = k_{\mathrm{T}}\sqrt{D/\rho_{\mathrm{P}} h}$. Using this expression and the dispersion relation $\Omega/k_{\mathrm{T}} = c_{\mathrm{P}}$, one can easily rewrite the radiation condition (7.111) as $c_{\mathrm{P}} > c$. Thus, if the phase velocity of bending waves in the plate is higher than the speed of the acoustic waves in the fluid, the plate will radiate sound. It should be remembered that the above analysis was performed for a flexible surface of infinite extent. In the case of a *finite* surface, there is acoustic radiation even below the cut-off condition, because the acoustic short-circuit cannot occur near the boundaries.

When acoustic radiation takes place, vibration energy of the surface propagates into the infinite fluid medium surrounding the surface in the form of sound. The surface obviously loses energy continuously, and as a result, its motion is damped. This is known as *radiation damping*, as discussed previously. If the surface is to maintain its motion, it has to be forced. In the analysis presented above in this section, the radiation damping effect was not observed, because we excluded the effect of the surrounding fluid on the motion of the surface. A metallic plate vibrating in air may be hardly affected by the motion of the surrounding air. Hence, the dynamics of the plate is practically uncoupled from the dynamics of the air. In such a case, the effect of radiation damping can be observed only over large time scales. However, for the case of a plate vibrating in water, for example, one cannot neglect the effect of the dynamics of the water on the plate. The effect of the fluid pressure has to be included in the equation of motion of the plate as an external pressure term (see [4]).

In order to include the effect of radiation damping for harmonic motion of a surface, say a membrane, we introduce a radiation reaction pressure term in the equation of motion as

$$\mu w_{,tt} - T\nabla^2 w = -p.$$

The radiation pressure can be conveniently related to the membrane kinematics through the impedance of the fluid. The impedance of the fluid, as seen by the membrane, is given by

$$\mathcal{Z} = \frac{\mathscr{A}[\overline{p}]}{\mathscr{A}[v_z]} = -\rho_0 \frac{\mathscr{A}[\psi_{,t}]}{\mathscr{A}[\psi_{,z}]}\bigg|_{z=0},$$

where ψ is in complex time-harmonic form and $\mathscr{A}[\cdot]$ denotes the complex amplitude. In the case of radiation from the membrane, using (7.102), we obtain

$$\mathcal{Z} = \frac{\rho_0 \Omega}{k_z}.$$

Now, we make the approximation that the transverse wave speed in the membrane remains unaffected by the surrounding fluid. Hence, using (7.107) in the definition of k_z in (7.99), one obtains

$$k_z = \frac{\Omega}{c_{\text{M}} c}\sqrt{c_{\text{M}}^2 - c^2}.$$

Since $c_{\text{M}} > c$ for radiation, $\mathcal{Z}$ is real (i.e., resistive), and we can express the pressure at the surface as $p = \mathcal{Z} w_{,t}$. Therefore, taking into account the radiation damping of a harmonically vibrating membrane, one obtains the equation of motion of the membrane as

$$\mu w_{,tt} + d w_{,t} - T\nabla^2 w = 0.$$

where

$$d = \frac{2\rho_0 c_{\text{M}} c}{\sqrt{c_{\text{M}}^2 - c^2}}.$$

Note that a factor of two has been taken in the expression of d to include the effect of fluid loading from both sides of the membrane. When $c_{\text{M}} \gg c$, it is easily observed that $d \approx 2\rho_0 c$. When $c_{\text{M}} < c$, $\mathcal{Z}$ is reactive (imaginary), and hence there is no radiation. In that case, one obtains an added inertia in the membrane due to the fluid. A similar analysis can be performed for a plate.

7.1.8 Waves in wave guides

Wave guides are structured media that can guide a traveling wave (or wave energy) from one point to another. For example, a stethoscope uses a wave guide for transmitting acoustic signals, picked up by the chestpiece, to the ear. Here we consider two simple geometries of wave guides and study the propagation of acoustic waves through them.

7.1.8.1 Wave guide with a rectangular cross-section

Consider an infinitely long linear wave guide with a rectangular cross-section, as shown in Figure 7.9. The reference frame is fixed, with the x-axis directed along the wave guide, as shown in the figure. We study the propagation of acoustic waves through the fluid inside the guide.

Consider the velocity potential corresponding to a wave traveling in the positive x-axis direction of the form

$$\psi(x, y, z, t) = (Ae^{-ik_y y} + Be^{ik_y y})(Ce^{-ik_z z} + De^{ik_z z})e^{i(-k_x x+\omega t)}. \tag{7.112}$$

The condition of zero normal velocity at the walls of the guide yields the corresponding boundary conditions as

$$v_y(x, 0, z, t) \equiv 0, \qquad v_y(x, a, z, t) \equiv 0, \tag{7.113}$$

$$v_z(x, y, 0, t) \equiv 0, \qquad v_z(x, y, b, t) \equiv 0. \tag{7.114}$$

Then, using the definition

$$v_y = \psi_{,y} = ik_y(-Ae^{-ik_y y} + Be^{ik_y y})(Ce^{-ik_z z} + De^{ik_z z})e^{i(-k_x x+\omega t)}, \tag{7.115}$$

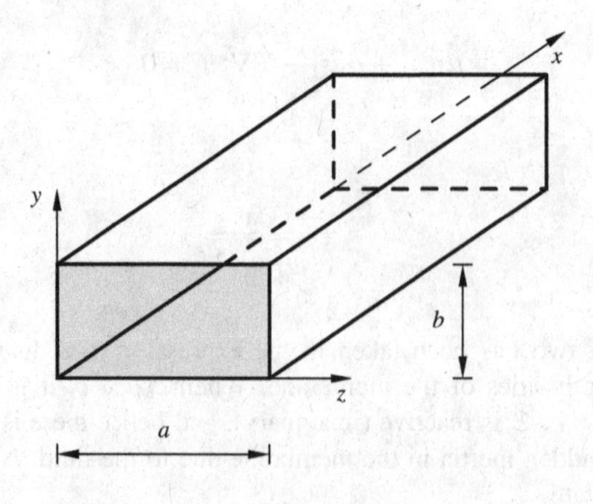

Figure 7.9 A rectangular wave guide

and substituting in the conditions in (7.113) yields

$$A = B \qquad \text{and} \qquad \sin k_y a = 0. \tag{7.116}$$

Similarly, from the conditions on v_z in (7.114), we get

$$C = D \qquad \text{and} \qquad \sin k_z b = 0. \tag{7.117}$$

From the second equalities in (7.113) and (7.114), we have

$$k_y = \frac{m\pi}{a} \qquad \text{and} \qquad k_z = \frac{n\pi}{b}. \tag{7.118}$$

Hence, for the mode (m, n),

$$\psi_{(m,n)}(x, y, z, t) = \left[\hat{\psi}_{(m,n)} \cos\frac{m\pi y}{a} \cos\frac{n\pi z}{b}\right] e^{i(-k_x x + \omega t)}. \tag{7.119}$$

Substituting this in the wave equation yields the dispersion relation

$$k_x^2 + \frac{m^2\pi^2}{a^2} + \frac{n^2\pi^2}{b^2} - \frac{\omega^2}{c^2} = 0. \tag{7.120}$$

It is evident that only the mode $(m, n) = (0, 0)$ is non-dispersive, while all other modes are dispersive. The dispersion relation for a square wave guide is shown in Figure 7.10. The phase velocity of waves in the wave guide is then given by

$$c_P = \frac{\omega}{k_x} = \frac{c}{k_x}\sqrt{k_x^2 + \frac{m^2\pi^2}{a^2} + \frac{n^2\pi^2}{b^2}}. \tag{7.121}$$

The group velocity is obtained as

$$c_G = \frac{d\omega}{dk_x} = \frac{c k_x}{\sqrt{k_x^2 + \frac{m^2\pi^2}{a^2} + \frac{n^2\pi^2}{b^2}}}. \tag{7.122}$$

Therefore, one can easily shown that $c_P\, c_G = c^2$.

Consider the propagation of an acoustic wave of frequency ω through the wave guide. If $\omega < \omega^c$, where

$$\omega^c_{(m,n)} := c\sqrt{\frac{m^2\pi^2}{a^2} + \frac{n^2\pi^2}{b^2}}, \tag{7.123}$$

the wave number k_x of the acoustic wave calculated from (7.120) becomes imaginary. Hence, the wave solution (7.119) for mode (m, n) becomes evanescent, and cannot propagate.

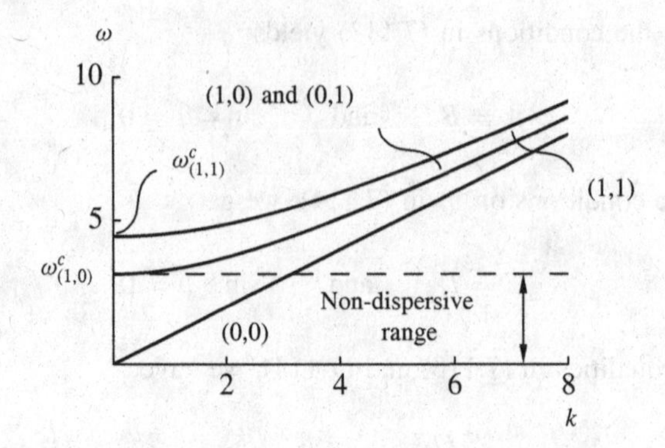

Figure 7.10 Dispersion relation of a square wave guide

However, higher modes may still propagate. The frequency $\omega^c_{(m,n)}$ is hence known as the *cut-off frequency* for the mode (m, n) of the wave guide. The cut-off frequencies are indicated in Figure 7.10 for the modes shown. There is no cut-off frequency for the mode $(0, 0)$.

If an acoustic signal consisting of a band of frequencies is to be transmitted through a wave guide, dispersion is to be avoided, so that the signal is not distorted. The frequency band of interest must then be made to fall within the non-dispersive frequency range shown in Figure 7.10. Therefore, the highest frequency of the band, $\omega_{\max}$, must satisfy

$$\omega_{\max} < \min[\omega^c_{(1,0)}, \omega^c_{(0,1)}]$$

$$\Rightarrow \frac{\omega_{\max}}{c} < \min\left[\frac{\pi}{a}, \frac{\pi}{b}\right] \qquad \text{(using (7.123))}. \tag{7.124}$$

In the non-dispersive range, $\omega_{\max}/c = k_{\max} = 2\pi/\lambda_{\min}$. Hence, in terms of the wavelength, the condition for no dispersion can be written from (7.124) as

$$\max[a, b] < \lambda_{\min}/2.$$

In other words, one must choose a wave guide with dimensions a and b smaller than half the shortest wavelength to be transmitted.

7.1.8.2 Wave guide with a circular cross-section

Next consider the more important case of a circular wave guide of radius R, as shown in Figure 7.11. Let the velocity potential in this case be

$$\psi(r, \phi, z, t) = F(r)\mathrm{e}^{im\phi}\mathrm{e}^{i(-k_z z+\omega t)}, \tag{7.125}$$

where m is an integer. Substituting (7.125) in the wave equation in the cylindrical polar coordinates (7.85) yields

$$F''(r) + \frac{1}{r}F'(r) + \left(\alpha^2 - \frac{m^2}{r^2}\right)F(r) = 0, \tag{7.126}$$

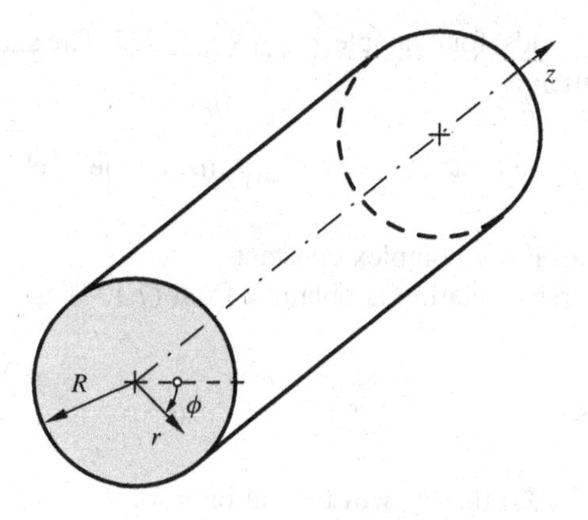

Figure 7.11 A circular wave guide

where

$$\alpha^2 := \frac{\omega^2}{c^2} - k_z^2. \tag{7.127}$$

The solution of the Bessel differential equation (7.126) can be written in the form (see, for example, [3])

$$F(r) = AJ_m(\alpha r) + BY_m(\alpha r). \tag{7.128}$$

From the finiteness condition at $r = 0$, we must have $B = 0$ in (7.128). Therefore, the velocity potential (7.125) can now be written as

$$\psi(r, \phi, z, t) = \hat{\psi} J_m(\alpha r) e^{im\phi} e^{i(-k_z z + \omega t)}. \tag{7.129}$$

Using (7.129) in the boundary condition $v_r(R) = \psi_{,r}|_{r=R} \equiv 0$ yields

$$J'_m(\alpha R) = 0. \tag{7.130}$$

The first few roots of (7.130) are obtained as $\alpha_{(0,1)}R = 0.0$, $\alpha_{(0,2)}R = 3.8317$, $\alpha_{(1,1)}R = 1.8412$, $\alpha_{(1,2)}R = 5.3314$, and so on. Using the approximate representation

$$J_m(x) \approx \sqrt{\frac{2}{\pi x}} \cos\left(x - (2m + 1)\frac{\pi}{4}\right), \qquad \text{for } x \gg 1 \tag{7.131}$$

in (7.130), one can obtain approximate values of $\alpha_{(m,n)}R$ as

$$\alpha_{(m,n)}R \approx [2m + 4n - 3]\frac{\pi}{4}.$$

This approximation yields, for example, $\alpha_{(0,2)}R = 3.927$. The solution for the velocity potential can now be written as

$$\psi_{(m,n)}(r, \phi, z, t) = C_{(m,n)} J_m(\alpha_{(m,n)} r) e^{im\phi} e^{i(-k_z z + \omega t)}, \tag{7.132}$$

where $C_{(m,n)}$ is an arbitrary complex constant.

Finally, the dispersion relation is obtained from (7.127) as

$$k_z^2 = \frac{\omega^2}{c^2} - \alpha_{(m,n)}^2. \tag{7.133}$$

It is evident from (7.133) that k_z will be real only for

$$\omega > \omega_{(m,n)}^c := c\alpha_{(m,n)}, \tag{7.134}$$

where $\omega_{(m,n)}^c$ is the cut-off frequency for the mode (m, n). Below the frequency $\omega_{(m,n)}^c$, there will be no transmission in the mode (m, n). Since $\alpha_{(0,1)} = 0$, the circular wave guide is non-dispersive for the mode $(0, 1)$ and dispersive for all higher modes. If one wants to avoid dispersion in a frequency band of interest, one must select the dimensions of the wave guide such that the first cut-off frequency $\omega_{(1,1)}^c$ lies above the maximum frequency of the considered frequency band, as was demonstrated for rectangular wave guides in Section 7.1.8.1.

7.1.9 Acoustic waves in a slightly viscous fluid

In the previous sections, we considered acoustic waves in inviscid fluid media. In this section we study the propagation (or attenuation) of waves when the medium has slight viscosity. To keep the analysis simple, the most important assumption we make is that the entropy generation due to the slight internal dissipation is negligible. Further, as before, we also neglect thermal diffusion. More detailed analysis (including thermal diffusion) can be found in [5].

To study the effect of viscosity, we begin with the linearized Navier–Stokes equation

$$\frac{\partial \mathbf{v}}{\partial t} = -\frac{1}{\rho_0}\nabla \overline{p} + \nu \nabla^2 \mathbf{v} + \frac{\nu}{3}\nabla(\nabla \cdot \mathbf{v}), \tag{7.135}$$

where $\nu := \mu/\rho_0$ is the kinematic viscosity, μ is the viscosity, and we assume μ to be small. The conditions used for linearization remain the same as in Section 7.1.1. The linearized equation of continuity and the equation of state are given by, respectively,

$$\frac{\partial \overline{\rho}}{\partial t} + \rho_0 \nabla \cdot \mathbf{v} = 0, \tag{7.136}$$

$$\overline{p} = c^2 \overline{\rho}. \tag{7.137}$$

Substituting the expression of $\overline{\rho}$ from (7.137) in (7.136) yields

$$\frac{\partial \overline{p}}{\partial t} + c^2 \rho_0 \nabla \cdot \mathbf{v} = 0, \tag{7.138}$$

and differentiating this with respect to time gives

$$\frac{\partial^2 \overline{p}}{\partial t^2} + c^2 \rho_0 \nabla \cdot \frac{\partial \mathbf{v}}{\partial t} = 0. \tag{7.139}$$

Substituting for the local acceleration $\partial \mathbf{v}/\partial t$ from (7.135) in (7.139), one obtains

$$\frac{\partial^2 \overline{p}}{\partial t^2} + c^2 \rho_0 \left[-\frac{1}{\rho_0} \nabla^2 \overline{p} + \nu \nabla^2 (\nabla \cdot \mathbf{v}) + \frac{\nu}{3} \nabla^2 (\nabla \cdot \mathbf{v}) \right] = 0$$

or

$$\frac{\partial^2 \overline{p}}{\partial t^2} + c^2 \rho_0 \left[-\frac{1}{\rho_0} \nabla^2 \overline{p} + \frac{4\nu}{3} \nabla^2 (\nabla \cdot \mathbf{v}) \right] = 0$$

or

$$\frac{\partial^2 \overline{p}}{\partial t^2} + c^2 \rho_0 \left[-\frac{1}{\rho_0} \nabla^2 \overline{p} - \frac{4\nu}{3\rho_0 c^2} \nabla^2 \frac{\partial \overline{p}}{\partial t} \right] = 0 \qquad \text{(using (7.138))}$$

or

$$\frac{\partial^2 \overline{p}}{\partial t^2} - c^2 \nabla^2 \overline{p} - \frac{4\nu}{3} \nabla^2 \frac{\partial \overline{p}}{\partial t} = 0. \tag{7.140}$$

This is the governing differential equation for acoustic pressure waves in a slightly viscous medium. As can be easily recognized in (7.140), we have the wave equation with an additional viscosity term which is responsible for attenuation, as will be shown below. It may be mentioned that no velocity potential can be defined here, since the fluid motion is rotational in general. However, for plane waves, and purely spherical and cylindrical waves one can easily show that $\nabla \times \mathbf{v} \equiv \mathbf{0}$. In these cases, therefore, one can write $\mathbf{v} = \nabla \psi$, and the wave equation with viscosity is obtained as

$$\frac{\partial^2 \psi}{\partial t^2} - c^2 \nabla^2 \psi - \frac{4\nu}{3} \nabla^2 \frac{\partial \psi}{\partial t} = 0.$$

Consider a propagating plane harmonic pressure wave represented by

$$\overline{p} = \overline{p}_0 e^{i(k\hat{\mathbf{n}} \cdot \mathbf{r} - \omega t)}. \tag{7.141}$$

Substituting this in (7.140) yields the dispersion relation

$$\omega^2 - c^2k^2 + i\frac{4\nu}{3}\omega k^2 = 0, \tag{7.142}$$

from where the k is obtained in terms of the frequency as

$$k^2 = \frac{\omega^2}{c^2 - i\frac{4\nu}{3}\omega}$$
$$\Rightarrow k \approx \pm\frac{\omega}{c}\left[1 + i\frac{2\nu}{3c^2}\omega\right]. \tag{7.143}$$

It is evident from (7.143) that viscosity makes the medium dispersive.

Consider a plane harmonic pressure wave traveling along the direction of the positive x-axis. Substituting the expression of k for the positive-traveling wave in (7.141), and writing $\hat{\mathbf{n}} \cdot \mathbf{r} = x$, we obtain

$$\overline{p} = \overline{p}_0 \mathrm{e}^{-\frac{2\nu\omega^2}{3c^3}x} \mathrm{e}^{i\frac{\omega}{c}(x-ct)}.$$

Thus, the pressure amplitude falls exponentially as the wave propagates. Further, the higher the frequency, the higher is the attenuation of the wave.

7.2 SURFACE WAVES IN INCOMPRESSIBLE LIQUIDS

The remaining part of this chapter deals with the dynamics of waves on the surface of incompressible liquids. It is known from the study of fluid mechanics that liquid surfaces are subject to surface tension forces. However, in the following analysis, we assume that the surface wave amplitude is much smaller than the wavelength. The surface tension forces due to the small surface curvature may then be neglected in comparison to the gravity force, which is much larger on the earth.

7.2.1 Dynamics of surface waves

Consider an inviscid incompressible homogeneous body of liquid with a free surface, as shown in Figure 7.12. Assume a coordinate system such that the x-y-plane coincides with the undisturbed (reference) surface of the liquid, as shown in the figure. Euler's equation of motion and the continuity equation can be written as, respectively, (see, for example, [1])

$$\rho\mathbf{v}_{,t} + \rho(\mathbf{v}\cdot\nabla)\mathbf{v} = -\rho g\hat{\mathbf{k}} - \nabla p = -\nabla\overline{p}, \tag{7.144}$$

$$\nabla\cdot\mathbf{v} = 0, \tag{7.145}$$

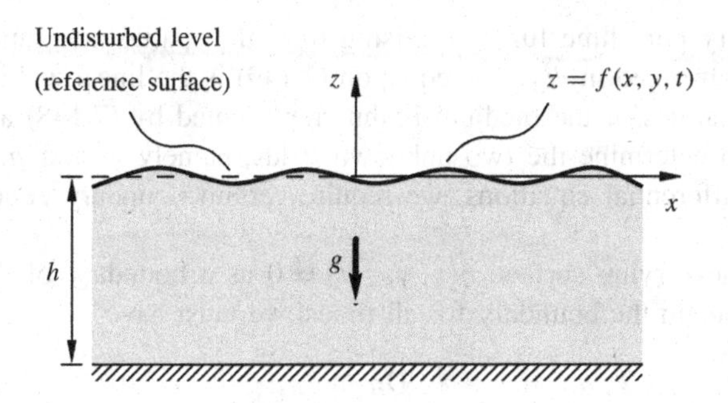

Figure 7.12 Surface waves on a liquid surface

where g is the acceleration due to gravity, $\hat{\mathbf{k}}$ is the unit vector along the z-axis direction, and

$$\overline{p} := \rho g z + p. \tag{7.146}$$

Since we consider linear dynamics only, the non-linear second term on the left-hand side of (7.144) will be dropped. Assuming that the motion of the fluid is irrotational, one can express the velocity vector $\mathbf{v}$ as

$$\mathbf{v} = \nabla \psi, \tag{7.147}$$

where ψ is a space–time-dependent scalar velocity potential function. Then, using (7.147) in (7.145), we obtain the Laplace equation

$$\nabla \cdot \nabla \psi = 0$$

or

$$\nabla^2 \psi = 0. \tag{7.148}$$

Using (7.147) in (7.144) yields

$$(\nabla \psi)_{,t} + \frac{1}{\rho} \nabla \overline{p} = \mathbf{0}$$

or

$$\nabla \left(\psi_{,t} + \frac{\overline{p}}{\rho} \right) = \mathbf{0}$$

$$\Rightarrow \psi_{,t} + g z + \frac{p}{\rho} = 0, \tag{7.149}$$

where any arbitrary pure time function arising from the spatial integration step in (7.149) is assumed to be absorbed in $\psi_{,t}$. The equation (7.149) is the linearized unsteady Bernoulli equation. The dynamics of the medium is thus represented by (7.148) and (7.149), which must be solved to determine the two unknown fields, namely ψ and p. In order to solve the two partial differential equations, we require certain boundary conditions, which are discussed next.

Consider a time-varying surface $\eta(x, y, z, t) = 0$ as a boundary of the medium. Since this is going to remain the boundary for all times, we must have

$$\frac{\mathrm{D}\eta}{\mathrm{D}t} = 0$$

or

$$\eta_{,t} + \mathbf{v} \cdot \nabla\eta = 0$$

or

$$\eta_{,t} + \nabla\psi \cdot \nabla\eta = 0. \tag{7.150}$$

We will now consider two special types of boundaries, as follows:

1. **Rigid boundary**
   In this case, the condition (7.150) reduces to

$$\left[\nabla\psi \cdot \nabla\eta\right]_{\eta=0} = 0.$$

   As an example, consider the bottom surface of a liquid body, as shown in Figure 7.12. The bottom surface is represented by $\eta(x, y, z, t) = z + h = 0$. Therefore, the boundary condition is given by

$$\left[\nabla\psi \cdot \hat{\mathbf{k}}\right]_{z=-h} = 0$$
$$\Rightarrow \psi_{,z}\big|_{z=-h} = 0.$$

2. **Free surface**
   For a free surface given by $z = f(x, y, t)$, we have $\eta(x, y, z, t) = z - f(x, y, t)$. Therefore, (7.150) yields

$$-f_{,t} + \left[\left(\psi_{,x}\hat{\mathbf{i}} + \psi_{,y}\hat{\mathbf{j}} + \psi_{,z}\hat{\mathbf{k}}\right) \cdot \left(-f_{,x}\hat{\mathbf{i}} - f_{,y}\hat{\mathbf{j}} + \hat{\mathbf{k}}\right)\right]_{\eta=0} = 0$$
$$\Rightarrow -f_{,t} + \left[-\psi_{,x}f_{,x} - \psi_{,y}f_{,y} + \psi_{,z}\right]_{\eta=0} = 0.$$

   Dropping all non-linear terms in the above condition, yields the linearized kinematic boundary condition at the free surface as

$$\psi_{,z}\big|_{\eta=0} - f_{,t} = 0.$$

In the presence of surface waves, pressure variations are created on the *reference surface* of the liquid. Note that, to keep the analysis linear, we must specify the pressure boundary condition on the reference surface, rather than the (deformed) liquid surface (on which the pressure is actually constant). This pressure variation is specified as a pressure (or dynamic) boundary condition on the (reference) surface $z = 0$. Taking the equation of the free surface as $z = f(x, y, t)$, the pressure on the surface $z = 0$ can be expressed as

$$p\big|_{z=0} = \rho g f$$

$$\Rightarrow -\rho\psi_{,t}\big|_{z=0} = \rho g f \qquad \text{(using (7.149))}$$

or

$$\psi_{,t}\big|_{z=0} + gf = 0. \tag{7.151}$$

To summarize, the dynamics of surface waves is governed by the Laplace equation (7.148) along with the kinematic boundary condition (7.150) and the free-surface dynamic boundary condition (7.151). The pressure field is determined from the linearized unsteady Bernoulli equation (7.149).

7.2.2 Sloshing of liquids in tanks

As an application of the formulation presented in the previous section, let us consider the problem of sloshing of liquids in partially filled tanks. This is an important issue in many situations, such as in overhead tanks on buildings or structures, and tankers transporting liquids. We consider here only some of the elementary aspects of the problem in the Cartesian and cylindrical coordinate systems. A more sophisticated analysis must also include the effects of structural flexibility and of the motion of the container on the dynamics of sloshing, as well as large-amplitude motion of the liquid (see [6]).

7.2.2.1 Sloshing in cuboidal tanks

Consider a cuboidal tank of length a and width b, filled with liquid up to a height h, as shown in Figure 7.13. The coordinate system is shown in the figure. It is easy to see that the boundary conditions here are the zero normal-velocity conditions on the walls and the bottom, and the kinematic and dynamic conditions on the free surface. Therefore, the equations governing the dynamics of the liquid can be written as

$$\psi_{,xx} + \psi_{,yy} + \psi_{,zz} = 0, \tag{7.152}$$

$$p = -\rho g z - \rho\psi_{,t}, \tag{7.153}$$

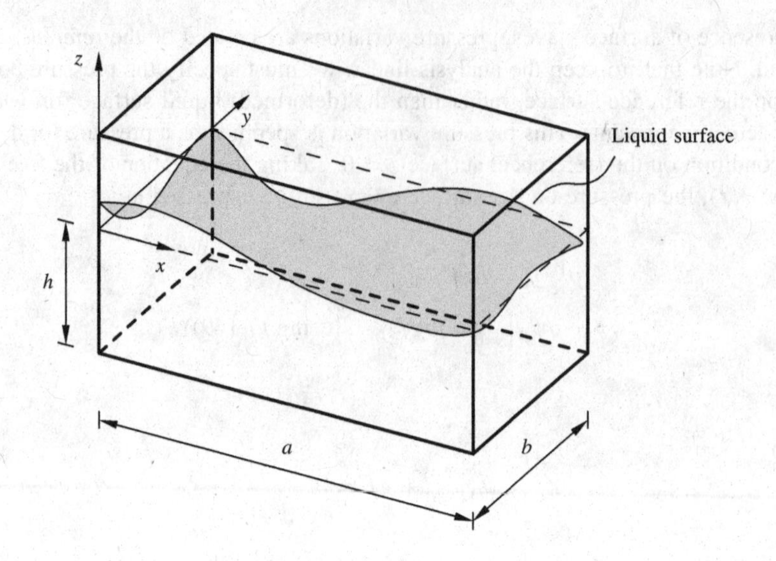

Figure 7.13 Sloshing in a partially filled cuboidal tank

along with the boundary and free-surface conditions:

$$\psi_{,x}\big|_{x=0,a} = 0, \tag{7.154}$$

$$\psi_{,y}\big|_{y=0,b} = 0, \tag{7.155}$$

$$\psi_{,z}\big|_{z=-h} = 0, \tag{7.156}$$

$$\psi_{,z}\big|_{z=0} - f_{,t} = 0, \tag{7.157}$$

$$\psi_{,t}\big|_{z=0} + gf = 0, \tag{7.158}$$

where (7.154) and (7.155) represent the zero normal-velocity condition on the walls, (7.156) is the zero normal-velocity condition at the bottom, and (7.157) and (7.158) are, respectively, the kinematic and dynamic conditions on the free surface $z = f(x, y, t)$. Differentiating (7.158) with respect to time, and substituting for $\partial f/\partial t$ from (7.157) yields

$$\left[\psi_{,tt} + g\psi_{,z}\right]_{z=0} = 0. \tag{7.159}$$

Assume a separable solution of (7.152) in the form

$$\psi(x, y, z, t) = X(x)Y(y)Z(z)\mathrm{e}^{i\omega t}. \tag{7.160}$$

Substituting this solution form in (7.152), and dividing by $X(x)Y(y)Z(z)$, yields

$$\frac{1}{X}\frac{\mathrm{d}^2X}{\mathrm{d}x^2} + \frac{1}{Y}\frac{\mathrm{d}^2Y}{\mathrm{d}y^2} + \frac{1}{Z}\frac{\mathrm{d}^2Z}{\mathrm{d}z^2} = 0. \tag{7.161}$$

It is evident that (7.161) can be satisfied if and only if each of the three terms is a constant, i.e.,

$$\frac{X_{,xx}}{X} = -\alpha^2, \qquad \frac{Y_{,yy}}{Y} = -\beta^2, \qquad \text{and} \qquad \frac{Z_{,zz}}{Z} = \gamma^2. \tag{7.162}$$

such that

$$\gamma^2 = \alpha^2 + \beta^2. \tag{7.163}$$

Note that we have taken the first two separation constants (associated with x and y) with negative sign, and the third separation constant (associated with z) with positive sign. This choice will simplify matching all the boundary conditions. We have from (7.162)

$$\begin{aligned} X(x) &= A_1 \mathrm{e}^{i\alpha x} + A_2 \mathrm{e}^{-i\alpha x}, \\ Y(y) &= B_1 \mathrm{e}^{i\beta y} + B_2 \mathrm{e}^{-i\beta y}, \\ Z(z) &= C_1 \mathrm{e}^{\gamma z} + C_2 \mathrm{e}^{-i\gamma z}, \end{aligned}$$

and the solution of the velocity potential $\psi(x, y, z, t)$ is obtained as

$$\psi(x, y, z, t) = (A_1 \mathrm{e}^{i\alpha x} + A_2 \mathrm{e}^{-i\alpha x})(B_1 \mathrm{e}^{i\beta y} + B_2 \mathrm{e}^{-i\beta y})(C_1 \mathrm{e}^{\gamma z} + C_2 \mathrm{e}^{-i\gamma z})\mathrm{e}^{i\omega t}. \tag{7.164}$$

Substituting this solution of the velocity potential in (7.154) yields the conditions at $x = 0$ and $x = a$ as

$$\begin{aligned} A_1 i\alpha - A_2 i\alpha &= 0, \\ A_1 i\alpha \mathrm{e}^{i\alpha a} - A_2 i\alpha \mathrm{e}^{-i\alpha a} &= 0 \end{aligned}$$

$$\Rightarrow \begin{bmatrix} 1 & -1 \\ \mathrm{e}^{i\alpha a} & -\mathrm{e}^{-i\alpha a} \end{bmatrix} \begin{Bmatrix} A_1 \\ A_2 \end{Bmatrix} = 0. \tag{7.165}$$

For non-trivial solutions in A_1 and A_2, we must have

$$\mathrm{e}^{i\alpha a} - \mathrm{e}^{-i\alpha a} = 0$$

or

$$\begin{aligned} &\sin \alpha a = 0, \\ &\Rightarrow \alpha_m = \frac{m\pi}{a}, \qquad m = 1, 2, \ldots, \infty. \end{aligned} \tag{7.166}$$

For these values of α, (7.165) yields $A_1 = A_2$. Similarly, using (7.164) in (7.155), one can obtain

$$\beta_n = \frac{n\pi}{b}, \qquad n = 1, 2, \ldots, \infty, \tag{7.167}$$

and $B_1 = B_2$. Using these conditions in (7.164) yields

$$\psi(x, y, z, t) = A \cos \frac{m\pi x}{a} \cos \frac{n\pi y}{b} (C_1 e^{\gamma_{(m,n)} z} + C_2 e^{-\gamma_{(m,n)} z}) e^{i\omega t}, \tag{7.168}$$

where

$$\gamma_{(m,n)} = \sqrt{\frac{m^2\pi^2}{a^2} + \frac{n^2\pi^2}{b^2}} \quad \text{(from (7.163))}. \tag{7.169}$$

Substituting (7.168) in (7.156), we have

$$C_1 \gamma_{(m,n)} e^{-\gamma_{(m,n)} h} - C_2 \gamma_{(m,n)} e^{\gamma_{(m,n)} h} = 0. \tag{7.170}$$

Using (7.168) in (7.159) yields on rearrangement

$$(g\gamma_{(m,n)} - \omega^2) C_1 - (g\gamma_{(m,n)} + \omega^2) C_2 = 0. \tag{7.171}$$

One can combine (7.170) and (7.171) as

$$\begin{bmatrix} e^{-\gamma_{(m,n)} h} & -e^{\gamma_{(m,n)} h} \\ g\gamma_{(m,n)} - \omega^2 & g\gamma_{(m,n)} + \omega^2 \end{bmatrix} \begin{Bmatrix} C_1 \\ C_2 \end{Bmatrix} = 0. \tag{7.172}$$

Now, for non-trivial solutions in C_1 and C_2, we must have from (7.172)

$$-e^{-\gamma_{(m,n)} h} (g\gamma_{(m,n)} + \omega^2) + e^{\gamma_{(m,n)} h} (g\gamma_{(m,n)} - \omega^2) = 0$$

$$\Rightarrow \omega_{(m,n)}^2 = g\gamma_{(m,n)} \tanh \gamma_{(m,n)} h. \tag{7.173}$$

This gives the sloshing frequency for the sloshing mode (m, n). For $\gamma_{(m,n)} h \gg 1$, one can use the approximation $\tanh \gamma_{(m,n)} h \approx 1$ to obtain the approximate sloshing frequencies as

$$\omega_{(m,n)} \approx \sqrt{g\gamma_{(m,n)}} = \sqrt{g} \left[\frac{m^2\pi^2}{a^2} + \frac{n^2\pi^2}{b^2} \right]^{1/4}.$$

From (7.170), one can write

$$C_2 = e^{-2\gamma h} C_1,$$

and then the velocity potential for the sloshing mode (m, n) can be written as

$$\psi_{(m,n)}(x, y, z, t) = A_{(m,n)} \cos \frac{m\pi x}{a} \cos \frac{n\pi y}{b} \cosh[\gamma_{(m,n)}(z + h)] e^{i\omega t}, \tag{7.174}$$

where a constant factor $e^{-\gamma_{(m,n)} h}$ has been absorbed in $A_{(m,n)}$.

One can now compute the free surface of the liquid by taking the real part of (7.158) as

$$
\begin{aligned}
z = f(x, y, t) &= -\mathscr{R}\left[\frac{i\omega}{g} A \cos\frac{m\pi x}{a} \cos\frac{n\pi y}{b} \cosh[\gamma_{(m,n)} h] e^{i\omega t}\right], \\
&= \frac{\omega}{g} A \cos\frac{m\pi x}{a} \cos\frac{n\pi y}{b} \cosh[\gamma_{(m,n)} h] \sin\omega t,
\end{aligned}
$$

and the pressure from (7.153) as

$$
\begin{aligned}
p &= \mathscr{R}\left[-\rho g z - i\omega\rho A \cos\frac{m\pi x}{a} \cos\frac{n\pi y}{b} \cosh[\gamma_{(m,n)}(z+h)] e^{i\omega t}\right] \\
&= -\rho g z + \omega\rho A \cos\frac{m\pi x}{a} \cos\frac{n\pi y}{b} \cosh[\gamma_{(m,n)}(z+h)] \sin\omega t.
\end{aligned}
$$

The first few sloshing modes of the free surface are shown in Figure 7.14.

7.2.2.2 Sloshing in cylindrical tanks

In order to study the case of sloshing in a cylindrical tank, as shown in Figure 7.15, we represent the governing equations and boundary conditions, in cylindrical polar coordinates. They can be written as

$$\psi_{,rr} + \frac{1}{r}\psi_{,r} + \frac{1}{r^2}\psi_{,\theta\theta} + \psi_{,zz} = 0, \tag{7.175}$$

$$p = -\rho g z - \rho\psi_{,t}, \tag{7.176}$$

$$\psi_{,r}\big|_{r=a} = 0, \tag{7.177}$$

$$\psi_{,z}\big|_{z=-h} = 0, \tag{7.178}$$

$$\psi_{,z}\big|_{z=0} - f_{,t} = 0, \tag{7.179}$$

$$\psi_{,t}\big|_{z=0} + g f = 0, \tag{7.180}$$

where a is the radius of the tank and h is the height of the liquid in the tank; (7.177) and (7.178) represent, respectively, the zero normal-velocities on the wall and the bottom, and (7.179) and (7.180) are, respectively, the kinematic and dynamic conditions on the free surface. Differentiating (7.180) once with respect to time, and eliminating $\partial f/\partial t$ from (7.179), yields

$$\left[\psi_{,tt} + g\psi_{,z}\right]_{z=0} = 0. \tag{7.181}$$

Substituting a separable function of the form

$$\psi(r, \theta, z, t) = R(r)\Theta(\theta)Z(z)e^{i\omega t}, \tag{7.182}$$

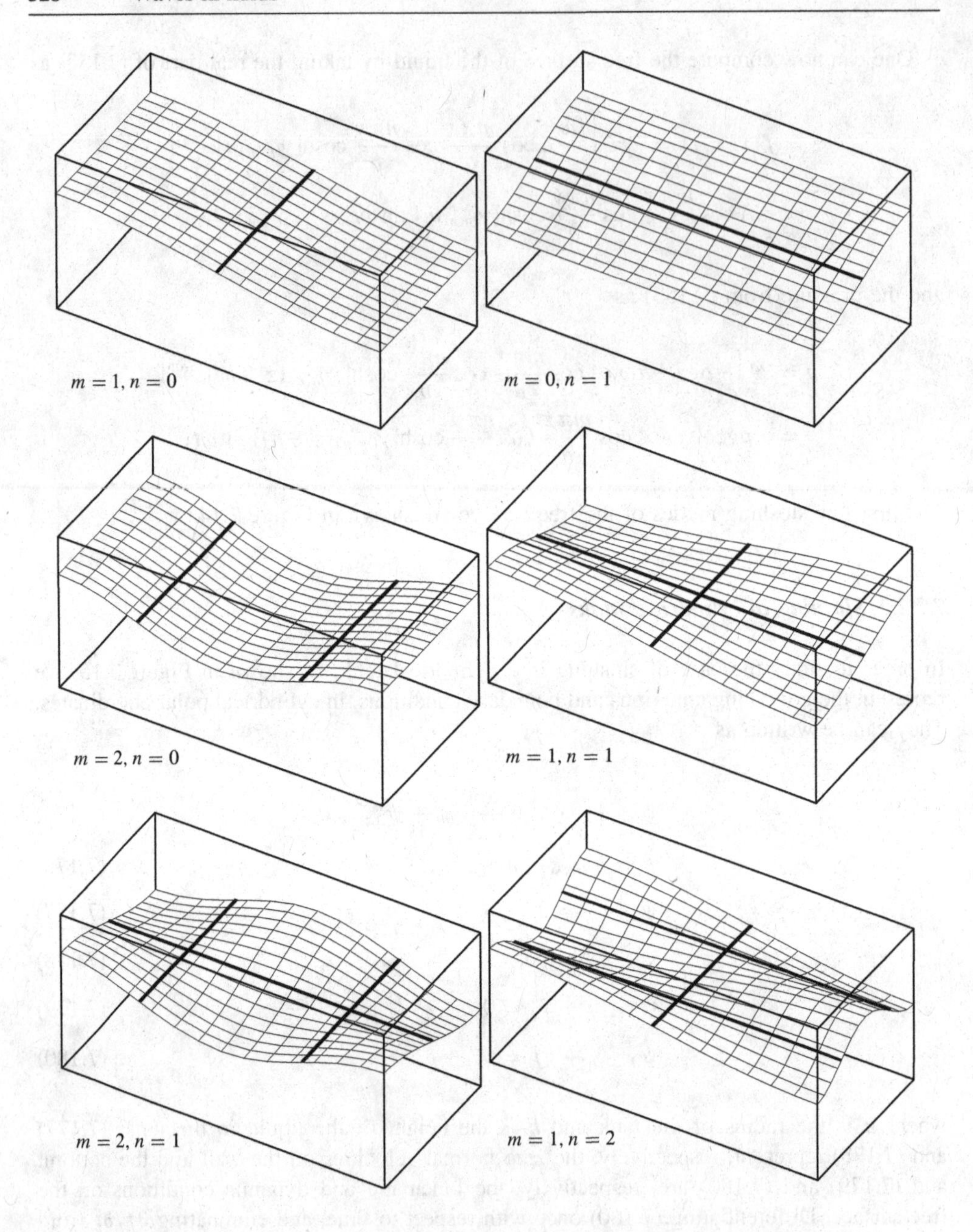

Figure 7.14 Sloshing modes of the free surface in a partially filled cuboidal tank

in (7.175), one obtains on rearrangement

$$\frac{1}{R}\left(\frac{\mathrm{d}^2 R}{\mathrm{d}r^2}+\frac{1}{r}\frac{\mathrm{d}R}{\mathrm{d}r}\right)+\frac{1}{r^2}\left(\frac{1}{\Theta}\frac{\mathrm{d}^2\Theta}{\mathrm{d}\theta^2}\right)+\frac{1}{Z}\frac{\mathrm{d}^2 Z}{\mathrm{d}z^2}=0. \tag{7.183}$$

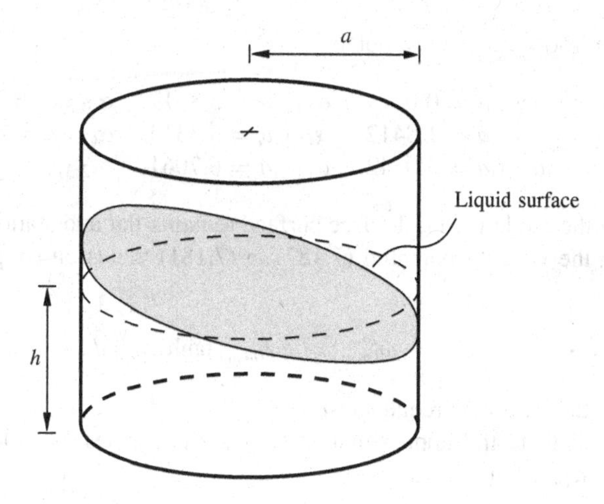

Figure 7.15 Sloshing in a partially filled cylindrical tank

It is evident that (7.183) can be satisfied if and only if

$$\frac{\mathrm{d}^2 R}{\mathrm{d}r^2} + \frac{1}{r}\frac{\mathrm{d}R}{\mathrm{d}r} + \left(\alpha^2 - \frac{\nu^2}{r^2}\right) R = 0, \tag{7.184}$$

$$\frac{\mathrm{d}^2 \Theta}{\mathrm{d}\theta^2} + \nu^2 \Theta = 0, \tag{7.185}$$

$$Z_{,zz} - \alpha^2 Z = 0, \tag{7.186}$$

where α and ν are the separation constants. For a cylindrical tank, periodicity of the solution in θ implies $\nu = m = 0, \pm 1, \ldots, \pm\infty$. The solution of the Bessel differential equation (7.184) then is of the form $R(r) = A_1 J_m(\alpha r) + A_2 Y_m(\alpha r)$. For finiteness of the solution at $r = 0$, we must have $A_2 = 0$. Next, solving (7.186) along with the boundary condition (7.178) gives

$$Z(z) = B\mathrm{e}^{-\alpha h} \cosh[\alpha(z + h)].$$

Using these expressions in (7.182), one can rewrite the velocity potential in the form

$$\psi(r, \theta, z, t) = A J_m(\alpha r)\mathrm{e}^{-\alpha h} \cosh[\alpha(z + h)]\mathrm{e}^{im\theta}\mathrm{e}^{i\omega t}. \tag{7.187}$$

Substituting (7.187) in the boundary condition (7.177), we have the characteristic equation

$$J'_m(\alpha a) = 0. \tag{7.188}$$

The first few roots $\alpha_{(m,n)}a$ are obtained as

$$\begin{aligned}
&\alpha_{(0,1)}a = 0.0, && \alpha_{(0,2)}a = 3.8317, && \alpha_{(0,3)}a = 7.0156,\\
&\alpha_{(1,1)}a = 1.8412, && \alpha_{(1,2)}a = 5.3314, && \alpha_{(1,3)}a = 8.5363,\\
&\alpha_{(2,1)}a = 3.0542, && \alpha_{(2,2)}a = 6.7061, && \alpha_{(2,3)}a = 9.9695.
\end{aligned}$$

Note that in the mode (0, 1), the free surface remains flat and stationary (a trivial solution). Substituting the velocity potential (7.187) in (7.181) yields on simplification the frequency equation

$$\omega^2_{(m,n)} = g\alpha_{(m,n)} \tanh \alpha_{(m,n)}h, \tag{7.189}$$

which gives the sloshing frequencies.

We can calculate the approximate sloshing frequencies as follows. Using (7.131), we approximate $\alpha_{(m,n)}a$ as

$$\alpha_{(m,n)}a \approx [2m + 4n - 3]\frac{\pi}{4}.$$

For $\alpha_{(m,n)}h \gg 1$, we also approximate the frequency equation (7.189) as $\omega^2_{(m,n)} \approx g\alpha_{(m,n)}$. Therefore, we have the approximate sloshing frequencies as

$$\omega_{(m,n)} \approx \sqrt{[2m + 4n - 3]\frac{\pi}{4}\frac{g}{a}}.$$

For every eigenfrequency $\omega_{(m,n)}$ with $m \neq 0$, there are two independent eigenfunctions given by

$$\begin{aligned}
W_{\rm C}^{(m,n)} &= J_m(\alpha_{(m,n)}r)\cosh[\alpha_{(m,n)}(z+h)]\cos m\theta,\\
W_{\rm S}^{(m,n)} &= J_m(\alpha_{(m,n)}r)\cosh[\alpha_{(m,n)}(z+h)]\sin m\theta.
\end{aligned}$$

Thus, there are two degenerate modes corresponding to each eigenfrequency $\omega_{(m,n)}$, $m \neq 0$. The first few sloshing mode-shapes of the free surface are shown in Figure 7.16.

7.2.3 Surface waves in a channel

Consider a channel formed by two boundaries at $y = 0$ and $y = b$, as shown in Figure 7.17. The governing equations for the surface waves in the channel are given by

$$\psi_{,xx} + \psi_{,yy} + \psi_{,zz} = 0, \tag{7.190}$$

$$p = -\rho g z - \rho\psi_{,t}, \tag{7.191}$$

$$\psi_{,y}\big|_{y=0,b} = 0, \tag{7.192}$$

$$\psi_{,z}\big|_{z=-h} = 0, \tag{7.193}$$

$$\psi_{,z}\big|_{z=0} - f_{,t} = 0, \tag{7.194}$$

$$\psi_{,t}\big|_{z=0} + gf = 0. \tag{7.195}$$

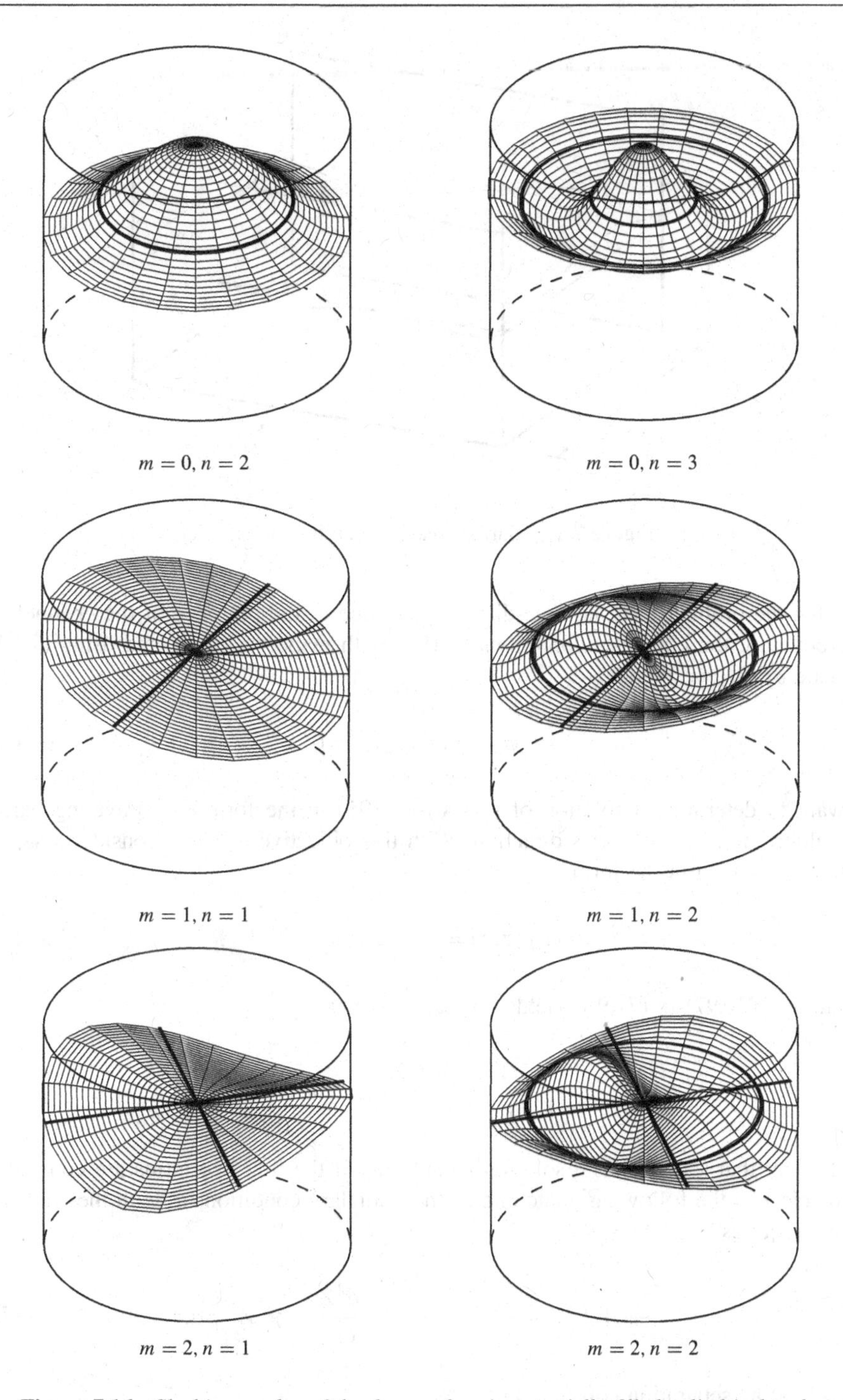

Figure 7.16 Sloshing modes of the free surface in a partially filled cylindrical tank

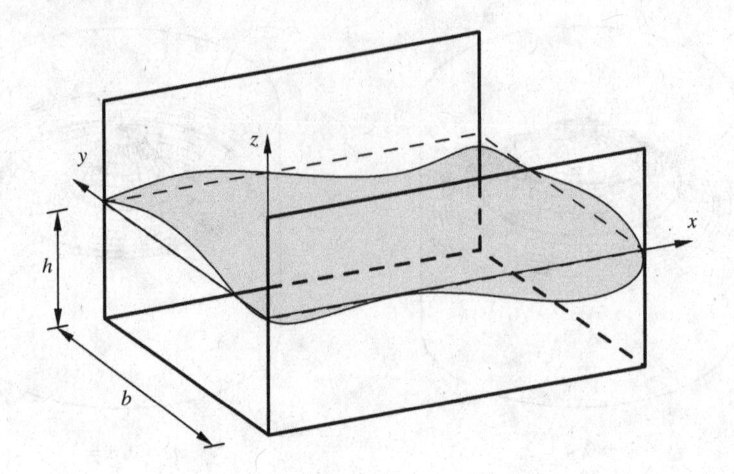

Figure 7.17 Surface waves in an infinite channel

Note that the problem here is three-dimensional, due to the presence of the channel walls. Differentiating the boundary conditions (7.195) with respect to time, and using (7.194) to eliminate $\partial f/\partial t$, we obtain

$$\left[\psi_{,tt} + g\psi_{,z}\right]_{z=0} = 0. \tag{7.196}$$

We want to determine a solution of (7.190)–(7.195) in the form of a traveling harmonic wave along the positive x-axis direction. With this objective in view, consider a separable solution of (7.190) in the form

$$\psi(x, y, z, t) = Y(y)Z(z)\mathrm{e}^{i(kx-\omega t)}. \tag{7.197}$$

Substituting (7.197) in (7.190) yields on rearrangement

$$\frac{1}{Y}\frac{\mathrm{d}^2Y}{\mathrm{d}y^2} + \frac{1}{Z}\frac{\mathrm{d}^2Z}{\mathrm{d}z^2} - k^2 = 0. \tag{7.198}$$

It is evident that (7.198) has a solution if and only if the first two terms are constants. As will be clear in the following, matching of the boundary conditions at both the walls of the channel requires

$$\frac{\mathrm{d}^2Y}{\mathrm{d}y^2} + \beta^2 Y = 0 \quad \text{and} \quad \frac{\mathrm{d}^2Z}{\mathrm{d}z^2} - \gamma^2 Z = 0, \tag{7.199}$$

where β is the separation constant, and

$$\gamma^2 = \beta^2 + k^2. \tag{7.200}$$

The solutions of $Y(y)$ and $Z(z)$ may be written as

$$Y(y) = A\cos\beta y + B\sin\beta y \qquad \text{and} \qquad Z(z) = Ce^{\gamma z} + De^{-\gamma z},$$

and hence, (7.197) is of the form

$$\psi(x, y, z) = (A\cos\beta y + B\sin\beta y)(Ce^{\gamma z} + De^{-\gamma z})e^{i(kx-\omega t)}. \tag{7.201}$$

Applying the boundary condition (7.192) yields

$$B = 0, \tag{7.202}$$

$$\sin\beta b = 0 \qquad \Rightarrow \qquad \beta_n = \frac{n\pi}{b}, \qquad n = 0, 1, \ldots, \infty. \tag{7.203}$$

Using the boundary conditions in (7.193), one obtains

$$D = Ce^{-2\gamma_n h}, \tag{7.204}$$

where

$$\gamma_n = \sqrt{\frac{n^2\pi^2}{b^2} + k^2}. \tag{7.205}$$

Using (7.202)–(7.204) in (7.201), we obtain the velocity potential

$$\psi(x, y, z, t) = A_n \cos\frac{n\pi y}{b}\cosh\gamma_n(z+h)e^{i(kx-\omega t)}. \tag{7.206}$$

Finally, substituting (7.206) in (7.196) yields on simplification the dispersion relation

$$\omega^2 - g\gamma_n \tanh\gamma_n h = 0, \tag{7.207}$$

where ω_n represents the circular frequency of the nth mode. The dispersion relation is shown in Figure 7.18 for different modes. It is observed that all modes are dispersive. Further, the lowest mode ($n = 0$) is always propagating, while the higher modes ($n \geq 1$) have a cut-off frequency below which they become evanescent. The nth cut-off frequency ω_n^c can be obtained by substituting (7.205) in (7.207), and setting $k = 0$ as

$$\omega_n^c = \sqrt{g\frac{n\pi}{b}\tanh\frac{n\pi h}{b}}, \quad n \geq 1.$$

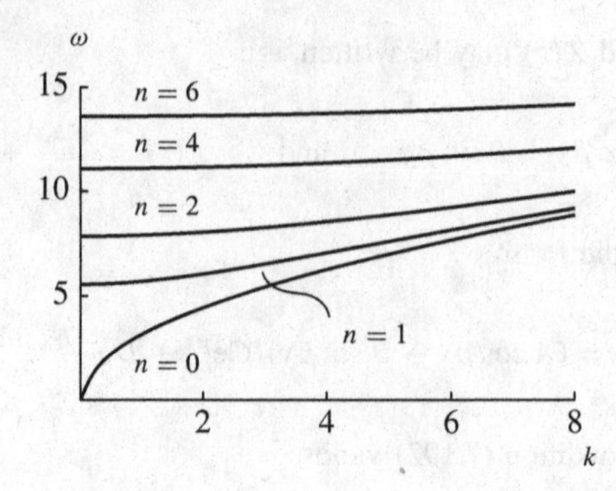

Figure 7.18 Dispersion relation for surface waves in a channel

EXERCISES

7.1 Determine the acoustic eigenfrequencies of a rectangular room with *acoustically hard* walls of length a, width b, and height h.

7.2 Determine the acoustic eigenfrequencies inside a half-cylindrical warehouse with *acoustically hard* walls of radius a and length l, as shown in Figure 7.19.

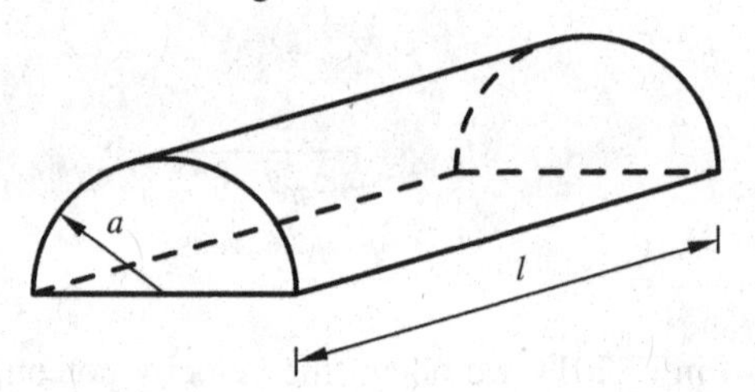

Figure 7.19 Exercise 7.2

7.3 Plane acoustic waves are incident at an angle on a stretched membrane of infinite extent. Determine the intensity of the wave transmitted across the membrane. Take the mass per unit area of the membrane μ, tension T, density of air ρ_0, and sound speed in air c.

7.4 A cylinder of radius R oscillates harmonically with a radial velocity $v = Ae^{i\omega t}$ in a fluid of density ρ and sound speed c.

(a) Determine the impedance of the medium as seen by the cylinder, and plot the variation of the radiation damping and radiation reactance coefficients with wave number.

(b) Calculate the intensity of the cylindrical waves generated at any radius $r > R$.

7.5 Determine the dispersion relation for a wave guide of semi-circular cross-section of radius a.

7.6 The chestpiece of an acoustic stethoscope is intended to pick up low-intensity sound waves in the frequency range 20 Hz to 20 kHz, and transmit them to the ear through a circular wave guide. Determine the upper bound on the radius R of the circular wave guide required to transmit the acoustic signals without dispersion. Assume $c = 340$ m/s.

7.7 Two circular wave guides of radii a and b are connected, as shown in Figure 7.20. A plane harmonic acoustic wave corresponding to the velocity potential $\psi_{\mathrm{I}} = Be^{i(kx-\omega t)}$ is incident on the junction from the left as shown. Determine the intensity of the wave transmitted across the junction.

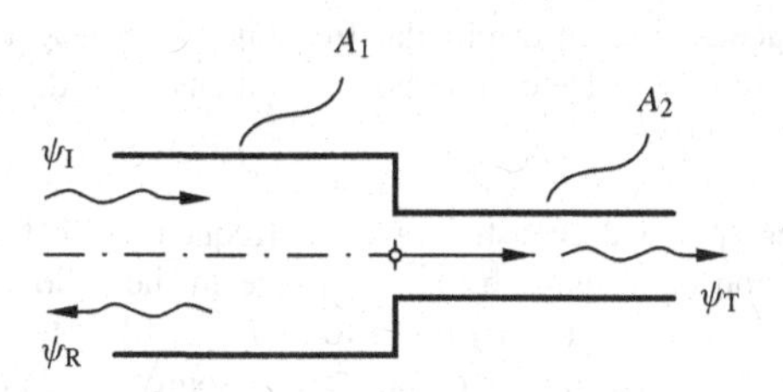

Figure 7.20 Exercise 7.7

7.8 A large stretched circular membrane lies in the x-y-plane with the origin at its center, and is surrounded by a compressible fluid.

(a) Assuming that the membrane is vibrating in one of its eigenmodes without being loaded by the fluid, determine the acoustic waves set up above the membrane surface.
(b) Also determine the radiation condition and the radiation damping coefficient for the membrane.

7.9 Let the phase velocity of transverse waves on an infinite surface be c_{T}. Assuming that $c_{\mathrm{T}} > c$, where c is the velocity of sound in air, determine the amplitude and wavelength of plane waves radiated into the air above the surface when a traveling wave of amplitude A_{T} and wave number k_{T} propagates on the surface.

7.10 Determine the coincidence frequency of a steel plate of thickness 2 mm vibrating in air. Take the sound speed in air as $c = 340$ m/s.

7.11 A stretched circular membrane completely covers one end of a semi-infinite circular wave guide of radius a.

(a) Assuming that the membrane is vibrating in one of its eigenmodes without being loaded by the fluid, determine the acoustic waves set up inside the wave guide.
(b) Determine the condition for no acoustic radiation through the wave guide.
(c) At what average rate must energy be supplied to the membrane to maintain its motion in a particular mode when there is acoustic radiation?

7.12 Determine the damping coefficient for a large rectangular plate vibrating harmonically and radiating sound of circular frequency Ω.

7.13 A cylinder of radius R oscillates radially uniformly along its length with circular frequency ω in a fluid of density ρ, sound speed c. Determine the added mass and damping coefficients for the cylinder.

7.14 A rigid cylinder of radius a, surrounded by a fluid, oscillates harmonically with a small amplitude A in a direction perpendicular to its axis, and uniformly along its length with circular frequency ω. Take ρ as the density of the fluid and c as the sound speed.

(a) Show that the velocity boundary condition for the fluid, on the surface of the cylinder is given by $\psi_{,r}(R, \phi, t) = A\cos\phi\, e^{i\omega t}$, where $\psi(r, \phi, t)$ is the velocity potential for the fluid, and ϕ is the angle measured from the line of motion of the cylinder.

(b) Determine the waves set up in the surrounding fluid. (*Hint*: Assume $\psi(r, \phi, t) = R(r) \cos\phi \, e^{i\omega t}$.)
(c) Determine the impedance of the fluid per unit length of the cylinder, and identify the radiation damping and added mass coefficients (per unit length).

7.15 A plane harmonic acoustic wave of circular frequency ω propagates through a fluid of density ρ, sound speed c, and viscosity η. Determine how the intensity of the wave varies as a function of the distance traveled by the wave.

7.16 A source is emitting spherical acoustic waves of frequency 5 kHz in air. Determine the sound pressure level as a function of distance from the source in the following two cases: (a) the air is considered inviscid, and (b) taking viscosity of air to be $\eta = 1.81 \times 10^{-5}$ Ns/m. For the above cases, also calculate the required average power output of the source so that it can be heard at a sound pressure level of 50 dB at a radius of 100 m. Assume the velocity of sound in air $c = 340$ m/s and the density of air $\rho_0 = 1.2$ kg/m^3.

7.17 A rectangular tank filled with water to a height h has a layer of oil of thickness d on top. Determine the sloshing frequencies of the system.

7.18 Determine the eigenfrequencies of sloshing of a liquid in the tank of constant depth whose top view is as shown in Figure 7.21.

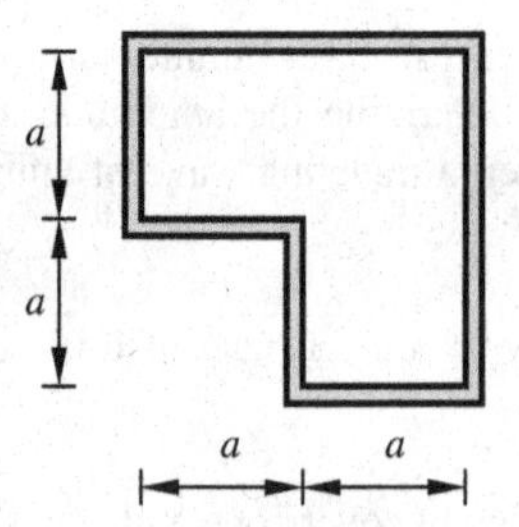

Figure 7.21 Exercise 7.18

7.19 Determine the sloshing eigenfrequencies and eigenfunctions in an annular tank of inner radius a and outer radius b.

7.20 Investigate the propagation of interface waves between two half-spaces of liquids, as shown in Figure 7.22. (*Hint*: Velocity potential far from the interface (i.e., $z \to \pm\infty$) is zero.)

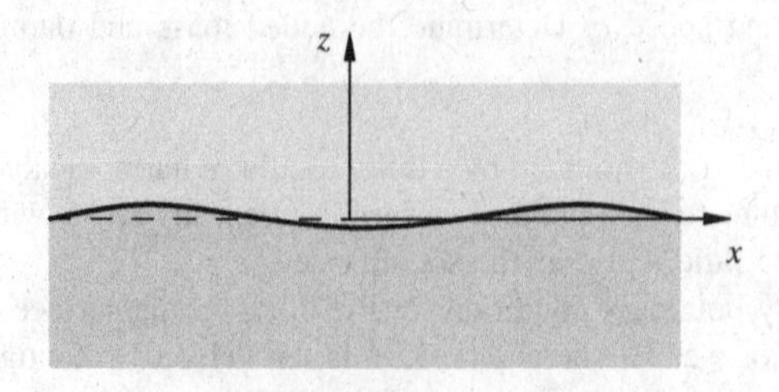

Figure 7.22 Exercise 7.20

7.21 In Exercise 7.20, if the upper layer of liquid is of thickness d, investigate the propagation of surface waves and determine the dispersion relation.

REFERENCES

[1] Kundu, P.K., and Cohen, I.M., *Fluid Mechanics*, 2e, Academic Press, San Diego, California, 2002.
[2] Goldstein, H., Poole, C.P., and Safko, J.L., *Classical Mechanics*, 3e, Addison-Wesley, Boston, 2002.
[3] Kreyszig, E., *Advanced Engineering Mathematics*, 2e, Wiley Eastern Pvt. Ltd., New Delhi, 1969.
[4] Hagedorn, P., A Note on the Vibrations of Infinite Elastic Plates in Contact with Water, *J. of Sound and Vibration*, 175(2), 1994, pp. 233–240.
[5] Morse, P.M., and Ingard, K.U., *Theoretical Acoustics*, McGraw-Hill Book Co., New York, 1968.
[6] Ibrahim, R.A., *Liquid Sloshing Dynamics: Theory and Applications*, Cambridge University Press, Cambridge, 2005.

8

Waves in elastic continua

In this chapter, we consider vibrations and waves in three-dimensional homogeneous and isotropic elastic continua. This study is important primarily for two reasons. First, any elastic body is essentially an elastic continuum with boundaries. Therefore, the accurate dynamics of elastic bodies and their exact solutions can be obtained only if we treat them as bounded elastic continua. The second reason lies in our interest in knowing the internal structure of materials and in evaluating material properties. An important technical application of such a study is in non-destructive testing and evaluation of materials. Other applications are in geophysics and seismology. However, this chapter is intended to give only an elementary introduction to the topic of waves in elastic continua.

8.1 EQUATIONS OF MOTION

Consider the free-body diagram of an infinitesimal element of a homogeneous and isotropic elastic continuum, as shown in Figure 8.1. In this figure, the stresses on the invisible surfaces have been omitted for clarity. Further, we have followed the convention of representing the x, y, and z in the subscripts by 1, 2, and 3, respectively, as shown in the figure. The coordinate axis x, y, and z are represented by x_1, x_2, and x_3, respectively, and σ_{ij} represents the stress component in the direction of the j-axis on a surface with normal along the i-axis. As will be evident in the following, this convention is very convenient for compact representations, and will be followed throughout this chapter.

Let the displacement of an infinitesimal element be denoted by $u_1(x_1, x_2, x_3, t)$, $u_2(x_1, x_2, x_3, t)$, and $u_3(x_1, x_2, x_3, t)$ along the coordinate directions x_1, x_2, and x_3, respectively. Then, applying Newton's second law to the infinitesimal element, the equations of motion can be written as

$$\rho\frac{\partial^2 u_1}{\partial t^2} - \left(\frac{\partial\sigma_{11}}{\partial x_1} + \frac{\partial\sigma_{12}}{\partial x_2} + \frac{\partial\sigma_{13}}{\partial x_3}\right) = 0, \tag{8.1}$$

$$\rho\frac{\partial^2 u_2}{\partial t^2} - \left(\frac{\partial\sigma_{21}}{\partial x_1} + \frac{\partial\sigma_{22}}{\partial x_2} + \frac{\partial\sigma_{23}}{\partial x_3}\right) = 0, \tag{8.2}$$

$$\rho\frac{\partial^2 u_3}{\partial t^2} - \left(\frac{\partial\sigma_{31}}{\partial x_1} + \frac{\partial\sigma_{32}}{\partial x_2} + \frac{\partial\sigma_{33}}{\partial x_3}\right) = 0, \tag{8.3}$$

Vibrations and Waves in Continuous Mechanical Systems P. Hagedorn and A. DasGupta

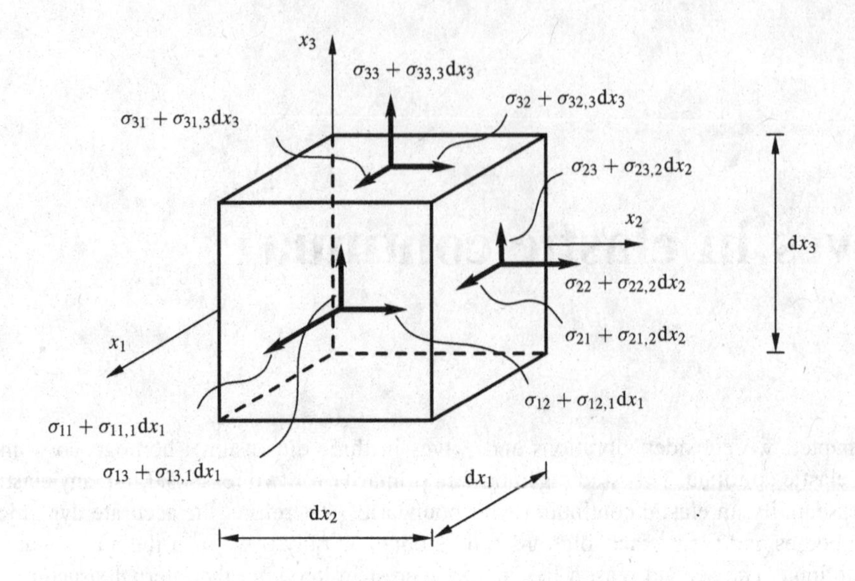

Figure 8.1 Stresses on an infinitesimal element of an elastic continuum

where we have neglected body forces. For simplicity, let us represent (8.1)–(8.3) in a compact form as

$$\rho u_{i,tt} - \sum_{j=1}^{3} \sigma_{ij,j} = 0, \qquad i = 1, 2, 3, \tag{8.4}$$

where $\sigma_{ij,j} := \partial\sigma_{ij}/\partial x_j$. The equations of motion (8.4) can also be derived easily using the variational formulation

$$\int_{t_1}^{t_2} [\delta \mathcal{T} + \overline{\delta \mathcal{W}}]\,\mathrm{d}t = 0, \tag{8.5}$$

where $\mathcal{T}$ is the kinetic energy and $\overline{\delta \mathcal{W}}$ is the infinitesimal virtual work done by the elastic forces. These are expressed as

$$\mathcal{T} = \frac{1}{2}\int_{\mathrm{V}} \rho \sum_{i=1}^{3} \dot{u}_i{}^2\,\mathrm{dV} \qquad \text{and} \qquad \overline{\delta \mathcal{W}} = \int_{\mathrm{S}} \sum_{i=1}^{3} \left(\sum_{j=1}^{3} \sigma_{ij}\hat{\mathbf{n}}_j \right) \delta u_i\,\mathrm{d}S, \tag{8.6}$$

where $\hat{\mathbf{n}}$ is the unit normal to the infinitesimal surface element dS. Substituting the expressions from (8.6) in (8.5), taking the variation, and using the Gauss divergence theorem (see, for example, [1]), leads to

$$\int_{t_1}^{t_2} \left[\int_{\mathrm{V}} \rho \sum_{i=1}^{3} \dot{u}_i \delta\dot{u}_i\,\mathrm{dV} + \int_S \sum \left(\sum_{j=1}^{3} \sigma_{ij}\hat{\mathbf{n}}_j \right) \delta u_i \mathrm{dS} \right] \mathrm{d}t = 0$$

$$\Rightarrow \int_{t_1}^{t_2} \int_V \left[\sum_{i=1}^{3} \left(-\rho \ddot{u}_i + \sum_{j=1}^{3} \sigma_{ij,j} \right) \delta u_i \right] \mathrm{d}V \, \mathrm{d}t = 0,$$

which again yields the equations of motion (8.4).

The task now is to express the stress tensor components σ_{ij} in terms of the displacement field components u_i. From elementary theory of elasticity, it is known that Hooke's law can be written as (see, for example, [2])

$$\sigma_{ij} = \lambda \left(\sum_{k=1}^{3} \epsilon_{kk} \right) \delta_{ij} + 2\mu \epsilon_{ij}, \tag{8.7}$$

where λ and μ are known as *Lamé parameters*. In terms of Young modulus E and Poisson ratio ν, the Lamé parameters are given by

$$\lambda = \frac{E\nu}{(1-2\nu)(1+\nu)} \quad \text{and} \quad \mu = \frac{E}{2(1+\nu)}. \tag{8.8}$$

The strain–displacement relations can be written as

$$\epsilon_{ij} = \frac{1}{2} \left(u_{i,j} + u_{j,i} \right). \tag{8.9}$$

Then, using (8.9) in (8.7), one obtains the stress–displacement relations as

$$\sigma_{ij} = \lambda \left(\sum_{k=1}^{3} u_{k,k} \right) \delta_{ij} + \mu \left(u_{i,j} + u_{j,i} \right)$$

$$\Rightarrow \sigma_{ij} = \lambda (\nabla \cdot \mathbf{u}) \delta_{ij} + \mu \left(u_{i,j} + u_{j,i} \right). \tag{8.10}$$

With (8.10), one can write

$$\sum_{j=1}^{3} \sigma_{ij,j} = \lambda \sum_{j=1}^{3} \left[(\nabla \cdot \mathbf{u})_{,j} \delta_{ij} \right] + \mu \sum_{j=1}^{3} \left[u_{i,jj} + u_{j,ij} \right]$$

$$\Rightarrow \sum_{j=1}^{3} \sigma_{ij,j} = \lambda (\nabla \cdot \mathbf{u})_{,i} \delta_{ij} + \mu \left[\nabla^2 u_i + (\nabla \cdot \mathbf{u})_{,i} \right]. \tag{8.11}$$

Using (8.11) in (8.4) yields the equations of motion

$$\rho \mathbf{u}_{,tt} - (\lambda + \mu) \nabla (\nabla \cdot \mathbf{u}) - \mu \nabla^2 \mathbf{u} = 0, \tag{8.12}$$

where $\mathbf{u} = (u_1, u_2, u_3)^{\mathrm{T}}$. This equation is known as *Navier's equation* for an elastic continuum.

Using a theorem due to Helmholtz (see, for example, [3]), one can decompose the displacement vector field $\mathbf{u}$ as

$$\mathbf{u}(x_1, x_2, x_3, t) = \mathbf{u}_{\mathrm{L}}(x_1, x_2, x_3, t) + \mathbf{u}_{\mathrm{S}}(x_1, x_2, x_3, t), \tag{8.13}$$

such that

$$\nabla \times \mathbf{u}_{\mathrm{L}} = 0 \qquad \text{and} \qquad \nabla \cdot \mathbf{u}_{\mathrm{S}} = 0. \tag{8.14}$$

The vector field $\mathbf{u}_{\mathrm{L}}$ is irrotational, and represents longitudinal displacement (or dilatoric deformations) of the continuum. On the other hand, $\mathbf{u}_{\mathrm{S}}$ is a divergence-free vector field, and hence preserves volume. This is characteristic of shear deformations (or deviatoric deformations). Therefore, the displacement field $\mathbf{u}_{\mathrm{S}}$ represents the shearing motion of the continuum. A traveling wave in an elastic continuum with displacements of the type $\mathbf{u}_{\mathrm{L}}$ is known as a *Primary* or P-wave. On the other hand, a traveling wave consisting of displacements of the type $\mathbf{u}_{\mathrm{S}}$, is termed a *Secondary* or S-wave. Physically, P- and S-waves correspond, respectively, to pressure and shear waves in an elastic continuum.

Using (8.13) in (8.12), one can write

$$\begin{aligned} &\rho(\mathbf{u}_{\mathrm{L},tt} + \mathbf{u}_{\mathrm{S},tt}) - (\lambda + \mu)\nabla[\nabla \cdot (\mathbf{u}_{\mathrm{L}} + \mathbf{u}_{\mathrm{S}})] - \mu\nabla^2(\mathbf{u}_{\mathrm{L}} + \mathbf{u}_{\mathrm{S}}) = 0 \\ \Rightarrow\ &\rho(\mathbf{u}_{\mathrm{L},tt} + \mathbf{u}_{\mathrm{S},tt}) - (\lambda + \mu)\nabla(\nabla \cdot \mathbf{u}_{L}) - \mu\nabla^2(\mathbf{u}_{\mathrm{L}} + \mathbf{u}_{\mathrm{S}}) = 0. \end{aligned} \tag{8.15}$$

Taking the divergence of (8.15), and simplifying using the properties

$$\nabla \cdot \mathbf{v}_{,tt} = (\nabla \cdot \mathbf{v})_{,tt}, \qquad \nabla \cdot \nabla^2\mathbf{v} = \nabla^2(\nabla \cdot \mathbf{v}),$$

and (8.14), we get

$$\begin{aligned} &\rho(\nabla \cdot \mathbf{u}_{\mathrm{L},tt}) - (\lambda + 2\mu)\nabla^2(\nabla \cdot \mathbf{u}_{\mathrm{L}}) = 0 \\ \Rightarrow\ &\nabla \cdot [\rho\mathbf{u}_{\mathrm{L},tt} - (\lambda + 2\mu)\nabla^2\mathbf{u}_{\mathrm{L}}] = 0. \end{aligned} \tag{8.16}$$

Further, using (8.16) in (8.15), one can also write

$$\nabla \cdot [\rho\mathbf{u}_{\mathrm{S},tt} - \mu\nabla^2\mathbf{u}_{\mathrm{S}}] = 0. \tag{8.17}$$

Next, taking the curl of (8.15), and simplifying using the properties

$$\nabla \times \mathbf{v}_{,tt} = (\nabla \times \mathbf{v})_{,tt}, \qquad \nabla \times \nabla^2\mathbf{v} = \nabla^2(\nabla \times \mathbf{v}),$$

and (8.14), one obtains

$$\rho(\nabla \times \mathbf{u}_{\mathrm{S},tt}) - \mu\nabla^2(\nabla \times \mathbf{u}_{\mathrm{S}}) = 0$$
$$\Rightarrow \nabla \times [\rho\mathbf{u}_{\mathrm{S},tt} - \mu\nabla^2\mathbf{u}_{\mathrm{S}}] = 0. \tag{8.18}$$

Further, we can also write

$$\nabla \times [\rho\mathbf{u}_{\mathrm{L},tt} - (\lambda + 2\mu)\nabla^2\mathbf{u}_{\mathrm{L}}] = 0. \tag{8.19}$$

Now, from (8.16) and (8.19) it is observed that the divergence and curl of the same vector quantity is zero. A similar observation is also made from (8.17) and (8.18). Therefore, it may be said that the individual vector quantities must be zero, i.e.,

$$\rho\mathbf{u}_{\mathrm{L},tt} - (\lambda + 2\mu)\nabla^2\mathbf{u}_{\mathrm{L}} = 0, \tag{8.20}$$
$$\rho\mathbf{u}_{\mathrm{S},tt} - \mu\nabla^2\mathbf{u}_{\mathrm{S}} = 0. \tag{8.21}$$

One can also rewrite (8.20) and (8.21) in the form of the standard wave equation as, respectively,

$$\mathbf{u}_{\mathrm{L},tt} - c_{\mathrm{L}}^2\nabla^2\mathbf{u}_{\mathrm{L}} = 0, \tag{8.22}$$
$$\mathbf{u}_{\mathrm{S},tt} - c_{\mathrm{S}}^2\nabla^2\mathbf{u}_{\mathrm{S}} = 0, \tag{8.23}$$

where

$$c_{\mathrm{L}} = \sqrt{\frac{\lambda + 2\mu}{\rho}} \quad \text{and} \quad c_{\mathrm{S}} = \sqrt{\frac{\mu}{\rho}}, \tag{8.24}$$

are, respectively, the longitudinal and the shear wave speeds. Using (8.8), the two wave speeds can also be represented as

$$c_{\mathrm{L}} = \sqrt{\frac{E}{\rho(1+\nu)}\left(\frac{1-\nu}{1-2\nu}\right)} \quad \text{and} \quad c_{\mathrm{S}} = \sqrt{\frac{E}{2\rho(1+\nu)}}.$$

The ratio of the two wave speeds $\kappa := c_{\mathrm{L}}/c_{\mathrm{S}}$ is obtained as

$$\kappa = \sqrt{\frac{2(1-\nu)}{1-2\nu}}. \tag{8.25}$$

It may be observed that $\kappa > 1$, i.e., the longitudinal wave speed is higher than the shear wave speed. For example, for $\nu = 1/3$, we have $c_{\mathrm{L}} = \sqrt{3}c_{\mathrm{S}}$.

8.2 PLANE ELASTIC WAVES IN UNBOUNDED CONTINUA

Let us first consider elastic waves in one dimension, say along the x_1-axis. The displacement fields $\mathbf{u}_L(x_1, t)$ and $\mathbf{u}_S(x_1, t)$ must satisfy the curl and divergence conditions (8.14). From the curl condition, we have

$$u_{L2,1} = 0 \quad \text{and} \quad u_{L3,1} = 0$$
$$\Rightarrow \mathbf{u}_L = (u_{L1}, 0, 0)^T.$$

On the other hand, the divergence condition yields

$$\mathbf{u}_{S1,1} = 0 \quad \Rightarrow \quad \mathbf{u}_S = (0, u_{S2}, u_{S3})^T.$$

Thus, the longitudinal wave or P-wave has particle motion only along the direction of propagation, while the shear wave or S-wave consists of particle motion along the two directions perpendicular to the direction of propagation. The shear waves can therefore be a vertically polarized SV-wave with particle motion along the x_2-axis or a horizontally polarized SH-wave with particle motion along the x_3-axis. Of course, any other direction of polarization orthogonal to the x_1-axis is also possible, and will be a combination of an SV and an SH wave. The P-wave and the SV-wave are visualized in Figure 8.2. The x_1-x_2-plane (containing the P- and SV-waves) is known as the *plane of incidence.*

Now, consider an arbitrary direction of wave propagation in a two-dimensional plane, as shown in Figure 8.3. Let us represent this direction by the unit vector $\hat{\mathbf{n}} = (\cos\theta, \sin\theta, 0)^T$. Then, any plane longitudinal harmonic wave solution of (8.22) can be written as

$$\mathbf{u}_L(x_1, x_2, t) = A_L \hat{\mathbf{n}}\, e^{i(k_L \hat{\mathbf{n}}\cdot\mathbf{r} - \omega t)} = A_L \hat{\mathbf{n}}\, e^{ik_L(x_1 \cos\theta + x_2 \sin\theta - c_L t)}, \tag{8.26}$$

where A_L is the (complex) amplitude, k_L is the wave number, $\mathbf{r} = (x_1, x_2, x_3)^T$, ω is the frequency, and $c_L = \omega / k_L$ is the phase speed of the wave. The product $\mathbf{k}_L := k_L \hat{\mathbf{n}}$ is known

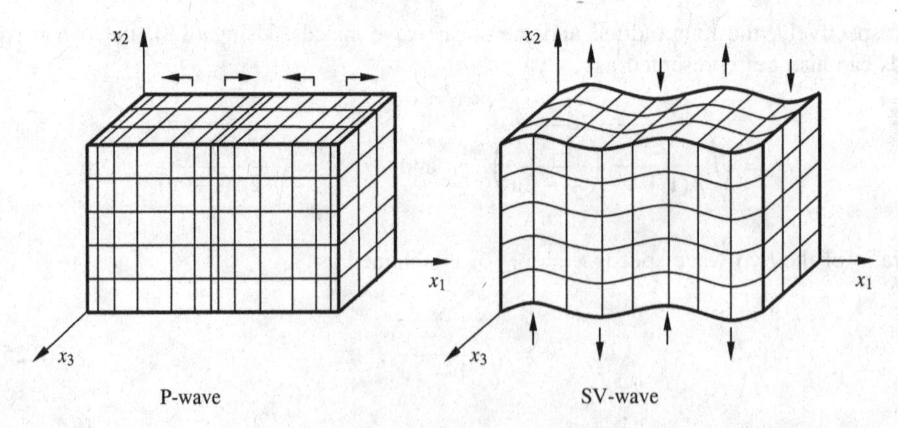

Figure 8.2 Motion of a section of a continuum due to P- and SV-waves traveling in the x_1-axis direction

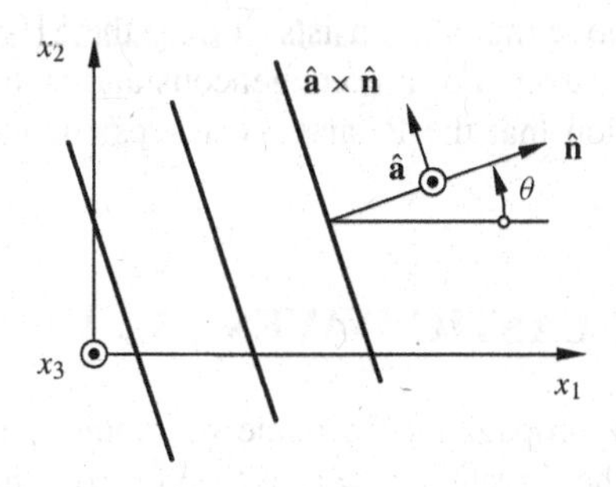

Figure 8.3 Representation of plane waves in a two-dimensional plane

as the *wave vector*. It can be easily checked that this displacement field satisfies the curl condition in (8.14), and is therefore irrotational.

The shear waves SV and SH can be represented as follows. Consider the unit vector $\hat{\mathbf{a}} = (0, 0, 1)^{\mathrm{T}}$ along the x_3-axis, as shown in Figure 8.3. Then $\hat{\mathbf{a}}$ and $\hat{\mathbf{a}} \times \hat{\mathbf{n}} = (-\sin\theta, \cos\theta, 0)^{\mathrm{T}}$ represent the two directions perpendicular to the direction of propagation, as shown in Figure 8.3. Now, one can represent harmonic shear waves as

$$\begin{aligned} \mathbf{u}_{\mathrm{S}}(x_1, x_2, t) = A_{\mathrm{V}} \hat{\mathbf{a}} \times \hat{\mathbf{n}}\, \mathrm{e}^{ik_{\mathrm{S}}(x_1 \cos\theta + x_2 \sin\theta - c_{\mathrm{S}} t)} \\ + A_{\mathrm{H}} \hat{\mathbf{a}}\, \mathrm{e}^{ik_{\mathrm{S}}(x_1 \cos\theta + x_2 \sin\theta - c_{\mathrm{S}} t)}, \end{aligned} \tag{8.27}$$

where A_{V} and A_{H} are the (complex) wave amplitudes, and $c_{\mathrm{S}} = \omega / k_{\mathrm{S}}$. The first term in (8.27) is the SV-wave, while the second term is the SH-wave. It can be easily checked from (8.27) that both SV- and SH-waves satisfy the divergence condition in (8.14), and hence are volume-preserving.

Consider the simultaneous existence of the longitudinal and transverse plane waves in an elastic space. Then, the total displacement field can be written using (8.26) and (8.27) as

$$\begin{aligned} \mathbf{u}(x_1, x_2, t) &= \mathbf{u}_{\mathrm{L}} + \mathbf{u}_{\mathrm{S}} \\ &= A_{\mathrm{L}} \hat{\mathbf{n}}\, \mathrm{e}^{ik_{\mathrm{L}}(x_1 \cos\theta + x_2 \sin\theta - c_{\mathrm{L}} t)} \\ &\quad + A_{\mathrm{V}} \hat{\mathbf{a}} \times \hat{\mathbf{n}}\, \mathrm{e}^{ik_{\mathrm{S}}(x_1 \cos\theta + x_2 \sin\theta - c_{\mathrm{S}} t)} \\ &\quad + A_{\mathrm{H}} \hat{\mathbf{a}}\, \mathrm{e}^{ik_{\mathrm{S}}(x \cos\theta + y \sin\theta - c_{\mathrm{S}} t)}. \end{aligned} \tag{8.28}$$

The components of the displacement vector field $\mathbf{u}$ can be written from (8.28) as

$$\begin{aligned} u_1 &= A_{\mathrm{L}} \cos\theta\, \mathrm{e}^{ik_{\mathrm{L}}(x_1 \sin\theta + x_2 \cos\theta - c_{\mathrm{L}} t)} \\ &\quad - A_{\mathrm{V}} \sin\theta\, \mathrm{e}^{ik_{\mathrm{S}}(x_1 \sin\theta + x_2 \cos\theta - c_{\mathrm{S}} t)}, \\ u_2 &= A_{\mathrm{L}} \sin\theta\, \mathrm{e}^{ik_{\mathrm{L}}(x_1 \sin\theta + x_2 \cos\theta - c_{\mathrm{L}} t)} \\ &\quad + A_{\mathrm{V}} \cos\theta\, \mathrm{e}^{ik_{\mathrm{S}}(x_1 \sin\theta + x_2 \cos\theta - c_{\mathrm{S}} t)}, \\ u_3 &= A_{\mathrm{H}} \mathrm{e}^{ik_{\mathrm{S}}(x_1 \sin\theta + x_2 \cos\theta - c_{\mathrm{S}} t)}. \end{aligned}$$

It may be observed from the above that u_3 consists of only the SH-wave and is completely decoupled from the P- and SV-waves. For a homogeneous and isotropic elastic continuum, this has the important implication that the P- and SV-waves cannot be excited by an SH-wave, and vice versa.

8.3 ENERGETICS OF ELASTIC WAVES

Let us now discuss the energy propagation in plane harmonic elastic waves. Consider a harmonic P-wave traveling in the direction $\hat{\mathbf{n}} = (\cos\theta, \sin\theta, 0)^{\mathrm{T}}$, and represented by

$$\begin{aligned}\mathbf{u}_{\mathrm{L}} &= \mathscr{R}[A\hat{\mathbf{n}}\,\mathrm{e}^{ik(\hat{\mathbf{n}}\cdot\mathbf{r}-c_{\mathrm{L}}t)}] \\ &= A\hat{\mathbf{n}}\cos[k(\hat{\mathbf{n}}\cdot\mathbf{r}-c_{\mathrm{L}}t)].\end{aligned} \tag{8.29}$$

The kinetic energy density can be written as

$$\begin{aligned}\hat{\mathcal{T}} &= \frac{1}{2}\rho\mathbf{u}_{\mathrm{L},t}\cdot\mathbf{u}_{\mathrm{L},t} \\ &= \frac{1}{2}\rho A^2 k_{\mathrm{L}}^2 c_{\mathrm{L}}^2 \sin^2[k_{\mathrm{L}}(\hat{\mathbf{n}}\cdot\mathbf{r}-c_{\mathrm{L}}t)] \\ &= \frac{1}{2}\rho A^2\omega^2 \sin^2[k_{\mathrm{L}}(\hat{\mathbf{n}}\cdot\mathbf{r}-c_{\mathrm{L}}t)].\end{aligned}$$

The average kinetic energy density (kinetic energy per unit wavelength) is then computed as

$$\langle\hat{\mathcal{T}}\rangle = \frac{k_{\mathrm{L}}}{2\pi}\int_0^{2\pi/k_{\mathrm{L}}} \hat{\mathcal{T}}\,\mathrm{d}\xi = \frac{1}{4}\rho A^2\omega^2, \tag{8.30}$$

where ξ is measured along the direction of propagation.

The potential energy density, which is the elastic strain energy per unit volume, is obtained from theory of elasticity as (see, for example, [3])

$$\begin{aligned}\hat{\mathcal{V}} &= \frac{1}{2}(\sigma_{11}\epsilon_{11}+\sigma_{22}\epsilon_{22}+\sigma_{33}\epsilon_{33})+\sigma_{12}\epsilon_{12}+\sigma_{23}\epsilon_{23}+\sigma_{13}\epsilon_{13} \\ &= \frac{1}{2}\lambda\left(\sum_{j=1}^{3}\epsilon_{jj}\right)^2+\mu\left(\sum_{j=1}^{3}\epsilon_{jj}^2\right)+2\mu(\epsilon_{12}^2+\epsilon_{23}^2+\epsilon_{13}^2),\end{aligned} \tag{8.31}$$

where we have used the strain–displacement relation (8.9). Using (8.29) in (8.9) yields

$$\begin{aligned}&\epsilon_{11} = -Ak_{\mathrm{L}}\cos^2\theta\ \sin[k_{\mathrm{L}}(\hat{\mathbf{n}}\cdot\mathbf{r}-c_{\mathrm{L}}t)], \\ &\epsilon_{12} = \epsilon_{21} = -Ak_L\cos\theta\ \sin\theta\ \sin[k_L(\hat{\mathbf{n}}\cdot\mathbf{r}-c_Lt)], \\ &\epsilon_{22} = -Ak_{\mathrm{L}}\sin^2\theta\ \sin[k_{\mathrm{L}}(\hat{\mathbf{n}}\cdot\mathbf{r}-c_{\mathrm{L}}t)], \\ &\epsilon_{13} = 0, \qquad \epsilon_{23} = 0, \qquad \text{and} \qquad \epsilon_{33} = 0.\end{aligned}$$

Using these expressions in (8.31), one obtains on simplification

$$\begin{aligned}\hat{\mathcal{V}} &= \frac{1}{2}(\lambda + 2\mu)A^2 k_L^2 \sin^2[k_L(\hat{\mathbf{n}} \cdot \mathbf{r} - c_L t)] \\ &= \frac{1}{2}\rho A^2 \omega^2 \sin^2[k_L(\hat{\mathbf{n}} \cdot \mathbf{r} - c_L t)],\end{aligned}$$

where we have used the relation

$$k_L^2 = \frac{\omega^2}{c_L^2} = \frac{\rho\omega^2}{\lambda + 2\mu}.$$

The average potential energy density (potential energy per unit wavelength) is then obtained as

$$\langle \hat{\mathcal{V}} \rangle = \frac{k_L}{2\pi} \int_0^{2\pi/k_L} \hat{\mathcal{V}} \, d\xi = \frac{1}{4}\rho A^2 \omega^2, \tag{8.32}$$

where, as before, ξ is measured along the direction of propagation. Now, using (8.30) and (8.32), the average total energy density of longitudinal harmonic plane waves can be written as

$$\langle \hat{\mathcal{E}} \rangle = \langle \hat{\mathcal{T}} \rangle + \langle \hat{\mathcal{V}} \rangle = \frac{1}{2}\rho A^2 \omega^2. \tag{8.33}$$

Next, we compute the energy flux (or power) crossing unit area of the continuum. Let the stress tensor at any point in the continuum be represented by the symmetric matrix

$$[\sigma] = \begin{bmatrix} \sigma_{11} & \sigma_{12} & \sigma_{13} \\ \sigma_{21} & \sigma_{22} & \sigma_{23} \\ \sigma_{31} & \sigma_{32} & \sigma_{33} \end{bmatrix},$$

Define the vector of power flow per unit area, or the intensity vector as

$$\mathcal{I} := -[\sigma]\mathbf{u}_{L,t},$$

where the negative sign follows from the convention that compressive stresses are negative. Then, the power flow per unit area across a surface perpendicular to the direction of propagation of the wave is obtained as

$$\mathcal{I}_n = -\hat{\mathbf{n}}^T[\sigma]\mathbf{u}_{L,t}. \tag{8.34}$$

The velocity vector is obtained from (8.29) as

$$\mathbf{u}_{L,t} = Ak_L\hat{\mathbf{n}} \sin[k_L(\hat{\mathbf{n}} \cdot \mathbf{r} - c_L t)], \tag{8.35}$$

and the stresses can be computed using (8.10) as

$$\sigma_{11} = -(\lambda + 2\mu\cos^2\theta)Ak_L \sin[k_L(\hat{\mathbf{n}}\cdot\mathbf{r} - c_L t)],$$
$$\sigma_{12} = -2\mu\cos\theta\,\sin\theta\, Ak_L \sin[k_L(\hat{\mathbf{n}}\cdot\mathbf{r} - c_L t)],$$
$$\sigma_{22} = -(\lambda + 2\mu\sin^2\theta)Ak_L \sin[k_L(\hat{\mathbf{n}}\cdot\mathbf{r} - c_L t)],$$
$$\sigma_{13} = 0, \qquad \sigma_{23} = 0, \qquad \text{and} \qquad \sigma_{33} = 0.$$

Finally, the energy flux per unit area in the direction of propagation of the wave is obtained from (8.34) and (8.35) as

$$\begin{aligned}\mathcal{I}_n &= (\lambda + 2\mu)A^2k_L^2 c_L \sin^2[k_L(\hat{\mathbf{n}}\cdot\mathbf{r} - c_L t)] \\ &= \rho A^2\omega^2 c_L \sin^2[k_L(\hat{\mathbf{n}}\cdot\mathbf{r} - c_L t)],\end{aligned}$$

from where the average power flow per unit area can be computed as

$$\langle\mathcal{I}_n\rangle = \frac{1}{2}\rho A^2\omega^2 c_L. \tag{8.36}$$

Comparing (8.36) and (8.33), we obtain

$$\langle\mathcal{I}_n\rangle = c_L\langle\hat{\mathcal{E}}\rangle.$$

Thus, energy of longitudinal waves propagates at the wave speed c_L. The energetics of shear waves can be studied similarly.

The specific impedance of an elastic medium can be defined as follows. It is known from theory of elasticity that the stresses on any surface with a unit surface normal $\hat{\mathbf{n}}$ is given by $[\sigma]\hat{\mathbf{n}}$, and the normal stress on the surface is obtained as $\hat{\mathbf{n}}^T[\sigma]\hat{\mathbf{n}}$. Then, for a harmonic P-wave, the specific impedance of the medium in the normal direction is defined by

$$\mathcal{Z}_L := -\frac{\mathscr{A}[\hat{\mathbf{n}}^T[\sigma]\hat{\mathbf{n}}]}{\mathscr{A}[\hat{\mathbf{n}}^T\mathbf{u}_{L,t}]}, \tag{8.37}$$

where $\mathbf{u}_{L,t}$ is the velocity vector field of the harmonic wave, $\mathscr{A}[\cdot]$ denotes the complex amplitude, and the negative sign follows from the convention that compressive stresses are negative. It may be mentioned that all calculations in (8.37) are required to be carried out using the complex harmonic wave representation. It may be checked that the impedance $\mathcal{Z}_L$ calculated for the harmonic wave field $\mathbf{u}_L = A\hat{\mathbf{n}}\,e^{ik(\hat{\mathbf{n}}\cdot\mathbf{r}-c_L t)}$ is given by $\mathcal{Z}_n = \rho c_L$. One can similarly define the impedance of the medium to shear waves.

8.4 REFLECTION OF ELASTIC WAVES

Let us now consider the interaction of elastic plane waves with plane boundaries. Assuming that a boundary B lies in the x_1-x_3-plane, some of the possible boundary conditions that occur are as follows:

1. **Free boundary**
   In this case all the stresses on the surface must be zero. Therefore, we have $\sigma_{12}|_{\mathrm{B}} = \sigma_{22}|_{\mathrm{B}} = 0$.
2. **Fixed boundary with a rigid body**
   When the continuum boundary is rigidly fixed, the displacements are zero on the boundary, i.e., $\mathbf{u}|_{\mathrm{B}} = 0$.
3. **Boundary fixed with another elastic half-space**
   In this case, the displacements, and the normal and shear stresses at the boundary of both spaces must match. Hence, we must have $\mathbf{u}^{\mathrm{I}}\big|_{\mathrm{B}} = \mathbf{u}^{\mathrm{II}}\big|_{\mathrm{B}}$ and $\sigma^{\mathrm{I}}_{2i}\big|_{\mathrm{B}} = \sigma^{\mathrm{II}}_{2i}\big|_{\mathrm{B}}$, $i = 1, 2, 3$, where the superscripts I and II distinguish the two half-spaces.

To keep the discussion simple, in the following, we only consider reflection of harmonic plane waves from a free surface.

8.4.1 Reflection from a free boundary

We consider the following three cases of reflection from a free boundary. The first is the reflection of an incident P-wave, the second considers an incident SV-wave, and finally, the case of an incident SH-wave is discussed.

8.4.1.1 Incident P-wave

Consider a plane P-wave incident at an angle θ_{L0} with the normal to the free boundary $x_2 = 0$, as shown in Figure 8.4. The reflected waves can consist of, in general, both longitudinal and shear waves in different directions. Hence, the total wave field may be represented as

$$\begin{aligned}\mathbf{u}(x_1, x_2, t) = {} & A_{\mathrm{L0}}\hat{\mathbf{n}}_{\mathrm{L0}}\mathrm{e}^{ik_{\mathrm{L0}}(x_1 \sin\theta_{\mathrm{L0}} + x_2 \cos\theta_{\mathrm{L0}} - c_{\mathrm{L}}t)} \\ & + A_{\mathrm{L}}\hat{\mathbf{n}}_{\mathrm{L}}\mathrm{e}^{ik_{\mathrm{L}}(x_1 \sin\theta_{\mathrm{L}} - x_2 \cos\theta_{\mathrm{L}} - c_{\mathrm{L}}t)} \\ & + A_{\mathrm{V}}\hat{\mathbf{a}} \times \hat{\mathbf{n}}_{V}\mathrm{e}^{ik_{\mathrm{S}}(x_1 \sin\theta_{\mathrm{V}} - x_2 \cos\theta_{\mathrm{V}} - c_{\mathrm{S}}t)} \\ & + A_{\mathrm{H}}\hat{\mathbf{a}}\mathrm{e}^{ik_{\mathrm{S}}(x_1 \sin\theta_{\mathrm{H}} - x_2 \cos\theta_{\mathrm{H}} - c_{\mathrm{S}}t)},\end{aligned} \tag{8.38}$$

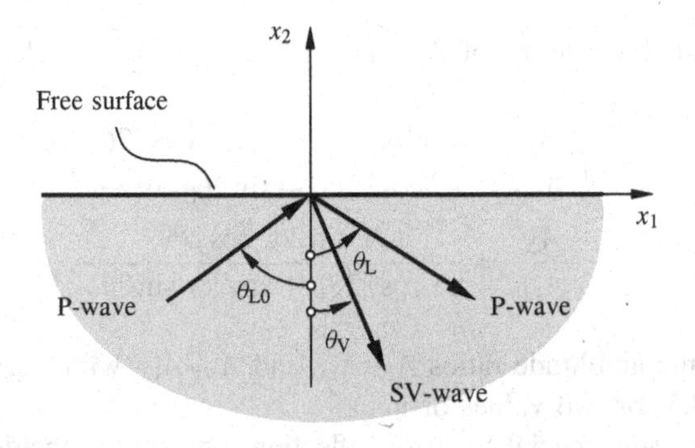

Figure 8.4 Reflection of a P-wave from a free boundary

where A_{L0} is the amplitude of the incident P-wave, A_L, A_V and A_H are, respectively, the amplitudes of the reflected P-, SV- and SH-waves, and

$$\hat{\mathbf{n}}_{L0} = (\sin\theta_{L0}, \cos\theta_{L0}, 0)^T, \qquad \hat{\mathbf{n}}_L = (\sin\theta_L, -\cos\theta_L, 0)^T,$$
$$\mathbf{n}_V = (\sin\theta_V, -\cos\theta_V, 0)^T, \qquad \mathbf{n}_H = (\sin\theta_H, -\cos\theta_H, 0)^T,$$

are, respectively, the directions of propagation of the incident P-wave and the reflected P-, SV- and SH-waves. It may be observed immediately from the u_3 component in (8.38) that $A_H = 0$. Thus, as mentioned before, the SH-wave cannot be excited by reflection of a P-wave from a free surface.

The boundary conditions for the free surface are

$$\sigma_{12}\big|_{x_2=0} = 0 \quad \Rightarrow \quad [u_{1,2} + u_{2,1}]\big|_{x_2=0} = 0. \tag{8.39}$$

$$\sigma_{22}\big|_{x_2=0} = 0 \quad \Rightarrow \quad [(\lambda + 2\mu)(u_{1,1} + u_{2,2}) - 2\mu u_{1,1}]\big|_{x_2=0} = 0. \tag{8.40}$$

It can be immediately concluded that these boundary conditions are identically satisfied if and only if

$$k_{L0}\sin\theta_{L0} = k_L\sin\theta_L = k_S\sin\theta_V \quad \text{and} \quad c_L k_{L0} = c_L k_L = c_S k_S$$
$$\Rightarrow k_{L0} = k_L, \quad \theta_{L0} = \theta_L \quad \text{and} \quad \frac{\sin\theta_L}{\sin\theta_V} = \frac{k_S}{k_L} = \frac{c_L}{c_S} = \kappa. \tag{8.41}$$

For a given angle of incidence $\theta_{L0} = \theta_L$, one can calculate the angle of the reflected SV-wave θ_V from the last equation in (8.41).

Substituting (8.38) in (8.39) and using (8.41) yields on simplification

$$A_{L0}\sin 2\theta_L - A_L\sin 2\theta_L - \kappa A_V\cos 2\theta_V = 0. \tag{8.42}$$

Similarly, from the condition (8.40), one obtains

$$A_{L0}\kappa\cos 2\theta_V + A_L\kappa\cos 2\theta_V - A_V\sin 2\theta_V = 0. \tag{8.43}$$

Solving for A_L and A_V in terms of A_{L0} from (8.42) and (8.43) yields

$$\frac{A_L}{A_{L0}} = \frac{\sin 2\theta_L\sin 2\theta_V - \kappa^2\cos^2 2\theta_V}{\kappa^2\cos^2 2\theta_V + \sin 2\theta_L\sin 2\theta_V},$$
$$\frac{A_V}{A_{L0}} = \frac{2\kappa\sin 2\theta_L\cos 2\theta_V}{\kappa^2\cos^2 2\theta_V + \sin 2\theta_L\sin 2\theta_V}.$$

The variations of the amplitude ratios A_L/A_{L0} and A_V/A_{L0} with angle of incidence θ_L are shown in Figure 8.5 for two values of ν.

Let us consider some special cases of reflection. For normal incidence, i.e., $\theta_L = 0$, we obtain $A_L/A_{L0} = -1$ and $A_V/A_{L0} = 0$. Thus, the incident P-wave is totally reflected, and

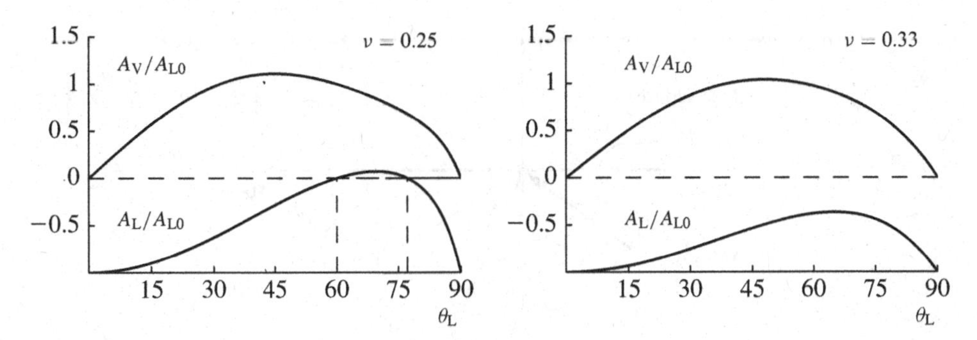

Figure 8.5 Amplitude ratios of reflection of a P-wave incident at θ_L on a free boundary for two values of Poisson ration ν

no SV-wave is generated. It may be checked that there is no phase change in this reflection. On the other hand, if θ_L is such that

$$\sin 2\theta_L \sin 2\theta_V - \kappa^2 \cos^2 2\theta_V = 0, \tag{8.44}$$

then $A_L/A_{LO} = 0$ and $A_V/A_{LO} = \kappa \cot 2\theta_V$. In this case, therefore, the P-wave is completely converted to a SV-wave. This phenomenon is termed *mode conversion*. It may be observed from Figure 8.5 that, for $\nu = 0.25$, there are two such incident angles, namely $\theta_L = 60^\circ$ and $\theta_L = 77.2^\circ$ at which mode conversion takes place. There is no mode conversion for $\nu = 0.33$.

8.4.1.2 Incident SV-wave

Next consider an SV-wave incident on a free surface at an angle θ_{SO}, as shown in Figure 8.6. In general, both longitudinal and transverse waves are generated. Hence, the total wave field can be represented by

$$\begin{aligned}\mathbf{u}(x_1, x_2, t) = {} & A_{VO}\hat{\mathbf{n}}_{SO} e^{ik_{SO}(x_1 \sin\theta_{VO} + x_2 \cos\theta_{VO} - c_S t)} \\ & + A_L \hat{\mathbf{n}}_L e^{ik_L(x_1 \sin\theta_L - x_2 \cos\theta_L - c_L t)} \\ & + A_V \hat{\mathbf{a}} \times \hat{\mathbf{n}}_S e^{ik_S(x_1 \sin\theta_V - x_2 \cos\theta_V - c_S t)} \\ & + A_H \hat{\mathbf{a}} e^{ik_S(x_1 \sin\theta_H - x_2 \cos\theta_H - c_S t)}.\end{aligned} \tag{8.45}$$

Writing out the u_3 component immediately yields $A_H = 0$.

The boundary conditions are given by (8.39)–(8.40). As in the previous case, it can be easily concluded that the boundary conditions are identically satisfied if and only if

$$k_{SO} \sin\theta_{VO} = k_L \sin\theta_L = k_S \sin\theta_V \quad \text{and} \quad c_S k_{SO} = c_L k_L = c_S k_S$$

$$\Rightarrow k_{SO} = k_S, \qquad \theta_{VO} = \theta_V \quad \text{and} \quad \frac{\sin\theta_L}{\sin\theta_V} = \frac{k_S}{k_L} = \frac{c_L}{c_S} = \kappa. \tag{8.46}$$

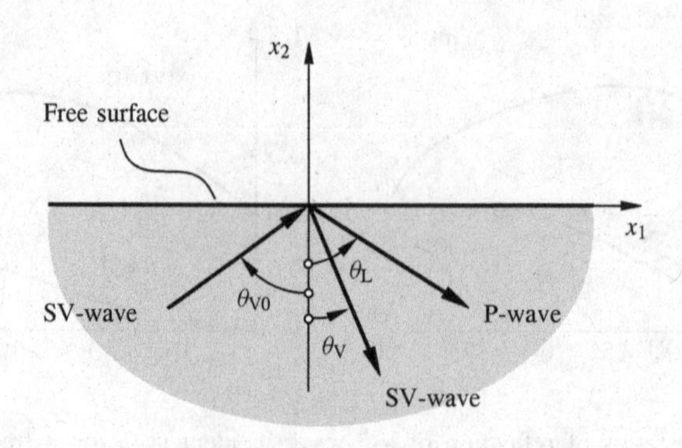

Figure 8.6 Reflection of an SV-wave from a free boundary

The last equation in (8.46) yields the angle of the reflected P-wave θ_L for a given incident angle $\theta_{V0} = \theta_V$. However, since $c_L > c_S$, there exists a critical incident angle θ_V^C such that $\kappa \sin\theta_V^C = 1$, and hence, from (8.46), $\theta_L = 90°$. For angles of incidence greater than this critical value of θ_V, we have a reflected P-wave traveling along the x-axis whose amplitude decreases exponentially along the negative y-axis. Such waves are known as *inhomogeneous waves*.

Substituting (8.45) in (8.39) and using (8.46) yields on simplification

$$A_{V0}\kappa \cos 2\theta_V + A_L \sin 2\theta_L + \kappa A_V \cos 2\theta_V = 0. \tag{8.47}$$

Similarly, the condition (8.40) yields

$$A_{V0} \cos 2\theta_V + A_L\kappa \cos 2\theta_V - A_V \sin 2\theta_V = 0. \tag{8.48}$$

Solving for A_L and A_V in terms of A_{V0} from (8.47) and (8.48) yields

$$\frac{A_L}{A_{V0}} = \frac{-2\kappa \sin 2\theta_V \cos 2\theta_V}{\kappa^2 \cos^2 2\theta_V + \sin 2\theta_L \sin 2\theta_V},$$

$$\frac{A_V}{A_{V0}} = \frac{\sin 2\theta_L \sin 2\theta_V - \kappa^2 \cos^2 2\theta_V}{\kappa^2 \cos^2 2\theta_V + \sin 2\theta_L \sin 2\theta_V}.$$

If the SV-wave is incident normally, we have $A_L/A_{V0} = 0$, and $A_V/A_{V0} = -1$, implying a total reflection with no phase change. When the condition (8.44) holds, $A_V/A_{V0} = 0$ and $A_L/A_{V0} = -(1/\kappa)\tan 2\theta_V$. Thus, we have the phenomenon of mode conversion from the incident SV-wave to a P-wave. However, the mode conversion will be observed only if it occurs at an incidence angle less than the critical angle of incidence θ_V^C discussed above in this section.

8.4.1.3 Incident SH-wave

Consider an incident SH-wave on a free surface. As observed in the above discussions, the SH-wave cannot excite the P- and SV-waves in a homogeneous and isotropic medium. Therefore, the reflected wave will comprise only a reflected SH-wave. Let us represent the complete wave field by

$$\mathbf{u}(x_1, x_2, t) = A_{\mathrm{H0}}\hat{\mathbf{a}}\mathrm{e}^{ik_{\mathrm{S}}(\hat{\mathbf{n}}_{\mathrm{H0}}\cdot\mathbf{r}-c_{\mathrm{S}}t)} + A_{\mathrm{H}}\hat{\mathbf{a}}\mathrm{e}^{ik_{\mathrm{S}}(\hat{\mathbf{n}}_{\mathrm{H}}\cdot\mathbf{r}-c_{\mathrm{S}}t)}, \tag{8.49}$$

where $\hat{\mathbf{n}}_{\mathrm{H0}} = (\sin\theta_{\mathrm{H0}}\cos\theta_{\mathrm{H0}}, 0)^T$ and $\hat{\mathbf{n}}_{\mathrm{H}} = (\sin\theta_{\mathrm{H}}, -\cos\theta_{\mathrm{H}}, 0)^{\mathrm{T}}$ are, respectively, the directions of the incident and reflected SH-waves.

The boundary conditions for the free surface are given by (8.39)–(8.40). It is immediately evident that (8.40) is identically satisfied by (8.49). Substituting (8.49) in (8.39) yields

$$\begin{aligned} &u_{3,2}\big|_{x_2=0} = 0 \\ &\Rightarrow A_{\mathrm{H0}}\cos\theta_{\mathrm{H0}}\mathrm{e}^{ik_{\mathrm{S}}(x_1\sin\theta_{\mathrm{H0}}-c_{\mathrm{S}}t)} - A_{\mathrm{H}}\cos\theta_{\mathrm{H}}\mathrm{e}^{ik_{\mathrm{S}}(x_1\sin\theta_{\mathrm{H}}-c_{\mathrm{S}}t)} = 0. \end{aligned} \tag{8.50}$$

The condition (8.50) is identically satisfied if and only if

$$\sin\theta_{\mathrm{H0}} = \sin\theta_{\mathrm{H}} \;\Rightarrow\; \theta_{\mathrm{H0}} = \theta_{\mathrm{H}}. \tag{8.51}$$

Then, we obtain from (8.50)

$$A_{\mathrm{H0}} = A_{\mathrm{H}}. \tag{8.52}$$

8.5 RAYLEIGH SURFACE WAVES

Under certain conditions, traveling waves can exist within a small depth from a free surface of an elastic continuum, while the bulk of the continuum remains almost undisturbed. Such waves are known as *Rayleigh waves*, and are useful in determining the surface and sub-surface characteristics of materials.

Consider an elastic half-space, as shown in Figure 8.7. Consider a general harmonic traveling wave directed along the positive x_1-axis, and composed of both longitudinal and vertical shear wave components of the form

$$\begin{aligned} \mathbf{u}(x_1, x_2, t) &= \mathbf{u}_{\mathrm{L}}(x_1, x_2, t) + \mathbf{u}_{\mathrm{S}}(x_1, x_2, t) \\ &= \mathbf{Y}_{\mathrm{L}}(x_2)\mathrm{e}^{i(kx_1-\omega t)} + \mathbf{Y}_{\mathrm{S}}(x_2)\mathrm{e}^{i(kx_1-\omega t)}, \end{aligned} \tag{8.53}$$

where $\mathbf{Y}_{\mathrm{L}}(x_2)$ and $\mathbf{Y}_{\mathrm{S}}(x_2)$ are two vector functions, and k and ω are, respectively, the wave number and frequency of the wave. Substituting the expressions of $\mathbf{u}_{\mathrm{L}}$ and $\mathbf{u}_{\mathrm{S}}$ from (8.53)

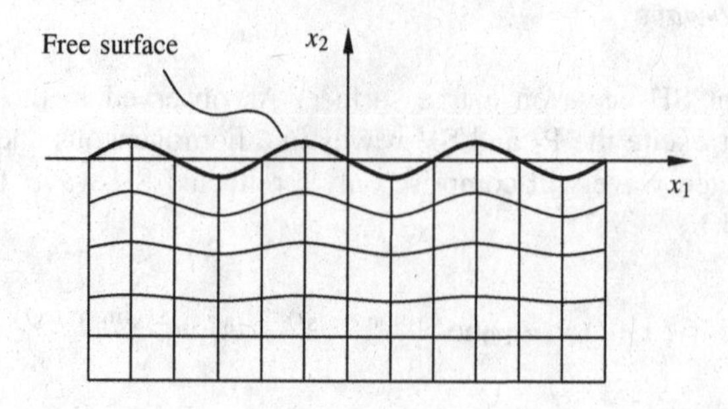

Figure 8.7 Motion of a section of a continuum due to Rayleigh surface waves

in the respective governing wave equations (8.22) and (8.23), we have

$$\mathbf{Y}_\mathrm{L}'' + \left(\frac{\omega^2}{c_\mathrm{L}^2} - k^2\right)\mathbf{Y}_\mathrm{L} = 0, \tag{8.54}$$

$$\mathbf{Y}_\mathrm{S}'' + \left(\frac{\omega^2}{c_\mathrm{S}^2} - k^2\right)\mathbf{Y}_\mathrm{S} = 0, \tag{8.55}$$

where prime denotes differentiation with respect to x_2. Now, we consider the case

$$\frac{\omega^2}{c_\mathrm{L}^2} - k^2 := -k_\mathrm{L}^2 < 0 \quad \text{and} \quad \frac{\omega^2}{c_\mathrm{S}^2} - k^2 := -k_\mathrm{S}^2 < 0. \tag{8.56}$$

Using these two definitions, the solutions of (8.54) or (8.55) can be represented in the general form as

$$\mathbf{Y}_J(x_2) = \mathbf{A}_J \mathrm{e}^{k_J x_2} + \mathbf{B}_J \mathrm{e}^{-k_J x_2},$$

where $J = \mathrm{L}$ or S. From the consideration of finiteness of the solution as $x_2 \to -\infty$, we must have $B_J = 0$. Therefore, the waveform (8.53) can be written as

$$\mathbf{u}(x_1, x_2, t) = (\mathbf{A}_\mathrm{L}\mathrm{e}^{k_\mathrm{L}x_2} + \mathbf{A}_\mathrm{S}\mathrm{e}^{k_\mathrm{S}x_2})\mathrm{e}^{i(kx_1 - \omega t)}. \tag{8.57}$$

Such a solution, if it exists, will clearly imply that $\mathbf{u} \to 0$ as $x_2 \to -\infty$. Thus, the wave motion will be prominent only on, and close to the surface $x_2 = 0$. Next, we must investigate the existence of a solution of the form (8.57).

The longitudinal and shear wave components in the solution (8.57) must satisfy the curl and divergence conditions (8.14). The curl condition yields

$$\nabla \times [\mathbf{A}_\mathrm{L}\mathrm{e}^{k_\mathrm{L}x_2}\mathrm{e}^{i(kx_1 - \omega t)}] = 0$$
$$\Rightarrow A_{\mathrm{L}3}k_L = 0, \quad ikA_{\mathrm{L}3} = 0, \quad \text{and} \quad ik_1A_{\mathrm{L}2} - k_\mathrm{L}A_{\mathrm{L}1} = 0. \tag{8.58}$$

Thus, we must have

$$A_{L3} = 0 \qquad \text{and} \qquad A_{L2} = \frac{k_L}{ik} A_{L1}. \tag{8.59}$$

Similarly, the divergence condition in (8.14) yields

$$\nabla \cdot [\mathbf{A}_S e^{k_S x_2} e^{i(kx_1 - \omega t)}] = 0$$
$$\Rightarrow A_{S2} = -\frac{ik}{k_S} A_{S1}. \tag{8.60}$$

Using (8.59) and (8.60) in (8.57) yields on rearrangement

$$u_1 = (A_{L1} e^{k_L x_2} + A_{S1} e^{k_S x_2}) e^{i(kx_1 - \omega t)}, \tag{8.61}$$
$$u_2 = \left(\frac{k_L}{ik} A_{L1} e^{k_L x_2} - \frac{ik}{k_S A_S 1} e^{k_S x_2} \right) e^{i(kx_1 - \omega t)}, \tag{8.62}$$
$$u_3 = 0. \tag{8.63}$$

The free boundary conditions can be written as

$$\sigma_{12}\big|_{x_2=0} = 0 \;\Rightarrow\; (u_{1,2} + u_{2,1})|_{x_2=0} = 0, \tag{8.64}$$
$$\sigma_{22}\big|_{x_2=0} = 0 \;\Rightarrow\; [(\lambda + 2\mu)(u_{1,1} + u_{2,2}) - 2\mu u_{1,1}]|_{x_2=0} = 0. \tag{8.65}$$

Substituting (8.61)–(8.62) in (8.64), we obtain

$$2k_L k_S A_{L1} + (k_S^2 + k^2) A_{S1} = 0. \tag{8.66}$$

Similarly, from (8.65), and using (8.24), we have

$$[(k^2 - k_L^2)\kappa^2 - 2k^2] A_{L1} - 2k^2 A_{S1} = 0, \tag{8.67}$$

where $\kappa = c_L / c_S$. Eliminating ω^2 from the two definitions in (8.56) yields

$$(k^2 - k_L^2)\kappa^2 = k^2 - k_S^2. \tag{8.68}$$

Using (8.68) in (8.67), one obtains

$$(k^2 + k_S^2) A_{L1} + 2k^2 A_{S1} = 0. \tag{8.69}$$

The two homogeneous equation (8.67) and (8.69) have non-trivial solution for A_{L1} and A_{S1} if and only if

$$4k_L k_S k^2 - (k^2 + k_S^2)^2 = 0$$

$$\Rightarrow \left(2k^2 - \frac{\omega^2}{c_S^2}\right)^4 - 16\left(k^2 - \frac{\omega^2}{c_L^2}\right)\left(k^2 - \frac{\omega^2}{c_S^2}\right)k^4 = 0 \qquad \text{(using (8.56))}$$

or

$$(2 - \xi^2)^4 - 16\left(1 - \frac{\xi^2}{\kappa^2}\right)(1 - \xi^2) = 0$$

$$\Rightarrow \xi^6 - 8\xi^4 + 8\left(3 - \frac{2}{\kappa^2}\right)\xi^2 - 16\left(1 - \frac{1}{\kappa^2}\right) = 0, \tag{8.70}$$

where $\kappa = c_L/c_S$ and

$$\xi = \frac{\omega}{c_S k}. \tag{8.71}$$

Since ω/k is the phase speed of the harmonic Rayleigh wave (8.57), ξ has the interpretation of a wave speed ratio.

For a given material, κ (given by (8.25)) is a constant, and roots of the polynomial (8.70) yield the possible values of ξ. However, not all such values are admissible for the dispersion relation (8.71) since they have to satisfy both the conditions in (8.56), i.e.,

$$\frac{\xi}{\kappa} - 1 < 0 \quad \text{and} \quad \xi - 1 < 0. \tag{8.72}$$

Since $c_L > c_S$ (i.e., $\kappa > 1$), the admissible values are those which satisfy the second condition in (8.72), i.e., $\xi < 1$. It can be shown that there is only one root of (8.70) which satisfies this condition. The variation of this root with ν is shown in Figure 8.8. The phase velocity of Rayleigh waves can be obtained from (8.71) as

$$c_R = \frac{\omega}{k} = c_S \xi. \tag{8.73}$$

Since $\xi < 1$, the speed of Rayleigh waves is less than the speed of shear waves (and hence also less than the speed of longitudinal waves).

For a Rayleigh wave, the locus of motion of any point of the continuum can be obtained by writing the real parts of (8.61) and (8.62) as

$$u_1 = a\cos(kx_1 - \omega t), \tag{8.74}$$

$$u_2 = b\sin(kx_1 - \omega t), \tag{8.75}$$

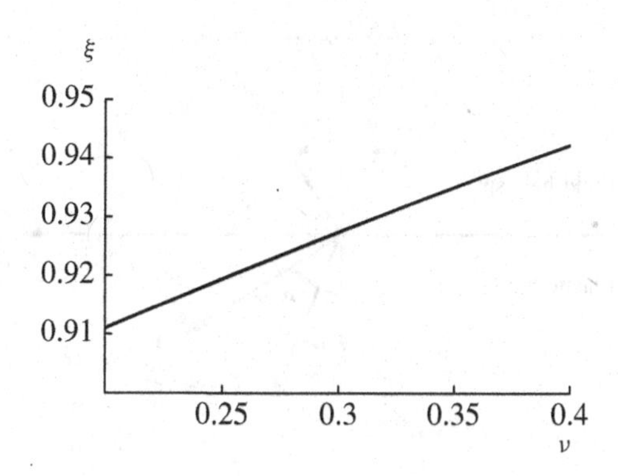

Figure 8.8 Variation of the admissible values of wave speed ratio ξ with Poisson ratio ν for Rayleigh waves

where a and b are constants for any point at a depth x_2. These are given by

$$a = A_{L1}e^{k_L x_2} + A_{S1}e^{k_S x_2},$$

$$b = \frac{k_L}{k}A_{L1}e^{k_L x_2} + \frac{k}{k_S}A_{S1}e^{k_S x_2}.$$

From (8.74) and (8.75), one can easily write

$$\frac{u_1^2}{a^2} + \frac{u_2^2}{b^2} = 1. \tag{8.76}$$

It may be noted that as $x_2 \to -\infty$, $a \to 0$, and $b \to 0$. Thus, all points of the continuum move on elliptic paths, and the ellipses shrink in size with increasing depth from the surface.

8.6 REFLECTION AND REFRACTION OF PLANAR ACOUSTIC WAVES

In this section, we consider an elastic half-space in contact with an inviscid compressible fluid, and study the waves generated in the elastic half-space by planar acoustic waves incident from the fluid on the interface. Situations similar to this occur in ultrasonic testing of materials and underwater acoustic sensing, and also in room acoustics.

Let a plane acoustic wave in the fluid be incident on the fluid–solid interface $x_2 = 0$ at an angle ϕ, as shown in Figure 8.9. This causes a reflected wave in the fluid, and in general, longitudinal and shear acoustic waves in the elastic half-space. Let the velocity potential for the fluid wave be represented by

$$\psi(x, y, t) = \psi_I e^{ik(x_1 \sin\phi - x_2 \cos\phi - ct)} + \psi_R e^{ik(x_1 \sin\phi + x_2 \cos\phi - ct)}, \tag{8.77}$$

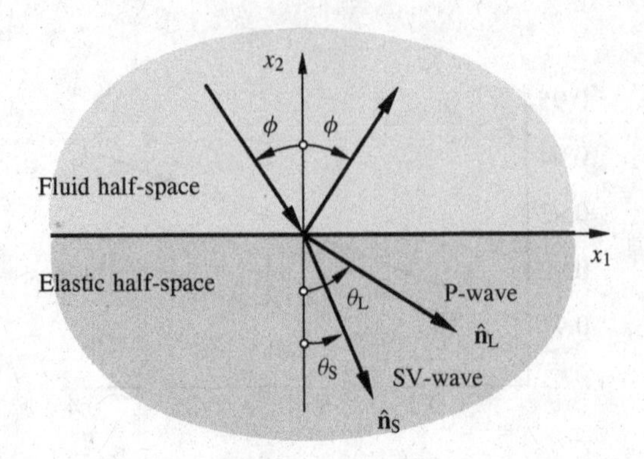

Figure 8.9 Reflection and refraction of sound at the boundary between fluid and elastic half-spaces

where ψ_I and ψ_R are constant amplitudes associated with the incident and reflected waves, respectively, c is the speed of acoustic waves in the fluid, and k is the wave number. The waves in the elastic half-space can be represented as

$$\mathbf{u}(x_1, x_2, t) = A_L \hat{\mathbf{n}}_L e^{ik_L(x_1 \sin\theta_L - x_2 \cos\theta_L - c_L t)} + A_S \hat{\mathbf{a}} \times \hat{\mathbf{n}}_S e^{ik_S(x_1 \sin\theta_S - x_2 \cos\theta_S - c_S t)}, \tag{8.78}$$

where $\hat{\mathbf{n}}_L = (\sin\theta_L, -\cos\theta_L, 0)^T$ and $\hat{\mathbf{n}}_S = (\sin\theta_S, -\cos\theta_S, 0)^T$. At the boundary, we have conditions on the velocity along the normal, the pressure (or the normal stress), and the shear stress. Since the fluid is assumed to be inviscid, there is no surface shear stress at the boundary of the elastic half-space. The velocity, pressure, and shear boundary conditions can be written as, respectively,

$$\psi_{,2}\big|_{x_2=0} = u_{2,t}\big|_{x_2=0}, \tag{8.79}$$

$$p = -\rho\psi_{,t}\big|_{x_2=0} = -\sigma_{22}\big|_{x_2=0}, \tag{8.80}$$

$$\sigma_{12}\big|_{x_2=0} = 0. \tag{8.81}$$

Substituting (8.77) and (8.78) in (8.79) yields

$$(-\psi_I + \psi_R) ik \cos\phi e^{ik(x_1 \sin\phi - ct)} = A_L \cos\theta_L i k_L c_L e^{ik_L(x_1 \sin\theta_L - c_L t)} - A_S \sin\theta_S i k_S c_S e^{ik_S(x_1 \sin\theta_S - c_S t)}. \tag{8.82}$$

If (8.82) is to be satisfied identically, we must have

$$k \sin\phi = k_L \sin\theta_L = k_S \sin\theta_S \quad \text{and} \quad kc = k_L c_L = k_S c_S$$

$$\Rightarrow k_L = k\frac{c}{c_L}, \quad k_S = k\frac{c}{c_S}, \quad \sin\theta_L = \frac{c_L}{c}\sin\phi, \quad \sin\theta_S = \frac{c_S}{c}\sin\phi. \tag{8.83}$$

These conditions then yield the wave numbers k_L and k_S, and the directions of the longitudinal and shear waves, θ_L and θ_S, respectively. Using (8.83) in (8.82), one can write

$$-\frac{c\cos\theta_L}{\cos\phi}A_L+\frac{c\sin\theta_S}{\cos\phi}A_S+\psi_R=\psi_I. \tag{8.84}$$

Similarly, from (8.80) and (8.81), we obtain, respectively,

$$-c_L\cos 2\theta_S A_L+c_S\sin 2\theta_S A_S-\psi_R=\psi_I, \tag{8.85}$$

$$c_S\sin 2\theta_L A_L+c_L\cos 2\theta_S A_S=0. \tag{8.86}$$

One can now easily solve for A_L, A_S, and ψ_R from the inhomogeneous linear equations (8.84)–(8.86).

There are some interesting observations here. When the angle of incidence of the acoustic wave in the fluid satisfies the condition $\sin\phi < c/c_L$, we have a reflected wave in the fluid, and P- and SV-waves in the elastic continuum. Thus, a fraction of the energy of the incident wave goes into the elastic half-space. For $\sin\phi = c/c_L$, we have from (8.83) $\theta_L = \pi/2$, implying that the P-wave travels parallel to the surface. It can be further observed from (8.86) that $A_S = 0$, i.e., the shear wave disappears. Thus, in this case we have only a P-wave in the elastic half-space. For $\sin\phi > c/c_L$, we observe from (8.83) that $\sin\theta_L > 1$. This implies that the P-wave becomes an inhomogeneous wave, and is restricted to being close to the surface. In the situation $\sin\phi > c/c_S$, both the P- and SV-waves become inhomogeneous, and hence we have only inhomogeneous waves in the elastic half-space. Since inhomogeneous waves cannot carry energy into the elastic half-space, in this case the incident wave is completely reflected back into the fluid without any loss of energy.

EXERCISES

8.1 An elastic half-space occupies the region $x_2 \in (-\infty, 0]$, as shown in Figure 8.10. The surface $x_2 = 0$ is rigidly fixed so as to prevent any motion of the points on the surface. Determine the reflected waves when (a) a plane harmonic P-wave is incident, and (b) a plane harmonic SV-wave is incident on the boundary. Determine the reflection coefficients as a function of the angle of incidence of the incident waves.

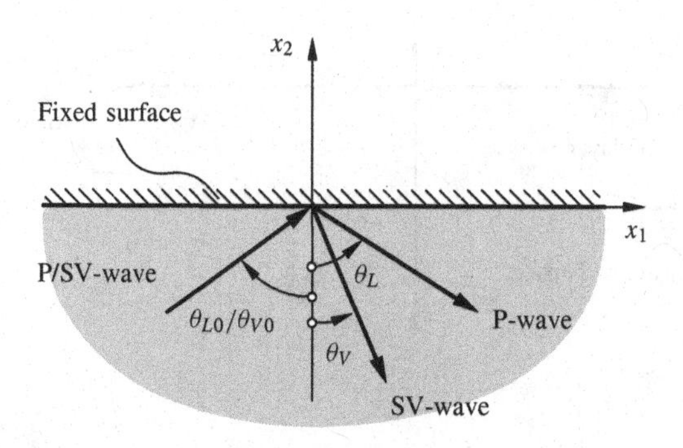

Figure 8.10 Exercise 8.1

8.2 Investigate the propagation of SH-waves through a wave guide, as shown in Figure 8.11. Determine the dispersion relation and cut-off frequencies for such waves.

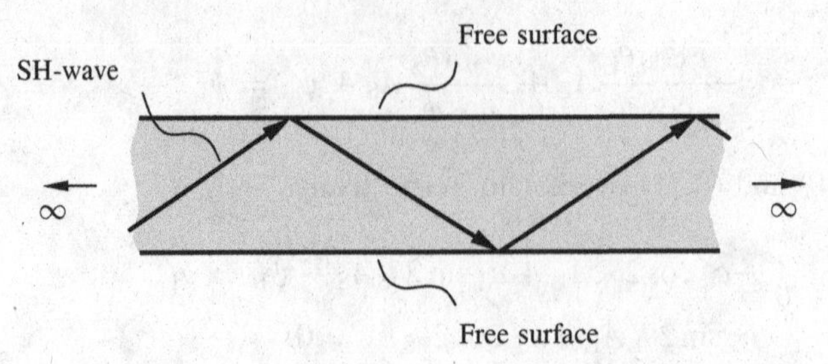

Figure 8.11 Exercise 8.2

8.3 Two elastic half-spaces are connected firmly with each other so as to prevent any slip between them. Investigate the existence and propagation of waves (combination of P- and SV-waves) that are restricted to a thin layer around the interface of the two half-spaces.

8.4 An elastic half-space is in contact with an ideal fluid. Investigate the propagation of interface waves that may exist within a thin layer of the interface of the fluid and the half-space. Such interface waves are known as *Stoneley waves* (see, for example, [4]).

8.5 An elastic half-space is connected firmly with a layer of another material with specific normal impedance $\mathcal{Z}_L$, and specific shear impedance $\mathcal{Z}_S$. Investigate the reflection of (a) an incident SH-wave, and (b) an incident P-wave from the surface. Determine the power lost in the reflection process as a function of the angle of incidence.

8.6 Two elastic half-spaces are connected firmly with each other so as to prevent any slip between them. A plane harmonic SH-wave is incident on the interface. Determine the transmitted and reflected waves as a function of the angle of incidence of the incident wave. Also determine the shear stress at the interface as a function of the angle of incidence.

8.7 An elastic half-space is firmly connected to a layer of another material of thickness d, as shown in Figure 8.12. Investigate the existence of surface SH-waves at the interface of the two materials. Such surface SH-waves are observed only in layered continua, and are known as *Love waves* (see, for example, [4]). Also determine the shear stress at the interface.

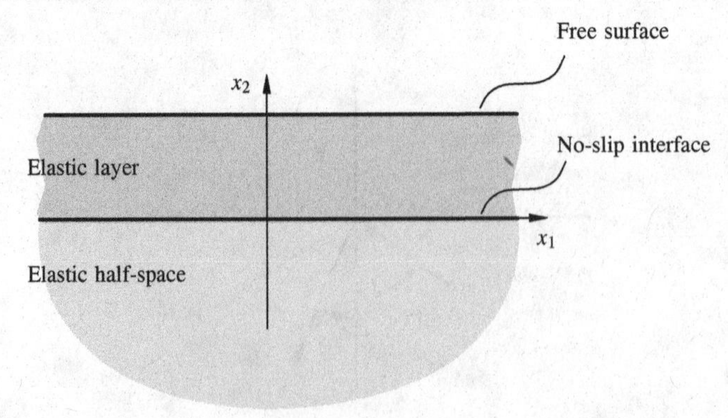

Figure 8.12 Exercise 8.7

8.8 An elastic half-space made of steel is in contact with water. A plane harmonic wave in the water is incident on the water–steel interface, as discussed in Section 8.6. Calculate and plot the amplitude ratios of the wave reflected back into the water, and the waves transmitted into the steel. How do the results change when water is replaced by air? Take $c_L = \sqrt{3}c_S = 5000$ m/s for steel, $c = 1500$ m/s for water and $c = 340$ m/s for air.

REFERENCES

[1] Kreyszig, E., *Advanced Engineering Mathematics*, 2e, Wiley Eastern Pvt. Ltd., New Delhi, 1969.
[2] Landau, L.D., and Lifshitz, E.M., *Theory of Elasticity*, 3e, Butterworth-Heinemann, Oxford, 1986.
[3] Fung, Y.C., *Foundations of Solid Mechanics*, Prentice-Hall, Inc., Englewood Cliffs, New Jersey, 1965.
[4] Achenbach, J.D., *Wave Propagation in Elastic Solids*, North-Holland Publishing Co., Amsterdam, 1973.

Appendix A

The variational formulation of dynamics

This appendix gives a brief introduction to the variational formulation of dynamics of continuous systems. More detailed discussions on the variational formulation of mechanics can be found in, for example, [1], and [2].

Consider for simplicity a one-dimensional continuous system with the field variable $w(x, t)$ which uniquely represents the configuration of the system at any time t. In the course of temporal evolution of the system, let the configurations at two time instants $t = t_1$ and $t = t_2$ be recorded as, respectively, $w(x, t_1)$ and $w(x, t_2)$. Assume that no information about the intermediate configurations is available. Now, the question is: Can we determine the intermediate configurations through which the system passed while going from $w(x, t_1)$ to $w(x, t_2)$? The answer to this question is provided by Hamilton's principle which states: *Of all the infinite paths available to a system between any two observed configurations, the system follows that path which extremizes the action A defined by*

$$A = \int_{t_1}^{t_2} \mathcal{L}\, \mathrm{d}t, \tag{A.1}$$

where $\mathcal{L} = \mathcal{T} - \mathcal{V}$ is known as the Lagrangian, and $\mathcal{T}$ and $\mathcal{V}$ are, respectively, the kinetic energy and potential energy expressions of the system at an arbitrary configuration. Extremization in mechanical systems can be construed as minimization. Thus, we will be searching over all possible paths, and the test that we have found the actual path is that any infinitesimal variation over that path should leave the value of A unchanged. This can be mathematically written as

$$\delta A = \delta \int_{t_1}^{t_2} \mathcal{L}\, \mathrm{d}t = \int_{t_1}^{t_2} \delta\mathcal{L}\, \mathrm{d}t = 0, \tag{A.2}$$

where $\delta(\cdot)$ is the infinitesimal variation operator which behaves very much like the total derivative operator, with the difference that $\delta(\cdot)$ does not vary the time. This difference should be obvious, since we are interested only in the infinitesimal variation of a path, and not time. Another important property of $\delta(\cdot)$ is that it commutes with any differential

Vibrations and Waves in Continuous Mechanical Systems P. Hagedorn and A. DasGupta

operator, i.e.,

$$\delta\left[\frac{\partial}{\partial x}(\cdot)\right] = \frac{\partial}{\partial x}[\delta(\cdot)]. \tag{A.3}$$

The Lagrangian is, in general, a function of the field variable, its time and space-derivatives, and time. Consider now a one-dimensional continuous system with a Lagrangian

$$\mathcal{L} = \int_0^l \hat{\mathcal{L}}(w, w_{,t}, w_{,x}, w_{,xx}, t)\,\mathrm{d}x,$$

where $\hat{\mathcal{L}}(\cdot)$ is known as the *Lagrangian density* of the system for obvious reasons, and l is the length of the continuum. Using the extremization condition (A.2), we obtain

$$\int_{t_1}^{t_2}\int_0^l \delta\hat{\mathcal{L}}(w, w_{,t}, w_{,x}, w_{,xx}, t)\,\mathrm{d}x\,\mathrm{d}t = 0,$$

or

$$\int_{t_1}^{t_2}\int_0^l \left[\frac{\partial\hat{\mathcal{L}}}{\partial w}\delta w + \frac{\partial\hat{\mathcal{L}}}{\partial w_{,t}}\delta w_{,t} + \frac{\partial\hat{\mathcal{L}}}{\partial w_{,x}}\delta w_{,x} + \frac{\partial\hat{\mathcal{L}}}{\partial w_{,xx}}\delta w_{,xx}\right]\mathrm{d}x\,\mathrm{d}t = 0.$$

Integrating by parts appropriately, one can obtain

$$\begin{aligned}\int_0^l \frac{\partial\hat{\mathcal{L}}}{\partial w_{,t}}\delta w\bigg|_{t_1}^{t_2}\mathrm{d}x + \int_{t_1}^{t_2}\left[\frac{\partial\hat{\mathcal{L}}}{\partial w_{,xx}}\delta w_{,x} + \left\{\frac{\partial\hat{\mathcal{L}}}{\partial w_{,x}} - \frac{\mathrm{d}}{\mathrm{d}x}\left(\frac{\partial\hat{L}}{\partial w_{,xx}}\right)\right\}\delta w\right]\bigg|_0^l \mathrm{d}t \\ + \int_{t_1}^{t_2}\int_0^l\left[\frac{\partial\hat{\mathcal{L}}}{\partial w} - \frac{\mathrm{d}}{\mathrm{d}t}\left(\frac{\partial\hat{\mathcal{L}}}{\partial w_{,t}}\right) - \frac{\mathrm{d}}{\mathrm{d}x}\left(\frac{\partial\hat{\mathcal{L}}}{\partial w_{,x}}\right) + \frac{\mathrm{d}^2}{\mathrm{d}x^2}\left(\frac{\partial\hat{\mathcal{L}}}{\partial w_{,xx}}\right)\right]\delta w\,\mathrm{d}x\,\mathrm{d}t = 0.\end{aligned} \tag{A.4}$$

Since we started with the objective of finding the intermediate configurations of the system between two configurations connected by a trajectory, we do not vary the initial and final configurations on the trajectory. Thus, $\delta w|_{t_1} = \delta w|_{t_2} \equiv 0$. Therefore, the first integral in (A.4) vanishes identically. Now, the remaining terms in (A.4) have to vanish for all the possible variations considered in Hamilton's principle. The second term in (A.4) contains the variations at the boundary only. Since these boundary variations can be held fixed independently of the variations at other points of the trajectory, the fundamental lemma of the calculus of variations tells us that the integrand of the third term in (A.4) must vanish. Thus,

$$\frac{\partial\hat{\mathcal{L}}}{\partial w} - \frac{\mathrm{d}}{\mathrm{d}t}\left(\frac{\partial\hat{\mathcal{L}}}{\partial w_{,t}}\right) - \frac{\mathrm{d}}{\mathrm{d}x}\left(\frac{\partial\hat{\mathcal{L}}}{\partial w_{,x}}\right) + \frac{\mathrm{d}^2}{\mathrm{d}x^2}\left(\frac{\partial\hat{\mathcal{L}}}{\partial w_{,xx}}\right) = 0, \tag{A.5}$$

which is the equation of motion of the system. Finally, the second integral vanishes, for example, if

$$\left.\frac{\partial \hat{\mathcal{L}}}{\partial w_{,xx}}\right|_{x=0} \equiv 0, \qquad \text{or} \qquad \delta w_{,x}\big|_{x=0} \equiv 0, \tag{A.6}$$

$$\left.\frac{\partial \hat{\mathcal{L}}}{\partial w_{,xx}}\right|_{x=l} \equiv 0, \qquad \text{or} \qquad \delta w_{,x}\big|_{x=l} \equiv 0, \tag{A.7}$$

$$\left.\frac{\partial \hat{\mathcal{L}}}{\partial w_{,x}} - \frac{\mathrm{d}}{\mathrm{d}x}\left(\frac{\partial \hat{\mathcal{L}}}{\partial w_{,xx}}\right)\right|_{x=0} \equiv 0, \qquad \text{or} \qquad \delta w\big|_{x=0} \equiv 0, \tag{A.8}$$

and

$$\left.\frac{\partial \hat{\mathcal{L}}}{\partial w_{,x}} - \frac{\mathrm{d}}{\mathrm{d}x}\left(\frac{\partial \hat{\mathcal{L}}}{\partial w_{,xx}}\right)\right|_{x=l} \equiv 0, \qquad \text{or} \qquad \delta w\big|_{x=l} \equiv 0. \tag{A.9}$$

These relations yield possible boundary conditions of the problem. Note, however, that Hamilton's principle only requires the *sum* of all the boundary terms to vanish, and not necessarily the different boundary terms individually (see the examples with special boundary conditions in Section 3.1.2). The solution of the equation of motion (A.5) along with the boundary conditions (A.6)–(A.9) yields the successive configurations of the system at any time instant starting from any arbitrary initial condition.

When a system is subject to generalized non-potential or external forces $Q(x,t)$, one can introduce them in Hamilton's principle through the virtual work expression $\overline{\delta\mathcal{W}} = Q(x,t)\delta w$. Hamilton's principle (A.2) is then written as

$$\int_{t_0}^{t_1} [\delta\mathcal{L} + \overline{\delta\mathcal{W}}]\,\mathrm{d}t = 0$$

$$\Rightarrow \int_{t_0}^{t_1} [\delta\mathcal{L} + Q(x,t)\delta w]\,\mathrm{d}t = 0. \tag{A.10}$$

This is, however, not strictly a variational principle, and is sometimes referred to as the *extended Hamilton's principle*. Following the steps discussed above, the equation of motion in this case can be easily obtained as

$$\frac{\partial \hat{\mathcal{L}}}{\partial w} - \frac{\mathrm{d}}{\mathrm{d}t}\left(\frac{\partial \hat{\mathcal{L}}}{\partial w_{,t}}\right) - \frac{\mathrm{d}}{\mathrm{d}x}\left(\frac{\partial \hat{\mathcal{L}}}{\partial w_{,x}}\right) + \frac{\mathrm{d}^2}{\mathrm{d}x^2}\left(\frac{\partial \hat{\mathcal{L}}}{\partial w_{,xx}}\right) + Q(x,t) = 0. \tag{A.11}$$

The boundary conditions (A.6)–(A.9), however, remain the same.

REFERENCES

[1] Landau, L.D., and Lifshitz, E.M., *Mechanics*, 3e, Butterworth-Heinemann, Oxford, 1976.
[2] Goldstein, H., Poole, C.P., and Safko, J.L., *Classical Mechanics*, 3e, Addison-Wesley, Boston, 2002.

Appendix B

Harmonic waves and dispersion relation

B.1 FOURIER REPRESENTATION AND HARMONIC WAVES

It is known from the theory of Fourier transforms (see, for example, [1]) that any sufficiently smooth square integrable function can be represented as

$$f(z) = \frac{1}{2\pi} \int_{-\infty}^{\infty} F(k) \mathrm{e}^{ikz} \, \mathrm{d}k, \tag{B.1}$$

where

$$F(k) = \int_{-\infty}^{\infty} f(z) \mathrm{e}^{-ikz} \, \mathrm{d}z,$$

and k is the Fourier variable. Hence, a traveling wave solution $f(x - ct)$ of the wave equation can be represented using (B.1) as

$$\begin{aligned} f(x - ct) &= \frac{1}{2\pi} \int_{-\infty}^{\infty} F(k) \mathrm{e}^{ik(x-ct)} \, \mathrm{d}k \\ &= \frac{1}{2\pi} \int_{-\infty}^{\infty} F(k) \mathrm{e}^{i(kx-\omega t)} \, \mathrm{d}k, \end{aligned} \tag{B.2}$$

where $\omega = ck$. The integrand in (B.2) can be easily recognized to be a harmonic traveling wave in complex notation. Thus, a general traveling wave can be represented as a linear superposition of harmonic traveling waves. This implies that wave propagation in linear continuous systems can be conveniently studied by studying the propagation of a harmonic wave solution of the form

$$w(x, t) = B \mathrm{e}^{i(kx - \omega t)}, \tag{B.3}$$

where $k = 2\pi/\lambda$ is defined as the wave number, λ is the wavelength, and ω is the circular frequency of the harmonic wave.

Consider an unforced taut string governed by the equation (see Chapter 1)

$$\rho A w_{,tt} - T w_{,xx} = 0, \tag{B.4}$$

where ρ is the density, A is the area of cross-section, and T is the tension. Substituting (B.3) in (B.4) yields

$$\begin{aligned} &D(\omega, k) B \mathrm{e}^{i(kx-\omega t)} = 0 \\ \Rightarrow\ &D(\omega, k) := -\rho A \omega^2 + T k^2 = 0, \end{aligned} \tag{B.5}$$

where $D(\omega, k) = 0$ is known as the *dispersion relation*. Thus, the dispersion relation provides, in terms of the properties of the medium (here, the string), a connection between the frequency and the wave number of the harmonic waves that can propagate in the medium.

A uniform Euler–Bernoulli beam is described by (see Chapter 3)

$$\rho A w_{,tt} + E I w_{,xxxx} = 0, \tag{B.6}$$

where E is the Young's modulus and I is the second moment of the area of cross-section. Substituting (B.3) in (B.6) yields the dispersion relation of the beam as

$$-\rho A \omega^2 + E I k^4 = 0. \tag{B.7}$$

Now, depending on the problem and the kind of analysis intended, there are two approaches, namely the *spatial framework* and the *temporal framework*. In the former, we first solve the dispersion relation for k in terms of ω. For example, in the case of a taut string

$$k = \pm \frac{\omega}{c}, \tag{B.8}$$

where $c = \sqrt{T/\rho A}$. Then, the complete harmonic wave solution may be written using (B.8) as

$$\begin{aligned} w(x, t) &= A_1 \mathrm{e}^{i(\omega x/c - \omega t)} + A_2 \mathrm{e}^{i(-\omega x/c - \omega t)} \\ &= [A_1 \mathrm{e}^{i\omega x/c} + A_2 \mathrm{e}^{-i\omega x/c}]\,\mathrm{e}^{-i\omega t}, \end{aligned} \tag{B.9}$$

where A_1 and A_2 are arbitrary constants. This solution is of the form $w(x, t) = W(x)\mathrm{e}^{-i\omega t}$, and hence suitable for modal analysis of finite continuous systems. Note that (B.9) represents the superposition of two counter-propagating harmonic waves of the form $A_1 \mathrm{e}^{i(kx-\omega t)}$ and $A_2 \mathrm{e}^{i(-kx-\omega t)}$, and can meet boundary condition requirements at some point in the medium. Hence (B.9) is also suitable for studying scattering of waves involving reflection and transmission of waves at boundaries and obstacles.

In the temporal framework, we solve the dispersion relation $D(\omega, k) = 0$ for ω in terms of k. For the string, this gives

$$\omega = \pm ck. \tag{B.10}$$

Therefore, the complete harmonic solution can be written as

$$w(x,t) = A_1 e^{ik(x-ct)} + A_2 e^{ik(x+ct)}, \tag{B.11}$$

where A_1 and A_2 are arbitrary constants. This harmonic wave solution has the form of the d'Alembert solution discussed in Chapter 2. This form of solution is suitable for initial value problems, or studying wave propagation in infinite continuous systems. For example, from the dispersion relation (B.7) of the Euler–Bernoulli beam, one can solve

$$\omega = \pm\left(\sqrt{\frac{EI}{\rho A}}\right) k^2 = \pm\beta k^2.$$

Then, the solution of the initial value problem for the beam with $w(x,0) = w_0(x)$ and $w_{,t}(x,0) = v_0(x)$ can be written as

$$w(x,t) = \frac{1}{2\pi}\int_{-\infty}^{\infty}\left[F(k)e^{i(kx-\beta k^2 t)} + G(k)e^{i(kx+\beta k^2 t)}\right] dk,$$

where $F(k)$ and $G(k)$ are obtained by solving

$$F(k) + G(k) = \int_{-\infty}^{\infty} w_0(x)e^{-ikx}\,dx$$

and

$$-F(k) + G(k) = \frac{1}{\beta k^2}\int_{-\infty}^{\infty} v_0(x)e^{-ikx}\,dx.$$

B.2 PHASE VELOCITY AND GROUP VELOCITY

Consider the harmonic traveling wave represented by (B.3). For a moving observer, this wave is described by

$$w(x,t) = Be^{i(kx(t)-\omega t)},$$

where $x(t)$ is the coordinate location of the observer at any time t. If the wave is to appear stationary to this moving observer, the phase $(kx(t) - \omega t)$ of the wave must not change with time, i.e.,

$$k\dot{x}(t) - \omega = 0 \quad\Rightarrow\quad \dot{x}(t) = \frac{\omega}{k} := c_P,$$

where the special velocity c_P is termed the *phase velocity* of the wave. Thus, c_P represents the speed at which the crests and troughs of a harmonic wave move. The ratio $c_P = \omega/k$ is easily obtained from the dispersion relation. For example, for a taut string $c_P = \omega/k = c$. In general, c_P can be a function of the wave number k. For example, for an Euler–Bernoulli beam, (B.7) yields $c_P = (\sqrt{EI/\rho A})k$. In that case, the medium is known as a *dispersive medium.* Thus, in a dispersive medium, harmonic waves of different wave numbers propagate at different speeds.

As mentioned previously, a general waveform can be considered to be composed of a number of harmonic wave components. When the waveform travels through a dispersive medium it will distort since the phase speeds of the harmonic components constituting the waveform are all different. Hence, we require a different definition of speed of the waveform in a dispersive medium, which is discussed next.

Consider the Fourier representation of a traveling wave in a dispersive medium of the form

$$f(x,t) = \frac{1}{2\pi}\int_{-\infty}^{\infty} F(k)\mathrm{e}^{i(kx-\omega(k)t)}\,\mathrm{d}k. \tag{B.12}$$

A wave is termed *narrow-band* when its spectrum is zero everywhere except over a narrow band of wave numbers, i.e., $F(k) = 0$ for $|k - k_0| > \epsilon$, as shown in Figure B.1. Such a waveform is an amplitude modulated wave like the one shown in Figure B.2, and is termed a wave packet. Such a wave packet has a functional representation of the form

$$\frac{1}{2\pi}\int_{-\infty}^{\infty} F(k)\mathrm{e}^{ikx}\,\mathrm{d}k = f_0(x)\mathrm{e}^{ik_0x}, \tag{B.13}$$

where $f_0(x)$ represents the modulating envelope, as shown in Figure B.2. It is of interest to determine the velocity of the profile of the wave packet since the energy of the packet is localized in the high-amplitude region.

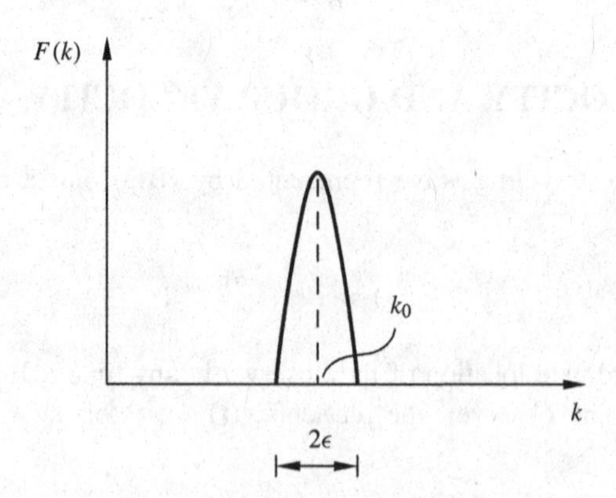

Figure B.1 Spectrum of a typical narrowband waveform in the wave number space

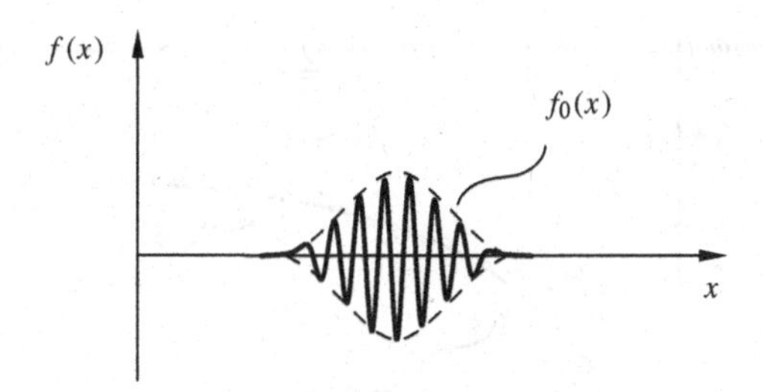

Figure B.2 Spatial signature of a typical narrow-band waveform

Let us take the central wave number as k_0, and write $k = k_0 + \xi$, where $\xi \in [-\epsilon, \epsilon]$. Then, one can easily show from (B.13) that

$$\frac{1}{2\pi}\int_{-\infty}^{\infty} F(k_0 + \xi)\mathrm{e}^{i\xi x}\,\mathrm{d}\xi = f_0(x). \tag{B.14}$$

Using this, one can now rewrite (B.12) as

$$f(x,t) = \frac{1}{2\pi}\int_{-\infty}^{\infty} F(k_0 + \xi)\mathrm{e}^{i[k_0 x + \xi x - \omega(k_0+\xi)t]}\,\mathrm{d}\xi$$

or

$$f(x,t) \approx \mathrm{e}^{i(k_0 x - \omega(k_0)t)}\frac{1}{2\pi}\int_{-\infty}^{\infty} F(k_0 + \xi)\mathrm{e}^{i\xi[x - \omega'(k_0)t]}\,\mathrm{d}\xi$$

or

$$f(x,t) = f_0[x - \omega'(k_0)t]\,\mathrm{e}^{i(k_0 x - \omega(k_0)t)} \qquad \text{(using (B.14))}$$

or

$$f(x,t) = f_0[x - c_{\mathrm{G}}t]\,\mathrm{e}^{ik_0(x - c_{\mathrm{P}}t)}, \tag{B.15}$$

where $c_{\mathrm{P}} = \omega(k_0)/k_0$, and

$$c_{\mathrm{G}} := \omega'(k_0) = \left.\frac{\mathrm{d}\omega(k)}{\mathrm{d}k}\right|_{k=k_0},$$

is the *group velocity* of the wave. It is evident from (B.15) that the amplitude envelope $f_0[\cdot]$ of a wave packet travels at the group velocity c_{G}, while the wave itself travels at the phase velocity given by $c_{\mathrm{P}} = \omega(k_0)/k_0$. It should be noted that the solution (B.15) is valid only

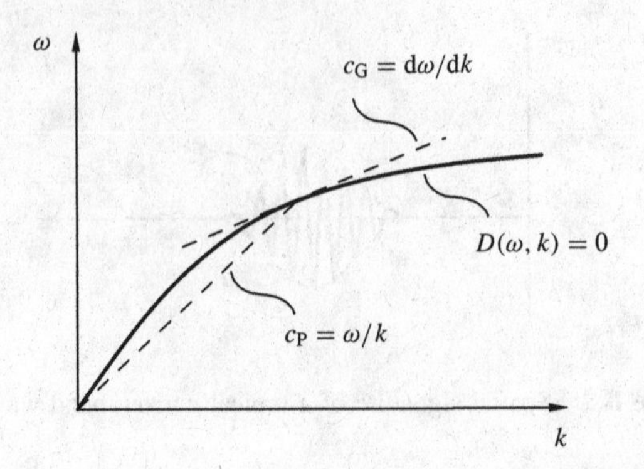

Figure B.3 The concept of phase and group velocities for a dispersive medium

for a small time due to the approximation involved in deriving it. The concepts of c_P and c_G in a dispersive medium are graphically explained in Figure B.3, where the slopes of the dashed lines yield the respective values.

REFERENCES

[1] Sneddon, I.N., *Fourier Transforms*, McGraw-Hill Book Co., New York, 1951.

Appendix C

Variational formulation for dynamics of plates

In deriving the equation of motion of a plate using the variational formulation, we will require the following two standard mathematical results:

$$a(x, y)\nabla \cdot \mathbf{v}(x, y) = \nabla(a\mathbf{v}) - (\nabla a) \cdot \mathbf{v}, \tag{C.1}$$

$$\int_{\mathrm{A}} \nabla \cdot \mathbf{v}(x, y)\, \mathrm{d}x\, \mathrm{d}y = \oint_{\mathrm{B}} \mathbf{v} \cdot \hat{\mathbf{n}}\, \mathrm{d}s, \tag{C.2}$$

where A is the domain of integration and B is the boundary enclosing the domain A, and $\hat{\mathbf{n}}$ is the unit outward normal to the infinitesimal line element ds on B, as shown in Figure C.1. The result (C.2) is the Gauss divergence theorem.

Consider a plate in the x-y-plane of the Cartesian coordinate system. Let $w(x, y, t)$ represent the transverse displacement field along the z-axis direction. The kinetic energy of a plate element can be represented by

$$\begin{aligned}\mathcal{T} &= \frac{1}{2}\int\int\int_{-h/2}^{h/2} \left[\rho w_{,t}^2 + \rho z^2 (w_{,xt}^2 + w_{,yt}^2)\right] \mathrm{d}z\, \mathrm{d}x\, \mathrm{d}y \\ &= \frac{1}{2}\int\int [\rho h w_{,t}^2 + I(w_{,xt}^2 + w_{,yt}^2)]\, \mathrm{d}x\, \mathrm{d}y,\end{aligned} \tag{C.3}$$

where $I = \rho h^3/12$ is the moment of inertia per unit area of the plate. The potential energy of the plate is given by the strain energy stored in the plate when it undergoes deformation. From linear theory of elasticity, the strain energy in this case can be written as (see [1])

$$\mathcal{V} = \frac{1}{2}\int\int\int_{-h/2}^{h/2} (\sigma_{xx}\epsilon_{xx} + \sigma_{yy}\epsilon_{yy} + 2\sigma_{xy}\epsilon_{xy})\, \mathrm{d}z\, \mathrm{d}y\, \mathrm{d}x.$$

Vibrations and Waves in Continuous Mechanical Systems P. Hagedorn and A. DasGupta

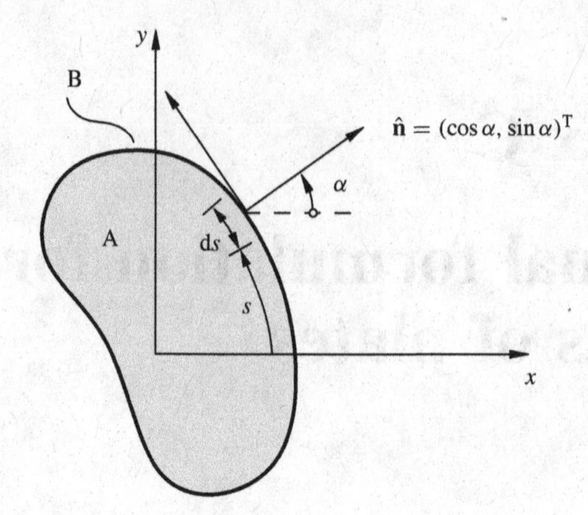

Figure C.1 Representation of unit tangent and normal vectors at a domain boundary

Using (5.4) and (5.6)–(5.8) in the above expression and simplifying yields

$$\begin{aligned}\mathcal{V} &= \frac{1}{2}\int\int_{\mathrm{A}} D[w_{,xx}^2 + w_{,yy}^2 + 2\nu w_{,xx}w_{,yy} + 2(1-\nu)w_{,xy}^2]\,\mathrm{d}x\,\mathrm{d}y \\ &= \frac{D}{2}\int\int_{\mathrm{A}} [(w_{,xx} + w_{,yy})^2 + 2(1-\nu)(w_{,xy}^2 - w_{,xx}w_{,yy})]\,\mathrm{d}x\,\mathrm{d}y,\end{aligned} \tag{C.4}$$

where

$$D = \frac{Eh^3}{12(1-\nu^2)}$$

is assumed to be independent of (x, y). In other words, the material properties E and ν, and the thickness h, from here onwards, are taken to be constant. For a more general case, one may take $D = D(x, y)$ and retain it inside the integral in (C.4).

Now, from Hamilton's principle, the dynamics of the plate must satisfy

$$\delta\int_{t_1}^{t_2} (\mathcal{T} - \mathcal{V})\,\mathrm{d}t = 0$$

or

$$\int_{t_1}^{t_2} \delta\mathcal{T}\,\mathrm{d}t - \int_{t_1}^{t_2} \delta\mathcal{V}_1\,\mathrm{d}t - \int_{t_1}^{t_2} \delta\mathcal{V}_2\,\mathrm{d}t = 0, \tag{C.5}$$

where

$$\mathcal{V}_1 = \frac{D}{2}\int\int (\nabla^2 w)^2\,\mathrm{d}x\,\mathrm{d}y,$$

$$\mathcal{V}_2 = D(1-\nu)\int\int (w_{,xy}^2 - w_{,xx}w_{,yy})\,\mathrm{d}x\,\mathrm{d}y.$$

We have partitioned the action integral into three terms for convenience.

Consider the first term in (C.5). Using the standard arguments of the variational formulation of dynamics, we can rewrite the first term as

$$\begin{aligned}
\int_{t_1}^{t_2} \delta T \, \mathrm{d}t &= \int_{t_1}^{t_2} \int\int [\rho h w_{,t}\, \delta w_{,t} + I(w_{,xt}\, \delta w_{,xt} + w_{,yt}\, \delta w_{,yt})] \, \mathrm{d}x \, \mathrm{d}y \, \mathrm{d}t \\
&= \int\int [\rho h w_{,t}\, \delta w + I(w_{,xt}\, \delta w_{,x} + w_{,yt}\, \delta w_{,y})]_{t_1}^{t_2} \, \mathrm{d}x \, \mathrm{d}y \\
&\quad + \int_{t_1}^{t_2} \int\int [-\rho h w_{,tt}\, \delta w - I(w_{,xtt}\, \delta w_{,x} + w_{,ytt}\, \delta w_{,y})] \, \mathrm{d}x \, \mathrm{d}y \, \mathrm{d}t \\
&= \int_{t_1}^{t_2} \int\int [-\rho h w_{,tt}\, \delta w - I\{(w_{,xtt}\, \delta w)_{,x} - w_{,xxtt}\, \delta w \\
&\qquad + (w_{,ytt}\, \delta w)_{,y} - w_{,yytt}\, \delta w\}] \, \mathrm{d}x \, \mathrm{d}y \mathrm{d}t \\
&= \int_{t_1}^{t_2} \int\int [-\rho h w_{,tt}\, \delta w - I\nabla \cdot (\delta w \nabla w_{,tt}) + I(\nabla^2 w_{,tt})\, \delta w] \, \mathrm{d}x \, \mathrm{d}y \, \mathrm{d}t.
\end{aligned}$$

Now, applying the Gauss divergence theorem (C.2) to the second term in the above expression, we have

$$\begin{aligned}
\int_{t_1}^{t_2} \delta T &= -\int_{t_1}^{t_2} \left[\oint I(\nabla w_{,tt}) \cdot \hat{\mathbf{n}}\, \delta w \, \mathrm{d}s + \int\int (\rho h w_{,tt} + I\nabla^2 w)\, \delta w \, \mathrm{d}x \, \mathrm{d}y \right] \mathrm{d}t \\
&= \int_{t_1}^{t_2} \left[-\oint I w_{,ntt}\, \delta w \, \mathrm{d}s + \int\int (\rho h w_{,tt} + I\nabla^2 w)\, \delta w \, \mathrm{d}x \, \mathrm{d}y \right] \mathrm{d}t, \qquad \text{(C.6)}
\end{aligned}$$

where we have used the definition

$$w_{,ntt} := \nabla w_{,tt} \cdot \hat{\mathbf{n}} = w_{,xtt} \cos\alpha + w_{,ytt} \sin\alpha.$$

Next, consider the integrand of the second term in (C.5). We can write

$$\begin{aligned}
\delta \mathcal{V}_1 &= D \int\int \nabla^2 w \nabla \cdot (\nabla \delta w) \, \mathrm{d}x \, \mathrm{d}y \\
&= D \int\int [\nabla \cdot (\nabla^2 w \nabla \delta w) - (\nabla \nabla^2 w) \cdot (\nabla \delta w)] \, \mathrm{d}x \, \mathrm{d}y \qquad \text{(using (C.1))} \\
&= D \int\int [\nabla \cdot (\nabla^2 w \nabla \delta w) - \nabla \cdot (\delta w \nabla \nabla^2 w) \\
&\qquad + \nabla^2 \nabla^2 w \, \delta w] \, \mathrm{d}x \, \mathrm{d}y \qquad \text{(using (C.1))} \\
&= D \oint [(\nabla^2 w \nabla \delta w) \cdot \hat{\mathbf{n}} - (\delta w \nabla \nabla^2 w) \cdot \hat{\mathbf{n}}] \, \mathrm{d}s \\
&\quad + D \int\int \nabla^4 w \, \delta w \, \mathrm{d}x \, \mathrm{d}y \qquad \text{(using (C.2))} \\
&= D \oint [\nabla^2 w \, \delta w_{,n} - \nabla^2 w_{,n}\, \delta w] \, \mathrm{d}s + D \int\int \nabla^4 w \, \delta w \, \mathrm{d}x \, \mathrm{d}y. \qquad \text{(C.7)}
\end{aligned}$$

The integrand of the third term in (C.5) can be written as

$$\begin{aligned}\delta\mathcal{V}_2 &= D(1-\nu)\int\!\!\int [2w_{,xy}\,\delta w_{,xy} - w_{,xx}\,\delta w_{,yy} - w_{,yy}\,\delta w_{,xx}]\,\mathrm{d}x\,\mathrm{d}y\\ &= D(1-\nu)\int\!\!\int [(w_{,xy}\,\delta w_{,y})_{,x} - w_{,xxy}\,\delta w_{,y} + (w_{,xy}\,\delta w_{,x})_{,y} - w_{,xyy}\,\delta w_{,x}\\ &\quad -(w_{,xx}\,\delta w_{,y})_{,y} + w_{,xxy}\,\delta w_{,y} - (w_{,yy}\,\delta w_{,x})_{,x} + w_{,xyy}\,\delta w_{,x}]\,\mathrm{d}x\,\mathrm{d}y\\ &= D(1-\nu)\int\!\!\int \nabla\cdot\mathbf{F}\,\mathrm{d}x\,\mathrm{d}y,\end{aligned} \tag{C.8}$$

where

$$\mathbf{F} = (F_x, F_y)^{\mathrm{T}} = [(w_{,xy}\,\delta w_{,y} - w_{,yy}\,\delta w_{,x}),\ (w_{,xy}\,\delta w_{,x} - w_{,xx}\,\delta w_{,y})]^{\mathrm{T}}.$$

Using (C.2) to (C.8) we get

$$\begin{aligned}\delta\mathcal{V}_2 &= D(1-\nu)\oint \mathbf{F}\cdot\hat{\mathbf{n}}\,\mathrm{d}s\\ &= D(1-\nu)\oint [(\cos\alpha w_{,xy} - \sin\alpha w_{,xx})\,\delta w_{,y}\\ &\quad + (\sin\alpha w_{,xy} - \cos\alpha w_{,yy})\,\delta w_{,x}]\,\mathrm{d}s.\end{aligned} \tag{C.9}$$

Now, one can relate the derivatives in the Cartesian and the $\hat{\mathbf{n}}$–$\hat{\mathbf{s}}$ systems as

$$\frac{\partial}{\partial x} = \cos\alpha\frac{\partial}{\partial n} - \sin\alpha\frac{\partial}{\partial s} \quad \text{and} \quad \frac{\partial}{\partial y} = \sin\alpha\frac{\partial}{\partial n} + \cos\alpha\frac{\partial}{\partial s}.$$

Using the above transformations one can rewrite (C.9) as

$$\begin{aligned}\delta\mathcal{V}_2 = D(1-\nu)\Bigg[&\oint [(\cos\alpha w_{,xy} - \sin\alpha w_{,xx})\sin\alpha\\ &+ (\sin\alpha w_{,xy} - \cos\alpha w_{,yy})\cos\alpha\,\delta w_{,n}\,\mathrm{d}s\\ &+ \oint [(\cos\alpha w_{,xy} - \sin\alpha w_{,xx})\cos\alpha\\ &-(\sin\alpha w_{,xy} - \cos\alpha w_{,yy})\sin\alpha]\,\delta w_{,s}]\,\mathrm{d}s\Bigg].\end{aligned} \tag{C.10}$$

Integrating by parts the second contour integral in the above, and remembering that the boundary term will be zero over a closed contour, we have

$$\begin{aligned}\delta\mathcal{V}_2 = D(1-\nu)\oint &(2\cos\alpha\sin\alpha w_{,xy} - \sin^2\alpha w_{,xx} - \cos^2\alpha w_{,yy})\,\delta w_{,n}\\ &-(\cos^2\alpha - \sin^2\alpha)w_{,xy} - \cos\alpha\sin\alpha(w_{,xx} - w_{,yy})_{,s}\,\delta w\,\mathrm{d}s.\end{aligned} \tag{C.11}$$

Substituting (C.6), (C.7), and (C.11) into (C.5) yields

$$\int_{t_1}^{t_2}\Big[-\oint D[\nabla^2 w_{,n} + D(1-\nu)\{\cos\alpha\sin\alpha(w_{,yy} - w_{,xx})$$

$$+ (\cos^2\alpha - \sin^2\alpha)w_{,xy}\}_{,s} + I w_{,ntt}]\,\delta w\,\mathrm{d}s$$

$$+ \oint D[\nabla^2 w + (1-\nu)(2\cos\alpha\sin\alpha w_{,xy}$$

$$- \cos^2\alpha w_{,yy} - \sin^2\alpha w_{,xx})]\,\delta w_{,n}\,\mathrm{d}s$$

$$+ \int_0^a\int_0^b\big[-\rho h w_{,tt} + I\nabla^2 w_{,tt} - D\nabla^4 w\big]\,\delta w\,\mathrm{d}x\,\mathrm{d}y\Big]\,\mathrm{d}t = 0.$$

The equation of motion is obtained from the last integral as

$$\rho h w_{,tt} - I\nabla^2 w_{,tt} + D\nabla^4 w = 0.$$

The boundary conditions are obtained as

$$\big[\nabla^2 w + (1-\nu)(2n_x n_y w_{,xy} - n_x^2 w_{,yy} - n_y^2 w_{,xx})\big] = 0$$

or

$$w_{,n} = 0, \qquad (x, y) \in \mathrm{B}, \tag{C.12}$$

and

$$-I w_{,ntt} - D[\nabla^2 w_{,n} + (1-\nu)\{n_x n_y(w_{,yy} - w_{,xx}) + (n_x^2 - n_y^2)w_{,xy}\}_{,s}] = 0$$

or

$$w = 0, \qquad (x, y) \in \mathrm{B}. \tag{C.13}$$

Let us now consider the boundary conditions for the special case of a rectangular Kirchhoff plate (that is, we set $I = 0$). The unit normals at one of the boundaries are shown in Figure C.2. In the case of the rectangular plate, (C.12) and (C.13) for the boundary $x = a$ yield, respectively,

$$D[\nabla^2 w - (1-\nu)w_{,yy}]_{x=a} = 0 \qquad \text{or} \qquad w_{,x}\big|_{x=a} = 0$$

and

$$D[(\nabla^2 w)_{,x} + (1-\nu)w_{,xyy}]_{x=a} = 0 \qquad \text{or} \qquad w\big|_{x=a} = 0.$$

The conditions at the other boundaries can be derived similarly.

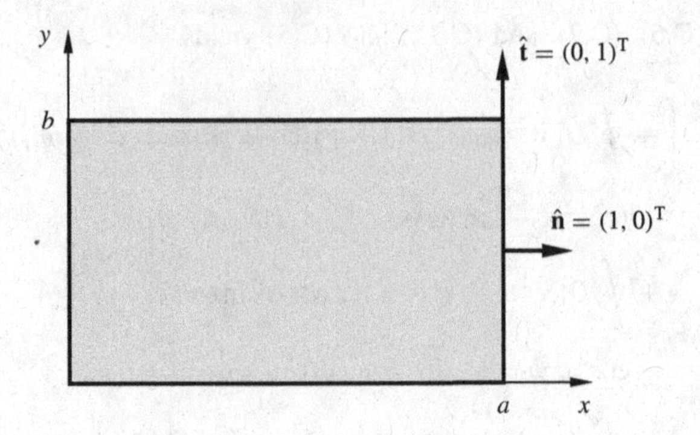

Figure C.2 Unit boundary normal and tangent vectors for a rectangular plate

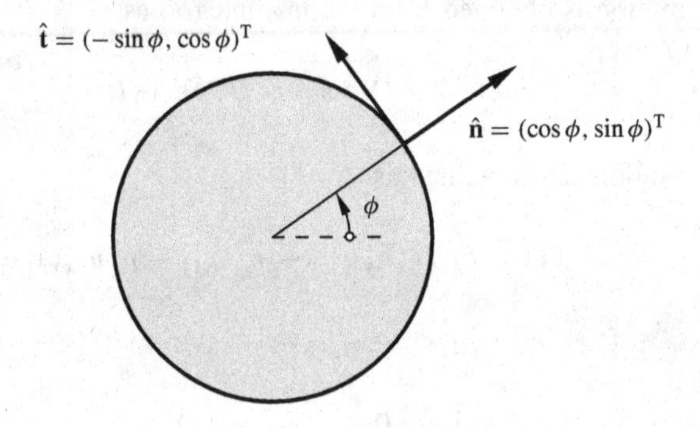

Figure C.3 Unit boundary normal and tangent vectors for a circular plate

For a circular plate, as shown in Figure C.3, the boundary conditions in the polar coordinates can be obtained from (C.12) and (C.13) using the operator transformations

$$\frac{\partial}{\partial x} = \cos\alpha \frac{\partial}{\partial r} - \sin\alpha \frac{\partial}{r\partial\phi} \quad \text{and} \quad \frac{\partial}{\partial y} = \sin\alpha \frac{\partial}{\partial r} + \cos\alpha \frac{\partial}{r\partial\phi}.$$

Using these transformations in (C.12) and (C.13) yields, respectively,

$$\left[\nabla^2 w - (1-\nu)\frac{1}{r}\left(w_{,r} + \frac{1}{r} w_{,\phi\phi}\right)\right]_{r=R} = 0 \quad \text{or} \quad w_{,r}\big|_{r=R} = 0,$$

and

$$\left[(\nabla^2 w)_{,r} + (1-\nu)\frac{1}{r}\left(\frac{1}{r} w_{,\phi\phi}\right)_{,r}\right]_{r=R} = 0 \quad \text{or} \quad w\big|_{r=R} = 0.$$

REFERENCES

[1] Landau, L.D., and Lifshitz, E.M., *Theory of Elasticity*, 3e, Butterworth-Heinemann, Oxford, 1986.

Index

Vibrations and Waves in Continuous Mechanical Systems P. Hagedorn and A. DasGupta